Nanophotonic Materials

Edited by
R. B. Wehrspohn, H.-S. Kitzerow,
and K. Busch

Related Titles

Prather, D. W., Sharkawy, A., Shouyuan, S. et al.

Photonic Crystals: Theory, Applications and Fabrication

Approx. 540 pages
2008
Hardcover
ISBN: 978-0-470-27803-1

Saleh, B. E. A., Teich, M. C.

Fundamentals of Photonics

1200 pages
2006
Hardcover
ISBN: 978-0-471-35832-9

Prasad, P. N.

Nanophotonics

432 pages
2004
Hardcover
ISBN: 978-0-471-64988-5

Busch, K. et al. (ed.)

Photonic Crystals

Advances in Design, Fabrication, and Characterization

380 pages
2004
ISBN: 978-3-527-40432-2

Nanophotonic Materials

Photonic Crystals, Plasmonics, and Metamaterials

Edited by
R. B. Wehrspohn, H.-S. Kitzerow, and K. Busch

WILEY-VCH Verlag GmbH & Co. KGaA

The Editors

Prof. Ralf B. Wehrspohn
Fraunhofer-Institute for Mechanics of Materials
Halle; Germany
ralf.wehrspohn@iwmh.fraunhofer.de

Prof. H.-S. Kitzerow
Department of Physical Chemistry FB13
University of Paderborn, Germany
heinz.kitzerow@upb.de

Prof. Kurt Busch
Institute of Solid State Physics
University of Karlsruhe, Germany
kurt@tkm.physik.uni-karlsruhe.de

Cover
Overlay of a 3D photonic crystal made from macroporous silicon, a simulation of a waveguide splitter and the lasing modes of a planar photonic crystals laser made from III-V-compound semiconductors (printed with kind permission of conImago, www.conimago.de).

Library of Congress Card No.: applied for

British Library Cataloguing-in-Publication Data
A catalogue record for this book is available from the British Library.

Bibliographic information published by the Deutsche Nationalbibliothek
Die Deutsche Nationalbibliothek lists this publication in the Deutsche Nationalbibliografie; detailed bibliographic data are available in the Internet at <http://dnb.d-nb.de>.

Typesetting Thomson Digital, Noida, India
Printing Strauss GmbH, Mörlenbach
Binding Litges & Dopf Buchbinderei GmbH, Heppenheim

Printed in the Federal Republic of Germany
Printed on acid-free paper

ISBN: 978-3-527-40858-0

Contents

Nanophotonic Materials: Photonic Crystals, Plasmonics, and Metamaterials. Edited by R.B. Wehrspohn, H.-S. Kitzerow, and K. Busch

ISBN: 978-3-527-40858-0

Preface

Stimulated by the pioneering work of Sajeev John and Eli Yablonovitsch in 1987, German research groups started in the early 1990s with theoretical and experimental work on 2D and 3D photonics crystals. This initial work was the basis of an application for a focused project on photonic crystals at the German Science Foundation (DFG) in 1999. In the last seven years, a consortium consisting of more than 20 German research groups worked together in the area of photonic crystals.

We started with linear, non-dispersive properties of purely dielectric 2D and 3D photonic crystals in the late 90s and developed the field of research step-by-step to non-linear and dispersive properties of dielectric photonic crystals including gain and/or losses. These properties where studied on different materials systems such as silicon, III-V-compound semiconductors, oxides and polymers as well as hybrid systems consisting of dielectric photonic crystals and liquid crystals. Applications of these systems were developed in the area of active photonic crystals fibres, functional optical components as well as sensors. Some of them have now even entered into industrial applications. During the funding period, some groups extended the initial focus to non-dielectric, dispersive materials such as metals and discussed the properties of periodic metallic structures (plasmonic crystals). After the groundbreaking work of John Pendry at the beginning of this century, resonances in dispersive structures with periodic permeability and permittivity (metamaterials) were studied as well. This was important in order to understand the difference of negative refraction in metamaterials and dielectric photonic crystals.

This special issue summarizes the work of those groups which were part of the focused German program. Other groups not funded by this project but by other grants joined the German initiative on photonic crystals. Workshops of all groups working in the field have been carried out at the German Physical Society meeting each year. Our work was also stimulated by the numerous Humboldt Awardees who visited Germany in the last seven years such as Sajeev John, Costas Soukoulis, Thomas Krauss and many others. We are also pleased that some of our close

Nanophotonic Materials: Photonic Crystals, Plasmonics, and Metamaterials. Edited by R.B. Wehrspohn, H.-S. Kitzerow, and K. Busch

ISBN: 978-3-527-40858-0

collaborators over the last six years have contributed to this issue: Dan Davidov's group from the Hebrew University, Israel, and Masanori Ozaki's group from Osaka University, Japan.

November 2007

R.B. Wehrspohn
H.-S. Kitzerow
K. Busch
Halle, Paderborn, and Karlsruhe

List of Contributors

A. Abdolvand
Martin-Luther University
Halle-Wittenberg
Institute of Physics
Friedemann-Bach-Platz 6
06108 Halle, Germany
abdolvand@physik.uni-halle.de
and
The University of Manchester
School of Mechanical, Aerospace and
Civil Engineering
Laser Processing Research Centre
Manchester M60 1QD
United Kingdom

M. V. Artemyev
Belarussian State University
Institute for Physico-Chemical
Problems
Minsk 220080
Belarus
artemyev@bsu.by

M. Augustin
Friedrich-Schiller-Universität Jena
Institute of Applied Physics/
»ultra optics«®
Max-Wien-Platz 1
07743 Jena
Germany

H. Bartelt
Institute of Photonic Technology
Albert-Einstein-Straße 9
07745 Jena
Germany
hartmut.bartelt@ipht-jena.de

M. Barth
Humboldt-Universität zu Berlin
Institute of Physics
Nano Optics Group
Hausvogteiplatz 5–7
10117 Berlin
Germany
michael.barth@physik.hu-berlin.de

C. Bauer
Max-Planck-Institut für
Festkörperforschung
(Solid State Research)
Heisenbergstr. 1
70569 Stuttgart
Germany
c.bauer@fkf.mpg.de

Nanophotonic Materials: Photonic Crystals, Plasmonics, and Metamaterials. Edited by R.B. Wehrspohn,
H.-S. Kitzerow, and K. Busch

ISBN: 978-3-527-40858-0

O. Benson
Humboldt-Universität zu Berlin
Institute of Physics
Nano Optics Group
Hausvogteiplatz 5–7
10117 Berlin
Germany
oliver.benson@physik.hu-berlin.de

R. Boucher
Institute for Photonic Technology Jena
Albert-Einstein Str. 9
07745 Jena
Germany
boucher@ipht-jena.de

K. Busch
Universität Karlsruhe
Institute of Thoeretical Solid State Physics
76128 Karlsruhe
Germany
kurt@tfp.uni-karlsruhe.de

S. Burger
Zuse Institute Berlin
Takustraße 7
14195 Berlin
Germany
burger@zib.de

B. N. Chichkov
Laser Center Hannover LZH
Hollertithallee 8
30419 Hannover
Germany
b.chichkov@lzh.de

D. N. Chigrin
University of Bonn
Physical Institute
Nussallee 12
53115 Bonn
Germany
chigrin@th.physik.uni-bonn.de

A. Christ
Max-Planck-Institut für Festkörperforschung
(Solid State Research)
Heisenbergstr. 1
70569 Stuttgart
Germany
a.christ@fkf.mpg.de

C. D'Andrea
Politecnico di Milano
Physics Department
Piazza L. da Vinci 32
20133 Milano
Italy
cosimo.dandrea@polimi.it

D. Davidov
The Hebrew University of Jerusalem
The Racah Institute of Physics
Jerusalem 91904
Israel
davidov@vms.huji.ac.il

P. Declerck
Fraunhofer ISC
Neunerplatz 2
97082 Würzburg
Germany

M. Eich
Hamburg University of Technology
Institute of Optical and Electronic Materials
Eissendorferstraße 38
21073 Hamburg
Germany

S. Essig
Universität Karlsruhe
Institute of Thoeretical Solid State Physics
76128 Karlsruhe
Germany
essig@tkm.physik.uni-karlsruhe.de

C. Etrich
Friedrich-Schiller-Universität Jena
Institute of Applied Physics/
»ultra optics«®
Max-Wien-Platz 1
07743 Jena
Germany
chris@pinet.uni-jena.de

F. Fleischhaker
University of Mainz
Institute of Organic Chemistry
Department of Chemistry, Pharmacy and Earth Science
Düsbergweg 10–14
55128 Mainz
Germany
fleischf@uni-mainz.de

A. Forchel
University of Würzburg
Technical Physics
Am Hubland
97074 Würzburg
Germany
forchel@physik.uni-wuerzburg.de

A. Frenkel
ANAFA –
Electromagnetic Solutions Ltd.
P.O.B. 5301
Kiriat Bialik 27000
Israel

L. Fu
University of Stuttgart
4th Physical Institute
Pfaffenwaldring 57
70569 Stuttgart
Germany
lf4@physik.uni-stuttgart.de

A. Fujii
Osaka University
Department of Electrical, Electronic and Information Engineering
2-1 Yamada-Oka, Suita
Osaka 565-0871
Japan

M. Gellner
University of Würzburg
Technical Physics
Am Hubland
97074 Würzburg
Germany

M. Gerken
Universität Karlsruhe (TH)
Light Technology Institute
Engesserstraße 13
76131 Karlsruhe
Germany
martina.gerken@lti.uni-karlsruhe.de

H. Giessen
4. Physikalisches Institut
Universität Stuttgart
Pfaffenwaldring 57
70569 Stuttgart
Germany
giessen@physik.uni-stuttgart.de

F. Glöckler
Universität Karlsruhe (TH)
Light Technology Institute
Engesserstraße 13
76131 Karlsruhe
Germany
felix.gloeckler@lti.uni-karlsruhe.de

M. Golosovsky
The Hebrew University of Jerusalem
The Racah Institute of Physics
Jerusalem 91904
Israel

H. Gräbeldinger
University of Stuttgart
4th Physical Institute
Pfaffenwaldring 57
70569 Stuttgart
Germany
h.graebeldinger@physik.
uni-stuttgart.de

H. Graener
Martin-Luther University
Halle-Wittenberg
Institute of Physics
Friedemann-Bach-Platz 6
06108 Halle
Germany
heinrich.graener@physik.uni-halle.de

H. Guo
University of Stuttgart
4th Physical Institute
Pfaffenwaldring 57
70569 Stuttgart
Germany
h.guo@physik.uni-stuttgart.de

W. Haase
Darmstadt University of Technology
Eduard- Zintl-Institute for Inorganic
and Physical Chemistry
Petersenstr. 20
64287 Darmstadt
Germany
haase@chemie.tu-darmstadt.de

W. Hergert
Martin-Luther University
Halle-Wittenberg
Institute of Physics
Friedemann-Bach-Platz 6
06108 Halle
Germany
wolfram.hergert@physik.uni-halle.de

J. Herrmann
Max Born Institute for Nonlinear Optics
and Short Pulse Spectroscopy
Max-Born-Str. 2a
12489 Berlin
Germany
jherrman@mbi-berlin.de

R. Houbertz
Fraunhofer ISC
Neunerplatz 2
97082 Würzburg
Germany
houbertz@isc.fhg.de

U. Hübner
Institute for Photonic Technology Jena
Albert-Einstein Str. 9
07745 Jena
Germany
huebner@ipht-jena.de

A. Husakou
Max Born Institute for Nonlinear Optics
and Short Pulse Spectroscopy
Max-Born-Str. 2a
12489 Berlin
Germany
husakou@mbi-berlin.de

R. Iliew
Friedrich-Schiller-Universität Jena
Institute of Condensed Matter Theory
and Solid State Optics
Max-Wien-Platz 1
07743 Jena
Germany
rumen.iliew@uni-jena.de

S. Kaiser
University of Stuttgart
1st Physical Institute
Pfaffenwaldring 57
70569 Stuttgart
Germany
kaiser@pi1.physik.uni-stuttgart.de

M. Kamp
University of Würzburg
Technical Physics
Am Hubland
97074 Würzburg
Germany
martin.kamp@physik.
uni-wuerzburg.de

J. Kirchhof
Institute of Photonic Technology
Albert-Einstein-Straße 9
07745 Jena
Germany
johannes.kirchhof@ipht-jena.de

O. Kiriyenko
Martin-Luther University
Halle-Wittenberg
Institute of Physics
Friedemann-Bach-Platz 6
06108 Halle
Germany
oleksey.kiriyenko@physik.uni-halle.de

H.-S. Kitzerow
University of Paderborn
Faculty of Science
Warburger Str. 100
33098 Paderborn
Germany
heinz.kitzerow@upb.de

E.-B. Kley
Friedrich-Schiller-Universität Jena
Institute of Applied Physics
Max-Wien-Platz 1
07743 Jena
Germany
kley@iap.uni-jena.de

R. Klose
Zuse Institute Berlin
Takustraße 7
14195 Berlin
Germany
klose@zib.de

J. Kobelke
Institute of Photonic Technology
Albert-Einstein-Straße 9
07745 Jena
Germany
jens.kobelke@ipht-jena.de

S. W. Koch
Philipps University
Department of Physics and Material
Sciences Center
Renthof 5
35032 Marburg
Germany
stephan.w.koch@physik.
uni-marburg.de

D. Konjhodzic
Max-Planck-Institut für
Kohlenforschung (Coal Research)
Kaiser-Wilhelm-Platz 1
45470 Mülheim an der Ruhr
Germany
denan@mpi-muelheim.mpg.de

J. Kroha
University of Bonn
Physical Institute
Nussallee 12
53115 Bonn
Germany
kroha@physik.uni-bonn.de

J. Kuhl
Max-Planck-Institut für
Festkörperforschung
(Solid State Research)
Heisenbergstr. 1
70569 Stuttgart
Germany
j.kuhl@fkf.mpg.de

B. Lange
Institute of Organic Chemistry
Department of Chemistry, Pharmacy
and Earth Science
University of Mainz
Düsbergweg 10–14
55128 Mainz
Germany
langeb@uni-mainz.de

A. V. Lavrinenko
Technical University of Denmark
Department of Communications,
Optics and Materials
COM-DTU/NanoDTU
Building 345V
2800 Kgs. Lyngby
Denmark
ala@com.dtu.dk

F. Lederer
Friedrich-Schiller-Universität Jena
Institute of Condensed Matter Theory
and Solid State Optics
Max-Wien-Platz 1
07743 Jena
Germany
dekanat@paf.uni-jena.de

H. Lehmann
Institute of Photonic Technology
Albert-Einstein-Straße 9
07745 Jena
Germany
hartmut.lehmann@ipht-jena.de

U. Lemmer
Universität Karlsruhe (TH)
Light Technology Institute
Engesserstraße 13
76131 Karlsruhe
Germany
uli.lemmer@lti.uni-karlsruhe.de

J. Leppert
Institute of Photonic Technology
Albert-Einstein-Straße 9
07745 Jena
Germany
jan.leppert@ipht-jena.de

N. Liu
University of Stuttgart
4th Physical Institute
Pfaffenwaldring 57
70569 Stuttgart
Germany
n.liu@physik.uni-stuttgart.de

A. Löffler
University of Würzburg
Technical Physics
Am Hubland
97074 Würzburg
Germany

A. Lorenz
University of Paderborn
Faculty of Science
Warburger Str. 100
33098 Paderborn
Germany
alorenz@mail.upb.de

Y. Lu
University of Mainz
Institute of Organic Chemistry
Düsbergweg 10–14
55099 Mainz
Germany

F. Marlow
Max-Planck-Institut für
Kohlenforschung (Coal Research)
Kaiser-Wilhelm-Platz 1
45470 Mülheim an der Ruhr
Germany
marlow@mpi-muelheim.mpg.de

Y. Matsuhisa
Osaka University
Department of Electrical, Electronic and
Information Engineering
Graduate School of Engineering
2-1 Yamada-Oka Suita
Osaka 565-0871
Japan
matsuhisa@ele.eng.osaka-u.ac.jp

H. Matthias
University of Paderborn
Faculty of Science
Warburger Str. 100
33098 Paderborn
Germany
heiner@upb.de

T. Meier
Philipps University
Department of Physics and Material
Sciences Center
Renthof 5
35032 Marburg
Germany
torsten.meier@physik.uni-marburg.de

B. M. Möller
University of Dortmund
Institute of Physics
Otto-Hahn-Straße 4
44227 Dortmund
Germany
bjoern.moeller@uni-dortmund.de

K. Mörl
Institute of Photonic Technology
Albert-Einstein-Straße 9
07745 Jena
Germany
klaus.moerl@ipht-jena.de

D. Nau
University of Bonn
Institute of Applied Physics
Wegelerstr. 8
53115 Bonn
Germany
dietmar.nau@iap.uni-bonn.de
and
University of Stuttgart
Physical Institute
Pfaffenwaldring 57
70569 Stuttgart
Germany

Y. Neve-Oz
The Hebrew University of Jerusalem
The Racah Institute of Physics
Jerusalem 91904
Israel
yairnz@pob.huji.ac.il

J. Niegemann
Universität Karlsruhe
Institute of Thoeretical Solid State
Physics
76128 Karlsruhe
Germany
niegeman@tfp.uni-karlsruhe.de

S. Nolte
Friedrich-Schiller-Universität Jena
Institute of Applied Physics
Max-Wien-Platz 1
07743 Jena
Germany
nolte@iap.uni-jena.de

A. Ovsianikov
Laser Center Hannover LZH
Hollertithallee 8
30419 Hannover
Germany
a.ovsianikov@lzh.de

M. Ozaki
Osaka University
Department of Electrical, Electronic and Information Engineering
Graduate School of Engineering
Osaka University
2-1 Yamada-Oka Suita
Osaka 565-0871
Japan
ozaki@ele.eng.osaka-u.ac.jp

B. Pasenow
Philipps University
Department of Physics and Material Sciences Center
Renthof 5
35032 Marburg
Germany
bernhard.pasenow@physik.uni-marburg.de

S. Passinger
Laser Center Hannover LZH
Hollertithallee 8
30419 Hannover
Germany
s.passinger@lzh.de

S. Peter
Universität Karlsruhe (TH)
Light Technology Institute
Engesserstraße 13
76131 Karlsruhe
Germany
sabine.peters@lti.uni-karlsruhe.de

F. Podgornov
Darmstadt University of Technology
Eduard-Zintl-Institute for Inorganic and Physical Chemistry
Petersenstr. 20
64287 Darmstadt
Germany
podgor@hrz2.hrz.tu-darmstadt.de
and
South Ural State University
Joint Nonlinear Optics Laboratory of the Institute of Electrophysics
(RAS, Ekaterinburg, Russia) and South Ural State University
Lenin Ave., 76
Chelyabinsk 454080
Russia

J. Pomplun
Zuse Institute Berlin
Takustraße 7
14195 Berlin
Germany
pomplun@zib.de

M. Reichelt
University of Arizona
Arizona Center for Mathematical Sciences
Tucson, AZ 85721
USA
reichelt@acms.arizona.edu

U. Röpke
Institute of Photonic Technology
Albert-Einstein-Straße 9
07745 Jena
Germany
ulrich.roepke@ipht-jena.de

V. Sandoghdar
Swiss Federal Institute of Technology (ETH)
Laboratory of Physical Chemistry
8093 Zurich
Switzerland
vahid.sandoghdar@ethz.ch

M. Scholz
University of Würzburg
Technical Physics
Am Hubland
97074 Würzburg
Germany

A. Schönhardt
University of Bremen
Institute of Environmental Physics
28359 Bremen
Germany
schoenhardt@iup.physik.uni-bremen.de

F. Schmidt
Zuse Institute Berlin
Takustraße 7
14195 Berlin
Germany
schmidt@zip.de

M. Schmidt
Hamburg University of Technology
Institute of Optical and Electronic Materials
Eissendorferstraße 38
21073 Hamburg
Germany

S. Schröter
Institute of Photonic Technology
Albert-Einstein Str. 9
07745 Jena
Germany
siegmund.schroeter@ipht-jena.de

K. Schuster
Institute of Photonic Technology
Albert-Einstein-Straße 9
07745 Jena
Germany
kay.schuster@ipht-jena.de

H. Schweizer
University of Stuttgart
4th Physical Institute
Pfaffenwaldring 57
70569 Stuttgart
Germany
h.schweizer@physik.uni-stuttgart.de

S. L. Schweizer
Institute of Physics
Martin-Luther University Halle-Wittenberg
06099 Halle
Germany

A. Schwuchow
Institute of Photonic Technology
Albert-Einstein-Straße 9
07745 Jena
Germany
anka.schwuchow@ipht-jena.de

J. Serbin
Laser 2000
Argelsrieder Feld 14
82234 Wessling
Germany
j.serbin@laser2000.de

S. Smolka
Humboldt-Universität zu Berlin
Institute of Physics
Nano Optics Group
Hausvogteiplatz 5–7
10117 Berlin
Germany
smolka@physik.hu-berlin.de

T. Stroucken
Philipps University
Department of Physics and Material Sciences Center
Renthof 5
35032 Marburg
Germany
tineke.stroucken@physik.uni-marburg.de

T. Sünner
University of Würzburg
Technical Physics
Am Hubland
97074 Würzburg
Germany
thomas.suenner@physik.uni-wuerzburg.de

S. Taccheo
Politecnico di Milano
Physics Department
Piazza L. da Vinci 32
20133 Milano
Italy
stefano.taccheo@polimi.it

L. Tkeshelashvili
Universität Karlsruhe
Institute of Thoeretical Solid State Physics
76128 Karlsruhe
Germany
lasha@tkm.physik.uni-karlsruhe.de

A. Tünnermann
Friedrich-Schiller-Universität Jena
Institute of Applied Physics
Max-Wien-Platz 1
07743 Jena
Germany
tuennermann@iap.uni-jena.de
and
Fraunhofer Institute for Applied Optics and Precision Engineering
Albert-Einstein-Straße 7
07745 Jena
Germany

W. Volksen
IBM Almaden Research Center
650 Harry Road
95120 San Jose
USA

S. Wackerow
Martin-Luther University Halle-Wittenberg
Institute of Physics
Friedemann-Bach-Platz 6
06108 Halle
Germany
stefan.wackerow@physik.uni-halle.de

R. B. Wehrspohn
Martin-Luther University
Halle-Wittenberg
Institute of Physics
06099 Halle
Germany
ralf.wehrspohn@physik.uni-halle.de

U. Woggon
University of Dortmund
Institute of Physics
Otto-Hahn-Straße 4
44227 Dortmund
Germany
ulrike.woggon@uni-dortmund.de

J. H. Wülbern
Hamburg University of Technology
Institute of Optical and Electronic Materials
Eissendorferstraße 38
21073 Hamburg
Germany
jan.wuelbern@tuhh.de

H. Yoshida
Osaka University
Department of Electrical, Electronic and Information Engineering
2-1 Yamada-Oka, Suita
Osaka 565-0871
Japan

R. Zentel
Institute of Organic Chemistry
Department of Chemistry, Pharmacy and Earth Science
University of Mainz
Düsbergweg 10–14
55128 Mainz
Germany
zentel@uni-mainz.de

T. Zentgraf
Max-Planck-Institut für Festkörperforschung (Solid State Research)
Heisenbergstr. 1
70569 Stuttgart
Germany
t.zentgraf@fkf.mpg.de

S. V. Zhukovsky
University of Bonn
Physical Institute
Nussallee 12
53115 Bonn
Germany
sergei@th.physik.uni-bonn.de

L. Zschiedrich
Zuse Institute Berlin
Takustraße 7
14195 Berlin
Germany
zschiedrich@zib.de

I
Linear and Non-linear Properties of Photonic Crystals

Nanophotonic Materials: Photonic Crystals, Plasmonics, and Metamaterials. Edited by R.B. Wehrspohn, H.-S. Kitzerow, and K. Busch

ISBN: 978-3-527-40858-0

1
Solitary Wave Formation in One-dimensional Photonic Crystals

Sabine Essig, Jens Niegemann, Lasha Tkeshelashvili, and Kurt Busch

1.1
Introduction

Periodically modulated dielectric structures exhibit the peculiar property that their multibranch dispersion relations may be separated by Photonic Band Gaps (PBGs) [1,2]. In the linear regime, optical waves with frequencies within these PBGs cannot propagate inside the sample and decay exponentially with distance. This gives rise to a number of novel physical phenomena that have significant potential for applications in telecommunication and all-optical information processing.

In the case of nonlinear periodic structures, the physics exhibits a much richer behavior [3]. For instance, due to the optical Kerr effect, i.e., an intensity-dependent refractive index, sufficiently intense electromagnetic pulses can locally tune the PBG. As a consequence, the system may become transparent to optical waves with frequencies in the (linear) band gaps. Moreover, in the presence of optical nonlinearities, modulational instabilities of these waves may occur. This leads to novel types of solitary excitations in PBG materials, the so-called gap solitons. Gap solitons have first been discovered by Chen and Mills [4], and are characterized by a central pulse frequency within a photonic band gap. Most notably, gap solitons can possess very low and even vanishing propagation velocities and, thus, lend themselves to various applications for instance in optical buffers and delay lines as well as in information processing. It should be noted that PBG materials allow solitary waves with carrier frequencies outside the band gaps, too. Such pulses are generally referred to as Bragg solitons [3]. To date, low group velocity Bragg solitons have been observed in fiber Bragg gratings [5,6]. These systems represent a very important class of (quasi) one-dimensional PBG materials which has already found many applications in various areas of photonics. In the following, we, therefore, concentrate on an analysis of fiber Bragg gratings. However, we would like to emphasize that owing to the universal nature of nonlinear wave propagation phenomena our results are of relevance for a number of physical

Nanophotonic Materials: Photonic Crystals, Plasmonics, and Metamaterials. Edited by R.B. Wehrspohn, H.-S. Kitzerow, and K. Busch

ISBN: 978-3-527-40858-0

systems ranging from two- or three-dimensional photonic crystals all the way to Bose–Einstein condensates in optical lattices.

The appropriate theoretical model to describe nonlinear wave phenomena in one-dimensional PBG materials is the so-called coupled mode theory [3]. In this model, forward and backward propagating waves are described through appropriate carrier waves that are modulated with corresponding (slowly varying) envelopes $E_{\pm}$. The resulting equations of motions for these envelopes are

$$i\frac{\partial E_+}{\partial z} + i\frac{\partial E_+}{\partial t} + E_- + |E_+|^2 E_+ + 2|E_-|^2 E_+ = 0, \tag{1.1a}$$

$$-i\frac{\partial E_-}{\partial z} + i\frac{\partial E_-}{\partial t} + E_+ + |E_-|^2 E_- + 2|E_+|^2 E_- = 0. \tag{1.1b}$$

Equations (1.1a) and (1.2b) are generally known as Nonlinear Coupled Mode Equations (NLCME). The NLCME are given in dimensionless variables, and the nonlinearity is assumed to be positive. For a detailed discussion of the NLCME, including their derivation directly from Maxwell's equations, we refer to Ref. [3]. Below, we will only give a brief synopsis of those results on the NLCME that pertain to our work on solitary wave formation described in Sections 1.2 and 1.3.

The coupled mode theory only accounts for one band gap located between two bands, an upper and a lower band, respectively. For systems such as fiber Bragg gratings, the NLCME provide a very accurate framework for studying nonlinear wave dynamics [5]. Although NLCME are nonintegrable, there are known exact solitary wave solutions to these equations which read as [3]

$$E_{\pm}(z,t) = \alpha\tilde{E}_{\pm}(z,t)e^{i\eta(\theta)}. \tag{1.2}$$

Here, $\tilde{E}_{\pm}$ represent the one-soliton solutions of the massive Thirring model which corresponds to the NLCMEs without self-phase modulations terms [3]. Explicitly, these one-soliton solutions read as

$$\tilde{E}_+ = \sqrt{\frac{1}{2}}\frac{1}{\Delta}\sin\delta\, e^{i\sigma}\sec h(\theta - i\delta/2), \tag{1.3a}$$

$$\tilde{E}_- = -\sqrt{\frac{1}{2}}\Delta\,\sin\delta\, e^{i\sigma}\sec h(\theta + i\delta/2), \tag{1.3b}$$

where we have introduced the abbreviations

$$\theta = \gamma(z - vt)\sin\delta, \qquad \sigma = \gamma(vz - t)\cos\delta, \tag{1.4}$$

as well as

$$\Delta = \left(\frac{1-v}{1+v}\right)^{\frac{1}{4}}, \qquad \gamma = \frac{1}{\sqrt{1-v^2}}. \tag{1.5}$$

Finally, α and $\eta(\theta)$ are given by

$$\frac{1}{\alpha^2} = 1 + \frac{1+\nu^2}{2(1-\nu^2)}, \qquad e^{i\eta} = \left(-\frac{e^{2\theta}+e^{\mp i\delta}}{e^{2\theta}+e^{\mp i\delta}}\right)^{\frac{2\nu}{3-\nu^2}}. \tag{1.6}$$

These solutions (1.2) of the NLCME depend on two independent parameters, the detuning δ and the scaled group velocity ν. The scaled group velocity can take on any value below the speed of light (i.e., $-1 \leq \nu \leq +1$) and may even vanish. The latter case corresponds to stationary solutions which have been found by Chen and Mills [4] through numerical experiments. The detuning δ of the soliton describes the value of the waves' central (carrier) frequency relative to the band edge and varies over a range $0 \leq \delta \leq \pi$. If δ is close to zero, the spectrum of the solitary wave is concentrated around the upper band edge (in the case of positive nonlinearity). The Bragg frequency associated with the center of the band gap corresponds to a detuning $\delta = \pi/2$. At this point, we would like to note that in terms of our dimensionless units the photonic band gap is located in the interval $[-1, 1]$. Equation (1.3) describes both gap and Bragg solitons. A stability analysis of these solitary wave solutions has shown that pulses with $\delta < \pi/2$ remain stable against small perturbations. However, for detuning $\delta > \pi/2$, the solitary waves become unstable and decay to radiation modes after a certain transient time [7].

In any realistic experimental situation, the exact soliton-shaped pulses cannot be launched directly. Instead, a formation of a solitary wave has to take place starting with an initial pulse with "distorted" shape that radiates away the excess energy into low amplitude linear modes. The influence of the PBG on this reshaping process is expected to be drastically increased for small values of the detuning. As alluded to above, for $\delta \rightarrow 0$ the spectrum of the pulse is mostly concentrated at the band edge where the group velocity of the eigenmodes is very small or even zero. Therefore, the linear waves (not to be confused with the linear eigenmodes) radiated from the initial nonlinear pulse will be strongly back-scattered by the PBG material. In turn, this will cause pronounced memory effects (also called non-Markovian effects; see the discussion in Section 1.3) in the soliton formation process quite similar to the atom-photon interaction processes in PBG materials [8,9]. This is the starting point of our analysis in the present paper. Owing to the non-integrability of the NLCME, we study this problem by means of a variational approach. We describe the details of this variational approach to the NLCME and how to include a coupling of the nonlinear radiation modes to linear losses in Sections 1.2 and 1.3, respectively. In Section 1.4, we report about results and comparison with numerically exact calculations. Finally, we summarize our findings in Section 1.5, where we also provide a perspective on improved treatments of losses within the NLCME.

1.2 Variational Approach to the NLCME

The variational approach to the NLCME is based on the Lagrangian density of the system in non-dimensional form

$$\begin{aligned}\mathcal{L} = \frac{i}{2}\Big[&E_+^*\left(\frac{\partial}{\partial t}+\frac{\partial}{\partial z}\right)E_+ - E_+\left(\frac{\partial}{\partial t}+\frac{\partial}{\partial z}\right)E_+^* \\ &+ E_-^*\left(\frac{\partial}{\partial t}-\frac{\partial}{\partial z}\right)E_- - E_-\left(\frac{\partial}{\partial t}-\frac{\partial}{\partial z}\right)E_-^*\Big] \\ &+\frac{1}{2}|E_+|^4+\frac{1}{2}|E_-|^4+2|E_+|^2|E_-|^2+E_+^*E_-+E_+E_-^*.\end{aligned} \tag{1.7}$$

The action corresponding to this Lagrangian density is invariant under temporal and spatial translations as well as under phase transformations. According to the Noether Theorem, every symmetry transformation generates a corresponding conserved quantity [10]. In our subsequent analysis, the mass and energy conservation laws are of major importance. The mass conservation law reads

$$\int_{-\infty}^{\infty}(|E_+|^2+|E_-|^2)\,\mathrm{d}z = \text{const}, \tag{1.8}$$

whereas the energy conservation law states

$$\begin{aligned}\int_{-\infty}^{\infty}\Bigg(&\frac{i}{2}\left[E_+^*\frac{\partial E_+}{\partial z}-E_+\frac{\partial E_+^*}{\partial z}-E_-^*\frac{\partial E_-}{\partial z}+E_-\frac{\partial E_-^*}{\partial z}\right] \\ &+\frac{1}{2}|E_+|^4+\frac{1}{2}|E_-|^4+2|E_+|^2|E_-|^2+E_+^*E_-+E_+E_-^*\Bigg)\mathrm{d}z = \text{const}.\end{aligned} \tag{1.9}$$

According to the general scheme for variational approach [11], we have to choose the trial functions for the forward and backward propagating modes of the NLCME that are tailored towards the problem under consideration. In the present case, our Ansatz must allow the pulse to relax to the exact gap soliton shape. For simplicity, we restrict ourselves to the evolution of stationary gap solitons, i.e., we choose the scaled group velocity v to be zero. Other situation can be treated analogously, albeit with considerably more complex notations. In addition, the trial functions must also contain a radiation part. Based on these considerations, we formulate the following Ansatz

$$E_+(z,t) = +\left[\eta_+(t)\,\mathrm{sec}\,h\left(\sin\delta_+(t)z-\frac{i}{2}\delta_+(t)\right)+ig_+(t)\right]\mathrm{e}^{-ia_+(t)}, \tag{1.10a}$$

$$E_-(z,t) = -\left[\eta_-(t)\,\mathrm{sec}\,h\left(\sin\delta_-(t)z+\frac{i}{2}\delta_-(t)\right)+ig_-(t)\right]\mathrm{e}^{-ia_-(t)} \tag{1.10b}$$

together with the initial conditions

$$E_+(z,t=0) = +\eta_0\,\mathrm{sec}\,h\left(\sin\delta_0-\frac{i}{2}\delta_0\right), \tag{1.11a}$$

$$E_-(z,t=0) = -\eta_0\,\mathrm{sec}\,h\left(\sin\delta_0+\frac{i}{2}\delta_0\right), \tag{1.11b}$$

where the initial amplitudes and detunings of the forward and backward propagating modes are chosen to be the same. The (time-dependent) variational parameters are the amplitudes $\eta_{\pm}$, the detunings $\delta_{\pm}$, the so-called shelf functions $g_{\pm}$, and the phases $a_{\pm}$. The first part of the trial function describes the initial pulse and the second part, consisting of the functions $g_{\pm}$, describes the nonlinear radiation modes. In our case, the shelf functions $g_{\pm}$ are of particular importance, since they allow the localized pulse to couple to linear radiation modes. Far away from the nonlinear pulse, the system dynamics is essentially linear so that the nonlinear extended modes can only contribute in a spatial region of finite extent around the pulse. As a consequence, the shelf should have a finite length l. Furthermore, in view of the symmetry between E_+ and E_- and inspired by numerically exact solutions of the NLCME, we assume

$$\eta_+ = \eta_- = \eta, \tag{1.12a}$$

$$\delta_+ = \delta_- = \delta, \tag{1.12b}$$

$$g_+ = g_- = g, \tag{1.12c}$$

$$a_+ = a_- = a, \tag{1.12d}$$

so that there are only four independent variational parameters in our problem.

Now, we insert the trial function given, Eqs. (1.10a) and (1.10b), into the Lagrangian density, Eq. 1.7, and carry out the integration over all space. As a result, we derive the effective Lagrangian of the system

$$\begin{aligned} L_{\text{eff}} = {} & 4\eta^2 \frac{\delta}{\sin^2\delta}\frac{da}{dt} - 2\pi\eta\frac{1}{\sin\delta}\frac{dg}{dt} - 4\eta^2\frac{1}{\sin\delta}(1-\delta\cot\delta) - 2\pi g\eta\frac{\cot\delta}{\sin\delta}\frac{d\delta}{dt} \\ & + 2\pi g\frac{1}{\sin\delta}\frac{d\eta}{dt} + 2g^2 l\frac{da}{dt} + 12\eta^4\frac{1}{\sin^3\delta}(1-\delta\cot\delta) + 4\eta^2 g^2\frac{1}{\sin\delta} \\ & + 3g^4 l + 8\eta^2 g^2\frac{\delta}{\sin^2\delta} - 4\eta^2\frac{1}{\sin\delta} - 2g^2 l. \end{aligned} \tag{1.13}$$

Using this expression, we obtain the following equations of motion for the four independent variational parameters η, δ, g, and a

$$\begin{aligned} \frac{d\eta}{dt} = \frac{g}{\pi}\Big(& -l\delta - 2\eta^2 - l\delta\cos^2\delta + 9\eta^2 l\delta\cot^2\delta - 4\eta^2\frac{\delta}{\sin\delta} + 3l\eta^2\frac{\delta}{\sin^2\delta} \\ & + l\sin\delta - g^2 l\sin\delta - 2l\delta\cos\delta + g^2 l\cos\delta + 3l\cos\delta\sin\delta \\ & - g^2 l\cos\delta\sin\delta + 8\eta^2\delta^2\frac{\cot\delta}{\sin\delta} + 4\eta^2\delta\cot\delta - 12\eta^2 l\cot\delta\Big) \end{aligned} \tag{1.14a}$$

$$\begin{aligned} \frac{d\delta}{dt} = \frac{2g}{\pi\eta}\Big(& 2\eta^2\delta - 6l\eta^2 + 6l\eta^2\delta\cot\delta + 4\eta^2\frac{\delta^2}{\sin\delta} - l\delta\sin\delta + lg^2\delta\sin\delta \\ & - l\delta\cos\delta\sin\delta + 2l\sin^2\delta - g^2 l\sin^2\delta\Big), \end{aligned} \tag{1.14b}$$

$$\frac{dg}{dt} = \frac{2\eta}{\pi}\left(-2+g^2+\frac{\delta^2}{\sin^2\delta}+6\eta^2\frac{1}{\sin^2\delta}-3\eta^2\frac{\delta^2}{\sin^4\delta}-\delta^2\cot^2\delta\right.$$
$$\left.+3\eta^2\delta^2\frac{\cot^2\delta}{\sin^2\delta}+2\delta\cot\delta-g^2\delta\cot\delta-6\eta^2\delta\frac{\cot\delta}{\sin^2\delta}\right). \tag{1.14c}$$

$$\frac{da}{dt} = -\frac{1}{2\sin\delta}\left(-\delta+6\eta^2\delta+\delta\cos 2\delta+4g^2\sin\delta-\sin 2\delta+g^2\sin 2\delta\right). \tag{1.14d}$$

In the above equations, the shelf length l is not fixed yet. In order to determine its value, we note that the differential Eqs. (1.14a)–(1.14d) exhibit a fixed point which corresponds to the exact stationary soliton solution of the NLCME

$$\eta = \frac{\sin\delta_{fp}}{\sqrt{3}}, \tag{1.15a}$$

$$\delta = \delta_{fp}, \tag{1.15b}$$

$$g = 0, \tag{1.15c}$$

$$a = t\cos\delta_{fp}. \tag{1.15d}$$

Moreover, the value of δ_{fp} can be calculated with the help of the energy conservation law. According to Eq. (1.9), the energy of the pulse described through Eqs. (1.10a) and (1.10b) is

$$\mathcal{E} = 4\eta^2\frac{1}{\sin\delta}(2-\delta\cot\delta)-12\eta^4\frac{1}{\sin^3\delta}(1-\delta\cot\delta)$$
$$-4\eta^2g^2\frac{1}{\sin\delta}-8\eta^2g^2\frac{\delta}{\sin^2\delta}+2g^2l-3g^4l. \tag{1.16}$$

Upon equating the energy at the fixed point $\varepsilon_{fp}=4\sin\delta_{fp}/3$ with the energy of the initial pulse, we obtain for δ_{fp}:

$$\delta_{fp} = \arcsin\left(3\eta_0^2\frac{1}{\sin\delta_0}(2-\delta_0\cot\delta_0)-9\eta_0^4\frac{1}{\sin^3\delta_0}(1-\delta_0\cot\delta_0)\right). \tag{1.17}$$

Now, in order to obtain the oscillation frequency ω of the variational parameters, we linearize Eq. (1.14) around δ_{fp}, leading to

$$\omega = \frac{2}{\sqrt{3}\pi}\sqrt{(3l-4\delta_{fp}-3l\cos\delta_{fp}-2\sin\delta_{fp})(2\delta_{fp}\cos\delta_{fp}+(\delta_{fp}^2-2)\sin\delta_{fp})}. \tag{1.18}$$

This allows us to finally deduce the length of the shelf by comparing this frequency with the actual oscillation frequency $\omega=1-\cos\delta_{fp}$ of the soliton, which represents the nonlinear frequency shift of the pulse [3]. Consequently, we obtain the length l of the shelf as

$$l = \frac{\pi^2}{4} \frac{1 - \cos \delta_{fp}}{2\delta_{fp} \cos \delta_{fp} + (\delta_{fp}^2 - 2) \sin \delta_{fp}} + \frac{4\delta_{fp} + 2 \sin \delta_{fp}}{3(1 - \cos \delta_{fp})}. \quad (1.19)$$

The above expression (1.19) for the shelf length diverges at $\delta_{fp} \approx 0.66\pi$ and it becomes even negative for larger values of the detuning. The reason of this singular behavior originates from the instabilities of the gap solitons in the lower half of the band gap. However, since we consider only the limit of small detunings, this unphysical behavior is of no relevance to our case.

1.3 Radiation Losses

To include the radiation loss in our calculation, we have to couple the nonlinear radiation modes to the linear modes of the system as discussed above. The linear modes represent the solutions of the linearized coupled mode equations (LCME)

$$i\frac{\partial E_+^r}{\partial z} + i\frac{\partial E_+^r}{\partial t} + E_-^r = 0, \quad (1.20a)$$

$$-i\frac{\partial E_-^r}{\partial z} + i\frac{\partial E_-^r}{\partial t} + E_+^r = 0. \quad (1.20b)$$

Here, $E_\pm^r$ denotes, respectively, the dispersive radiation in forward and backward propagating modes in the regions to the left and the right of the shelf. Then, the mass propagating to the right of the pulse is given by

$$\frac{d}{dt} \int_{l/2}^{\infty} (|E_+^r|^2 + |E_-^r|^2) dz = |E_+^r(z = l/2, t)|^2 - |E_-^r(z = l/2, t)|^2. \quad (1.21)$$

This equation describes the non-Markovian radiation dynamics of the solitary wave formation process alluded to above: Radiation travels away from the pulse further to the right (E_+^r), but may become back-reflected from the PBG material so that certain parts of it (E_-^r) return to the pulse. Eventually, only the difference of the masses contained in the modes associated with E_+^r and E_-^r can actually leave the pulse. Since the pulse is symmetric, the radiation propagating to the left exhibits the same behavior. Therefore, in order to allow the coupling of the nonlinear radiation modes to the linear radiation modes, we have to modify the mass conservation of our variational approach to allow for this effect according to

$$\frac{d}{dt}\left(4\eta^2 \frac{\delta}{\sin^2 \delta} + 2g^2 l\right) = -2\,|E_+^r(z = l/2, t)|^2 + 2\,|E_-^r(z = l/2,\, t)|^2. \quad (1.22)$$

On the r.h.s. of Eq. (1.22), only the radiation part occurs. The l.h.s. of Eq. (1.22) represents the mass conservation law within the variational approach which results from inserting the trial functions into Eq. (1.8). Alternatively, the same expression can be derived directly from Eq. (1.14).

Due to the fact that near the upper band edge we can reduce the NLCME to the Nonlinear Schrödinger equation (NLSE), we may now write [3]

$$\begin{pmatrix} E_+^r(z,t) \\ E_-^r(z,t) \end{pmatrix} = \frac{1}{\sqrt{2}}\left(a^r(z,t)\begin{pmatrix} 1 \\ -1 \end{pmatrix} - \frac{i}{2}\frac{\partial a^r(z,t)}{\partial z}\begin{pmatrix} 1 \\ 1 \end{pmatrix}\right) e^{-it}, \tag{1.23}$$

where a^r denotes the solution of the linearized NLSE. Upon inserting Eq. (1.23) into the r.h.s. of Eq. (1.21) we obtain

$$2\,|E_+^r(z=l/2,t)|^2 - 2\,|E_-^r(z=l/2,t)|^2 = -2\,\mathrm{Im}\,(a^{r*}a_z^r)|_{z=l/2}. \tag{1.24}$$

The corresponding calculation for the r.h.s. of Eq. (1.22) is identical to that presented in Ref. [12], where radiation losses for the NLSE – as opposed to the present analysis of the NLCME – have been investigated. As a result, the modified mass conservation equation of the NLCME takes the form (see Ref. [12])

$$\frac{\mathrm{d}}{\mathrm{d}t}\left(4\eta^2\frac{\delta}{\sin^2\delta} + 2g^2 l\right) = -2r\frac{\mathrm{d}}{\mathrm{d}t}\int_0^t \frac{r}{\sqrt{\pi(t-t)}}\,\mathrm{d}t, \tag{1.25}$$

where the height r of the shelf is defined as

$$r^2 = |a^r(z=l/2,t)|^2 = |E_+^r(z=l/2,t)|^2 + |E_-^r(z=l/2,t)|^2. \tag{1.26}$$

Following the argumentation of Ref. [12], the radiation losses may be introduced to the differential equation for the shelf g by adding to its r.h.s. the loss term $-2\beta g$

$$\begin{aligned}\frac{\mathrm{d}g}{\mathrm{d}t} = \frac{2\eta}{\pi}\Bigg(&-2+g^2+\frac{\delta^2}{\sin^2\delta}+6\eta^2\frac{1}{\sin^2\delta}-3\eta^2\frac{\delta^2}{\sin^4\delta}-\delta^2\cot^2\delta \\ &+3\eta^2\delta^2\frac{\cot^2\delta}{\sin^2\delta}+2\delta\cot\delta-g^2\delta\cot\delta-6\eta^2\delta\frac{\delta}{\sin^2\delta}\Bigg)-2\beta g.\end{aligned} \tag{1.27}$$

What remains is to derive an expression for the height r of the shelf and the loss coefficient β within our variational approach to the NLCME. The mass conservation can be rewritten as (recall that in our case $\delta \ll 1$)

$$\frac{\mathrm{d}}{\mathrm{d}t}\left(4\eta^2\frac{\delta}{\sin^2\delta} + 2\,\mathrm{l}g^2\right) \approx \frac{\mathrm{d}}{\mathrm{d}t}\left(4\frac{\eta^2}{\delta} + 2\,\mathrm{l}g^2\right). \tag{1.28}$$

Then, for small derivations of the parameters from the fixed point η_1 and δ_1, we obtain

$$\frac{\mathrm{d}}{\mathrm{d}t}\left(4\frac{\eta^2}{\delta} + 2\,\mathrm{l}g^2\right) \approx \frac{8}{3\delta_{\mathrm{fp}}}\left(1+\frac{32}{3\pi^2}\right)\frac{\mathrm{d}}{\mathrm{d}t}\left(2\eta_1^2 + 2\frac{9\pi^2}{64}g^2\right). \tag{1.29}$$

Moreover, the linearized Eq. (1.21) for g and η_1 give

$$\eta_1 = -\frac{r}{\sqrt{2}}\sin\phi, \tag{1.30a}$$

$$g = -\frac{8}{3\pi}\frac{r}{\sqrt{2}}\cos\phi, \tag{1.30b}$$

where the angle ϕ determines the (temporal) oscillatory behavior of the functions g and $\eta 1$. Finally, by combining Eqs. (1.25),(1.29) and (1.30) we obtain the decay factor β as

$$\beta = \left(\frac{3\pi}{8}\right)^2 \frac{r}{2r(0)\, l\sqrt{\pi t}}, \tag{1.31}$$

with

$$r^2 = \frac{3\,\delta_{fp}}{8}\frac{1}{1+\dfrac{32}{3\pi^2}}\left(\frac{4}{3}\,\delta_{fp} - 2\,\mathrm{l}g^2 - 4\eta^2\frac{\delta}{\sin^2\delta}\right). \tag{1.32}$$

This completes the calculation of the radiation losses and we have obtained a closed set of equations of motion, Eqs. (1.14a), (1.14b), (1.14d) and (1.27), together with Eqs. (1.17), (1.19), (1.31), and (1.32) as well as the initial conditions, Eq. (1.11a) and (1.11b).

1.4 Results

In order to compare the results of our variational approach with numerically exact results, we analyze the maximum intensity $I_{max}(t)$ of the pulse starting with the same initial conditions. Owing to the symmetry of the problem, this maximum always occurs at the origin and within the variational approach it is given as

$$I_{max}(t) = I(z=0,\, t) = 2\left|\frac{\eta}{\cos(\delta/2)} + ig\right|^2. \tag{1.33}$$

In Figure 1.1 we compare the evolution of a pulse (initial conditions $\delta_0 = 0.1\pi$ and $\eta_0 = 1.01\sin(0.1\pi)/\sqrt{3}$) within a numerically exact solution of the NLCME [13] (dashed blue line) and our variational approach (solid red line).

For relatively short times, the variational equations describe the wave dynamics rather accurately. However, for longer times the results become successively less accurate. This is a consequence of the approximations that have reduced the radiation losses of the NLCME model to those of the NLSE, i.e., Eq. (1.27). In other words, for long time scales, the NLSE model becomes an inaccurate approximation to the NLCME model, since the NLSE model does not properly account for the dispersion relation of the linear modes near the PBG. As a result, the expected non-Markovian effects are lost in this case. This is in contrast to the situation without losses, where the NLCME maps exactly to the NLSE for frequencies near the PBG.

In order to restore the correct quantitative behavior for long times, a more sophisticated treatment of losses has to be developed. Specifically, the LCME have to be solved without using the NLSE reduction.

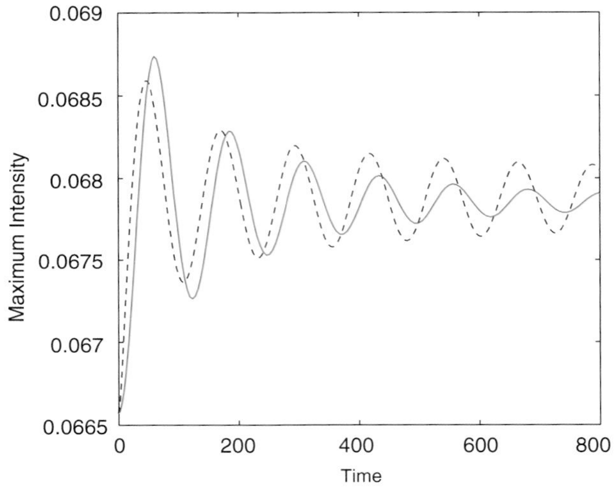

Figure 1.1 Comparison of the losses of the variational solution (solid line) with the direct numerical solution of the NLCME (dashed line). The maximum intensity $I_{max}(t)$ is plotted as a function of the dimensionless time for an initial pulse with $\delta_0 = 0.1\pi$ and $\eta_0 = 1.01\, sin\,(0.1\pi)/\sqrt{3}$.

1.5
Conclusions and Outlook

To summarize, we have presented the variational approach to relaxation of arbitrary pulses to exact solitary wave solutions in one-dimensional PBG materials within the NLCME model. For frequencies near the band edge, we have treated losses to linear radiation modes within the NLSE model. This approach allows a quantitatively correct description of the wave dynamics for not too long time scales. However, for long time scales, the results become less and less accurate, although still remaining qualitatively correct. This is somewhat surprising since naively one would expect that near band edges the NLCME results reduce to the NLSE results. In fact, this is the case for problems without radiation losses such as in the nonresonant collision of Bragg and/or gap solitons [14,15]. However, if the coupling to linear radiation modes is important, for long time scales the oversimplified dispersion relation inherent in the NLSE model fails to correctly take into account the non-Markovian radiation dynamics associated with strong Bragg scattering in nonlinear and/or coupled systems. Therefore, future progress has to be based on a more accurate treatment of these memory effects.

Acknowledgements

We acknowledge support by the Center for Functional Nanostructures (CFN) of the Deutsche Forschungsgemeinschaft within project A1.2. The research of L.T. and K.B.

is further supported by the DFG-Priority Program SPP 1113 "Photonic Crystals" under grant No. Bu-1107/6-1.

References

1 Joannopoulos, D. Meade, R.D. and Winn, J.N. (1995) *Photonic Crystals: Molding the Flow of Light*, Princeton University Press, Princeton, NJ.

2 Busch, K. *et al.* (2007) *Phys. Rep.* **444**, 101.

3 de Sterke C.M. and Sipe, J.E. (1994) in *Progress in Optics*, Vol. 33 (ed. E. Wolf), Elsevier Science, Amsterdam p. 203.

4 Chen W. and Mills, D.L. (1987) *Phys. Rev. Lett.* **58**, 160.

5 Eggleton, B.J. *et al.* (1996) *Phys. Rev. Lett.*, **76**, 1627.

6 Mok, J.T. *et al.* (2006) *Nature Physics*, **2**, 775.

7 Barashenkov, I.V. Pelinovsky, D.E. and Zemlyanaya, E.V. (1998). *Phys. Rev. Lett.* **80**, 5117.

8 Quang T. and John, S. (1994). *Phys. Rev. A* **50**, 1764.

9 Vats, N. John, S. and Busch, K. (2002). *Phys. Rev. A*, **65**, 043808.

10 Gelfand I.M. and Fomin, S.V. (1963). *Calculus of Variations*, Prentice-Hall, Englewood Cliffs, NJ.

11 Malomed, B.A. (2002)in *Progress in Optics* Vol. 43 (ed. E. Wolf,), Elsevier Science, Amsterdam, p. 71.

12 Kath W.L. and Smyth, N.F. (1995) *Phys. Rev. E*, **51**, 1484.

13 de Sterke, C.M. Jackson, K.R. and Robert, B.D. (1991) *J. Opt. Soc. Am. B*, **8**, 403.

14 Tkeshelashvili, L. Pereira, S. and Busch, K. (2004) *Europhys. Lett.*, **68**, 205.

15 Tkeshelashvili, L. *et al.* (2006) *Photon. Nanostruct.: Fundam. Appl.*, **4**, 75.

2
Microscopic Analysis of the Optical and Electronic Properties of Semiconductor Photonic-Crystal Structures

Bernhard Pasenow, Matthias Reichelt, Tineke Stroucken, Torsten Meier, and Stephan W. Koch

2.1
Introduction

The concept of using spatially-structured dielectric systems to control properties of the electromagnetic (EM) field and the light–matter interaction was proposed in 1987 by Yablonovitch [1] and by John [2]. In the following two decades substantial research has been performed in this field. For reviews see, e.g., Refs. [3–7]. Besides a wide range of interesting physical effects, photonic crystal (PhC) systems also offer promising opportunities for improved devices, see, e.g., Ref. [8]. Concurrent with the progress in material science and experimental research, also advanced theoretical tools have been developed and used to simulate and predict the properties of nanophotonic and nanoelectronic systems [3,5,6,9–11].

Combinations of PhCs and semiconductor nanostructures, called semiconductor photonic-crystal structures in this paper, offer a wide range of attractive features. On the one hand, semiconductor nano-structures can be grown with almost atomic precision and, on the other hand, they exhibit many interesting optical effects. For example, the excitonic resonances with large oscillator strengths [12–14] make it possible to reach the regime of strong light–matter coupling, as shown for the example of quantum dots in high-Q PhC cavities [15,16]. Systems of this kind can be used to explore fundamental questions in the area of quantum optics and may even contribute to the field of quantum information processing.

The quantum efficiency of well designed optoelectronic semiconductor devices approaches the fundamental radiative lifetime limit, and combinations of semiconductor nanostructures and photonic crystals allow for the possibility to optimize characteristics of light-emitting diodes and lasers [17–25]. In addition to the application potential, semiconductor photonic-crystal structures are also of interest in the context of fundamental physics. For example, it has already been demonstrated that the obtainable reduction of the spontaneous emission in PhC results in significant

Nanophotonic Materials: Photonic Crystals, Plasmonics, and Metamaterials. Edited by R.B. Wehrspohn, H.-S. Kitzerow, and K. Busch

ISBN: 978-3-527-40858-0

modifications of the exciton statistics and Coulomb many-particle correlations inside a semiconductor material [26].

In this paper, we review a recent theoretical approach for the description of the optical and electronic properties of spatially-inhomogeneous semiconductor photonic-crystal structures on a microscopic basis [9,10,27–29]. In some of our previous treatments [30–35], we have concentrated on longitudinal effects, in particular spatial inhomogeneities induced by dielectric shifts of the exciton resonance, whereas the transverse EM field has been assumed to be homogeneous. Here, we extend these studies and analyze the coupled dynamics of the field and the material excitations by evaluating self-consistently the Maxwell semiconductor Bloch equations (MSBE) including modifications of both the longitudinal EM field, i.e., modifications of the Coulomb interaction, and the transverse propagating EM field.

In our numerical approach, we solve Maxwells' equations in each time step using the finite-difference time-domain (FDTD) method [36]. The semiconductor Bloch equations (SBE) [12–14,37], which are used here in a real-space basis in order to describe spatially-inhomogeneous situations, are integrated using either the Runge–Kutta or the predictor-corrector algorithm [38]. This hybrid approach has successfully been applied to analyze optical absorption and gain spectra in the presence of quasi-equilibrium electron and hole populations [9] and the coherent wave-packet dynamics of photoexcited electron and hole populations in semiconductor photonic-crystal structures [10]. Appropriate variations of this theory have been used to investigate quantum dots in nanocavities [11], planar chiral structures [39,40], and nonlinear optical features of multi-quantum wells [41–43] and vertical-cavity surface-emitting laser structures (VCSEL) utilizing one-dimensional (1D) PhCs [44–46].

In Section 2.2 of this paper the semiclassical microscopic approach and the self-consistent analysis of the light–matter interaction in semiconductor photonic-crystal structures is summarized. Numerical results on the excitonic absorption in spatially-inhomogeneous situations, the coherent dynamics of excitonic wave packets, the decay of the coherent oscillations due to dephasing and relaxation processes, and density-dependent absorption and gain spectra in quasi-equilibrium situations are presented and discussed in Section 2.3. The most important results are briefly summarized in Section 2.4.

2.2
Theoretical Approach

In this section, we present a brief derivation of the real-space MSBE that describe the interactions between a classical light field and semiconductors within or in the vicinity of a spatially-inhomogeneous photonic-crystal environment. In general, the MSBE combine the wave equations for the optical field and the SBE [12–14,37] which are the microscopic equations of motions for the material excitations. Within such a semiclassical approach, the propagating transverse part of the EM field is treated via the wave equation, whereas the longitudinal part of the EM field is responsible for the Coulomb interaction between charged particles. In the vicinity of a spatially-structured dielectric environment, charged particles not only interact with each other directly but also indirectly via induced surface polarizations. A consistent approach should therefore treat these indirect particle–particle interaction on an equal level with the direct Coulomb interaction. As shown in Section 2.2.1, this can be achieved by separating the dielectric displacement into its transverse and longitudinal parts. Within this scheme, the transverse components of the dielectric displacement are determined by Maxwell's equations in which the optical material polarization enters as a source term. The static part of the dielectric displacement modifies the particle–particle interaction and leads to a generalized Coulomb potential as shown in Section 2.2.1.2. This generalized Coulomb potential enters in the Hamiltonian that governs the dynamics of photoexcited electrons and holes in the semiconductor material, which is introduced in Section 2.2.2. Using this Hamiltonian, we evaluate the Heisenberg equations of motion for the optical material excitations in real space, see Section 2.2.3, where also the inclusion of nonradiative dephasing and relaxation is discussed.

2.2.1 Spatially-Inhomogeneous Maxwell Equations in Semiconductor Photonic-Crystal Structures

The propagation of EM waves in macroscopic material is governed by the following set of Maxwell equations

$$\begin{aligned}
&\text{a)}\nabla\cdot\boldsymbol{D}=\rho, \quad \text{b)}\ \nabla\times\boldsymbol{H}-\frac{\partial}{\partial t}\boldsymbol{D}=j \quad \text{with} \quad \boldsymbol{D}=\varepsilon_0\varepsilon(\boldsymbol{r})\boldsymbol{E},\\
&\text{c)}\nabla\cdot\boldsymbol{B}=0, \quad \text{d)}\ \nabla\times\boldsymbol{E}+\frac{\partial}{\partial t}\boldsymbol{B}=0 \quad \text{with} \quad \boldsymbol{B}=\mu_0\boldsymbol{H}.
\end{aligned} \tag{2.1}$$

Here, $\boldsymbol{E}$ and $\boldsymbol{H}$ are the macroscopic electric and magnetic fields, $\boldsymbol{D}$ and $\boldsymbol{B}$ are the dielectric displacement and magnetic induction fields, and ρ and $\boldsymbol{j}$ are the free charges and currents, respectively. For the constitutive relations $\boldsymbol{D}=\boldsymbol{D}[\boldsymbol{E}]$ and $\boldsymbol{B}=\boldsymbol{B}[\boldsymbol{H}]$ we have made a linear ansatz and restricted ourselves to nonmagnetic media, i.e., use $\mu\equiv 1$. Within this ansatz, the PhC structure is described via a periodically-varying scalar dielectric function $\varepsilon(\boldsymbol{r})$, which is assumed to be local and frequency independent. Resonant frequency-dependent contributions are included in the material charge and current densities which are, as is shown below, computed explicitly from the corresponding equations of motion of the material excitations. As is well known, the charge and current densities couple to the scalar and vector potential, respectively. These potentials are introduced with the aid of the homogeneous Maxwell equations as $\boldsymbol{B}=\nabla\times\boldsymbol{A}$ and $\boldsymbol{E}=-(\partial\boldsymbol{A}/\partial t)-\nabla\phi$. Inserting the potentials into the inhomogeneous Maxwell Eq. (2.1a) and (2.1b), we obtain

$$\begin{aligned}
&\nabla\cdot\varepsilon(\boldsymbol{r})\left(\frac{\partial}{\partial t}\boldsymbol{A}+\nabla\phi\right)=-\rho/\varepsilon_0 \quad \text{and}\\
&\nabla\times\nabla\times\boldsymbol{A}+\frac{\varepsilon(\boldsymbol{r})}{c^2}\frac{\partial^2}{\partial t^2}\boldsymbol{A}=-\frac{\varepsilon(\boldsymbol{r})}{c^2}\nabla\frac{\partial}{\partial t}\phi+\mu_0\boldsymbol{j}.
\end{aligned} \tag{2.2}$$

Equation (2.2) shows that the scalar potential is coupled to the vector potential and the charge density. In homogeneous media, the scalar potential can be eliminated from the equations of motion with the aid of the Coulomb gauge $\nabla \cdot \boldsymbol{A} = 0$. In the presence of a spatially-varying dielectric function we can use the generalized Coulomb gauge $\nabla \varepsilon(\boldsymbol{r}) \cdot \boldsymbol{A} = 0$ [47,48] to divide the dielectric displacement into its transverse $\boldsymbol{D}_{\mathrm{T}} = -\varepsilon(\boldsymbol{r})\partial/\partial t\, \boldsymbol{A}$ and longitudinal $\boldsymbol{D}_{\mathrm{L}} = -\varepsilon(\boldsymbol{r})\nabla\phi$ parts. The propagating part of the EM field is associated with this transverse part of the dielectric displacement, whereas the longitudinal part leads to the generalized static Coulomb interaction.

2.2.1.1 Transverse Part: Self-Consistent Solution of the Maxwell Semiconductor Bloch Equations

Within our semiclassical description, the transverse part of the dielectric displacement and the magnetic field obey the coupled wave equations

$$\nabla \times \frac{\boldsymbol{D}_{\mathrm{T}}}{\varepsilon_0 \varepsilon(r)} + \frac{\partial}{\partial t}\boldsymbol{B} = 0 \quad \text{and} \quad \nabla \times \boldsymbol{B} - \mu_0 \frac{\partial}{\partial t}\boldsymbol{D}_{\mathrm{T}} = \mu_0 \boldsymbol{j}_{\mathrm{T}}, \tag{2.3}$$

where $\boldsymbol{j}_{\mathrm{T}}$ is the transverse part of the current, which appears as a source term for the EM field. As has been shown Refs. [10,28], in semiconductor photonic-crystal structures the time derivative of the macroscopic polarization is one of its possible sources, i.e.,

$$\boldsymbol{j}_T = \frac{\partial}{\partial t}\boldsymbol{P}. \tag{2.4}$$

The dynamics of the polarization is determined by the Heisenberg equations of motion which include the interaction with the classical light field, see Section 2.2.3. The Maxwell equations have to be solved self-consistently together with the microscopic equations for the semiconductor excitations. Therefore, we combine the finite difference time domain (FDTD) method [36] for the transversal fields with an predictor-corrector algorithm [38] which is used for the time propagation of the microscopic semiconductor polarizations and electron and hole densities.

2.2.1.2 Longitudinal Part: The Generalized Coulomb Interaction

Within our treatment, the field energy associated with the longitudinal part of the dielectric displacement results in the Coulomb interaction among charged particles. Due to the spatial variation of $\varepsilon(\boldsymbol{r})$, the Coulomb potential $V^{\mathrm{C}}(\boldsymbol{r}, \boldsymbol{r}')$ has to be introduced as the solution of the generalized Poisson equation for a single electron at position $\boldsymbol{r}'$, i.e., a charge density of $\rho(\boldsymbol{r}, t) = \delta(\boldsymbol{r} - \boldsymbol{r}')$

$$-\nabla \cdot [\varepsilon(\boldsymbol{r})\nabla V^{\mathrm{C}}(\boldsymbol{r}, \boldsymbol{r}')] = \delta(\boldsymbol{r} - \boldsymbol{r}')/\varepsilon_0, \quad \text{with} \quad \phi(\boldsymbol{r}, t) = \int \mathrm{d}^3 r' V^{\mathrm{C}}(\boldsymbol{r}, \boldsymbol{r}')\,\rho(\boldsymbol{r}', t). \tag{2.5}$$

Inserting the definition of $\boldsymbol{D}_{\mathrm{L}}$ and $\nabla \cdot \boldsymbol{D}_{\mathrm{L}} = \rho$ into the expression for the field energy,

$$\hat{H}_{\mathrm{C}} = \frac{1}{2}\int \mathrm{d}^3 r \frac{\boldsymbol{D}_{\mathrm{L}} \cdot \boldsymbol{D}_{\mathrm{L}}}{\varepsilon_0 \varepsilon(\boldsymbol{r})} = \frac{1}{2}\int \mathrm{d}^3 r \phi(\boldsymbol{r})\rho(\boldsymbol{r}) = \frac{1}{2}\int \mathrm{d}^3 r \int \mathrm{d}^3 r' \rho(\boldsymbol{r}) V^{\mathrm{C}}(\boldsymbol{r}, \boldsymbol{r}')\rho(\boldsymbol{r}') \tag{2.6}$$

is obtained [30,31].

Since we consider situations close to a structured dielectric environment the generalized Coulomb potential $V_{12}^{\mathrm{C}} \equiv V^{\mathrm{C}}(\boldsymbol{r}_1, \boldsymbol{r}_2)$ depends explicitly on the two coordinates of the interacting carriers and not just on the relative distance as in the homogeneous case.

For a piecewise constant dielectric function, i.e., $\varepsilon(\boldsymbol{r}) = \varepsilon_i$ if $\boldsymbol{r} \in D_i$, the generalized Coulomb potential describing the interaction between unit charges can be written as [30]

$$\begin{aligned} V^{\mathrm{C}}(\boldsymbol{r}, \boldsymbol{r}') &= \frac{1}{4\pi\varepsilon_0} \left[\frac{1}{\varepsilon(\boldsymbol{r}')} \frac{1}{|\boldsymbol{r}-\boldsymbol{r}'|} - \sum_{i<j} \left(\frac{1}{\varepsilon_i} - \frac{1}{\varepsilon_j} \right) \int_{\partial D_{ij}} \mathrm{d}a'' \frac{1}{|\boldsymbol{r}''-\boldsymbol{r}|} \boldsymbol{n}_i \cdot \boldsymbol{D}_l(\boldsymbol{r}'', \boldsymbol{r}') \right] \\ &= V_0(\boldsymbol{r}, \boldsymbol{r}') + \delta V(\boldsymbol{r}, \boldsymbol{r}'), \end{aligned} \tag{2.7}$$

where ∂D_{ij} denotes the interface which separates regions D_i and D_j of different dielectric constants, $\boldsymbol{n}_i$ is the unit vector normal to the surface at $\boldsymbol{r}''$ pointing out of region D_i into D_j, and $\boldsymbol{D}_l(\boldsymbol{r}'', \boldsymbol{r}')$ is that part of the dielectric displacement field, which is connected to the scalar potential via $\boldsymbol{D}_l(\boldsymbol{r}, \boldsymbol{r}') = -\varepsilon(\boldsymbol{r}) \nabla V^{\mathrm{C}}(\boldsymbol{r}, \boldsymbol{r}')$. Equation (2.7) shows that V^{C} is given by the sum of two contributions. The first term, $V_0(\boldsymbol{r}, \boldsymbol{r}') \propto [\varepsilon(\boldsymbol{r}') |\boldsymbol{r} - \boldsymbol{r}'|]^{-1}$ is the usual Coulomb interaction which is screened with the local value of the dielectric constant. The second contribution, $\delta V(\boldsymbol{r}, \boldsymbol{r}')$ in Eq. (2.7), originates from induced polarizations at the interfaces between the regions of different dielectric constants [30,31,34].

The dependence of the generalized Coulomb potential V^{C} on the center-of-mass coordinate obeys the same symmetry properties as the dielectric function. Except for some rather simple geometries, like, e.g., two dielectric half spaces separated by a plane or a single sphere embedded in a material of different dielectric constant [49], V^{C} cannot be obtained analytically. It is, however, possible to evaluate V^{C} numerically using an integral equation for the dielectric displacement $\boldsymbol{D}_l$ at the interfaces $\partial \boldsymbol{D}_{ij}$. This equation can be obtained by applying $\boldsymbol{n}_i(\boldsymbol{r}) \cdot \nabla$ to Eq. (2.7), where $\boldsymbol{n}_i(\boldsymbol{r})$ is the unit vector normal to the surface at $\boldsymbol{r}$. Denoting the normal component of the dielectric displacement by $D_{n_i}(\boldsymbol{r}, \boldsymbol{r}') = \boldsymbol{n}_i(\boldsymbol{r}) \cdot \boldsymbol{D}_l(\boldsymbol{r}, \boldsymbol{r}')$, one obtains

$$4\pi D_{n_i}(\boldsymbol{r}, \boldsymbol{r}') = \boldsymbol{n}_i(\boldsymbol{r}) \cdot \frac{\boldsymbol{r}-\boldsymbol{r}'}{|\boldsymbol{r}-\boldsymbol{r}'|^3} - \lim_{\gamma \to 0_+} \frac{\varepsilon_j - \varepsilon_i}{\varepsilon_j} \int_{\partial D_{ij}} \mathrm{d}a'' \boldsymbol{n}_i(\boldsymbol{r}) \cdot \frac{\boldsymbol{r}_\gamma - \boldsymbol{r}''}{|\boldsymbol{r}_\gamma - \boldsymbol{r}''|^3} D_{n_i}(\boldsymbol{r}'', \boldsymbol{r}'), \tag{2.8}$$

with $\boldsymbol{r}_\gamma = \boldsymbol{r} - \gamma \boldsymbol{n}_i(\boldsymbol{r})$ [32,34]. To obtain $\boldsymbol{D}_n$, Eq. (2.8) can be solved by matrix inversion on a grid in real space. Inserting this solution into Eq. (2.7) allows one to determine the generalized Coulomb potential V^{C} in the regions of interest.

2.2.2
Hamiltonian Describing the Material Dynamics

The Hamiltonian $\hat{H}$ which governs the optical and electronic properties of semiconductors and semiconductor nanostructures consists of three additive terms. The first term $\hat{H}_0$ contains the single-particle band structure and the Coulomb self-energies, the second one $\hat{H}_I$ denotes the interaction of the semiconductor with the classical EM field, and the third one $\hat{H}_C$ describes the many-body Coulomb interaction among charged particles including the modifications due to the dielectric structuring discussed above.

The effective single-particle part of the Hamiltonian reads [9,10,30]

$$\hat{H}_0 = \int d\boldsymbol{r}_1 \left(\hat{c}_1^+ \left[E_G - \frac{\hbar^2 \nabla_1^2}{2m_e} \right] \hat{c}_1 + \hat{d}_1^+ \left[-\frac{\hbar^2 \nabla_1^2}{2m_h} \right] \hat{d}_1 \right)$$

$$\int + \frac{e^2}{2} \int d\boldsymbol{r}_1 \delta V_{11} (\hat{c}_1^+ \hat{c}_1 + \hat{d}_1^+ \hat{d}_1). \tag{2.9}$$

Here, E_G is the band-gap energy, m_e and m_h are the effective masses of the electrons and holes, respectively, and $e^2 \delta V_{11}/2 \equiv e^2 \delta V(\boldsymbol{r}_1,\boldsymbol{r}_1)/2$ are the electron and hole self-energies. Even though the self-interaction due to the bulk part of the Coulomb interaction is unphysical and has been removed, the presence of a spatially-structured dielectric environment leads to physically meaningful interactions of the charges with their self-induced surface polarizations [50–52]. Consequently, besides the kinetic energies, first term of Eq. (2.9), H_0 also contains additionally spatially-varying electron and hole potentials.

In dipole approximation, the light–matter interaction is given by [9,10,30]

$$\hat{H}_I = -\int d\boldsymbol{r}_1 \boldsymbol{E}_1^T(t) \cdot \mu (\hat{d}_1 \hat{c}_1 + \hat{c}_1^+ \hat{d}_1^+), \tag{2.10}$$

where $\boldsymbol{E}^T$ is the transversal part of the optical field and μ is the interband dipole matrix element which is taken to be real. $\hat{d}_1 \hat{c}_1 (\hat{c}_1^+ \hat{d}_1^+)$ describes the local interband coherence which corresponds to destroying (creating) an electron–hole pair at position $\boldsymbol{r}_1$. Using the microscopic polarizations $\hat{p}_{12} = \langle \hat{d}_1 \hat{c}_2 \rangle$, the total optical polarization of the material system reads

$$\boldsymbol{P} = \langle \hat{\boldsymbol{P}} \rangle = \int d\boldsymbol{r}_1 \boldsymbol{P}(\boldsymbol{r}_1) = \int d\boldsymbol{r}_1 \mu (p_{11} + p_{11}^*). \tag{2.11}$$

This expression is used as the source term for the wave Eq. (2.3).

The many-body Coulomb part of the Hamiltonian is given by [9,10,30]

$$\hat{H}_C = \frac{e^2}{2} \int d\boldsymbol{r}_1 \int d\boldsymbol{r}_2 V_{12}^C (\hat{c}_1^+ \hat{c}_2^+ \hat{c}_2 \hat{c}_1 + \hat{d}_1^+ \hat{d}_2^+ \hat{d}_2 \hat{d}_1 - 2 \hat{c}_1^+ \hat{d}_2^+ \hat{d}_2 \hat{c}_1). \tag{2.12}$$

Here, the first two terms represent the repulsive interaction among electrons and holes, respectively, whereas the third term denotes the attractive interaction between an electron and a hole. Since we consider situations close to a structured dielectric environment the matrix element for the Coulombic interactions is the generalized potential $V_{12}^C \equiv V^C(\boldsymbol{r}_1, \boldsymbol{r}_2)$.

2.2.3
Semiconductor Bloch Equations in Real Space

The dynamics of the semiconductor system is described by the equations of motion for the relevant quantities that describe the material excitations [12]. The equation of motion for the expectation value of an arbitrary operator $\mathcal{O} = \langle \mathcal{O} \rangle$ is the Heisenberg equation, i.e.,

$$i\hbar \frac{\partial}{\partial t}\mathcal{O}(t) = \langle \left[\hat{\mathcal{O}}, \hat{H}\right] \rangle. \tag{2.13}$$

Whereas the commutators with $\hat{H}_0$ and $\hat{H}_I$ lead to a set of closed equations of motion on the single-particle level, i.e., the optical Bloch equations, the many-particle part of the Hamiltonian, $\hat{H}_c$, introduces a coupling to an infinite hierarchy of correlation functions [12–14,53]. Due to $\hat{H}_c$, two-operator expectation values like $\langle \hat{d}_1 \hat{c}_2 \rangle$ couple to four-operator expectation values like $\langle c_3^\dagger d_1 c_2 c_3 \rangle$, etc. Since below we are mainly interested in the space-dependence of excitonic resonances and to keep the numerical requirements within reasonable numerical limits, we restrict our present analysis to the level of the time-dependent Hartree–Fock approximation [12], where the four-operator expectation values are split in products of two-operator expectation values.

Using the Hartree–Fock factorization, we obtain a closed set of coupled equations of motion that determine the dynamics of the expectation values of all two-operator quantities. The real-space SBE in time-dependent Hartree–Fock approximation read [10]

$$\begin{aligned} i\hbar\frac{\partial}{\partial t}\left[p_{12}-p_{12}|_{\text{corr}}\right] = & \left[E_G - \frac{\hbar^2\nabla_1^2}{2m_h} - \frac{\hbar^2\nabla_2^2}{2m_e} + \frac{e^2}{2}(\delta V_{11}+\delta V_{22}) - e^2 V_{12}^C - e^2\Delta_{12}^{eh}\right] p_{12} \\ & + e^2\int d^\delta \boldsymbol{r}_3 (V_{13}^C - V_{32}^C)(n_{32}^e p_{13} - n_{31}^h p_{32}) - \mu\cdot(\boldsymbol{E}_1\delta_{12} - \boldsymbol{E}_1 n_{12}^e - \boldsymbol{E}_2 n_{21}^h), \end{aligned} \tag{2.14}$$

$$\begin{aligned} i\hbar\frac{\partial}{\partial t}\left[n_{12}^e - n_{12}^e|_{\text{corr}}\right] = & \left[\frac{\hbar^2}{2m_e}(\nabla_1^2 - \nabla_2^2) - \frac{e^2}{2}(\delta V_{11} - \delta V_{22}) - e^2\Delta_{12}^{eh}\right] n_{12}^e \\ & + e^2\int d^\delta \boldsymbol{r}_3 (V_{13}^C - V_{32}^C)(n_{13}^e n_{32}^e + p_{31}^* p_{32}) + \mu\cdot(\boldsymbol{E}_1 p_{12} - \boldsymbol{E}_2 p_{21}^*), \end{aligned} \tag{2.15}$$

with

$$\Delta_{12}^{eh} = \int d^\delta r_3 (V_{13}^C - V_{32}^C)(n_{33}^e - n_{33}^h) \text{ and } \Delta_{12}^{he} = -\Delta_{12}^{eh}, \tag{2.16}$$

where δ is the dimensionality of the system. The equation for the hole density n_{12}^h is not shown explicitly, since it can be obtained from Eq. (2.15) by replacing e $\leftrightarrow$ h and $p_{ij} \rightarrow p_{ji}$. In Eqs. (2.14), (2.15), the terms denoted by |corr formally represent all many-body correlations that are beyond the time-dependent Hartree–Fock limit [12–14,53]. In the analysis presented here, these correlation terms are either neglected or treated at a phenomenological level.

Within a phenomenological description the nonradiative decay of the interband polarization, often called dephasing, is assumed to be exponential and is described by adding $-i\hbar p_{12}/T_2$ to the right hand side of Eq. (2.14). In the coherent limit, the decrease of the carrier populations and intraband coherences is induced purely by the finite lifetime of the photoexcited carriers. Thus, in this limit the carrier populations and intraband coherences decay with the time constant $T_2/2$ [14]. In reality, however, the coherent limit is often not well suited to describe the dynamics of electron–hole excitations in semiconductors. Typically, the populations and intraband coherences do not vanish on a time scale similar to the dephasing of the optical polarization, but rather become incoherent and approach quasi-equilibrium distributions in the respective bands. On a phenomenological level, this thermalization process can be modeled by adding $-i\hbar(n_{12}^{\mathrm{e}}-n_{12}^{\mathrm{e,eq}})/T_1$ to the right hand side of Eq. (2.15) where $n_{12}^{\mathrm{e,eq}}$ denotes the populations and intraband coherences in quasi-thermal equilibrium [10].

The single-particle self-energies δV appear as potentials in the homogeneous parts of the equations of motion, Eqs. (2.14), (2.15). For the electron–hole interband coherence p_{12}, the homogeneous part of the equation of motion is furthermore influenced by the electron–hole Coulomb attraction $-e^2 V_{12}^{\mathrm{C}}$, which gives rise to excitonic effects. Additionally, integrals over the generalized Coulomb potential V^{C} and products of p's and n's appear in Eqs. (2.14), (2.15) and these equations of motion contain sources representing the driving by the electric field.

Equations (2.14), (2.15) together with the FDTD equations for the EM field allow for a self-consistent description of the dynamical evolution of coupled light and material systems, where the field is driven by the material polarization that is in turn driven by the electric field. This set of equations may be solved for arbitrary field intensities. It contains many-body Coulomb effects and can be used to investigate high-intensity effects like, e.g., Rabi-flopping. Due to the self consistency of the solution, radiative decay processes are included automatically, yielding the correct radiative decay rates for the polarization and the carrier populations even within a semiclassical description.

2.2.3.1 **Low-Intensity Limit**

In order to eliminate density-dependent shifts of the single-particle energies and to prevent rapid dephasing and relaxation due to carrier–carrier scattering, the numerical results presented below are obtained assuming incident laser beams of weak intensities. Therefore, one can describe the light–matter coupling perturbatively and classify the material excitations according to their order in the optical field.

Let us assume that the semiconductor is in its ground state before the optical excitation, i.e., the electron and hole populations as well as the intraband and interband coherences vanish initially. In this case, the linear optical properties of a semiconductor are determined by the equation of motion for the first-order electron–hole interband coherence $p_{12}^{(1)}$ [10,31]

$$i\hbar\frac{\partial}{\partial t}p_{12}^{(1)} = \left[E_{\mathrm{G}} - \frac{\hbar^2\nabla_1^2}{2m_{\mathrm{h}}} - \frac{\hbar^2\nabla_2^2}{2m_{\mathrm{e}}} + \frac{e^2}{2}(\delta V_{11}+\delta V_{22}) - e^2 V_{12}^{\mathrm{C}}\right]p_{12}^{(1)} - \mu\cdot\boldsymbol{E}_1\delta_{12}. \tag{2.17}$$

By diagonalizing the homogeneous part of Eq. (2.17), one can obtain the energies of the excitonic resonances ε_x and the corresponding eigenfunctions Ψ_x ($\boldsymbol{r}_1$, $\boldsymbol{r}_2$). For excitation with a homogeneous light field, i.e., $\boldsymbol{E}_1 \equiv \boldsymbol{E}(\boldsymbol{r}_1)$ is constant in space, the oscillator strength of each excitonic state is proportional to $\mu^2|\int d^{\delta}r\Psi_X(\boldsymbol{r},\boldsymbol{r})|^2$, i.e., to the absolute square of the electron–hole overlap, since the field generates electrons and holes at the same position in space, see Eqs. (2.10) and (2.17). For an inhomogeneous excitation the spatial overlap of the polarization eigenfunctions and the light field redistributes the absorption strengths among the excitonic states.

In second order in the light field, carrier populations and intraband coherences $n_{12}^{e(2)}$ are generated. This process is described by [10]

$$
\begin{aligned}
i\hbar\frac{\partial}{\partial t}n_{12}^{e(2)} &= \left[\frac{\hbar^2}{2m_e}(\nabla_1^2-\nabla_2^2)-\frac{e^2}{2}\delta V_{11}+\frac{e^2}{2}\delta V_{22}\right]n_{12}^{e(2)} \\
&\quad+\int d^{\delta}r_3 e^2(V_{13}^C-V_{32}^C)(p_{31}^{(1)})^{*}p_{32}^{(1)}+\mu\cdot(\boldsymbol{E}_1 p_{12}^{(1)}-\boldsymbol{E}_2(p_{21}^{(1)})^{*}).
\end{aligned}
\tag{2.18}
$$

By diagonalizing the homogeneous part of Eq. (2.18), one can determine the electron and hole eigenstates. Using the spatial periodicity induced by the PhC, the dispersions $\varepsilon_{\boldsymbol{k}}^{\nu}$, where ν labels the mini bands, and the corresponding Bloch-type eigenfunctions $\Phi_{\boldsymbol{k}}^{\nu}(\boldsymbol{r})$ can be computed [9,10]. In quasi-equilibrium, the states in this single-particle basis are populated according to thermal distributions, i.e., $n_{\boldsymbol{k}}^{\nu}$ is given by the Fermi–Dirac function F. The total density n at temperature T can be expressed as

$$
n=\sum_{\boldsymbol{k},\nu} n_{\boldsymbol{k}}^{\nu}=\sum_{\boldsymbol{k},\nu} F(\varepsilon_{\boldsymbol{k}}^{\nu},T,\mu), \tag{2.19}
$$

where $n_{\boldsymbol{k}}^{\nu}$ denotes the population of state $\Phi_{\boldsymbol{k}}^{\nu}$ with energy $\varepsilon_{\boldsymbol{k}}^{\nu}$. Since the total density n depends on the chemical potential μ, $\mu(n,T)$ needs to be determined self-consistently. Having obtained $n_{\boldsymbol{k}}^{\nu}$, we transform back to real space since this is numerically advantageous for performing the dynamic calculations [9,10].

Note, that due to the spatial integrals which appear as a consequence of the many-body Coulomb interaction it requires a lot more effort to numerically solve Eq. (2.18) than the equation for the linear polarization, Eq. (2.17), which contains no spatial integrals. However, if the dynamics is fully coherent and one considers only terms up to second order in the field, i.e., in the coherent $\chi^{(2)}$-limit, it is actually not necessary to solve the combined set of Eqs. (2.17), (2.18). Using the equations of motion, one can easily verify that in this limit the carrier populations and intraband coherences are determined by the interband coherence via the sum rule

$$
n_{12}^{e(2)}=\int d^{\delta}r_3 p_{32}^{(1)}(p_{31}^{(1)})^{*}, \tag{2.20}
$$

i.e., they are given by spatial integrals over products of linear polarizations [10,14].

2.3
Numerical Results

2.3.1
Semiconductor Photonic-Crystal Structure

Numerical solutions of the MSBE for semiconductor photonic-crystal structures typically require a considerable amount of computer time and memory. On the one hand, in general situations a three-dimensional (3D) space discretization is necessary for FDTD solutions of Maxwell's equations. Since the optical wave length and the photonic structure have to be resolved with suitable accuracy, such evaluations have to be performed with a high number of grid points. On the other hand, due to the generalized Coulomb interaction and the coupling to spatially-inhomogeneous light fields, the analysis of the material excitations has to be performed taking into account both the relative and the center-of-mass coordinates. Thus, the SBE have to be evaluated for spatially-inhomogeneous situations where different length scales have to be resolved since the exciton Bohr radius is typically at least one order of magnitude smaller than the optical wave length.

In order to keep the complexity within reasonable limits, we chose for the analysis presented here a model system which consists of a 1D array of dielectric slabs ($\varepsilon = 13$) which extend in z-direction and are separated by air ($\varepsilon = 1$), see Figure 2.1. The substrate below this dielectric structure is made of the same material as the dielectric slabs. Light

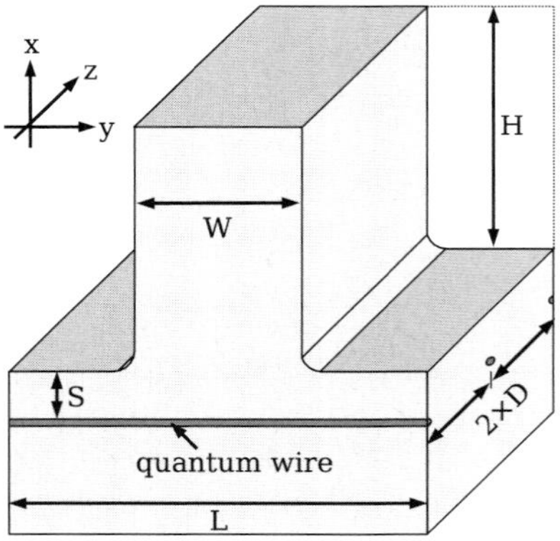

Figure 2.1 Schematical drawing of the considered semiconductor photonic-crystal structure. A simple PhC is modeled as a periodic 1D array of dielectric slabs ($\varepsilon = 13$) which are separated by air ($\varepsilon = 1$). The length L of the unit cell in y-direction is 180 nm. The substrate below the PhC is made of the same material as the dielectric slabs. It surrounds the array of parallel quantum wires which lies $S = 2.6$ nm underneath the PhC. The distance between adjacent wires, which is also the length of the unit cell in z-direction, is $D = 18$ nm. For the width W of the slabs we use 90 nm. For comparison we introduce also two homogeneous reference systems. The situations $W = 0$ and $W = L = 180$ nm are denoted by half-space and homogeneous case, respectively, as is explained in the text. The height H of the slabs is always 700 nm. Taken from Ref. [10].

propagating in this structure may create photoexcitations in an array of parallel semiconductor quantum wires, which extend in y-direction perpendicular to the slabs and are separated from the PhC structure by the distance S. In y-direction the unit cell with length L is repeated periodically. In addition, periodic boundary conditions (PBC) are also used in z-direction with period D, which is the distance between adjacent quantum wires. The parameters used in the numerical calculations are as follows: The length of the unit cell in y-direction is $L = 180$ nm and the height of the slabs is $H = 700$ nm. A number of different values are used for the width W of the slabs. The parallel wires are separated by $D = 18$ nm and the distance to the PhC is $S = 2.6$ nm.

Due to the light propagation through the dielectric structure, the optical field is spatially-varying along the quantum wires. Since the wires are oriented perpendicular to the dielectric slabs, the generalized Coulomb interaction varies periodically along the wires. Our model system thus includes both a space-dependence of the EM field and space-dependent modifications of the semiconductor properties.

For later purposes, we introduce two reference systems. The limit $W = L$, is referred to as homogeneous case, since the dielectric-air interface is far away ($S + H = 702.6$ nm) from the quantum wires and thus the modifications of the Coulomb interaction are negligibly small. The second limit, $W = 0$, is denoted as the half-space case in which the planar air-dielectric interface is very close ($S = 2.6$ nm) to the wire array and therefore the Coulomb interaction is significantly modified. Due to the planar interfaces, the generalized Coulomb potential is homogeneous with respect to the center-of-mass coordinate along the quantum wire for both these cases.

In our calculations, we solve Eq. (2.8) for a single dielectric slab, insert the solution in Eq. (2.7) to obtain the modified Coulomb interaction, and add the resulting δV for the two nearest slabs, i.e., perform a superposition of the Coulomb modifications of three non-interacting slabs. Numerical tests have justified this approximation if the distance between the slabs is not too small. The edges between the dielectric substrate and the slabs have been smoothed as shown in Figure 2.1. This avoids numerical problems when solving Eq. (2.8) using the Nystrom method [38] and takes into account that realistic PhCs made by etching techniques have no sharp edges.

Figure 2.2 shows the Coulomb modifications $\delta V(r_2,r_1)$ in the quantum wires for three fixed, representative positions $r_2 = -90$ nm (a), $r_2 = -45$ nm (b), and $r_2 = 0$ nm (c) as function of position r_1. Positions r_1 underneath the 90 nm wide dielectric slabs are indicated by the gray shading. The Coulomb modifications δV are small for positions r_2 directly underneath the slabs, e.g., case (c), since in this situation the distance to the dielectric-air interfaces of the photonic structure is large. δV is bigger in the regions between the slabs, as can be seen in case (a), because in this situation the distance to the dielectric-air interfaces is small. Between these extremal values, there is a quite sharp transition which takes place within a few nanometers directly under the edges of the slabs, i.e., at $r_2 \approx \pm 45$ nm in Figure 2.2. The single-particle potential, see dashed line in Figure 2.2, shows this sharp transition and follows the periodicity of the PhC. For short distances $|r_1 - r_2|$, the magnitude and space dependence of the modified Coulomb interaction calculated for $r_2 = -90$ nm, i.e., case (a) in Figure 2.2, differ only marginally from those which arise for the half-space case ($W = 0$). The curves (a) and (b) show a strong decrease of the Coulomb

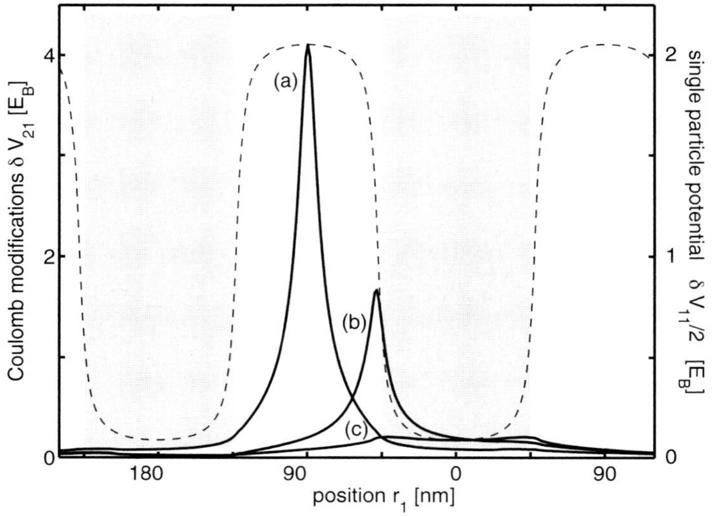

Figure 2.2 Modification of the Coulomb potential $\delta V(r_2, r_1) = \delta V_{21}$ (solid) in the quantum wires for positions $r_2 = -90$ nm (a), $r_2 = -45$ nm (b), and $r_2 = 0$ nm (c) as function of position r_1. The single-particle potential $\delta V(r_1, r_1)/2 = \delta V_{11/2}$ is also displayed (dashed). The dielectric slabs of the PhC are $W = 90$ nm wide and $H = 700$ nm high. Positions r_1 underneath the slabs are indicated by the gray areas. The energy unit is the 3D exciton binding energy of GaAs $E_B = 4.25$ meV. Taken from Ref. [10].

modifications with increasing distance between r_1 and r_2. For positions r_2 underneath the slab, i.e., case (c), δV depends only weakly on the distance if it is smaller than about half the slab width and decreases significantly if the distance exceeds $W/2$.

The dashed line in Figure 2.2 shows that the spatially-periodically-varying dielectric environment introduces a periodic single-particle potential $\delta V(r_1, r_1)/2$ for the electron and holes, with minima underneath the dielectric slabs. This single-particle potential influences the optical and electronic properties of the quantum wire via electron and hole confinement effects and corresponds to a periodic modulation of the effective band gap. Furthermore, the excitonic properties are altered as a result of the modified space-dependent electron–hole attraction.

2.3.2
Linear Excitonic Absorption

In this section, the excitonic resonances in the linear absorption spectra are obtained by solving the linear polarization equation, Eq. (2.17). We assume the incoming external light field to be a plane wave (PW) propagating in negative x-direction. The incident electric field is linearly polarized in y-direction, i.e., parallel to the extension of the quantum wires. The y- and z-components of the electric and magnetic field, E_y and H_z, respectively, have slowly varying Gaussian envelopes and oscillate in time with central frequencies close to the band gap frequency, i.e., $E_G/\hbar$. The linear spectra have been computed using the net energy flux through the boundaries of our FDTD simulation space

$$\Delta\left(\frac{\partial}{\partial t}E\right) = \int d\sigma \boldsymbol{n}\cdot\boldsymbol{S}, \tag{2.21}$$

where $\boldsymbol{S}=\boldsymbol{E}\times\boldsymbol{H}$ is the Poynting vector. The net flux contains all information about the absorbed or gained energy per unit time. In spectrally-resolved experiments, the net flux is measured over all times and analyzed in frequency space

$$\Delta E = \int dt\, \Delta\left(\frac{\partial}{\partial t}E\right) = \int d\omega\alpha(\omega)I_0(\omega), \tag{2.22}$$

where

$$\alpha(\omega) = \frac{1}{I_0(\omega)}\int d\sigma(\boldsymbol{E}^*(\boldsymbol{r},\omega)\times\boldsymbol{H}(\boldsymbol{r},\omega)+\text{c.c.})\cdot\boldsymbol{n} \tag{2.23}$$

and $I_0(\omega)$ is the intensity of the incoming light field. The absorption spectra shown below have been obtained by computing $\alpha(\omega)$ from Eq. (2.23).

The semiconductor parameters used in the following are $\mu = 3.5e\text{Å}\boldsymbol{e}_y$ for the dipole matrix element, $m_h/m_e = 4$ and $m_e = 0.066m_0$ for the electron and hole masses, and $E_G = 1.42$ eV for the energy gap. Considering a dielectric constant of $\varepsilon = 13$, these parameters result in a 3D exciton binding energy of $E_B = 4.25$ meV and a Bohr radius of $a_B \approx 13$ nm In most of the calculations, nonradiative homogeneous broadening is modeled by introducing a decay rate of $\gamma = \hbar/T_2 = 1\ meV$ in the equation of motion of the interband polarization. The Coulomb potential for the 1D wires has been regularized using $V_0 = 1/(|r|+a_0)$ [54,55]. The regularization parameter is chosen as $a_0 = 0.16a_B$. Except for changes of the nonradiative decay time T_2 in Sections 2.3.3 and 2.3.4, these parameters are kept constant in the following, except for Section 2.3.5 where a different structure is analyzed.

To obtain the results presented in the following, the FDTD calculations are performed on a grid with a spatial resolution of 5 nm and a temporal resolution of $dt = dx/(2c) \simeq 8.3\times10^{-18}$ s [36]. The SBE have to be solved with a resolution smaller than the exciton Bohr radius a_B, therefore, we use 1.3 nm, i.e., $\approx a_B/10$. The self-consistent solution of the MSBE is done in the following scheme [9,10]: With the electric field at time t, the magnetic field at $t + dt/2$ and the polarization at $t + dt$ are computed. The polarizations at t and $t + dt$ are used to determine its time derivative at $t + dt/2$ which together with the magnetic field allows us to evaluate the electric field at $t + dt$. Then these steps are repeated.

For the case of a semiconductor quantum well placed close to a planar dielectric-air interface, image charge effects cause a shift of the single-particle energies, i.e., the band gap shifts to higher energies. Since the electron–hole attraction is increased close to air, the exciton binding energy increases as well [30,52]. In semiconductor photonic-crystal structures both, the band gap and excitonic binding energy, are space dependent [31,32]. The band gap variation induces potential valleys underneath the dielectric slabs which give rise to confined single-particle and exciton states, see insets in Figure 2.3. This potential affect the linear absorption spectra and cause the

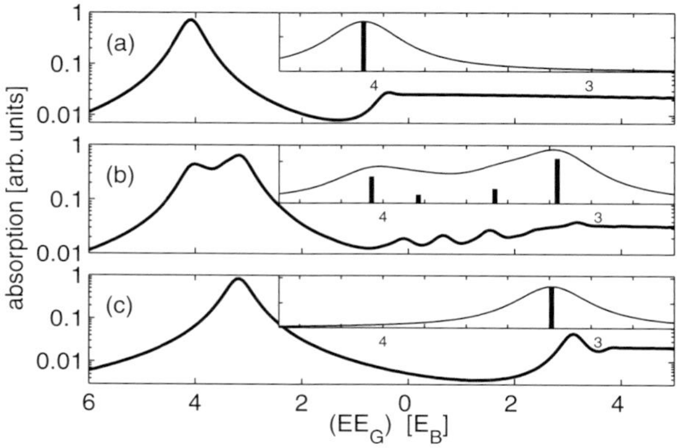

Figure 2.3 Excitonic linear absorption spectra for dielectric slabs of widths (a) $W = 180$ nm (homogeneous case), (b) $W = 90$ nm, and (c) $W = 0$ nm (half-space case), respectively, on a logarithmic scale. The height of the slabs is 700 nm in all cases. The insets show the decomposition of the excitonic resonances into the contributing bound excitonic states and their oscillator strengths on a linear scale. The calculations have been performed using a damping of $\gamma = 1$ meV for the interband polarization and a quantum wire length of five unit cells with PBC. After Ref. [10].

double-peaked 1s-exciton resonance visible in Figure 2.3(b). To improve the visibility of the continuum absorption, the spectra are plotted on a logarithmic scale. For comparison, also the homogeneous and the half-space cases are shown in Figure 2.3 (a) and (c), respectively. Comparing Figure 2.3(b) with (a) and (c) shows that the lower exciton energy agrees with the position of the exciton in the homogeneous case, whereas the upper one corresponds to the half-space case.

Additional information on the optically active exciton resonances is shown in the insets of Figure 2.3. Diagonalizing the linear polarization equation, Eq. (2.17), for one unit cell with PBC yields the energetic positions of the excitonic states. Since the optical field at the quantum wire is spatially varying, the oscillator strengths of the different resonances have been computed by fitting the self-consistently evaluated linear absorption spectra by a sum of Lorentzian curves $\sum_i A_i/((E-\varepsilon_{X,i})^2+\gamma^2)$, where $\gamma = 1$ meV is the decay constant used in the equation of motion for the interband polarization. The insets of Figure 2.3 demonstrates that except for the homogeneous and the half-space cases, (a) and (c), respectively, which are dominated by a single excitonic peak, in a general situation as considered in (b) more than one excitonic resonances contribute to the absorption.

The logarithmic plots of the linear absorption show that the band gap appears at $0E_B$ and $\approx 4E_B$ for the homogeneous and the half-space case, respectively, see Figure 2.3(a) and (c). Thus the exciton binding energy, i.e., the energetic distance between the lowest exciton resonance and the onset of the continuum, increases from $\approx 4E_B$ in the homogeneous case to $\approx 7.2E_B$ in the half-space case. Qualitatively similar results have been obtained for quantum wells close to two-dimensional (2D) PhCs [31,32].

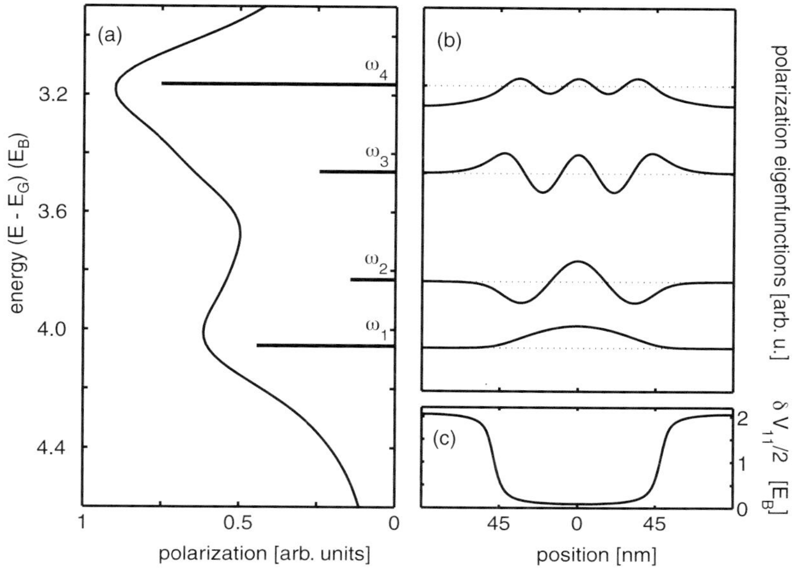

Figure 2.4 (a) Linear excitonic absorption spectrum for the parameters considered in Figure 3.3(b). The lines indicate the spectral positions and the oscillator strengths of the contributing excitonic resonances. (b) Eigenfunctions of the interband polarization obtained by diagonalizing Eq. (3.17). Shown is the spatial variation of the polarization eigenfunction for equal electron and hole positions, i.e., $\Psi_{X,i}(r_1, r_1)$. The dotted lines indicate the eigenenergies $\Psi_{X,i}$ and correspond to the zero polarization axes. (c) Corresponding single-particle potential induced by the spatially-varying dielectric environment. After Ref. [10].

For a more detailed understanding of the excitonic resonances, we analyze in Figure 2.4 also the polarization eigenfunctions, which belong to the excitonic states of Figure 2.3(b). These real polarization eigenfunctions $\Psi_X(r_1,r_2)$ have been obtained by diagonalization Eq. (2.17) for one unit cell with PBC. Shown in Figure 2.4(b) is the spatial variation of the eigenfunction for equal electron and hole positions, i.e., $\Psi_X(r_1,r_1)$. The three lowest states are localized in the potential valley underneath the dielectric slabs, see Figure 2.4(b) and (c). They look similar to usual quantum mechanical eigenfunctions of a particle which is confined in a box-shaped potential and show an increasing number of nodes with increasing energy. Due to its higher energy, the fourth state has strong contributions for positions in between the dielectric slabs, which explains its half-space like character.

2.3.3 Coherent Wave Packet Dynamics

So far, we have focused on the linear optical properties of the system, e.g., the excitonic resonances, their oscillator strengths, resonance energies, and the space-dependent eigenfunctions. In this section, we investigate the intricate coherent wave packet dynamics of the electron density after resonant excitation of the excitonic resonances. The absorption spectra shown in the previous section have been computed using five unit cells of the quantum wire array with PBC. This number of unit cells is required to obtain a converged continuum absorption. Numerical test have shown that for resonant excitation at the excitonic resonances it is justified to reduce the system to one unit cell with PBC since continuum effects do not contribute to the dynamics and the excitonic resonance is already stable for this system size.

The densities are computed for excitation with laser pulses of weak intensities up to second order ($\chi^{(2)}$) in the light–matter interaction. Here, we focus on the fully coherent dynamics and therefore neglect nonradiative dephasing and relaxation processes, i.e., use infinite relaxation and dephasing times $T_1, T_2 \rightarrow \infty$. The coherent electron density is obtained by solving Eq. (2.17) and using Eq. (2.20). Figure 2.5 shows the spatio-temporal dynamics of the electron density $n_{11}^{e(2)}$ after excitation with a Gaussian pulse of 2 ps full width at half maximum (FWHM) duration of the pulse envelope and a central frequency which is tuned to the four energetically lowest excitonic resonances shown in Figure 2.4. For excitation at the three lowest resonances, Figure 2.5(a)–(c), the electron density is basically concentrated at spatial

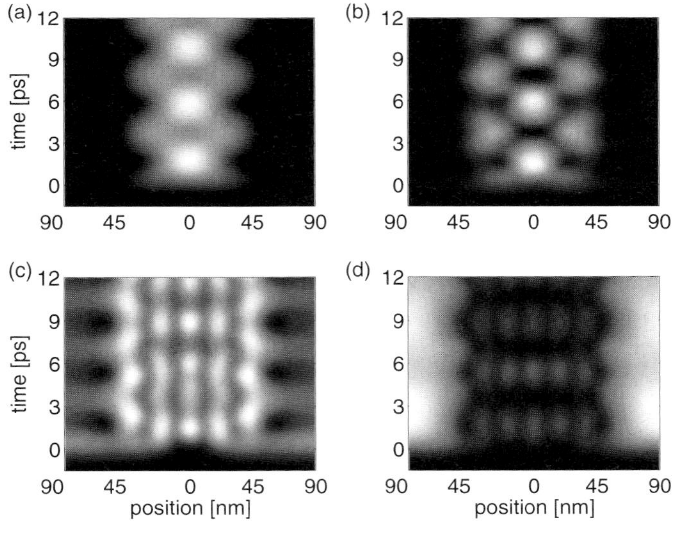

Figure 2.5 Contour plots showing the spatiotemporal dynamics of the coherent electron density $n_{11}^{e(2)}$, see Eq. (3.20), along one quantum wire unit cell. The system is excited by a Gaussian pulse of 2 ps duration (FWHM). The central frequencies of the pulses, ωL, are tuned to the four lowest exciton states: (a) $\hbar\omega_L = E_G - 4.05E_B$, (b) $\hbar\omega_L = E_G - 3.83E_B$, (c) $\hbar\omega_L = E_G - 3.46E_B$, and (d) $\hbar\omega_L = E_G - 3.16E_B$, respectively. The dielectric slabs are $W = 90$ nm wide and $H = 700$ nm high. The calculations have been performed assuming a fully coherent situation, i.e., nonradiative dephasing and thermalization have not been considered ($T_1, T_2 \rightarrow \infty$). White corresponds to the maximal density and black to zero density in each plot. Taken from Ref. [10].

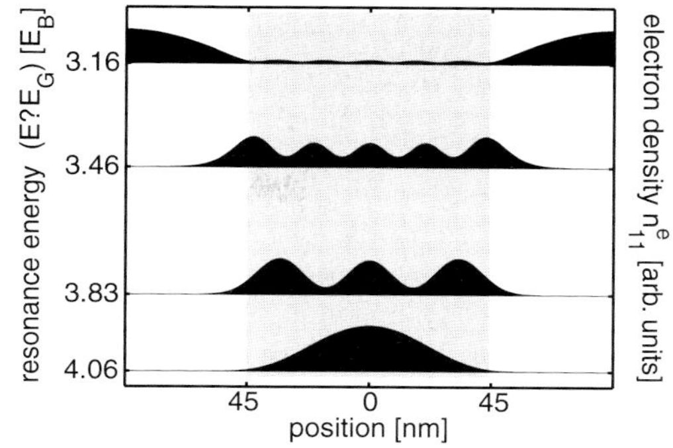

Figure 2.6 Coherent electron densities calculated using Eq. (2.20) and the polarization eigenfunctions shown in Figure 2.4(b). On the left side the energetic positions of the resonances are shown. The vertically displaced lines indicate zero density. Positions underneath the dielectric slabs are marked gray. After Ref. [10]

positions underneath the dielectric slabs, i.e., between ±45 nm. Since the spectral width of the incident electric field of $0.3E_B$ (FWHM of field intensity) is comparable to the energetic spacing between the resonances, the density is not constant as function of time. The pulses generate a coherent superposition of the exciton transitions which leads to wave packet dynamics. However, comparing Figure 2.5(a)–(c) with the electron densities corresponding to the three lowest resonances, see Figure 2.6, shows that for each case the resonantly excited exciton give the strongest contributions to the density. When exciting at the fourth excitonic resonance, Figure 2.5(d), the electron density is concentrated underneath the air regions of the PhC. In this case, the density dynamics corresponds essentially to a coherent superposition of the fourth and third excitonic resonances, see Figure 2.6.

By using spectrally-narrower, i.e., temporally-longer laser pulses, it is possible to selectively excite single exciton resonances. In the limiting cases of very narrow pulses one therefore obtains these static photoexcited densities as shown in Figure 2.6.

2.3.4
Wave Packet Dynamics with Dephasing and Relaxation

As explained above, exciting semiconductor nanostructures with a short optical laser pulses generates a coherent optical material polarization. With increasing time, this polarization decays due to a variety of processes. Depending on the relevant physical mechanisms and the excitation conditions, typical dephasing times can vary between many picoseconds or just a few femtoseconds. Radiative decay due to the finite lifetime of the excited states always contributes to the decay of the optical polarization. However, in semiconductors the dephasing is typically dominated by the interaction with phonons, by the many-body Coulomb interaction, or sometimes by disorder [56]. Simultaneously with the dephasing of the polarization, the initially coherently-excited carrier distributions change their nature and gradually become incoherent. Due to the interaction with phonons and Coulomb scattering among the electrons and holes, these incoherent populations approach thermal quasi-equilibrium distributions in the course of time.

For spatially-homogeneous systems, it is possible to describe dephasing and relaxation at a microscopic level [12,56]. For spatially-inhomogeneous systems, such an analysis is much more complicated. For example, the evaluation of Coulomb scattering processes in the presence of disorder are computationally very demanding and can be performed only for very small systems, see, e.g., Ref. [57]. Therefore, we describe these processes here on a phenomenological level. As discussed in Section 2.2.3, the nonradiative decay of the polarization is modeled by a dephasing time T_2 and the carrier populations approach a quasi-equilibrium Fermi–Dirac distribution with the relaxation time T_1.

Figure 2.7 shows the electron quasi-equilibrium density, i.e., $n_{11}^{e,eq}$, for a small average density of $n_0 = 0.001/a_B$ (dotted line) at a temperature of 50 K. Since the thermal energy $k_B T \approx 1 E_B$ is smaller than the depth of the single-particle potential $\delta V(r_1,r_1)/2$ of $\approx 2E_B$, see Figure 2.4(c), the quasi-equilibrium distribution is strongly concentrated in the regions of low potential energy, i.e., underneath the dielectric slabs. The computed spatio-temporal dynamics of the electron density including dephasing and relaxation is visualized in Figure 2.8. Compared to Figure 2.5 we have used the same structural parameters and excitation conditions and only included the incoherent processes by using relaxation and nonradiative dephasing times of $T_1 = T_2 = 6$ ps. These values are reasonable considering that excitonic transitions are excited with pulses of weak intensities. Figure 2.8 clearly demonstrates that immediately after the excitation the densities exhibit signatures of coherent wave packet dynamics, similar to Figure 2.5. Due to relaxation and dephasing, this wave

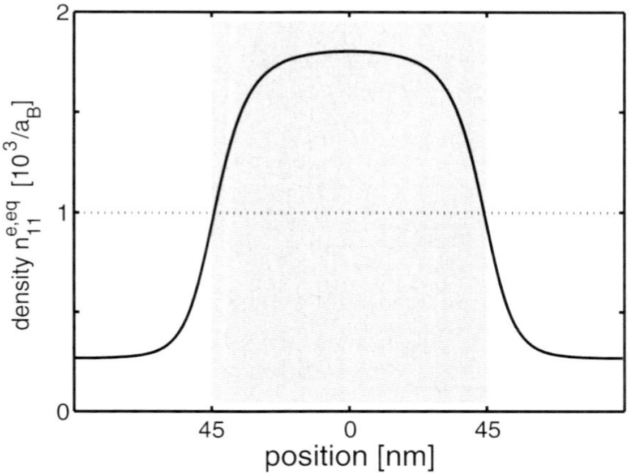

Figure 2.7 Space-dependent quasi-equilibrium electron carrier densities, i.e., $n_{11}^{e;eq}$, for an average density of $n_0 = 0.001/a_B$ (dotted) at a temperature of 50 K. Positions underneath the dielectric slabs are marked gray. After Ref. [10].

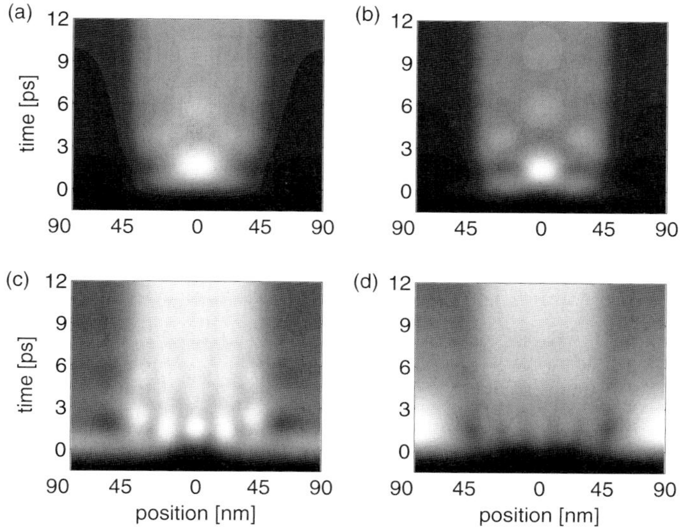

Figure 2.8 Contour plots showing the spatio-temporal dynamics of the electron density $n_{11}^{e(2)}$ along one quantum wire unit cell. The system is excited by a Gaussian pulse of 2 ps duration (FWHM). The central frequencies of the pulses, ωL, are tuned to the four lowest exciton states: (a) $\hbar\omega_L = E_G - 4.05E_B$, (b) $\hbar\omega_L = E_G - 3.83E_B$, (c) $\hbar\omega_L = E_G - 3.46E_B$, and (d) $\hbar\omega_L = E_G - 3.16E_B$, respectively. The dielectric slabs are $W = 90$ nm wide and $H = 700$ nm high. Nonradiative dephasing and relaxation processes have been included using $T_1 = T_2 = 6$ ps at a temperature of 50 K. Except for the dephasing and relaxation, the material parameters and the excitation conditions are the same as in Figure 2.5. White corresponds to the maximal density and black to zero density in each plot. Taken from Ref. [10].

packet dynamics is damped with increasing time. In the limit of long times, i.e., $t \gg T_1, T_2$, the electron population approaches a quasi-equilibrium Fermi–Dirac distribution, cp. Figure 2.7. Therefore, regardless of the excitation conditions which determine the position of the initially generated density, the electrons eventually accumulate in the regions of low potential energy, i.e., underneath the dielectric slabs. This localization of the carriers makes such structures interesting for possible applications in laser structures. The fact that, in the regions of high carrier density, population inversion can be reached at a lower overall density may lead to a reduction of the laser threshold [9,34] as is shown in the next section.

2.3.5
Quasi-Equilibrium Absorption and Gain Spectra

So far, we have focussed our attention on the optical and electronic properties of the inhomogeneous excitonic resonances. These states appear as a consequence of the interaction of the semiconductor electrons with the PhC via the self-induced surface polarizations. In the preceding section, we have shown how the electron densities evolve in the presence of dephasing and relaxation. These quasi-equilibrium static electron and hole densities are used to compute the corresponding absorption and gain spectra by solving Eq. (2.14) with the Runge–Kutta method [38] for a weak incident light field. The computed spectra demonstrate how the linear absorption changes in the presence of spatially-inhomogeneous quasi-equilibrium electron and hole distributions [9]. For these calculations the PhC structure is realized as a rectangular 2D array of air cylinders which is surrounded by a dielectric medium, see inset of Figure 2.9 and Ref. [9] for further details. Although such a structure is different from the one shown in Figure 2.1, the general properties of the excitonic resonances, quasi-equilibrium densities, etc., are very similar for both cases.

Figure 2.9 demonstrates that the two excitonic peaks which appear as a consequence of the structured dielectric environment in the linear absorption spectra have a very different density-dependent behavior. At small carrier densities, the absorption of both peaks is smaller than the absorption at vanishing density. This bleaching occurs due to phase-space-filling effects which are caused by quasi-equilibrium electron and hole distributions $n^{e,eq}$ and $n^{h,eq}$. However, the bleaching of the energetically lower resonance is much stronger than that of the higher resonance. This difference is due to the spatial variation of the carrier populations. The population is biggest in the regions underneath the dielectric material, see Figure 2.7(b). Since these positions are associated with the lower resonance, its absorption is strongly reduced due to the large $n^{e,eq}$ and $n^{h,eq}$. The energetically higher resonance shows

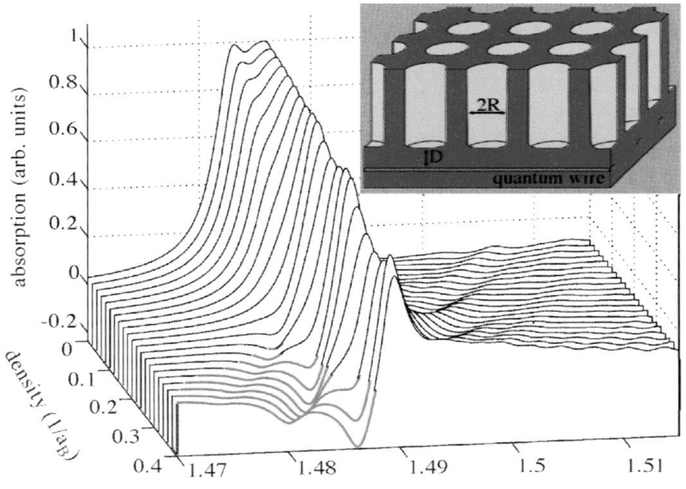

Figure 2.9 (Density-dependent absorption/gain spectra for an array of quantum wires that is separated by $D = 0.2a_B$ from the PhC with air cylinders of radius $R = 2.65a_B$, see inset. The presence of negative absorption, i.e., optical gain, is highlighted by changing the lines from black to red. In the calculations a decay rate for the polarization has been used which results in a homogeneous broadening of 1.7 meV FWHM. After Ref. [9].

less bleaching since it is associated with positions underneath the air cylinders where $n^{e,eq}$ and $n^{h,eq}$ are small.

With increasing carrier density, the bleaching of the 'low density' peak continues, see Figure 2.9. The absorption of the 'high density' resonance, however, vanishes at a certain carrier density. At this density, the sum of the electron and hole populations at the positions underneath the dielectric cylinders is so large that no further absorption is possible. When the density is increased further, the absorption becomes negative, i.e., optical gain is achieved. In this case, the incoming light field is amplified via stimulated emission from the highly-excited material system [9].

The modulations of the absorption which are visible in Figure 2.9 energetically above the exciton resonances and in the gain region are caused by the presence of confined states due to the single-particle potential. The band gap renormalization [12] leads to spectral shifts of these resonances as function of the density.

If one computes density-dependent absorption/gain spectra for wires that are homogeneously surrounded by dielectric material ($R = 0$) only one resonance is present and the carrier population in the wires is homogeneous in space. To obtain optical gain in the homogeneous system a total density of about $0.3/a_B$ is required. However, for the inhomogeneous structure, Figure 2.9, gain is already present for a density of $0.24/a_B$. Our results therefore demonstrate that the spatial accumulation of the population in the regions of low self energy may lead to a significant reduction of the gain threshold which, for the structure considered here, amounts to 20% [9].

2.4 Summary

We have briefly reviewed a theoretical approach which is well suited to describe optoelectronic properties of spatially-inhomogeneous semiconductor photonic-crystal structures on a fully microscopic basis. Self-consistent solutions of the Maxwell semiconductor Bloch equations including modifications of both the longitudinal and the transversal EM field are obtained numerically. As examples we discuss the excitonic absorption in spatially-inhomogeneous situations, the coherent dynamics of excitonic wave packets, the decay of the coherent oscillations due to dephasing and relaxation processes, and density-dependent absorption and gain spectra in quasi-equilibrium situations. In the future, our versatile microscopic scheme will be applied to a variety of other systems and excitation configurations.

Acknowledgements

This work is supported by the Deutsche Forschungsgemeinschaft (DFG) through the Schwerpunktprogramm "Photonische Kristalle" (SPP 1113). T.M. thanks the DFG for support via a Heisenberg fellowship (ME 1916/1). We acknowledge the John von Neumann Institut für Computing (NIC), Forschungszentrum Jülich, Germany, for grants for extended CPU time on their supercomputer systems.

References

1 Yablonovitch, E. (1987) *Phys. Rev. Lett.*, **58**, 2059.

2 John, S. (1987) *Phys. Rev. Lett.*, **58**, 2486.

3 Joannopoulos, J.D., Meade, R.D. and Winn, J.N. (1995) *Photonic Crystals: Molding the Flow of Light*, Princeton University Press, Princeton.

4 Sakoda, K. (2001) Optical Properties of Photonic Crystals, in: Springer Series in Optical Sciences, Vol. 80.

5 Soukoulis, C. (ed.) (2001) *Photonic Crystals and Light Localization in the 21st Century*, Kluwer Academic Publishers, Boston.

6 Busch, K., Lölkes, S., Wehrspohn, R.B. and Föll, H., (eds) (2004) *Photonic Crystals – Advances in Design, Fabrication and Characterization*, Wiley-VCH, Berlin.

7 Prasad, P.N. (2004) *Nanophotonics*, Wiley, Hoboken.

8 Benisty, H., Lourtioz, J.-M., Chelnokov, A., Combrié, S. and Checoury, X. (2006) *Proc. IEEE*, **94**, 997.

9 Reichelt, M., Pasenow, B., Meier, T., Stroucken, T. and Koch, S.W. (2005) *Phys. Rev. B*, **71**, 035346.

10 Pasenow, B., Reichelt, M., Stroucken, T., Meier, T. and Koch, S.W. (2005) *Phys. Rev. B*, **71**, 195321.

11 Dineen, C., Förstner, J., Zakharian, A.R., Moloney, J.V. and Koch, S.W. (2005) *Opt. Express*, **13**, 4980.

12 Haug, H. and Koch, S.W. (2004) *Quantum Theory of the Optical and Electronic Properties of Semiconductors*, 4th ed., World Scientific, Singapore.

13 Schäfer, W. and Wegener, M. (2002) *Semiconductor Optics and Transport Phenomena*, Springer, Berlin.

14 Meier, T., Thomas, P. and Koch, S.W. (2007) *Coherent Semiconductor Optics: From Basic Concepts to Nanostructure Applications*, Springer, Berlin.

15 Reithmaier, J.P., Sek, G., Loffler, A., Hofmann, C., Kuhn, S., Reitzenstein, S., Keldysh, L.V., Kulakovskii, V.D., Reinecke, T.L. and Forchel, A. (2002) *Nature*, **432**, 197.

16 Yoshie, T., Scherer, A., Hendrickson, J., Khitrova, G., Gibbs, H.M., Rupper, G., Ell, C., Shchekin, O.B. and Deppe, D.G. (2004) *Nature*, **432**, 200.

17 Labilloy, D., Benisty, H., Weisbuch, C., Krauss, T.F., De La Rue, R.M., Bardinal, V., Houdré, R., Oesterle, U., Cassagne, D. and Jouanin, C. (1997) *Phys. Rev. Lett.*, **79**, 4147.

18 Painter, O., Lee, R.K., Scherer, A., Yariv, A., O'Brien, J.D., Dapkus, P.D. and Kim, I. (1999) *Science*, **284**, 1819.

19 Imada, M., Noda, S., Chutinan, A., Tokuda, T., Murata, M. and Sadaki, G. (1999) *Appl. Phys. Lett.*, **75**, 316.

20 Boroditsky, M., Krauss, T.F., Coccioli, R., Vrijen, R., Bhat, R. and Yablonovitch, E. (1999) *Appl. Phys. Lett.*, **75**, 1036.

21 Hwang, J.-K., Ryu, H.-Y., Song, D.-S., Han, I.-Y., Song, H.-W., Park, H.-G., Lee, Y.-H. and Jang, D.-H. (2000) *Appl. Phys. Lett.*, **76**, 2982.

22 Erchak, A.A., Ripin, D.J., Fan, S., Rakich, P., Joannopoulos, J.D., Ippen, E.P., Petrich, G.S. and Kolodziejski, L.A. (2001) *Appl. Phys. Lett.*, **78**, 563.

23 Ryu, H.-Y., Hwang, J.-K., Song, D.-S., Han, I.-Y. and Lee, Y.-H. (2001) *Appl. Phys. Lett.*, **78**, 1174.

24 Noda, S., Yokoyama, M., Imada, M., Chutinan, A. and Mochizuki, M. (2001) *Science* **293**, 1132.

25 Song, D.-S., Kim, S.-H., Park, H.-G., Kim, C.-K. and Lee, Y.-H. (2001) *Appl. Phys. Lett.*, **80**, 3901.

26 Kira, M., Hoyer, W., Stroucken, T. and Koch, S.W. (2001) *Phys. Rev. Lett.*, **87**, 176401.

27 Reichelt, M. (2005) *Dissertation*, Philipps-University Marburg, http://archiv.ub.uni-marburg.de/diss/z2005/0123/pdf/dmr.pdf.

28 Pasenow, B. (2006) *Dissertation*, Philipps-University Marburg, http://archiv.ub.uni-

marburg.de/diss/z2006/0584/pdf/dbp.pdf.

29 Meier, T., Duc, H.T., Reichelt, M., Pasenow, B., Stroucken, T. and Koch, S.W. (2006) John von Neumann Institute für Computing, Jülich, Germany, NIC Series, Vol. 32 219.

30 Stroucken, T., Eichmann, R., Banyai, L. and Koch, S.W. (2002) *J. Opt. Soc. Am. B*, **19**, 2292.

31 Eichmann, R., Pasenow, B., Meier, T., Stroucken, T., Thomas, P. and Koch, S.W. (2003) *Appl. Phys. Lett.*, **82**, 355.

32 Eichmann, R., Pasenow, B., Meier, T., Thomas, P. and Koch, S.W. (2003) *phys. stat. sol. (b)*, **238**, 439.

33 Eichmann, R. (2002) *Dissertation*, Philipps-University Marburg, http://archiv.ub.uni-marburg.de/diss/z2003/0173/pdf/dre.pdf.

34 Meier, T. and Koch, S.W. (2004) in *Photonic Crystals – Advances in Design, Fabrication and Characterization*, (eds K. Busch, S. Lölkes, R.B. Wehrspohn, and H. Föll), Wiley-VCH, Berlin. pp. 43–62.

35 Meier, T., Pasenow, B., Thomas, P. and Koch, S.W. (2004) John von Neumann Institute für Computing, Jülich, Germany, NIC Series Vol. 20 261.

36 Taflove, A. (1995) *Computational Electrodynamics – The Finite Difference Time-Domain Method*, Artech-House, Boston.

37 Lindberg, M. and Koch, S.W. (1988) *Phys. Rev. B*, **38**, 3342.

38 Press, W.H., Teukolsky, S.A., Vettering, W.T. and Flannery, B.P. (2002) *Numerical Recipes in C++*, Cambridge University Press, Cambridge.

39 Krasavin, A.V., Schwanecke, A.S., Zheludev, N.I., Reichelt, M., Stroucken, T., Koch, S.W. and Wright, E.M. (2005) *Appl. Phys. Lett.*, **86**, 201105.

40 Reichelt, M., Koch, S.W., Krasavin, A.V., Moloney, J.V., Schwanecke, A.S., Stroucken, T., Wright, E.M. and Zheludev, N.I. (2006) *Appl. Phys. B*, **84**, 97.

41 Prineas, J.P., Zhou, J.Y., Kuhl, J., Gibbs, H.M., Khitrova, G., Koch, S.W. and Knorr, A. (2002) *Appl. Phys. Lett.*, **81**, 4332.

42 Nielsen, N.C., Kuhl, J., Schaarschmidt, M., Förstner, J., Knorr, A., Koch, S.W., Khitrova, G., Gibbs, H.M. and Giessen, H. (2004) *Phys. Rev. B*, **70**, 075306.

43 Schaarschmidt, M., Förstner, J., Knorr, A., Prineas, J.P., Nielsen, N.C., Kuhl, J., Khitrova, G., Gibbs, H.M., Giessen, H. and Koch, S.W. (2004) *Phys. Rev. B*, **70**, 233302.

44 Meier, T., Sieh, C., Koch, S.W., Lee, Y.-S., Norris, T.B., Jahnke, F., Khitrova, G. and Gibbs, H.M. (2004) *Optical Microcavities*, (eds K. Vahala), World Scientific, Singapore. pp. 239–317.

45 Pasenow, B., Reichelt, M., Stroucken, T., Meier, T., Koch, S.W., Zakharian, A.R. and Moloney, J.V. (2005) *J. Opt. Soc. Am. B*, **22**, 2039.

46 Thränhardt, A., Meier, T., Reichelt, M., Schlichenmaier, C., Pasenow, B., Kuznetsova, I., Becker, S., Stroucken, T., Hader, J., Zakharian, A.R., Moloney, J.V., Chow, W.W. and Koch, S.W. (2006) *J. Non-Cryst. Solids*, **352**, 2480.

47 Vogel, W. and Welsch, D.G. (1994) *Quantum Optics*, Akademie Verlag, Berlin.

48 Tip, A. (1997) *Phys. Rev. A*, **56**, 5022.

49 Banyai, L. and Koch, S.W. (1993) *Semiconductor Quantum Dots*, World Scientific, Singapore.

50 Keldysh, L.V. (1988) *Superlattices Microstruct.*, **4**, 637.

51 Tran Thoai, D.B., Zimmermann, R., Grundmann, M. and Bimberg, D. (1990) *Phys. Rev. B*, **42**, 5906.

52 Kulik, L.V., Kulakovskii, V.D., Bayer, M., Forchel, A., Gippius, N.A. and Tikhodeev, S.G. (1996) *Phys. Rev. B*, **54**, R2335.

53 Meier, T. and Koch, S.W. (2001) *Ultrafast Physical Processes in Semiconductors special issue of Series Semiconductors and Semimetals*, Vol. 67 Academic Press, 231–313.

54 Loudon, R. (1976) *Am. J. Phys.*, **44**, 1064.

55 Banyai, L., Galbraith, I., Ell, C. and Haug, H. (1987) *Phys. Rev. B*, **36**, 6099.

56 Koch, S.W., Meier, T., Jahnke, F. and Thomas, P. (2000) *Appl. Phys. A*, **71**, 511.

57 Varga, I., Thomas, P., Meier, T. and Koch, S.W. (2003) *Phys. Rev. B*, **68**, 113104.

3
Functional 3D Photonic Films from Polymer Beads

Birger Lange, Friederike Fleischhaker, and Rudolf Zentel

3.1
Introduction

Artificial opals represent a special sub-class of photonic crystals (PC), which are formed by self-assembly from inorganic or organic monodisperse colloids. Thus they are also called colloidal photonic crystals (CPC). Originally, they were prepared – like natural opals – from silica, but later on opals from polymer received a lot of attention, due to their easy accessibility and the possibility for post-processing. For reviews see the following references [1–3]. This article focuses especially on the chemistry performed in CPCs from polymers to obtain functionality.

Surfactant free emulsion polymerization is the method of choice for the preparation of monodisperse polymer colloids that can be self-assembled to artificial polymer opals or colloidal photonic crystals (CPC) [3–5]. For this process water insoluble monomer is dispersed in water and polymerized with the help of a charged water soluble radical initiator (mostly peroxodisulfate). The charged sulfate radicals initiate the polymerization. The oligomers act as detergents and form micelles as soon as their concentration grows above the critical micelle concentration. Further polymerization proceeds inside these micelles in analogy to the classical form of emulsion polymerization. Slight modification of the polymerization conditions allows for the synthesis of monodisperse polymer colloids from various acrylate monomers (see Figure 3.1). Compared to their inorganic equivalents, polymer colloids offer the advantage that they allow for facile tailoring of many material properties [3,6,7], such as glass transition temperature or refractive index by simple use of differently functionalized monomers as starting material. The use of chemically reactive monomers allows for chemical modification of the spheres after crystallization, while crosslinking can be employed to increase the thermal and mechanical stability as well as resistance towards solvents.

The size of the polymer colloids can be changed by varying the monomer to water ratio (see Figure 3.2) [5]. This is possible since the number of growing micelles (to become the colloids later on) is determined by the critical micelle concentration in

Nanophotonic Materials: Photonic Crystals, Plasmonics, and Metamaterials. Edited by R.B. Wehrspohn, H.-S. Kitzerow, and K. Busch

ISBN: 978-3-527-40858-0

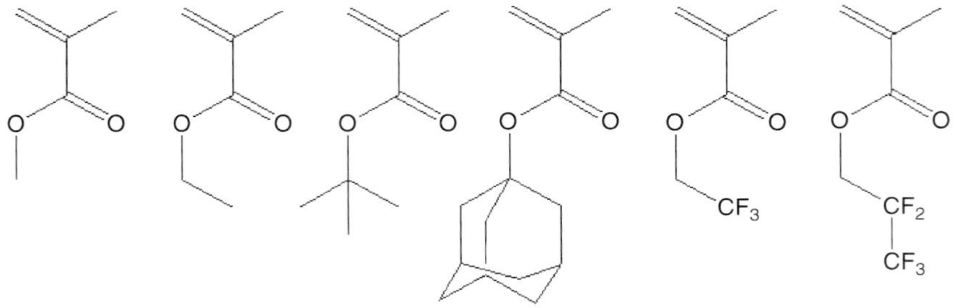

Figure 3.1 Different acrylate monomers, which can be polymerized successfully resulting in monodisperse polymer colloids.

water at a very early stage of the emulsion polymerization. The micelle growth is limited by the amount of available monomer supply that consequently determines the final size of the resulting colloids.

Inorganic colloids are generally attractive, because they offer increased thermal stability and resistance towards organic solvents. Also, they may possess higher refractive indices, as well as semiconducting or fluorescent properties and interesting optical properties in combination with metallo-dielectric structures. For a long time however, only monodisperse silica particles were available. Recently this has changed and fabrication techniques of monodisperse colloids from TiO_2 [8], ZnS [9], selenium or wismut [10] as well as from various metallodielectrics [11,12] have been developed.

It is preferably to crystallize thin films of the colloidal dispersion on flat substrates. This leads to thin opaline films, which present a 3D PC in its 2D limit and in which

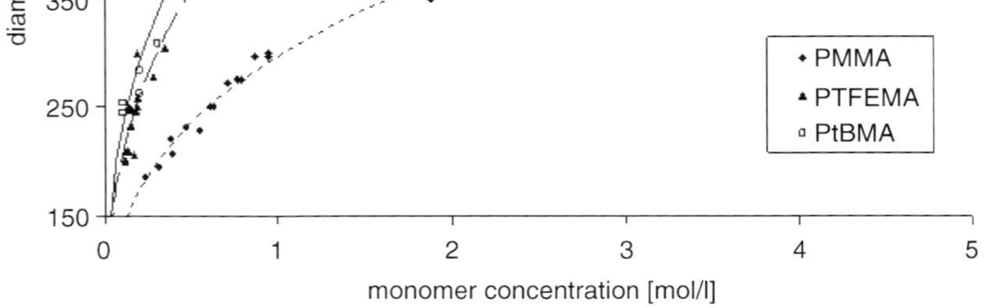

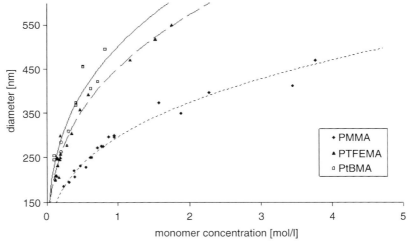

Figure 3.2 Dependence of sphere diameter on the initial monomer to water ratio for three different monomer types; PMMA: polymethylmethacrylate, PTFEMA: polytrifluorethylmethacrylate, PtBMA: poly-*t*-butylmethacrylate [5].

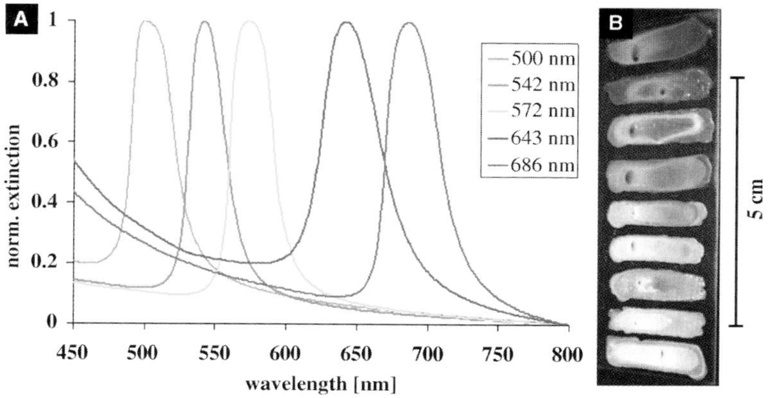

Figure 3.3 (A) Transmission spectra showing the Bragg-peaks of polymer CPC films, consisting of different sphere sizes and (B) corresponding photographs.

the [111]-axis of the fcc packed opal is oriented perpendicular to the substrate. In the simplest version the crystallization of polymer opals is performed by spreading a colloidal dispersion on a flat horizontal substrate [1,4]. Thereafter the dispersion is dried slowly (time varying from over night to one week). In this way large substrates (several cm^2) can be covered by an opaline film (with a thickness of 1–100 μm) of optically homogenous appearance. Figure 3.3 shows pictures of such films and the corresponding UV-spectra. Different colors are obtained by crystallizing colloids of different size.

Optical microscopy shows cracks, which separate the crystallites and which appear due to volume shrinkage during drying. However, large crystals can be obtained (several 100 μm^2), if the film is dried slowly [4,13–15]. Despite crack formation, all crystallites have a similar orientation of the lattice planes and the [111]-plane is oriented parallel to the substrate. The problem of crack formation can be reduced by applying acoustic noise during crystallization [16]. It can be eliminated by crystallizing on a fluid matrix like liquid gallium [3,17].

The preparation of opaline films of homogeneous and controlled thickness is possible by vertical crystallization in a meniscus moving with controlled speed. Experimentally this is either achieved by lowering the liquid level by slow evaporation of the dispersion agent [18,19] (mostly used for SiO_2 colloids suspended in ethanol) or by slowly lifting the substrate [5,20] with a speed of some 100 nm/s. This latter process is commonly applied for polymers, which are dispersed in water (slow evaporation rate). As a result, high quality fcc-packed opaline films of very homogenous thickness can be obtained. They show excellent optical properties, even if the measurement is integrated over areas of several mm^2, with a high virtual absorption due to reflection [3]. Figure 3.4 shows the SEM picture of a corresponding opal.

In addition, the crystallization in a moving meniscus allows for stacking opals or inverted opals of different sphere size or material (see Figure 3.5) [3,7,19–21]. Such architectures are known as heterostructures. The difference in lattice constants and

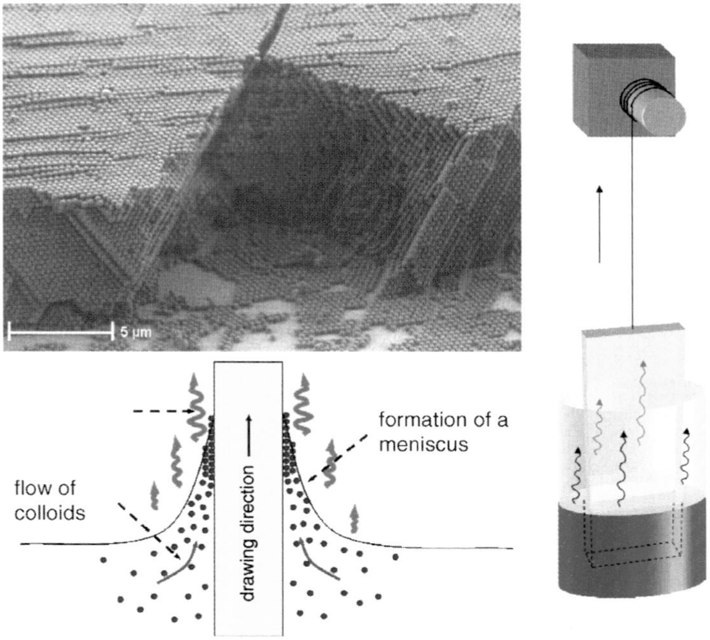

Figure 3.4 Schemes of vertical crystallization, showing the formation of an opaline structure in a moving meniscus and a SEM image of a polymer opal.

material properties of the single stacked opal types may lead to superposition of several stop bands along one probing direction. Scanning electron microscopy investigations show a sharp interface separating both films (see Figure 3.5) [3,21], even if the lattice constant is incommensurable. Also co-crystallization of differently

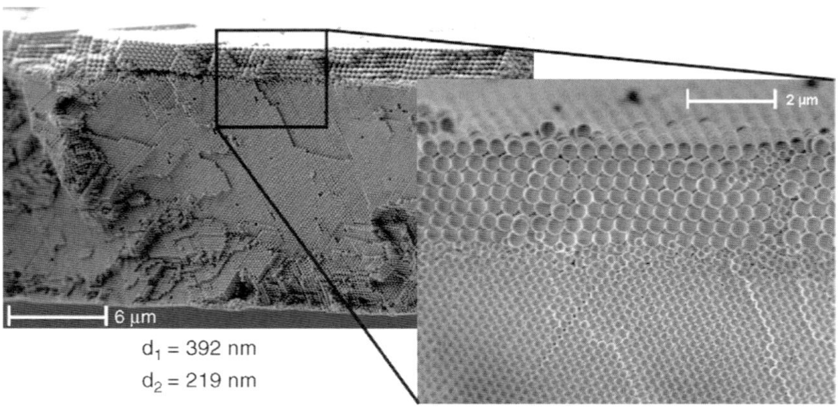

Figure 3.5 Heterostructure of PMMA spheres of 219 nm and 392 nm crystallized in a two step process [7].

sized spheres is possible when performing crystallization in a moving meniscus. Here, the intercalation of smaller spheres into the octahedral and tetrahedral voids of a larger sphere fcc packing can be achieved [22]. In this way bimodal and even trimodal colloidal crystals are available.

An alternative route to the preparation of opaline films of homogenous and well controlled thickness is the crystallization in specially designed packing cells [23]. In cases when the packing quality of the opal is less important and attention is turned to color impression, opals can also be prepared by spraying the colloidal dispersion onto a substrate. Suitable substrates are especially porous materials such as paper [24].

While the packing of monodisperse colloids hardly depends on the material they consist of, new chemistry is needed to realize functional opals and their functional replica. The search for functional opals is thereby especially focused on four topics:

(I) The use of opals and their replica as "simple" coloring agents,

(II) the controlled incorporation of fluorescent materials into opals to study the influence of the photonic band structure on emission, accompanied with the intention to realize thresholdless lasing in such systems,

(III) the replication with different materials to extend the chemical and physical properties of opaline photonic crystals, especially focused on temperature and chemical resistance and

(IV) most importantly the controlled incorporation of defects.

3.2 Opals as Coloring Agents

The creation of chemicals for coloration has been an important part of industrial chemistry since its infancy more than 100 years ago. Usually colorants and color pigments rely on absorption of electromagnetic radiation in the visible spectrum. An alternative possibility to the creation of colors is selective reflection by nanostructured materials. This phenomenon is realized in nature in the wings of various butterflies and in beetles, for example, and it is present in "natural" and "artificial" opals. Although it is easily possible to crystallize brilliantly-colored opaline films from colloidal dispersion on large-area substrates in the laboratory, it is difficult to employ this method for colored coatings: The control of drying (crystallization) conditions on substrates with arbitrary topology is crucial to achieve brilliant reflection colors but unfortunately not straightforward. Also, the requirements for a coating are not fulfilled by an opaline film due to its porous structure (accessibility of the holes in the fcc structure).

There are different methods do obtain the effect pigments, one is to blend a clear coating with opaline flakes and handle them like a coating [25]. Another possibility is to use spraying of concentrated colloidal dispersions to form photonic crystal films on a given substrate and subsequently cover them with a protective layer [24].

3.2.1
Opal Flakes as Effect Pigments in Clear Coatings

For the use of opaline flakes in a clear coating, a method to adjust the refractive index contrast between colloids and filled interstitials and to prevent swelling of the colloids in the coating is required. (1) The refractive index of most polymers is rather close to that of transparent coatings. Hence, the refractive index contrast responsible for the brilliant diffraction colors is low and the optical effect rather poor. (2) Polymer colloids swell and sometimes dissolve if dispersed in solvent-based transparent coatings. (3) The coating application often requires temperatures well above 100 °C.

One possibility to solve these problems is the use of fluorinated monomers for the preparation of monodisperse spheres [25]. Due to their low refractive index (1.44 and lower) the contrast to the transparent coating remains satisfactory. The poor solubility of the corresponding polymer in organic solvents that can be additionally improved by cross-linking reduces swelling of the colloids strongly. A high crosslinking density gives rise to thermal stabilities up to decomposition temperatures above 250 °C. A photograph of such film flakes dispersed as effect pigments in a clear coating is shown in Figure 3.6.

3.2.2
Opaline Effect Pigments by Spray Induced Self-Assembly

Effect pigments are usually prepared as flakes before being actually transferred onto the substrate. During the creation of fine pattern (structures) on substrates, as it is done in an extreme case by ink-jet printing, the resolution is limited by the size of the sprayed liquid droplets. Consequently flakes (edge length from 50 μm to 1000 μm) used as effect pigments may easily exceed the desired size of the aerosol droplets.

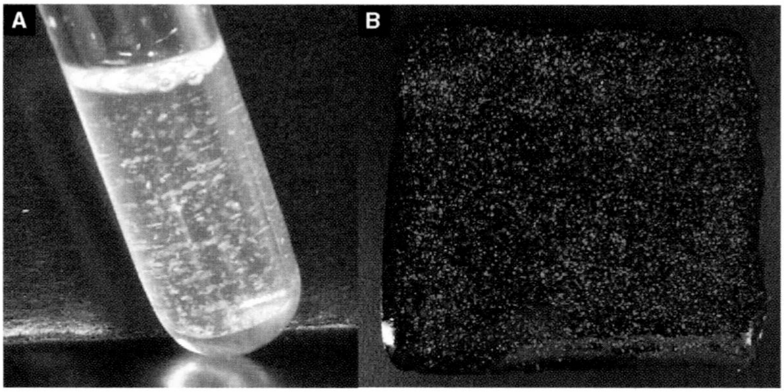

Figure 3.6 Crystal flakes of fluorinated methacrylates as effect pigments (A) dispersed in a clear coating; (B) coated on a substrate after photopolymerization of the transparent coating [25].

This is a general problem since the dimension of the effect pigments have to be a multiple of the wavelength of visible light and cannot be reduced. The flakes' size is necessary to achieve both the interference of electromagnetic radiation leading to the desired photonic effects and a flat arrangement of the flakes on the substrate. An alternative approach to get large effect pigments is to transfer the pigments' building blocks onto the substrate and to let them self assemble afterwards. Self-assembling of opaline materials on porous substrates like paper requires a very fast crystallization. Since crystallization time has a direct impact on the crystal quality, fast drying only gives satisfactory results (brilliant colors) if the used colloids are highly monodisperse. For most applications the use of water as a dispersion agent is a convenient solution. To achieve crystallization from water in a short time it is in addition necessary to use high colloid concentrations of about 40 vol%. This opens the possibility to apply effect pigments to porous substrates not by spraying of the rather large effect pigments themselves, but by spraying of their building blocks, which self-assemble later on. The feasibility of this approach is presented in Figure 3.7 for monodisperse PMMA spheres. This process also tolerates additives used for ink-jet printing [24].

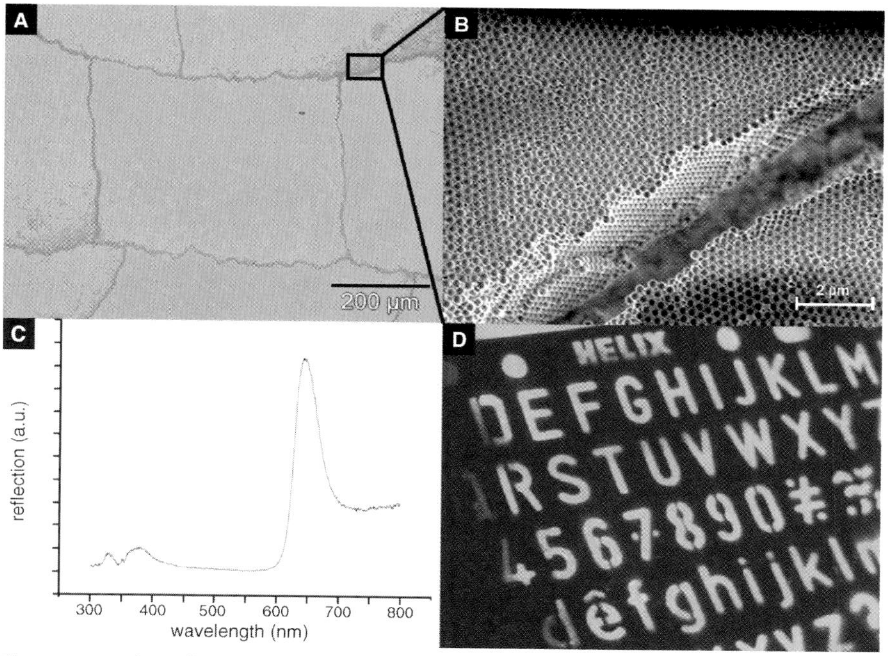

Figure 3.7 Opaline effect pigment from 300 nm PMMA spheres applied with an air brush: (A) Microscope image of large domains; (B) SEM magnification showing the fcc lattice of the opal; (C) reflectance spectrum of the same sample, showing a Bragg-peak at 640 nm (red); (D) photograph of a colloidal dispersion sprayed on a black standard paper through a mask with a resolution of ~1 mm [24].

3.3
Loading of Opals with Highly Fluorescent Dyes

The loading of opals with fluorescent dyes is of interest to be studied for various reasons. From a general point of view the localization of light emitters in CPC based architectures allows us to investigate the influence of the photonic band structure on the photoluminescence properties. In addition the feedback mechanism inside the opal can give rise to lasing or enhanced stimulated emission [26–33]. Loading of the opaline structure with light emitters can be achieved – in the easiest approach – by infiltrating the opal voids with fluorescent materials from solution.

To realize more complex opal structures including defect structures it is necessary to incorporate the fluorophores directly into the matrix. Fluorescent dyes possessing sufficient water-solubility can be directly incorporated into the colloidal particles during the emulsion polymerization process [27]. Others can be incorporated by a swelling and deswelling process of the already synthesized colloids in water/ethanol mixtures [34]. However, highly stable fluorescent materials like perrylenes and especially quantum dots cannot be incorporated employing either of these methods. They require a two-stage preparation process as depicted in Figure 3.8 [26].

First highly fluorescent CdSe quantum dots, which are resistant to the required chemical and mechanical treatment, were synthesized (according to the SILAR

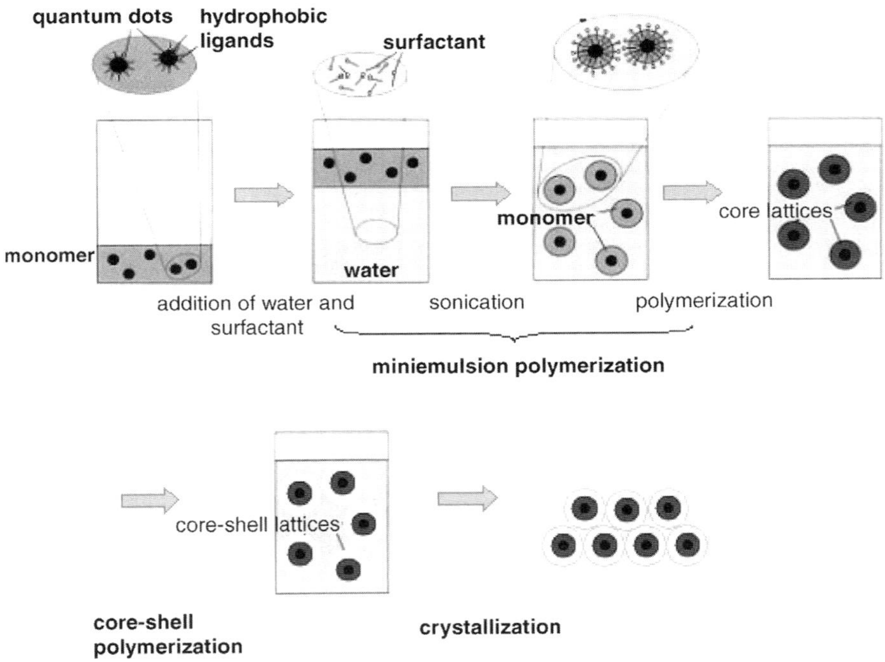

Figure 3.8 Schematic illustration of the incorporation of highly fluorescent CdSe quantum dots into the cores of core-shell colloids [26].

method [35]). They are then transferred into a styrene solution, out of which rather monodisperse core lattices were prepared by a modified miniemulsion process [36]. To increase the size of the core colloids and to increase their homogeneity in size they are subjected to a core-shell polymerization. As a result monodisperse colloids (diameter 200 nm to 400 nm) incorporating – in average – one CdSe quantum dot per sphere can be obtained. These colloids can be crystallized into opals combining the size dependent fluorescent properties of the quantum dots and the photonic band structure of an opal. The modification of the emission by the opaline matrix is shown by angular dependent measurements [26].

3.4 New Properties Through Replication

One possibility to extend the physical and chemical properties of opaline photonic crystals is the change of material, which can be accomplished during a replica formation. In this process the voids of a self-assembled colloidal photonic crystal (CPC) are infiltrated with a different material and the host opal is removed afterwards, leaving an inverse opal consisting of the new material. There are different infiltration methods including sol–gel processes or chemical vapor deposition from a variety of materials.

3.4.1 Increase of Refractive Index

As described above CPCs can be prepared from monodisperse silica or polymer (polymethacrylates or polystyrene derivates) spheres. These materials are limited in their refractive index (at least for high refractive indices) ranging from 1.4 to 1.6. One goal in optics is a photonic crystal with a complete photonic band gap that exists for inverse opal systems. Thus, one needs a high refractive index contrast: If the low index material is air ($n = 1$), the refractive index of the high index replication material has tobe at least 2.8. If a full band gap in the IR region is desired, silicon and germanium possessing high refractive indices in the IR-region ($n \approx 3.45$ or 4.0, respectively) are the materials of choice for replication. With silicon the full band gap has already been realized for the near IR wavelength region [37]. For the visible range no full band gap has been realized so far, due to the lack of transparent high refractive index materials in the visible wavelength region. A high refractive index contrast is still desirable to obtain high reflection along a stop band direction combined with intense reflection colors. Here, mostly TiO_2 ($n \approx 2.8$ nm at 600 nm, rutile form) and SnS_2 ($n \approx 3.2$ nm at 600 nm) are used (see Figure 3.9) [28,38–42]. An important factor besides the refractive index of the bulk material is the filling factor, which determines the effective refractive index. Thus it is essential to find a method that allows for high-degree infiltration of the CPC voids without destruction of the CPC template and for subsequent removal of the host matrix without damaging the replica frame. TiO_2 inverted opals are e.g. prepared by sol–gel infiltration techniques; silicon or SnS_2

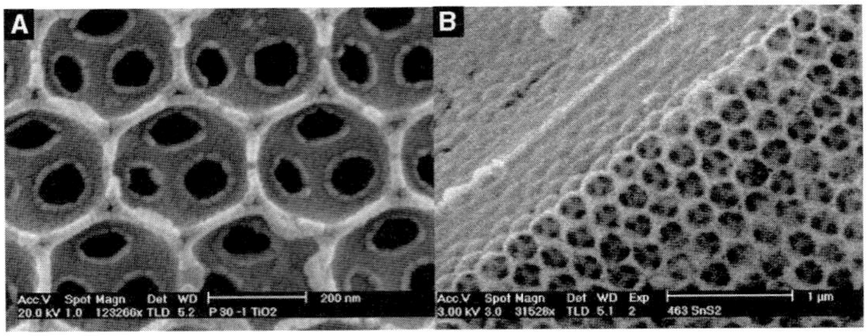

Figure 3.9 SEM images of replica structures from materials, which are transparent in the visible wavelength range (A) TiO_2 and (B) SnS_2 [38].

fillings are usually realized by chemical vapor deposition processes. In both methods the maximum filling fraction is not achieved. The infiltration process is followed by removal of the CPC template. HF and organic solvents are suitable agents for the dissolution of silica and polymer CPCs, respectively. Alternatively polymer templates can be removed by oxygen-plasma.

Figure 3.9 shows SEM images of inverted opals with stop gaps in the visible range. The air holes at the location of the original PMMA or silica spheres form the expected fcc-packed structure of the former opal template. If e.g. a sol–gel process is used, as in the case of TiO_2 replica, solvent is present during filling. This solvent – as well as the reactive groups, which are split-off during the reaction – occupy space. As a result the inverted opal is highly microporous (Figure 3.9A) and possesses a lower effective refractive index than the bulk material. Advantages of the microporosity and the resulting high surface area of inverted metaloxide CPCs can be found in other fields of application than pure optics, such as catalysis (see 4.3).

Inorganic replica can also be made by infiltration of opals with nanoparticles [43] or by filling through the gas phase. The problems remain, i.e. incomplete filling due to blocking of voids (gas phase filling) or volume loss caused by solvent evaporation (nanoparticles).

3.4.2 Robust Replica

Another problem of colloidal crystals made from polymers and their replica is often mechanical (scratch resistance, or adhesion to the substrate) and chemical (swelling or dissolving in organic solvents) stability. The mechanical stability of inverted CPCs is strongly related to a high infiltration degree. Complete infiltration of the CPC template voids can be achieved for example with a neat liquid monomer (applied without solvent), which can be transformed into a solid material by a polymerization process. Such a process has been successfully applied to obtain opal replica from classical polymers [44], they do, however, not fulfill the required demands with respect to mechanical and thermal stability. Replica formation from a liquid inorganic–organic hybrid-material like Ormocer [45] (this material combines acrylate groups for radical crosslinking with silanol groups to form silica) leads to nanostructured

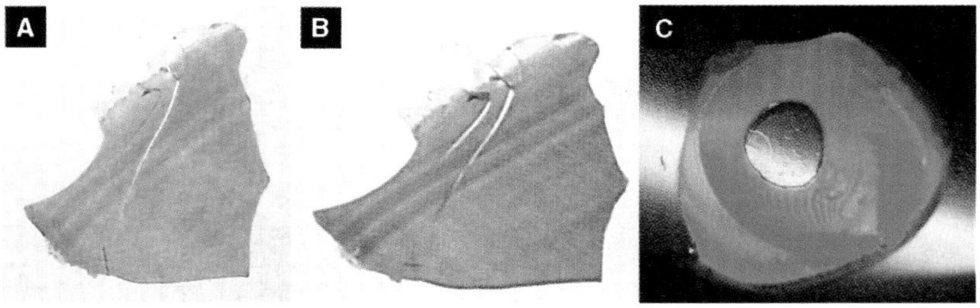

Figure 3.10 Ormocer inverse opal (A) free standing film – top view, (B) free standing film – tilted viewing angle, (C) replica on a glass substrate (2 cm × 2 cm) [46].

photonic crystals incorporating the outstanding mechanical and thermal properties of Ormocer (robustness and temperature stability up to 350 °C) [46].

For this purpose a polymer opal is used as a template and infiltrated with liquid Ormocer-oligomer. After polymerization of the resin the host opal is dissolved in THF and a high-quality inverse opal is obtained. It is possible to prepare large scale replica films (2 cm^2 size), which can be handled as free-standing films with a pair of tweezers (see Figure 3.10).

SEM images of the host opal template as well as the final high-quality Ormocer inverted opal are shown in Figure 3.11.

3.4.3
Inert Replica for Chemistry and Catalysis at High Temperatures

Inverse opals from inorganic materials such as TiO_2 or Al_2O_3 are also interesting candidates for enhanced catalysis due to their periodic structure and well defined large surface area. An example is the combination of the use of the periodic structure to reduce the velocity of photons due to multiple scattering with the photocatalytic activity of TiO_2. Slow photons in photonic crystals are shown to optically amplify the photoactivity of anatase TiO_2 in an inverse opal structure. An enhancement in TiO_2 absorption as a result of slow photons leads to a larger population of electron-hole pairs and faster degradation of organic molecules. A remarkable twofold enhancement is achieved when the energy of the slow photons is optimized with respect to the absorption edge of anatase [47].

This effect can also be used to increase the conversion efficiency of dye-sensitized TiO_2 photoelectrochemical cells. Higher conversion efficiency in the spectral range of 600–800 nm where the dye ($RuL_2(SCN)_2$) has a low extinction coefficient was observed at disordered titania structures and inverse opals with properly positioned stop bands, and at bilayer electrodes of photonic crystals and disordered layers coupled to nanocrystalline TiO_2 films. Multiple scattering events at disordered regions in the ordered and disordered inverse opals, and to a less extent localization of slow photons at the edges of a photonic gap, account for the improved light-harvesting behavior of these structures [48].

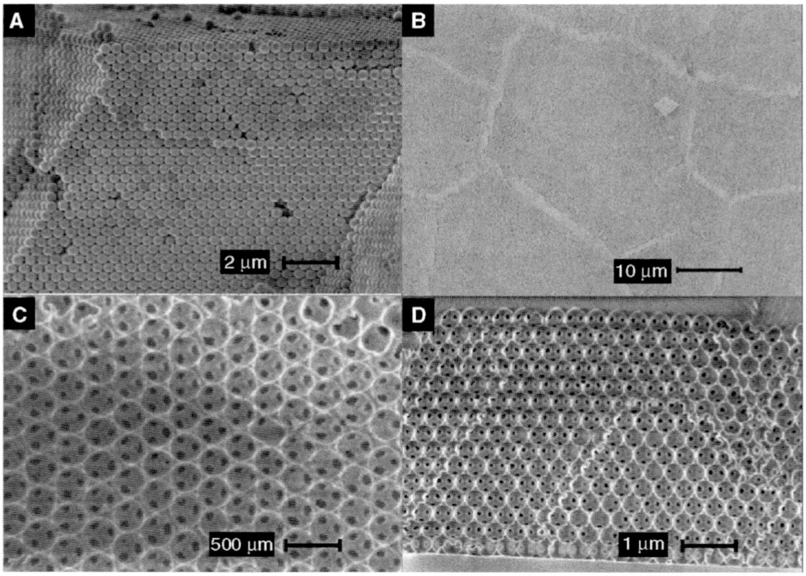

Figure 3.11 (A) SEM image of the colloidal crystal template formed by using 367 nm PMMA spheres, (B) SEM image of an inverse opal at low magnification showing that the cracks of the opal are filled with Ormocer in the respective inverse opal, (C) SEM image of the (111) facet of the inverse opal, (D) SEM image of the cross-section of the inverse opal [46].

Other applications just make use of the well defined large surface area of inverted CPCs e.g. in micro-structured reactors, which have shown significant advantages over conventional reactors due to their enhanced mass and heat transport and compactness [49]. However, using this new type of reactor for heterogeneously catalyzed gas phase reactions such as propane combustion, any maldistributed coating of catalysts still results in the formation of hot spots and negates the advantages of uniform temperature and concentration profiles in the microchannels. An even distribution of catalyst coating in the microchannel enhances the process intensification by micro-structured reactors. For this application first a PMMA opal was sedimented in the channel and transformed into an Al_2O_3 inverse opal afterwards. In a next step, a thin platinum layer was assembled by wet chemistry. This novel Pt-based catalyst with a periodic inverse opal in a microchannel reactor (see Figure 3.12) shows excellent conversion and stable activity for propane combustion at low temperatures [50].

3.5 Defect Incorporation into Opals

The unique ability of PCs to manipulate the transmission of light may lead to potential applications ranging from simple optical switches to an optical computer.

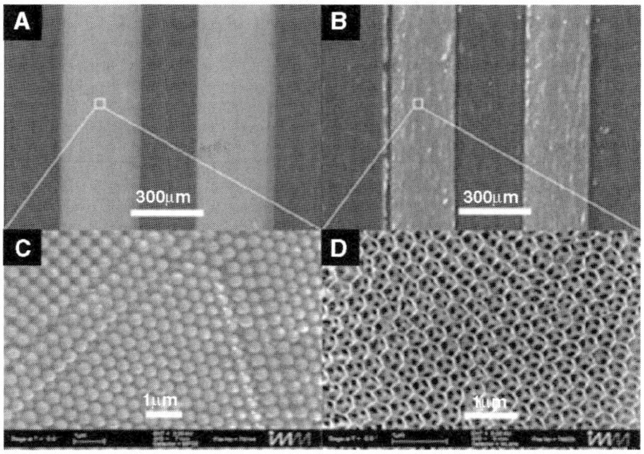

Figure 3.12 Optical micrographs of PMMA opals (A) and corresponding Al_2O_3 inverse opals (B) in the microchannels, and their SEM images: (C) for PMMA opals and (D) for Al_2O_3 inverse opals [50].

This requires the development of methods to incorporate designed defect structures into the crystal, though. The flow of photons in PCs is analogous to the flow of electrons in semiconductors. Analogously to the local doping in semiconductors that is the precondition for the creation of diodes, the controlled incorporation of defects is the precondition for the localization and guiding of photons in photonic crystals.

Realizing defects in PCs, that are built in a top-down method fashion, is rather straight forward and can be accomplished in one step during the fabrication of the matrix [51–57]. Achieving this objective by bottom-up self-assembly in 3D photonic crystals is only a recent development. One possibility to realize simple patterns is site selective crystal formation on structured substrates [58]. The generation of advanced defects in CPCs is a complex matter, but great progress has been gained in the development of novel processes for the incorporation of point [59,60], linear [34,61–64], planar [65–69] and 3D defects [70–75] during the last five years. Thus, large scale organization of monodisperse colloids can be combined with controlled defect generation. The appearance of two recent review articles [76,77] details the efforts being taken in this direction.

3.5.1
Patterning of the Opal Itself

Colloidal crystals from inorganic materials such as silica or from polystyrene are limited in terms of chemical modifications, which can be used for patterning. Colloids based on various polymethacrylates possess the advantage that the corresponding CPCs can be structured by electron beam writing [59,61] or UV-lithography [34,70] for example.

Electron beam lithography is a standard lithography technique widely used to create two-dimensional patterns if high resolution (<10 nm) and high versatility are required. Since PMMA is an electron beam resist, PMMA opals can be patterned by this technique as well. Structuring happens in a two-step process. First, the opal film is locally exposed to an electron beam, which leads to a change in the chemical structure of the methacrylate and enables development in a solvent. Subsequently, the exposed material is selectively removed. In conventional electron beam lithography, the resist is usually deposited on a substrate as a thin layer (typically below 300 nm). In this nearly two-dimensional geometry, the processed material is basically defined by the area that has been exposed to the scanning electron beam. Since in the case of an opal film, the thickness is typically 5 μm to 10 μm, this description is no longer valid. In fact, the volume of processed material depends on the 3D distribution of the electrons within the film. By varying the electron accelerating voltage the penetration depth of electrons can be manipulated [78] and it becomes possible to control the depth of the written structure. Optimization of the process allows the inscription of defect sites down to the size of an individual sphere, being approximately half a micrometer. Advantages with this method are that it employs standard processes for electron beam lithography, and that it is scalable to large areas. This process may be extended to the fabrication of buried single-site defects in self-assembled photonic crystals (see Figure 3.13B). This opens the possibility to prepare devices with single intragap defect lines, and is of interest for applications related to modification and control of directionality of scattered light (see Figure 3.13) [59].

As an alternating route to defects created by chemical modification, special photoprocessable opals can be processed with UV-lithography. Such UV-processable monodisperse colloids have been synthesized from the acid labile polymer poly-*t*-butylmethacrylate. They can be loaded with photoacid generator and crystallized into polymer opal photonic crystals. Irradiation with UV-light followed by baking and development with aqueous base allows for subsequent patterning of the opaline films and the introduction of tailored defects (see Figure 3.14) [34].

Defect structures can also be embedded into an opaline matrix, if a heterostructure of photosensitive and photostable spheres is used (see Figure 3.15) [70]. To stabilize

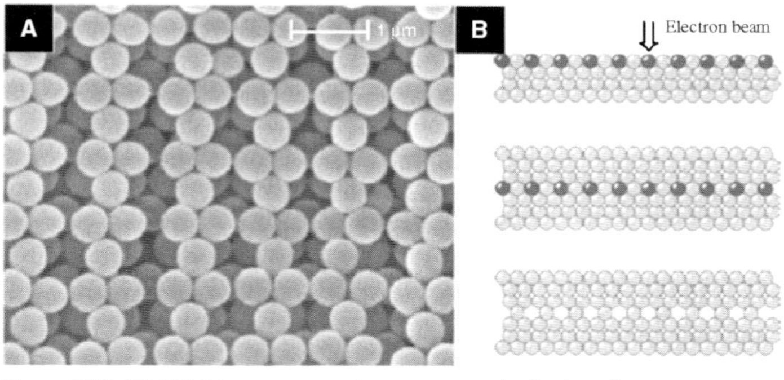

Figure 3.13 (A) SEM image presenting a rectangular lattice of point defects defined on the surface of a PMMA CPC (lattice parameter is 498 nm); (B) proposed process for embedding defects: 1. e-beam exposure, 2. growth of second CPC, 3. development of exposed regions [59].

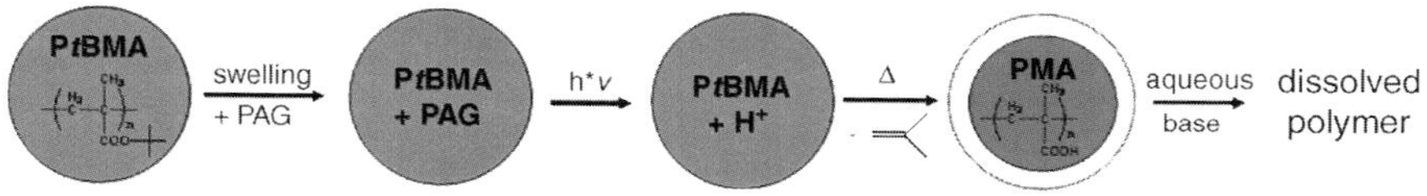

Figure 3.14 Scheme illustrating the preparation of photoprocessable polymer opals and the employed chemical reaction, which is an acid catalyzed saponification used for the realization of light-induced patterning. PtBMA: poly-*t*-butylmethacrylate, PAG: photoacid generator, PMA: polymethacrylic acid [34].

the system and to prevent the defects from collapsing, the photostable top-layer has to be mechanically stabilized. For this purpose the preparation of core-shell colloids with a reactive outer shell containing epoxy-groups is reported (see Figure 3.16), that allow for temperature induced chemical crosslinking [70,75].

3.5.2
Patterning of an Infiltrated Material

It is possible to extend e-beam or nanoimprint techniques to define linear and other 2D defects in CPCs [59]. A similar approach to define embedded linear extrinsic defects is to use conventional photolithography to pattern a photoresist deposited on a CPC. Following the assembly of another crystal on this structure and removal of the photoresist, buried linear air defects have been incorporated within CPCs. These linear defects have been suggested for use as waveguides [63,64]. The methods

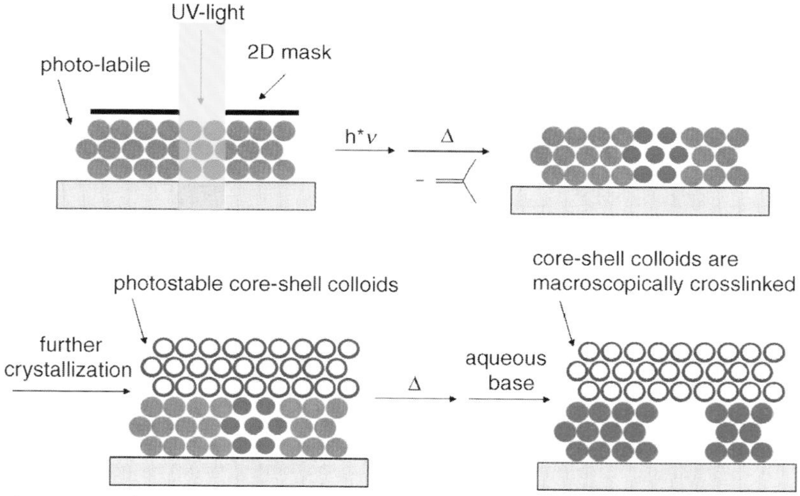

Figure 3.15 Scheme for the construction of a 3D defect via multilayer built-up of photo-labile PtBMA colloids and core-shell-colloids, which can be thermally crosslinked [70].

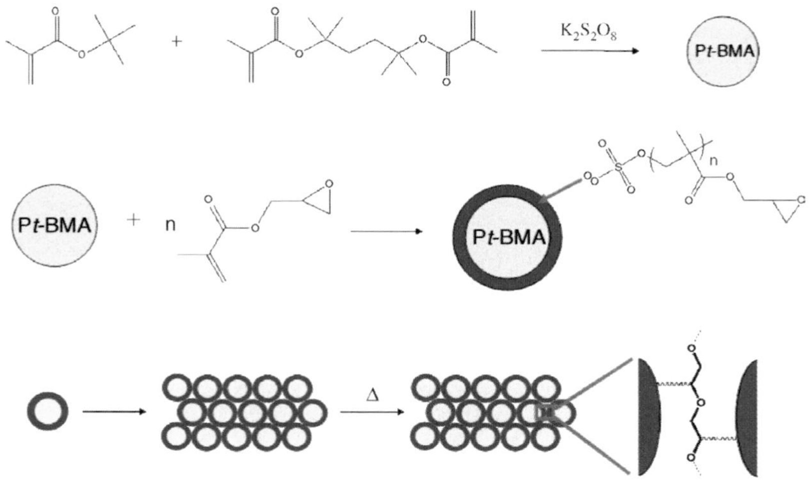

Figure 3.16 Synthesis of core-shell-colloids containing epoxy-functionalized methacrylate shells, which are able to perform a macroscopic crosslinking upon heat treatment [75].

presented so far require a complicated multi-step process in order to create well-defined 3D defects, because the modification is done at the surface of the opal. Thus a second opal has to be grown above or around the opal afterwards. The direct (one step) incorporation of 3D defects into the interior of an artificial opal requires methods to initiate 3D resolved chemical reactions inside an opal like confocal microscopy [71,72] or more preferably two-photon lithography [73,75].

Two-photon lithography, is a technique, which uses two-photon absorption to build complex 3D structures. The advantage of this technique is that outside the focal point, the incident light is not absorbed by the reaction medium. By tightly focusing a femtosecond laser beam into the resin, the photochemical processes occur only in close proximity to the focal point, allowing the fabrication of a 3D structure by directly writing 3D patterns. This unique spatial confinement is due to the fact that the simultaneous absorption rate of two photons has a quadratic dependence on laser power intensity [79–81].

By taking advantage of these processes, defect structures can be fabricated directly inside the opal with control over all three dimensions using a two-photon sensitive photopolymerizable resin. As a host system both opals [71] and inverse opals [75] can be used. Replica from Ormocer have the advantage of chemical and mechanical robustness and hence stays undamaged during the different process steps (see Figure 3.17) [75].

An important consideration for any photonic band gap application is the ability to convert the PC to a high-refractive-index structure that exhibits a complete photonic band gap. Generally, this is accomplished for CPCs through infiltration with a high-index material such as silicon at an elevated temperature, followed by removal of the silica, resulting in an inverse opal structure. Two-photon polymerization can be used

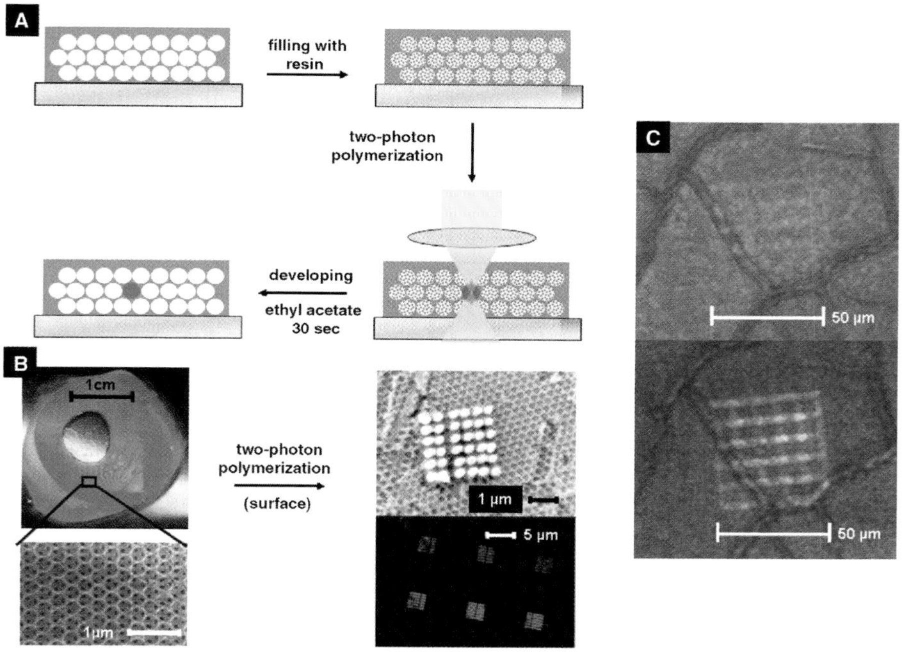

Figure 3.17 (A) Fabrication scheme of an inverse opal with 3D defects with two-photon lithography; (B) left: Digital photograph of an Ormocer replica and a SEM image showing a magnification of the inverse opal structure; right: SEM image of defects introduced on the surface of the replica with two photon lithography and corresponding image of a fluorescence microscope; (C) confocal microscope images showing the fluorescence of the embedded defects, when the focus is at the surface (top) the defect can hardly be seen, when the focus is inside the replica (bottom) the defect pattern can clearly be seen [75].

to deposit a silica hybrid photoresist within a silica-based opal. After chemical vapor deposition of silicon and HF etching embedded hollow waveguide structures within a high-refractive-index inverse opal were formed [72,82]. For success following this general procedure it is important that the materials for defect formation are stable at the requisite high temperatures (250–350 °C for silicon replica).

3.5.3
Chemistry in Defect Layers

Planar defect layers of different materials and with different chemical functionality have been embedded within CPCs by a number of bottom-up approaches. It has been shown that a monolayer of spheres sandwiched between two opal films made of spheres of different diameter can be prepared with the Langmuir-Blodgett technique and behaves as a two dimensional defect [5,83,84]. Spheres and various nanocrystalline aggregates that exceed the size of the CPC entrance windows have been incorporated as defect layers into CPCs by spincoating [85]. Also, a synthesis method that enables the integration of dielectric planar defects in inverted CPCs using chemical vapour deposition (CVD) has been reported [67,86]: A colloidal crystal template is controlled over-infiltrated by CVD to create a homogeneous surface layer. A second CPC is grown on top, fully infiltrated as well and the opal matrix removed afterwards.

Additionally, two bottom-up approaches have been developed that allow for the introduction of "smart" defect structures into CPC. In contrast to the previously described systems, these architectures can be actively addressed by various external stimuli. The defect is based on a functional thin film that is either prepared in a layer-by-layer self-assembly fashion and microcontact transfer printing or by spin-coating and sacrificial CPC infiltration. Thus, it can consist of a wide range of charged or non-charged polymers, biopolymers, dyes, quantum dots and other particles, to name a few [68,69,87–89]. This active element makes the materials interesting for applications as sensors, tunable filters or – if light emitters are incorporated – tunable CPC based laser sources.

In the first method (see Figure 3.18A), the functional thin film is layer-by-layer self-assembled onto a flat piece of polydimethylsiloxane, and the entire multilayer is microcontact transfer-printed onto the surface of a mechanically stabilized colloidal crystal. A second layer of CPC is then grown on top of this surface to form the embedded structural defect. In the second method (see Figure 3.18B), the defect layer is directly spin-coated on the surface of the bottom-CPC. To prevent penetration of the defect material into surrounding CPC voids during the spin-coating process, the bottom-CPC is melt-infiltrated with a sacrificial ribose filling. The sacrificial sugar filling is dissolved in water after spin-coating the defect film from all kind of hydrophobic solvents and the top-CPC is grown to complete the structure.

Optical spectra show a sharp transmission state within the photonic stopband, induced by the defect (see Figure 3.19). The position of the defect state wavelength can be actively tuned by varying the thickness and/or refractive index of the defect layer. The concept is illustrated in Figure 3.20.

CPCs with defect layers consisting of a photochemically active azobenzene based polymer [68], a redox active polyferrocenylsilane (PFS) metallopolymer [69] or a mechanically addressable thermoplastic elastomers [89] have been synthesized, for

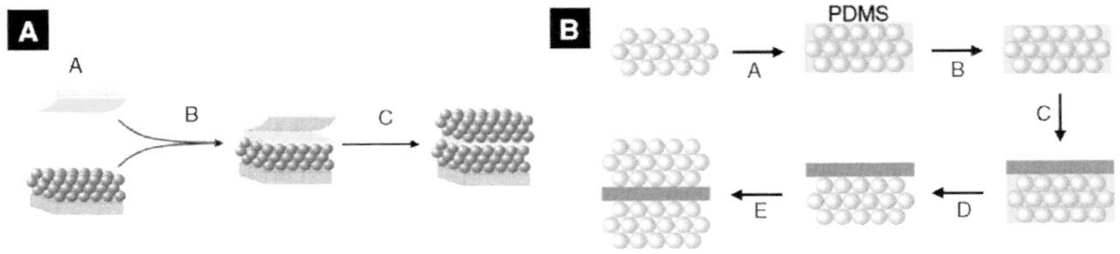

Figure 3.18 Preparation techniques for embedding "smart" defect layers into CPCs (schematic illustration). (A) Defect layer preparation based on layer-by-layer self-assembly and microcontact transfer printing [88]. (B) Defect preparation by spin-coating and sacrificial sugar infiltration of the bottom-CPC [89].

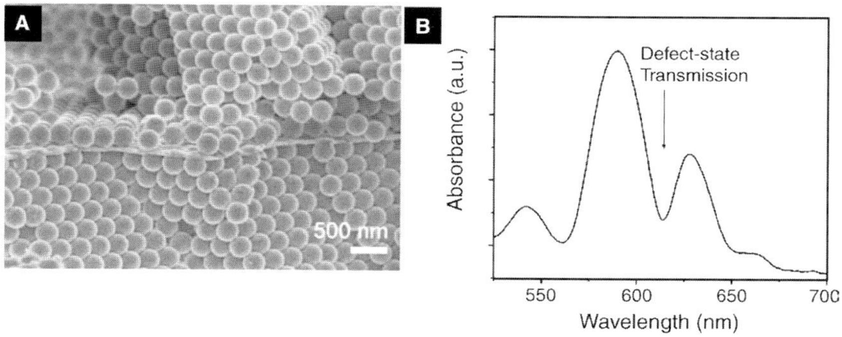

Figure 3.19 (A) Cross-sectional SEM image of a CPC with embedded addressable azobenzene-based defect layer [68]. Slight distortions in the structure result from the cleavage of the sample; (B) corresponding optical spectrum at normal incidence clearly showing Bragg-peak and intragap defect transmission state.

example. Active tuning of the defect state is induced by photochemical trans-cis isomerization and thermal backisomerization of the azo group in the azobenzene containing polymer defect CPC. The defect transmission state of PFS defect CPCs is dynamically switched by oxidizing and reducing the ferrocene units in the PFS polymer film. Oxidation leads to intercalation of the reduced oxidation agent as counterion into the defect layer, reduction to subsequent release, thus changing the optical thickness of the defect reversibly. CPCs incorporating an elastomeric defect

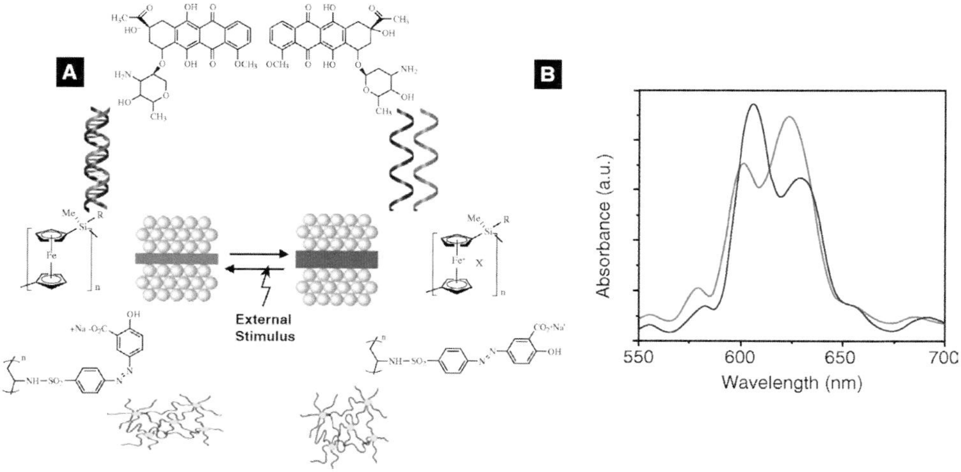

Figure 3.20 (A) Schematic illustration of various responsive defect materials inducing a change in the optical thickness of the defect layer when addressed by an external stimulus. (B) Optical spectra exemplifying a shift of the intragap defect mode resulting from a change in thickness and/or refractive index of the defect layer.

can simply be addressed by applying mechanical pressure leading to compression and decompression of the planar defect. Precise and reversible switching of the defect position is also possible by thermal cycling of polyelectrolyte multilayer CPC heterostructures [68].

In addition, the preparation of CPCs with functional biomolecular planar defect is reported. Biomacromolecules such as proteins and DNA are embedded as nanometer thin sheets in a CPC, while maintaining both the bioactivity of these molecules and the optical properties of the CPC. It is shown that defect CPCs also present a new class of materials to optically monitor various aspects of chemistry or biochemistry taking place in the functional defect layer, through precise shifts of the defect mode. DNA-based planar defect CPCs are used by example to demonstrate defect mode based optical monitoring of DNA conformational changes such as melting and annealing, as well as their enantioselctive interaction with a chiral anti-cancer drug [88]. Optical characterization requires no sophisticated instrumentation and it is performed with a simple fiber optics spectrometer attached to an optical microscope. The setup allows for real time measurements and a spot size of less than 2 μm in diameter can be probed. The developed bio-defect CPCs are consequently highly suitable for array-based analysis and biochip applications. Since CPCs can be used for chromatography, the presented bio-defect CPCs are also potential candidates for combining separation and biomonitoring in a single microstructured sample.

Figure 3.20 graphically summarizes various responsive defect layer materials inducing optical shifts of the defect mode when addressed by an external stimulus.

Acknowledgements

The authors of this paper would like to thank the DFG (Priority program: Photonic crystals) and the EU (PHAT) for financial support. F. Fleischhaker thanks the "Fonds der Chemischen Industrie" for a graduate scholarship and Prof. G.A. Ozin (Universtiy of Toronto) for the possibility to work in his lab.

References

1 Xia, Y., Gates, B., Yin, Y. and Lu, Y. (2000) *Adv. Mater.*, **12**, 693; and special volume devoted to CPCs, (2001) *Adv. Mater.*, **13**, 369–534.

2 Lopez, C. (2003) *Adv. Mater.*, **15**, 1679.

3 Egen, M., Zentel, R., Ferrand, P., Eiden, S., Maret, G. and Caruso, F. (2004) Preparation of 3D Photonic Crystals from Opals, in *Photonic Crystals – Advances in Design, Fabrication and Characterization* (eds K. Busch, S. Lölkes R.B. Wehrspon and H., Föll), Wiley-VCH, Weinheim, p. 109ff.

4 Müller, M., Zentel, R., Maka, T., Romanov, S.G. and Sotomayor Torres, C.M. (2000) *Chem. Mater.*, **12**, 2508.

5 Egen, M. and Zentel, R. (2004) *Macromol. Chem. Phys.*, **205**, 1479.

6 Egen, M. and Zentel, R. (2002) *Chem. Mater.*, **14**, 2176.

7 Egen, M., Voss, R., Griesebock, B., Zentel, R., Romanov, S.G. and Sotomayor Torres, C.M. (2003) *Chem. Mater.*, **15**, 3786.

8 Jiang, P. Bertone, J.F. and Colvin, V.L. (2001) *Science*, **291**, 453.
9 Velikov, K.P. and van Blaaderen, A. (2001) *Langmuir*, **17**, 4779.
10 Jeong, U., Wang, Y., Ibisate, M. and Xia, Y. (2005) *Adv. Funct. Mater.*, **15**, 1907.
11 Graf, C. and van Blaaderen, A. (2002) *Langmuir*, **18**, 524.
12 Wang, Y., Ibisate, M., Li, Z.Y. and Xia, Y. (2006) *Adv. Mater.*, **18**, 471.
13 Jiang, P., Bertone, J.F., Hwang, K.S. and Colvin, V.L. (1999) *Chem. Mater.*, **11**, 2132.
14 van Blaaderen, A., Ruel, R. and Wiltzius, P. (1997) *Nature*, **385**, 321.
15 Wong, S., Kitaev, V. and Ozin, G.A. (2003) *J. Am. Chem. Soc.*, **125**, 15589.
16 Greulich-Weber, S., von Rhein, E. and Bielawny, A. (2006) Control of Colloidal Crystal Growth by External Fields, in *Assembly at the Nanoscale – Toward Functional Nanostructured Materials, Mater. Res. Soc. Symp. Proc. 910E* (eds C.S. Ozkan, F. Rosei, G.P. Lopinski and Z.L. Wang), Warrendale, PA, 0901-Rb10-08.
17 Griesebock B., Egen, M. and Zentel, R. (2002) *Chem. Mater.*, **14**, 4023.
18 Vlasov, Y.A., Bo, X.Z., Sturm, J.C. and Norris, D.J. (2001) *Nature*, **414**, 289.
19 Rengarajan, R., Jiang, P., Larrabee, D.C., Colvin, V.L. and Mittleman, D.M. (2001) *Phys. Rev. B*, **64**, 205103.
20 Gu, Z.Z., Fujishima, A. and Sato, O. (2002) *Chem. Mater.*, **14**, 760.
21 Romanov, S.G., Egen, M., Zentel, R. and Sotomayor Torres, C.M. (2006) *Physica E*, **32**, 476.
22 Wang, J., Li, Q., Knoll, W. and Jonas, U. (2006) *J. Am. Chem. Soc.*, **128**, 15606.
23 Gates, B., Lu, Y., Li, Z.Y. and Xia, Y. (2003) *Appl. Phys. A*, **76**, 509.
24 Allard, D., Lange, B., Fleischhaker, F., Zentel, R. and Wulff, M. (2005) *Soft Mater.*, **3**, 121.
25 Egen, M., Braun, L., Zentel, R., Tännert, K., Frese, P., Reis, O. and Wulf, M. (2004) *Macromol. Mater. Eng.*, **289**, 158.
26 Fleischhaker, F. and Zentel, R. (2005) *Chem. Mater.*, **17**, 1346.
27 Romanov, S.G., Maka, T., Sotomayor Torres, C.M., Müller, M. and Zentel, R. (1999) *Appl. Phys. Lett.*, **75**, 1057.
28 Romanov, S.G., Maka, T., Sotomayor Torres, C.M., Müller, M. and Zentel, R. (2001) *Appl. Phys. Lett.*, **79**, 731.
29 Lodahl, P., Driel, A.F.v., Nikolaev, I.S., Irman, A., Overgaag, K., Vanmaekelbergh, D. and Vos, W.L. (2004) *Nature*, **430**, 656.
30 Gaponenko, S.V., Bogomolov, V.N., Petrov, E.P., Kapitonov, A.M., Yarotsky, D.A., Kalosha, I.I., Eychmüller, A.A., Rogach, A.L., McGilp, J., Woggon, U. and Gindele, F. (1999) *J. Lightwave Technol.*, **17**, 2128.
31 Blanco, A., López, C., Mayoral, R., Miguez, H., Meseguer, F., Mifsud, A. and Herrero, J. (1998) *Appl. Phys. Lett.*, **73**, 1781.
32 Wang, D., Rogach, A.L. and Caruso, F. (2003) *Chem. Mater.*, **15**, 2724.
33 Yoshino, K., Lee, S.B., Tatsuhara, S., Kawagishi, Y., Ozaki, M. and Zakhidov, A.A. (1998) *Appl. Phys. Lett.*, **73**, 3506.
34 Lange, B., Zentel, R., Ober, C.K. and Marder, S. (2004) *Chem. Mater.*, **16**, 5286.
35 Li, J., Wang, Y.W., Guo, W., Keay, J.C., Mishima, T.D., Johnson, M.B. and Peng, X. (2003) *J. Am. Chem. Soc.*, **125**, 12567.
36 Landfester, K. (2001) *Macromol. Rapid Commun.*, **22**, 896.
37 Blanco, A., Chomski, E., Grabtchak, S., Ibisate, M., John, S., Leonhard, S., López, C., Meseguer, F., Míguez, H., Mondla, J.P., Ozin, G.A., Toader, O. and van Driel, H.M. (2000) *Nature*, **405**, 437.
38 Müller, M., Zentel, R., Maka, T., Romanov, S.G. and Sotomayor Torres, C.M. (2000) *Adv. Mater.*, **12**, 1499.
39 Wijnhoven, J.E.G.J. and Vos, W.L. (1998) *Science*, **281**, 802.
40 Holland, B.T., Blanford, C.F., Do, T. and Stein, A. (1999) *Chem. Mater.*, **11**, 795.
41 Wijnhoven, J.E.G.J., Bechger, L. and Vos, W.L. (2001) *Chem. Mater.*, **13**, 4486.
42 Míguez, H., Chomski, E., García-Santamaría, F., Ibisate, M., John, S., López, C., Meseguer, F., Mondia, J.P., Ozin, G.A.,

Toader, O. and van Driel, H.M. (2001) *Adv. Mater.*, **13**, 1634.

43 Subramania, G., Constant, K., Biswas, R., Sigalas, M.M. and Ho, K. (2001) *Adv. Mater.*, **13**, 443.

44 Cassagneau, T. and Caruso, F. (2002) *Adv. Mater.*, **14**, 34.

45 Popall, M., Dabek, A., Robertsson, M.E., Valizadeh, S., Hagel, O.J., Büstrich, R., Nagel, R., Cergel, L., Lambert, D. and Schaub, M. (2000) *Mol. Cryst. Liq. Cryst.*, **354**, 123.

46 Lange, B., Wagner, J. and Zentel, R. (2006) *Macromol. Rapid Commun.*, **27**, 1746.

47 Chen, J.I.L., von Freymann, G., Choi, S.Y., Kitaev, V. and Ozin, G.A. (2006) *Adv. Mater.*, **18**, 1915.

48 Halaoui, L.I., Abrams, N.M. and Mallouk, T.E. (2005) *J. Phys. Chem. B*, **109**, 6334.

49 Hessel, V., Löwe, H., Müller, A. and Kolb, G. (2005) *Chemical Micro Process Engineering*, Wiley-VCH, p. 281.

50 Guan, G., Zapf, R., Kolb, G., Men, Y., Hessel, V., Löwe, H., Ye, J. and Zentel, R. (2007) *Chem. Commun.*, 260.

51 Noda, S., Tomoda, K., Yamamoto, N. and Chutinan, A. (2000) *Science*, **289**, 604.

52 Birner, A., Wehrspohn, R.B., Gosele, U.M. and Busch, K. (2001) *Adv. Mater.*, **13**, 377.

53 Qi, M.H., Lidorikis, E., Rakich, P.T., Johnson, S.G., Joannopoulos, J.D., Ippen, E.P. and Smith, H.I. (2004) *Nature*, **429**, 538.

54 Mertens, G., Wehrspohn, R.B., Kitzerow, H.S., Matthias, S., Jamois, C. and Gösele, U. (2005) *Appl. Phys. Lett.*, **87**, 1108.

55 Seet, K.K., Mizeikis, V., Matsuo, S., Juodkazis, S. and Misawa, H. (2005) *Adv. Mater.*, **17**, 541.

56 Scrimgeour, J., Sharp, D.N., Blanford, C.F., Roche, O.M., Denning, R.G. and Turberfield, A.J. (2006) *Adv. Mater.*, **12**, 1557.

57 Wong, S., Deubel, M., Perez-Willard, F., John, S., Ozin, G.A., Wegener, M. and von Freymann, G. (2006) *Adv. Mater.*, **18**, 265.

58 Ye, J., Zentel, R., Arpianen, S., Ahopelto, J., Jonsson, F., Romanov, S.G. and Sotomayor Torres, C.M. (2006) *Langmuir*, **22**, 7378.

59 Jonsson, F., Sotomayor Torres, C.M., Seekamp, J., Schniedergers, M., Tiedemann, A., Ye, J. and Zentel, R. (2005) *Microelectron. Eng.*, **78/79**, 429.

60 Yan, Q.F., Chen, A., Chua, S.J. and Zhao, X.S. (2005) *Adv. Mater.*, **17**, 2849.

61 Ferrand, P., Egen, M., Zentel, R., Seekamp, J., Romanov, S.G. and Sotomayor Torres, C.M. (2003) *Appl. Phys. Lett.*, **83**, 5289.

62 Juarez, B.H., Golmayo, D., Postigo, P.A. and López, C. (2004) *Adv. Mater.*, **16**, 1732.

63 Vekris, E., Kitaev, V., von Freymann, G., Perovic, D.D., Aitchison, J.S. and Ozin, G.A. (2005) *Adv. Mater.*, **17**, 1269.

64 Yan, Q.F., Zhou, Z.C., Zhao, X.S. and Chua, S.J. (2005) *Adv. Mater.*, **17**, 1917.

65 Jiang, P., Ostojic, G.N., Narat, R., Mittleman, D.M. and Colvin, V.L. (2001) *Adv. Mater.*, **13**, 389.

66 Wostyn, K., Zhao, Y.X., de Schaetzen, G., Hellemans, L., Matsuda, N., Clays, K. and Persoons, A. (2003) *Langmuir*, **19**, 4465.

67 Palacios-Lidon, E., Galisteo-Lopez, J.F., Juarez, B.H. and López, C. (2004) *Adv. Mater.*, **16**, 341.

68 Fleischhaker, F., Arsenault, A.C., Kitaev, V., Peiris, F.C. von Freymann, G., Manners, I., Zentel, R. and Ozin, G.A. (2005) *J. Am. Chem. Soc.*, **127**, 9318.

69 Fleischhaker, F., Arsenault, A.C., Wang, Z., Kitaev, V., Peiris, F.C., von Freymann, G., Manners, I., Zentel, R. and Ozin, G.A. (2005) *Adv. Mater.*, **17**, 2455.

70 Lange, B., Zentel, R. and Ober, C.K. (2006) *Poly. Prepr.*, **47**, 517.

71 Lee, W.M., Pruzinsky, S.A. and Braun, P.V. (2002) *Adv. Mater.*, **14**, 271.

72 Jun, Y.H., Leatherdale, C.A. and Norris, D.J. (2005) *Adv. Mater.*, **17**, 1908.

73 Pruzinsky, S.A. and Braun, P.V. (2005) *Adv. Funct. Mater.*, **15**, 1995.

74 Pruzinsky, S.A., García-Santamaría, F. and Braun, P.V. (2006) *PMSE Preprints*, **94**, 99.

75 Lange, B., Zentel, R., Jhaveri, S.J. and Ober, C.K. (2006) *Proc. SPIE*, 61821W/1.

76 Braun, P.V., Rinne, S.A. and García-Santamaría, F. (2006) *Adv. Mater.*, **18**, 2665.

77 Arsenault, A.C., Fleischhaker, F., von Freymann, G., Kitaev, V., Migues, H., Mihi, A., Tetreault, N., Vekris, E., Manners, I., Aitchison, S., Perovic, D. and Ozin, G.A. (2006) *Adv. Mater.*, **18**, 2779.

78 McCord, M.A. and Rooks, M.J. (1997) in *Handbook of Microlithography, Micromachining and Microfabrication*, Vol. 1 (ed. P. Rai-Choudhury), SPIE Optical Engineering Press, Bellingham, Washington, p. 139.

79 Tanaka, T., Sun, H.-B. and Kawata, S. (2002) *Appl. Phys. Lett.*, **80**, 312.

80 Yang, D., Jhaveri, S.J. and Ober, C.K. (2005) *MRS Bull.*, **30**, 976.

81 Zhou, W., Kuebler, S.M., Braun, K.L., Yu, T., Cammack, K.J., Ober, C.K., Perry, J.W. and Marder, S.R. (2002) *Science*, **296**, 1106.

82 Pruzinsky, S.A. (2006) Ph.D. thesis, University of Illinois at Urbana-Champaign, Urbana.

83 Zhao, Y., Wostyn, K., de Schaetzen, G., Clays, K., Hellemans, L., Persons, A., Szekeres, M. and Schoonheydt, R.A. (2003) *Appl. Phys. Lett.*, **82**, 3764.

84 Massé, P., Reculusa, S., Clays, K. and Ravaine, S. (2006) *Chem. Phys. Lett.*, **422**, 251.

85 Pozas, R., Mihi, A., Ocana, M. and Miguez, H. (2006) *Adv. Mater.*, **18**, 1183.

86 Tetreault, N., Mihi, A., Miguez, H., Rodriguez, I., Ozin, G.A., Meseguer, F. and Kitaev, V. (2004) *Adv. Mater.*, **16**, 346.

87 Tetreault, N., Arsenault, A.C., Mihi, A., Wong, S., Kitaev, V., Miguez, H. and Ozin, G.A. (2005) *Adv. Mater.*, **17**, 1912.

88 Fleischhaker, F., Arsenault, A.C., Peiris, F., Kitaev, V., Manners, I., Zentel, R. and Ozin, G.A. (2006) *Adv. Mater.*, **18**, 2387.

89 Fleischhaker, F., Arsenault, A.C., Schmidtke, J., Zentel, R. and Ozin, G.A. (2006) *Chem. Mater.*, **18**, 5640.

4
Bloch Modes and Group Velocity Delay in Coupled Resonator Chains

Björn M. Möller, Mikhail V. Artemyev, and Ulrike Woggon

4.1
Introduction

Structures built from coherently coupled optical microresonators currently attract much attention for guiding and bending light on the microscale, the construction of photonic circuits and slowing down light in the frame of optical computing [1,2]. The waveguides are here constituted by a sequence of resonators, whose mutual field overlaps lead to coherent interresonator coupling, thus opening a band for signal transmission [3]. Coupled-resonator structures can be realized in various ways, e.g. as closely arranged defects in photonic crystals (PhCs) [4–10], in which resonators are formed in the stop band of the PhC. Alternative realizations of coupled-resonator geometries are given by side-coupled resonators attached to waveguides [11] and by bottom-up approaches, in which individual dielectric microresonators are arranged periodically in lines and arrays in free space. Experimentally, coherent-coupling of individual resonators has been demonstrated first in the photonic molecule picture [12–16], which later extended to waveguide geometries [17–20].

Of particular interest for bottom-up approaches are microresonators with cylindrical [21], ring-like [19] or spherical [17,18,20] shapes. Since these structures intrinsically exhibit an isotropic mode pattern, interresonator coupling mediated by the evanescent component of individual resonator fields can occur for a wide range of possible coupled-resonator geometries and thus allow for the design of complex structures. Together with experimental progress, more realistic numerical studies emerged, which go beyond the conceptually intuitive assumption of infinite structures [22,23].

This report hence aims at the demonstration of coherent effects in one-dimensional chains of micronsized microspheres, the exploration of the mode structure in finite CROWs, the demonstration of the potential of coupled-resonator systems to slow down the group velocity of light and to explore the impact of size tuning in a CROW system.

Nanophotonic Materials: Photonic Crystals, Plasmonics, and Metamaterials. Edited by R.B. Wehrspohn, H.-S. Kitzerow, and K. Busch

ISBN: 978-3-527-40858-0

This report is organized as follows: With a spatially and spectrally resolved mode mapping technique described in Section 4.2 we demonstrate the formation of coherently coupled multiresonator fields in a one-dimensional coupled resonator chain (Section 4.3). The mode structure in *finite* coupled-resonator optical waveguides is discussed in Section 4.4 and a concise analytical model is offered. From experimentally obtained mode structures (Section 4.5) we explore coupled-resonator chains as optical delay lines for optical signals. Comparing experimental data with a theoretical analysis, we derive a slowing factor of $S = 31$. In Section 4.6 we additionally provide a concise model for the impact of tuned single resonator sizes on the resulting CROW modes; as a result, we explain the resulting coupled-resonator modes in terms of Bloch modes combined with mutual anticrossing relations.

4.2 Experiment

As building blocks for CROW structures exactly size-matched microspheres ($R = 1.4$ μm or $R = 2.25$ μm, respectively, obtained from Polysciences, Inc.) have been aligned in one-dimensional rows. The alignment is achieved by drying a microsphere suspension onto a quartz substrate: The presence of a sub-millimeter sized template (e.g. a glass rod) results in the formation of an one-dimensionally extended meniscus, into which the individual microspheres are dragged during evaporation of the solvent. After the drying process, ordered one-dimensional (1D) rows of microresonators are obtained (see Figure 4.1).

As a method to explore the CROW fields experimentally, the microresonators have been doped before arrangement [24,25] with a subsurface layer of CdSe nanocrystals prepared according to Refs. [26–28]. Excited by an argon-ion laser, the non-resonant

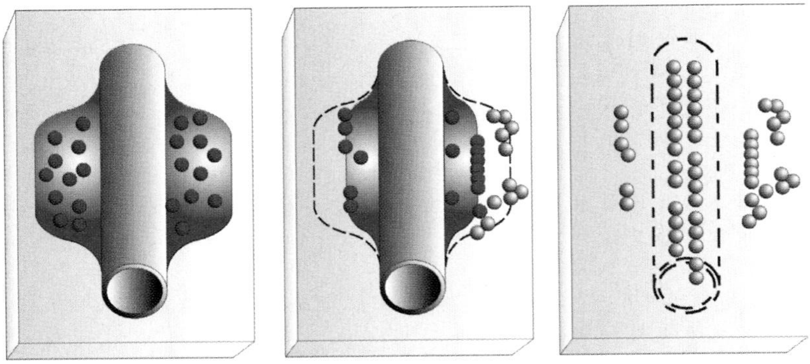

Figure 4.1 Self-organized arrangements of one-dimensional coupled-resonator structures. A suspension containing spherical microresonators is given onto a quartz substrate (left image) in the proximity of a small glass rod. During evaporation of the solvent (middle image), the microresonators are dragged into the meniscus of the suspension between glass rod and substrate. After complete solvent evaporation, ordered one-dimensional microresonator chains are obtained.

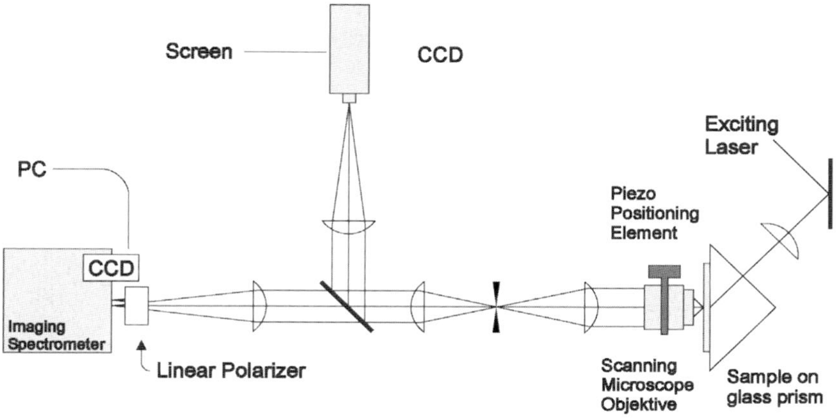

Figure 4.2 Experimental setup. The coupled-resonator samples are mounted onto a glass prism and excited through one of its back-facets by an argon-ion laser (488 nm). The PL emission is collected with a microscope objective and transferred to a spectrometer. Mode maps of the multi-resonator structures are achieved by spatially scanning the sample via a piezo-controlled movement of the objective.

nanocrystal emission is confined in the coupled-resonator structures. Thereby, the nanocrystal is used as an efficient CROW field probe for spectroscopic studies.

In order to explore the formation of coherently coupled light fields in 1D chains, we apply spatially and spectrally resolved mode mapping and microphotoluminescence spectroscopy. The quartz substrate containing the resonator chains is adhesively mounted onto a glass prism (see experimental scheme in Figure 4.2), thus providing an efficient way for incoupling light. An argon-ion laser beam is coupled into the prism through one of its back facets, thus evanescently exciting the nanocrystal doped samples. The non-resonant photoluminescence (PL) emission of the nanocrystals is transferred onto the entrance slit of an imaging spectrometer. The dispersed slit image is subsequently read out by a CCD chip for further image processing. Via piezo controlled movement of the microscope objective, the sample can be scanned in the spatial direction perpendicular to the slit.

4.3 Coherent Cavity Field Coupling in One-Dimensional CROWs

With this setup detailed in Section 4.2, spectra for any spatial position across the sample can be acquired. We use this technique to explore mode field modifications due to coherent resonator coupling by displaying the mode intensity of specific resonances for a whole two-dimensional image: Three exemplary mode maps for a multiresonator structure ($R = 2.25\,\mu\text{m}$) are plotted in Figure 4.3. In diagram a), the spectrally integrated PL emission is plotted in order to visualize the geometry of an end-raked coupled-resonator chain. In diagram b), the PL intensity at an energy window coinciding with a single resonator resonance is displayed for a

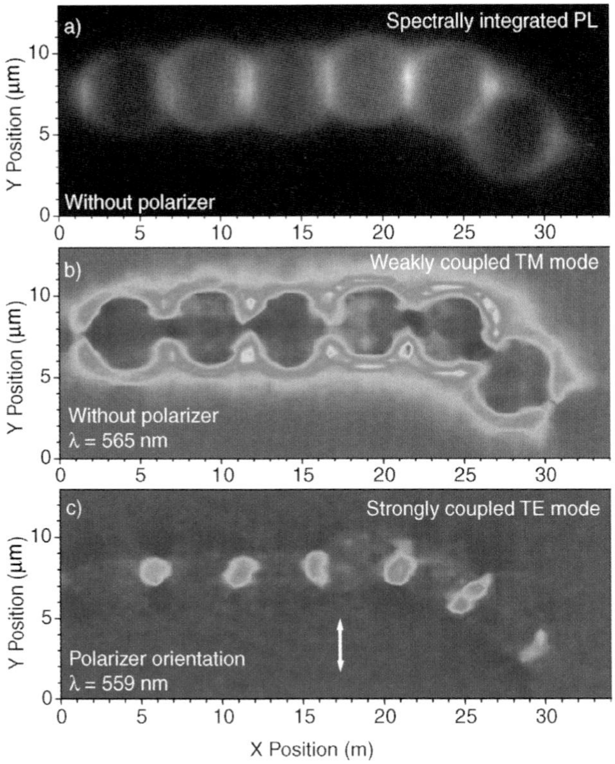

Figure 4.3 Mode maps of a six-resonator structure: (a) Integrated PL. (b) Mode map for a weakly coupled predominantly TE-polarized mode at 557 nm. Coherent coupling is evidenced by vanishing mode intensities at the microsphere intersection points. (c) Mode map for a strongly coupled predominantly TE-polarized mode at 559 nm. Coherent coupling is evidenced by enhanced mode intensities at the intersection regions.

predominantly TM-polarized mode at 565 nm. Here, modifications of the mode intensity profile are apparent: While for a single resonator the mapped mode intensity is confined to a ring along the circumference of the resonator (see Refs. [25,29]), we find significant differences in a coupled-resonator chain: At the intersection regions, the mode intensity almost completely vanishes. For predominantly TE-polarized mode resonances, a similar mode profile can be obtained.

A different mode map results for specific energies far-off the single resonator mode resonances: As illustrated in diagram c) for specific largely split-off energies still mode intensity can be oberved. In contrast to the image b), the mode intensity is here located exactly at those regions, where the modes in diagram b) display no intensity. The predominant TE-polarized character of this mode is verified with a linear polarizer inserted into the detection beam path [30].

As discussed in detail in Ref. [17], the modes can be classified into *strongly* and *weakly* coupled modes. In contrast to the strongly coupled modes, which exhibit comparatively large frequency splits and significant mode broadening, weakly coupled modes represent the weak coupling limit, which for slow-light experiments is of major importance. Thus we will focus both experimentally and theoretically on the mode structure and coherent effects of weakly coupled modes in what follows.

4.4 Mode Structure in Finite CROWs

For the exploration of CROW phenomena such as the group velocity of light inside a CROW, and the impact of size detuning, detailed insight into the mode structure in finite systems is required. Based on the description of infinite CROWs as introduced in Ref. [1], we first give a short summary of the specific characteristics concerning degenerate CROWs of finite size:

The field modes of an infinite CROW are described by a Bloch ansatz, in which the individual electric fields of the n-th resonator $E_\Omega(\boldsymbol{r} - nRe_z)$ are connected by Bloch waves with the wave number k_c

$$\boldsymbol{E}_{k_c}(\boldsymbol{r}, t) = E_0 \exp^{\eth\omega_{k_c}t} \sum_n \exp^{-\eth n k_c R} \boldsymbol{E}_\Omega(\boldsymbol{r} - nRe_z). \tag{4.1}$$

In this notation the parameter R labels the interresonator distance. The solution of Maxwell's equations for the coupled-resonator chain results in the following dispersion relation for the i-th wave number k_c^i

$$\omega_{k_c^i}^2 = \Omega^2[1 + 2\kappa \cos(k_c^i \cdot R)]. \tag{4.2}$$

The coupling constant κ is obtained by the overlap integrals of the electromagnetic field of neighboring resonator in the limit of next-neighbor coupling.

From the dispersion relation, the group velocity in a coupled-resonator optical waveguide immediately follows as its derivative:

$$v_g = \frac{d\omega_{k_c^i}}{dk_c^i} = -R\Omega\kappa \sin(k_c^i \cdot R). \tag{4.3}$$

Thus, the group velocity in a CROW scales directly with the coupling constant κ.

For a sufficiently large number of resonators N, for which the structure can be regarded as infinetely long, the wave number which solution Eq. (4.2) depends on reads

$$k_c^i = \frac{i \cdot \pi}{NR} \quad \text{with} \quad i = 1, \ldots, N. \tag{4.4}$$

For the calculation of coupled-cavity modes in finite structures, mainly numerical models [23] and transfer matrix approaches [22] have been applied. It might be tempting to argue, that finite structures would not show Bloch mode formation because of the lack of translational symmetry, which prevents the Bloch theorem from being valid. Nevertheless, we will show, that a Bloch wave interpretation of coupled-resonator modes is possible in a straight-forward manner.

In order to explore coupled modes in finite structures, we first interprete Eqs. (4.1) and (4.2) as the solution of an eigenvalue equation of infinite dimension (see also Ref. [31]). This eigenvalue equation can be obtained by writing the resonator field of the n-th resonator as the n-th vector component of a CROW field vector Ψ. After multiplying the squared dispersion relation Eq. (4.2) with Ψ, the cosine term in the dispersion relation can be rewritten into a term involving the $(n+1)$-th and $(n+1)$-th component of Ψ. In the relevant limit of weak interresonator coupling, the coupling constant κ is small compared to unity, and terms of higher order in κ can be omitted. Finally, the eigenvalue equation reads

$$\Psi_n + \kappa\Psi_{n-1} + \kappa\Psi_{n+1} = \frac{\omega^2_{k^i_c}}{\Omega^2}\Psi_n. \tag{4.5}$$

In a finite system, however, *surface* terms have to be included, so that the first (or last, respectively) resonator is coupled to one neighbour only. Therby, a finite system is described by a finite matrix equation

$$\begin{pmatrix} 1 & \kappa & 0 & \cdots \\ \kappa & 1 & \kappa & \ddots \\ & \ddots & \ddots & \ddots \\ \cdots & 0 & \kappa & 1 \end{pmatrix} \cdot \Psi = \frac{\omega^2_{k^i_c}}{\Omega^2}\Psi, \tag{4.6}$$

in which the coupling of the first and last resonators to their single neighbours is represented by the upper left and lower right corner.

This finite matrix equation is not solved with the ansatzes Eqs. (4.1), (4.2), (4.4) due to the surface terms. The solution vector for the electric cavity fields are instead given by the expression

$$\begin{pmatrix} \sin(1k^i_c R) \\ \vdots \\ \sin(nk^i_c R) \\ \vdots \\ \sin(Nk^i_c R) \end{pmatrix} \cdot \Psi = \frac{\omega^2_{k^i_c}}{\Omega^2}\Psi, \tag{4.7}$$

in which the wave number now reads

$$k^i_c = \frac{i \cdot \pi}{(N+1)R} \quad \text{with} \quad i = 1, \ldots, N. \tag{4.8}$$

With this modified wave number expression, a similar dispersion relation is obtained:

$$\frac{\omega_{k^i_c}}{\Omega} = \sqrt{1 + 2\kappa\cos(k^i_c R)} \approx 1 + \kappa\cos(k^i_c R). \tag{4.9}$$

Thus, the finiteness results in modified coupled-resonator light states which have an energy dependent succession of bright and dark states. A visualization of the light fields in finite structures is given in Figure 4.4. The mode structures are illustrated in

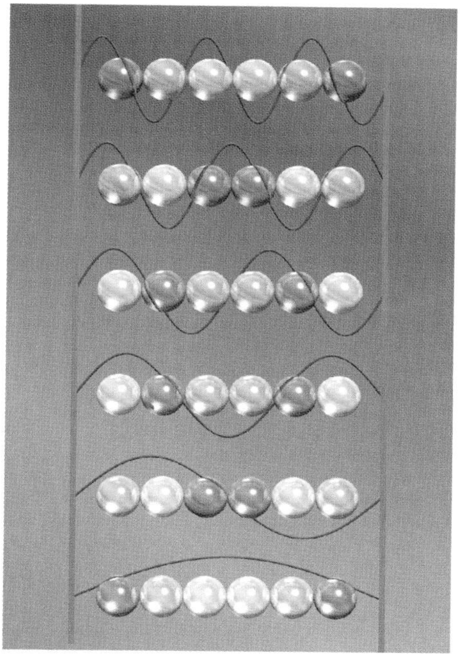

Figure 4.4 Mode structure in finite CROWs. For a six-resonator chain, the resulting cavity field pattern is displayed (in ascending order for increasing wave number) illustrating the wave-like nature of the collective modes. Solid lines indicate the strength of the electric field in the respective resonators. The sphere colors represent the corresponding intensity. The coupled-resonator modes evolve in a pattern of bright and dark resonator states. The mode structure does not depend on the magnitude of the coupling constant or the cavity geometry. The wavenumber after Eq. (4.8) results in common nodes of the envelope. As depicted in the scheme, these nodes are located one resonator diameter apart from the first or last resonator center, respectively.

ascending order for increasing wave numbers (or decreasing eigenenergy, respectively). Solid curves indicate the envelope of the electric field strengths in the respective resonator after Eq. (4.7). The strength of the electric field can be evaluated through the envelope value at the center of the resonator under consideration. The sphere colors indicate the brightness of the respective resonator and are deduced by squaring the numeric value of the electric field.

As can be noted from the illustration, the mode pattern for the succession of bright and dark states displays an energy dependent mirror symmetry, which can be utilized experimentally for the identification of coherently coupled modes. Additionally, according to the dispersion relation Eq. (4.9), the coupling constant can be obtained from the split of these Bloch modes. Thus, we will make use of this set of relations in Section 4.5 to identify the coupled modes from which the coupling constant and the slowing factor in a coupled-resonator structure can be derived.

The transition between coupled modes in infinite and finite structures can be simply deduced from the wave number expression. While the expression scales with

$1/N$ and $1/(N+1)$, respectively, the difference between the mode structures becomes negligible for sufficiently large number.

This modification of the wave numbers compared to those of infinite systems can be interpreted as an appropriate *scaling* of the lattice for finite coupled-resonator structures. Since the derivation of this result solely relies on the property of weak next-neighbor coupling without further accounting for cavity geometry or the specific coupling mechanism, it might serve for the description of a variety of related problems. For example, the validity of Eq. (4.8) has been confirmed recently for weakly coupled defects in a theoretical approach for a similar problem in *phononic* crystals (see Ref. [32]).

Beside finite and strictly periodic systems, coupled systems with locally varying coupling constants can readily covered by Eq. (4.6) as well. Hence, this formalism might serve as a tool for a fast evaluation of the effects of superlattices, which have been explored e.g. in metallodielectric one-dimensional photonic crystals (see Ref. [33]) or for the characterization of multiple-defect PhCs in chirp-compensating devices [34].

4.5
Slowing Down Light in CROWs

Possibly the most interesting application for coupled-resonator optical waveguides is their ability to slow down the group velocity of light. An impressive slowing down of light by a factor of 22.9 at the Bloch band center has been demonstrated recently by Poon et al. in a chain of coherently coupled microrings 120 μm in diameter [19]. In their experiment, a modulated laser is coupled into the microring chain and compared to a reference waveguide. The group delay has been determined with a lock-in technique measuring the phase lag between the signal and the reference.

Nevertheless, due to the chosen parameters, e.g. large cavity radius and small coupling constant, the spectral response in the transmission window prevents a direct connection of the slowing factors to the Bloch mode picture. Thereby, we follow a different approach. Making use of a way smaller microresonator size, CROWs can afford a significantly larger coupling constant while at the same time exhibiting a large slowing factor. This can be understood by expressing the slowing factor S in terms of the wavelength of the guided light λ, the cavity radius R:

$$S = \frac{c}{\nu_g^{\max}} = \frac{\lambda}{4\pi\kappa \cdot R}. \tag{4.10}$$

According to Eq. (4.10), the slowing factor remains unchanged if the product of the coupling constant and the cavity radius is kept constant.

If the coupling constant is chosen large enough to resolve the individual Bloch modes spectrally (i.e. the splitting due to coherent coupling is larger than the resonator line-widths) we can deduce the slowing factor directly from the spectral system response in the *Bloch mode regime*.

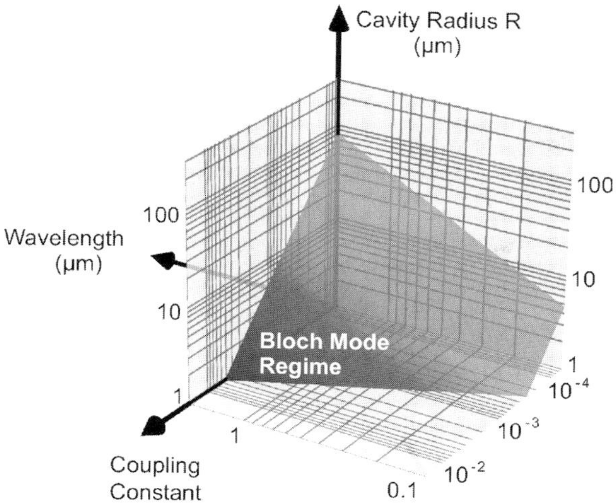

Figure 4.5 Possible parameter choices for a given slowing factor S. The coupling constant κ, cavity radius R and the wavelength λ of the guided light form a plane of constant slowing factor $S = 31$ in a triple logarithmic representation.

Hence, many different structures can be easily compared, which naturally stem from a large parameter space. Basically, coupled-resonator optical waveguides can be realized in a large parameter space [35–41]. A visualization of the trade-offs between the CROW parameters is shown in Figure 4.5. The parameters for a constant slowing factor ($S = 31$) constitute a plane in a triple logarithmic representation.

In order to unambiguously estimate the slowing-factor experimentally, we consider the smallest possible structure with alternating bright and dark resonator states and use the formalism of Section 4.4.

This system consists of three resonators ($R = 1.4$ µm) in a linear row and according to the discussion of Eq. (4.7), a symmetric spatio-spectral mode pattern indicating bright and dark states arises. The mode pattern of a linear three-resonator structure is evaluated experimentally in dependence of both the resonator number and the mode energy in Figure 4.6. For comparison, a theoretical calculation for a linear three-resonator chain after Eq. (4.7) is added.

For this particular structure, the observed frequency split is related to the coupling constant via Eq. (4.11). From comparison of experimental and theoretical data (Figure 4.6), a coupling constant of 1.1×10^{-3} is derived:

$$\kappa = \frac{\sqrt{2} \cdot \Delta E}{E_{\text{central}}} \approx \frac{\sqrt{2} \cdot \Delta\lambda}{\lambda_{\text{central}}} \approx 1.1 \times 10^{-3}. \quad (4.11)$$

From this result, the slowing factor (fraction of vacuum light speed and maximum light velocity in a given CROW) can be readily determined. With straight forward calculus (for details, see Ref. [45]), a slowing factor $S = 31$ can be obtained.

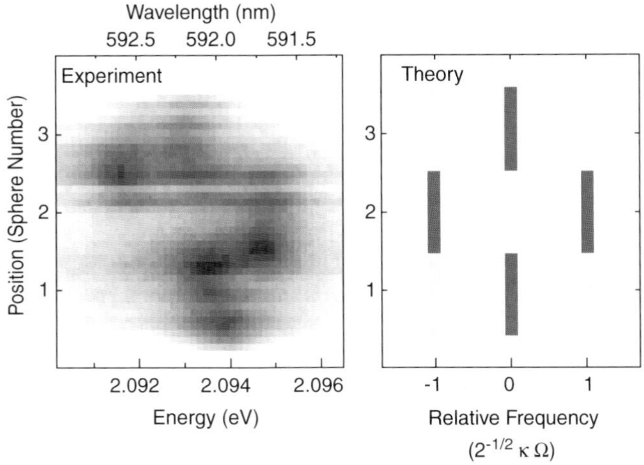

Figure 4.6 Mode pattern in a size-matched three-resonator system. The experimentally obtained mode splitting is resolved for each resonator (left) and compared to the theoretical model developed in Section 4.4. The mode splitting reveals a slowing-factor $S = 31$ (see text).

4.6
Disorder and Detuning in CROWs

In Section 4.4, we kept the single resonator mode frequencies degenerate for a discussion of the coupled-resonator mode patterns. In the following, we will exploit the modifications of coupled-resonator modes due to size tuning of individual resonators in a chain.

In the case of periodic photonic structures, the discussion of disorder effects is mainly focussed on random deviations from a strict periodicity. Typically, a parameter characterizing the degree of disorder is chosen and a number of configurations are modelled in accordance with this disorder parameter. Disorder is usually found to result in significant light localization within characteristic length scales, the so-called *Anderson localization* [42–44]. Connected with light localization, the modifications of the photonic density of states and the photonic band structure are of major importance.

For the case of size detuning in coupled-resonator optical waveguides which we discuss in this section, we take a related approach to size tuning. However, instead of modelling a statistical ensemble of size deviations, we explore the modifications of coherently coupled photon modes for a single defect, whose eigenfrequency is continuously tuned through the photonic band. Hence, we make use of the previously discussed model of finite CROWs.

In a similar way to that presented in Section 4.4, size detuning can be included into the matrix model. However, the model of a finite CROW with degenerate individual resonance frequencies Ω leads to a specific role of the frequency Ω in the corresponding equations. As discussed in a detailed manner in Ref. [45], CROWs built from resonators with different resonance frequencies $\Omega_1, \Omega_2, \ldots, \Omega_N$ can be efficiently modelled with a modified eigenvalue equation, in which all appearing frequencies contribute in a similar way.

$$\begin{pmatrix} \Omega_1^2 & \frac{\Omega_1^2+\Omega_2^2}{2}\cdot\kappa & 0 & \cdots \\ \frac{\Omega_1^2+\Omega_2^2}{2}\cdot\kappa & \Omega_2^2 & \frac{\Omega_2^2+\Omega_3^2}{2}\cdot\kappa & \ddots \\ 0 & \ddots & \ddots & \ddots \\ \cdots & 0 & \frac{\Omega_{n-1}^2+\Omega_n^2}{2}\cdot\kappa & \Omega_n^2 \end{pmatrix} \cdot \Psi_{\text{perturbed}} = \omega_{k_i}^2 \Psi_{\text{perturbed}}. \tag{4.12}$$

As one example of size tuning of a single resonator in a CROW, we consider a five resonator structure, in which the first four resonators obey degenerate resonance frequencies Ω_0. The fifth resonator's frequency is detuned across this frequency from $\Omega_0 - 3\kappa\Omega_0$ to $\Omega_0 + 3\kappa\Omega_0$. The evolving eigenfrequencies of the coupled system are displayed in Figure 4.7. For comparison, the Bloch mode frequencies of a degenerate four-resonator system are marked with dashed horizontal lines. The diagonal dashed line indicates the tuned resonance of the fifth resonator. As apparent from Figure 4.7, for detunings as large as several frequency scaled coupling constants $\kappa\Omega_0$, the system

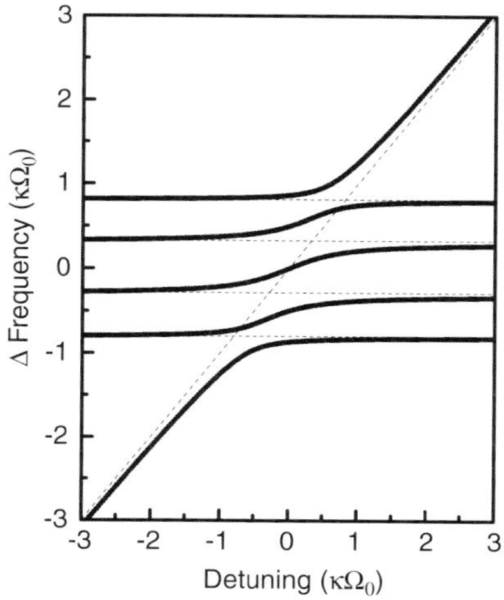

Figure 4.7 Evolving multiresonator modes for five resonators in dependence on the fifth's resonator tuned resonance. Dashed lines indicate Bloch modes of a four resonator system (horizontal) and the individual tuned resonance of the fifth resonator (diagonal), respectively. The tuned frequencies are given in units of frequency scaled coupling constants.

can be simply treated as a coherently coupled four-resonator system and an uncoupled single fifth resonator. For small detunings – when the fifth resonator crosses the four-resonator Bloch band – mutual anticrossing behavior can be observed: Whenever the frequency of the fifth resonator and one of the Bloch modes meet, avoided level crossing takes place and the modes bend over into each other. This behavior is a direct result from the coupled-oscillator analogy and thus nicely mimics related effects in solid state physics, e.g. the coupling of electronic resonances in electronic crystals. Specifically, these relations might serve as a rapid design tool for the required fine-tuning in crossing and intersecting CROW structures (see e. g. Ref. [46]) or for the connection of actively tuned [47] with otherwise insensitive structures.

4.7
Summary

Chains of coherently coupled microresonators have been studied both experimentally and theoretically. The evolution of individual microresonator modes into coherent waveguide modes has been demonstrated by means of microphotoluminescence mode mapping. A coupled-oscillator model is developed explaining the coupled-resonator mode patterns due to coherent photon mode coupling. An extension of this model has been shown covering structures of finite size. The potential of slowing-down the group velocity of light is evidenced experimentally. A slowing-factor of $S = 31$ has been deduced. A model for disorder in coupled-resonator optical waveguides is presented. Size detuning in CROWs has been explored in terms of anticrossing relations in close analogy to solid-state systems.

Acknowledgements

We cordially acknowledge the financial commitment of the DFG in the frame of the priority program SPP 1113 (Photonic Crystals).

References

1 Yariv, A., Xu, Y., Lee, R.K. and Scherer, A. (1999) *Opt. Lett.*, **24** (11), 711.
2 Stefanou N. and Modinos, A. (1998) *Phys. Rev. B*, **57** (19), 12127.
3 Bayindir, M., Temelkuran, B. and Ozbay, E. (2000) *Phys. Rev. Lett.*, **84**, 2140.
4 Christodoulides, D.N., Lederer, F. and Silberberg, Y. (2003) *Nature*, **424**, 817.
5 Richter, S., Hillebrand, R., Jamois, C., Zacharias, M., Gösele, U., Schweizer, S.L. and Wehrspohn, R.B. (2004) *Phys. Rev. B*, **70**, 193302.
6 Notomi, M., Yamada, K., Shinya, A., Takahashi, J., Takahashi, C. and Yokohama, I. (2001) *Phys. Rev. Lett.*, **87** (25), 253902.
7 Engelen, R.J.P., Sugimoto, Y., Wanatabe, Y., Korterik, J.P., Ikeda, N., van Hulst, N.F., Asakawa, K. and Kuipers, L. (2006) *Opt. Express*, **14** (4), 1685.

8 Vlasov, Y.A., O'Boyle, M., Hamann, H.F. and McNab, S.J. (2005) *Nature*, **438**, 65.

9 Altug H. and Vučković, J. (2005) *Appl. Phys. Lett.*, **86**, 111102.

10 Olivier, S., Smith, C., Rattier, M., Benisty, H., Weisbuch, C., Krauss, T., Houdré, R. and Oesterle, U. (2001) *Opt. Lett.*, **26** (13), 1019.

11 Heebner, J.E., Boyd, R.W. and Park, Q.-H. (2002) *Phys. Rev. E*, **65** (3), 036619.

12 Hara, Y., Mukaiyama, T., Takeda, K. and Kuwata-Gonokami, M. (2003) *Opt. Lett.*, **28**, 2437.

13 Mukaiyama, T., Takeda, K., Miyazaki, H., Jimba, Y. and Kuwata-Gonokami, M. (1999) *Phys. Rev. Lett.*, **82**, 4623.

14 Zhuk, V., Regelman, D.V., Gershoni, D., Bayer, M., Reithmaier, J.P., Forchel, A., Knipp, P.A. and Reinecke, T.L. (2002) *Phys. Rev. B*, **66**, 115302.

15 Möller, B.M., Woggon, U., Artemyev, M.V. and Wannemacher, R. (2004) *Phys. Rev. B*, **70**, 115323.

16 Rakovich, Y.P., Donegan, J.F., Gerlach, M., Bradley, A.L., Connolly, T.M., Boland, J.J., Gaponik, N. and Rogach, A. (2004) *Phys. Rev. A*, **70** , 051801(R).

17 Möller, B.M., Woggon, U. and Artemyev, M.V. (2005) *Opt. Lett.*, **30** (16), 2116.

18 Hara, Y., Mukaiyama, T., Takeda, K. and Kuwata-Gonokami, M. (2005) *Phys. Rev. Lett.*, **94**, 203905.

19 Poon, J.K.S., Zhu, L., DeRose, G.A. and Yariv, A. (2006) *Opt. Lett.*, **31** (4), 456.

20 Kanaev, A.V., Astratov, V.N. and Cai, W. (2006) *Appl. Phys. Lett.*, **88**, 111111.

21 Deng, S., Cai, W. and Astratov, V.N. (2004) *Opt. Express*, **12** (26), 6468.

22 Poon, J.K.S., Scheuer, J., Mookherjea, S., Paloczi, G.T., Huang, Y. and Yariv, A. (2004) *Opt. Express*, **12** (1), 90.

23 Chen Y. and Blair, S. (2004) *Opt. Express*, **12** (15), 3353.

24 Han, M., Gao, X., Su, J.Z. and Nie, S. (2001) *Nature Biotech.*, **19**, 631.

25 Möller, B., Woggon, U., Artemyev, M.V. and Wannemacher, R. (2003) *Appl. Phys. Lett.*, **83** (13), 2686.

26 Talapin, D.V., Rogach, A.L., Kornowski, A., Haase, M. and Weller, H. (2001) *Nano Lett.*, **1** (4), 207.

27 Hines M.A. and Guyot-Sionnest, P. (1996) *J. Phys. Chem.*, **100**, 468.

28 Murray, C.B., Norris, D.J. and Bawendi, M.G. (1993) *J. Am. Chem. Soc.*, **115**, 8706.

29 Woggon, U., Wannemacher, R., Artemyev, M.V., Möller, B., LeThomas, N., Anikeyev, V. and Schöps, O. (2003) *Appl. Phys. B*, **77**, 469.

30 Möller, B.M., Woggon, U., Artemyev, M.V. and Wannemacher, R. (2002) *Appl. Phys. Lett.*, **80** (18), 3253.

31 Möller, B., Woggon, U. and Artemyev, M.V. (2006) *J. Opt. A, Pure Appl. Opt.*, **8**, S113.

32 Sainidou, R., Stefanou, N. and Modinos, A. (2006) *Phys. Rev. B*, **74**, 172302.

33 Zentgraf, T., Christ, A., Kuhl, J., Gippius, N.A., Tikhodeev, S.G., Nau, D. and Giessen, H. (2006) *Phys. Rev. B*, **73**, 115103.

34 Belardini, A., Bosco, A., Leahu, G., Centini, M., Fazio, E., Sibilia, C., Bertolotti, M., Zhukovsky, S. and Gaponenko, S.V. (2006) *Appl. Phys. Lett.*, **89**, 031111.

35 Martńez, A., Garća, A., Sanchis, P. and Marť, J. (2003) *J. Opt. Soc. Am. A*, **20** (1), 147.

36 Mori D. and Baba, T. (2005) *Opt. Express*, **13** (23), 9398.

37 Darmawan S. and Chin, M.K. (2006) *J. Opt. Soc. Am. B*, **23** (5), 834.

38 Melloni, A., Costa, R., Monguzzi, P. and Martinelli, M. (2003) *Opt. Lett.*, **28** (17), 1567.

39 Yannopapas, V., Modinos, A. and Stefanou, N. (2002) *Phys. Rev. B*, **65** (23), 235201.

40 Guven K. and Ozbay, E. (2005) *Phys. Rev. B.*, **71**, 085108.

41 Povinelli, M.L., Johnson, S.G. and Joannopoulos, J.D. (2005) *Opt. Express*, **13** (18), 7145.

42 John, S. Phys. Rev. Lett., **58** (23), 2486.

43 Vlasov, Y.A., Kaliteevski, M.A. and Nikolaev, V.V. (1999) *Phys. Rev. B*, **60** (3), 1555.

44 Yannopapas, V., Modinos, A. and Stefanou, N. (2003) *Phys. Rev. B*, **68**, 193205.

45 Möller, B.M., Woggon, U. and Artemyev, M.V. (2007) *Phys. Rev. B*, **75**, 245327.

46 Jiao, Y., Mingaleev, S.F., Schillinger, M., Miller, D.A.B., Fan, S. and Busch, K. (2005) *IEEE Photonics Technol. Lett.*, **17** (9), 1875.

47 Mertens, G., Wehrspohn, R.B., Kitzerow, H.-S., Matthias, S., Jamois, C. and Gösele, U. (2005) *Appl. Phys. Lett.*, **87**, 241108.

5
Coupled Nanopillar Waveguides: Optical Properties and Applications

Dmitry N. Chigrin, Sergei V. Zhukovsky, Andrei V. Lavrinenko, and Johann Kroha

5.1
Introduction

Photonic crystals (PhCs) are known for offering unique opportunities for controling the flow of light by acting as waveguides, cavities, dispersive elements, etc. [1–4]. Photonic crystal waveguides (PCW) are one of the promising examples of PhCs applications at micron and sub-micron length-scales. They can be formed by removing one or several lines of scatterers from the PhC lattice (Figure 5.1a). PCW based on PhCs with different two-dimensional (2D) lattices of both air holes in a dielectric background and dielectric rods in air were reported [1–3]. Light confinement in PCW is obtained due to a complete photonic bandgap (PBG), in contrast to the standard guiding mechanism in a conventional dielectric waveguide (Figure 5.1b). It was theoretically predicted that a PhC waveguide can possess loss-free propagation as soon as a guiding mode falls into a complete PBG. However, progress in PhC research has revealed that losses are inevitable and sometimes might be rather high even in spite of broad PBG. Special optimization efforts are now intensively applied for decreasing optical losses and the results are quite promising [5,6].

At the same time, PBG guiding is not the only waveguiding mechanism in a PhC. Unique anisotropy of PhCs can cancel out the natural diffraction of the light, leading to the self-guiding of a beam in a non-channel PCW [7–9]. The common principle of index guiding (guiding due to total internal reflection) can be also found in periodic systems. It is rather straightforward if a waveguide is organized as a defect in a lattice of holes in a dielectric material. Then, the channel itself has higher index of refraction than the average index of the drilled or etched medium. Topologically inverted

Nanophotonic Materials: Photonic Crystals, Plasmonics, and Metamaterials. Edited by R.B. Wehrspohn, H.-S. Kitzerow, and K. Busch

ISBN: 978-3-527-40858-0

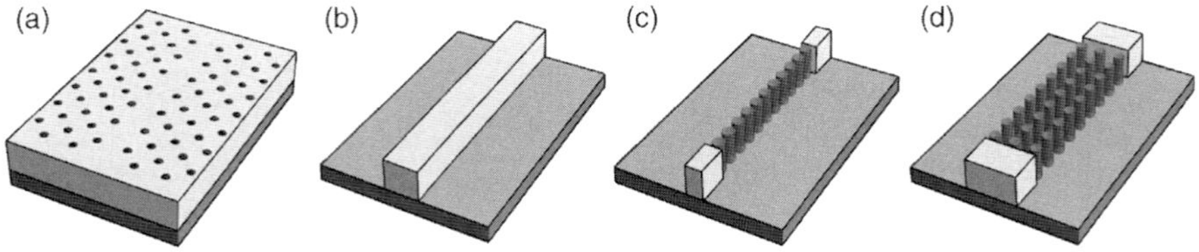

Figure 5.1 Optical waveguides: (a) photonic crystal waveguide, (b) dielectric waveguide, (c) nanopillar waveguide, and (d) coupled nanopillar waveguide.

systems like periodic arrays of rods or nanopillars placed in air can also provide waveguiding due to index difference [2]. However, fabrication of rod arrays on the nanoscale is a relatively difficult technological problem.

The recent progress in the fabrication of nanorod structures has proved the relevance of their study not only as a useful theoretical model. For example, two-dimensional (2D) silicon-on-insulator (SOI) pillar PhC have recently been fabricated and characterized [10]. Sandwich-like structures have also been successfully realized in $GaAs/Al_xO_y$ material system [11]. Membrane-like structures have been realized, based on polymer membranes incorporating Si rods [12]. Recently, various combinations of active materials inserted in single nanowires or arrays of nanopillars have been under attention as well [13]. It is important to point out that all of the above mentioned studies do not only present a successful practical realization of the pillar PhC structures, but also report transmission efficiencies and out-of-plane radiation losses comparable with the 2D PhC based on hole geometry.

A one-dimensional (1D) chain of rods placed at equal distance from one another (Figure 5.1c) possesses guiding properties as was shown by Fan *et al.* [2]. The fundamental mode of such a periodic nanopillar waveguide lies below the light line and below the first PBG corresponding to the 2D PhCs with a square lattice of the same rods. Guiding is due to total internal reflection. A better confinement of light can be achieved, if several 1D periodic chains are placed in parallel (Figure 5.1d) [14]. Such waveguides are called coupled nanopillar waveguides (CNPWs) and are designated as W*n*, where *n* is the number of parallel rows comprising the CNPW. In building a CNPW both the longitudinal and the transverse relative shift between individual waveguides can be arbitrary, and thus, a high flexibility in dispersion engineering can be achieved.

In this chapter, we review basic properties of coupled nanopillar waveguides and discuss their possible applications for integrated optics. In Section 5.2, a CNPW is introduced and possible ways to tune the CNPW dispersion are discussed. The transmission efficiency of 2D and 3D CNPWs is reported in Section 5.3. The route to improve the coupling between a nanopillar waveguide and an external dielectric waveguide (like an optical fiber) is discussed in Section 5.4 with respect to aperiodic NPWs. Possible applications of coupled periodic and aperiodic nanopillar waveguides are discussed in Section 5.5. Section 5.6 concludes the chapter.

5.2 Dispersion Engineering

5.2.1 Dispersion Tuning

In Ref. [2] it was shown that a single row of periodically placed dielectric rods is effectively a single-mode waveguide within a wide frequency range (Figure 5.2, left panel). It has a well confined fundamental mode. Attaching one, two or more identical W1 waveguides in parallel to the original one produces a coupled-waveguide structure [14]. It is well known in optoelectronics that this leads to the splitting of the original mode into n modes, where n is the number of coupled waveguides [15].

In Figure 5.2, dispersion diagrams for W1, W2, W3 and W4 CNPWs are shown. All rods are placed at the vertices of a square lattice. To model a CNPW dispersion we used the plane-wave expansion method (PWM) [16]. The supercell consists of one period in the z direction and 20 periods in the x direction, where n periods occupied by dielectric rods were placed in the center of the supercell. The waveguide is oriented along the z-axis (Figure 5.2). The calculations were performed for 2D structures and for TM polarization. The n modes of the CNPW are bound between the Γ–X and X–M projected bands of the corresponding infinite PhC of a 2D square lattice of rods (Figure 5.2, dashed lines) [14]. All modes are effectively localized within the waveguide region. Near the irreducible Brillouin zone (IBZ) boundary the dispersion is strongly affected by the system periodicity.

It is well known that by varying the filling factor, i.e. the rod radius, and the dielectric constant of the rods, one can tailor the frequency range and slope of the PhC

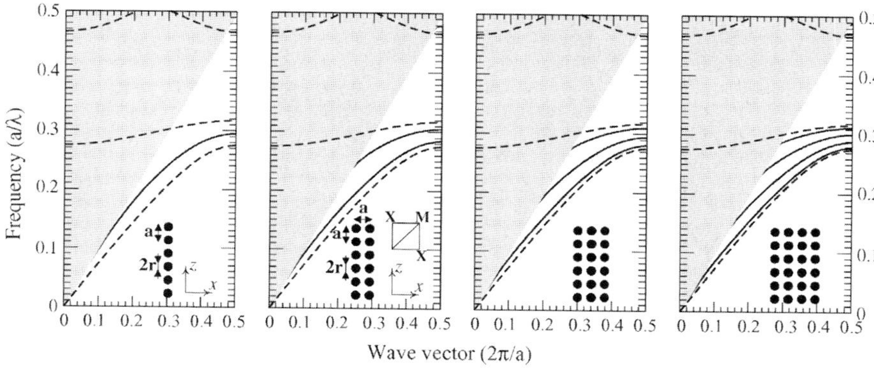

Figure 5.2 Dispersion diagrams for CNPWs with 1, 2, 3 and 4 rows. The insets show a sketch of the waveguides. The coordinate system together with the first quadrant of the first Brillouin zone for the square lattice with period a are also shown. The grey areas depict the continuum of radiated modes lying above the light line. Guided modes are shown as black solid lines. The projected band structure of the infinite 2D PhC is shown as dashed lines. Here $\varepsilon = 13.0$ and $r = 0.15a$.

bands. Taking into account that the CNPW modes are bound by Γ–X and X–M bands of the corresponding infinite 2D PhC, a proper frequency adjustment of nanopillar waveguide modes can be done by changing these two parameters. Decreasing the dielectric constant of the rods, while keeping their radius constant, pushes the bundle of n CNPW modes to higher frequencies. The modes shift towards lower frequencies, if the nanopillar radius increases, with fixed dielectric constant. In general, the mode tuning follows the rule: the larger the average refractive index of the system, the lower the mode frequencies [14].

Another option for tuning the mode dispersion of CNPW is to change the distance between individual waveguides, the transverse offset. Examples are shown in Figure 5.3 for two transverse offsets, $d = 0.5a$ (left) and $d = 2.0a$ (right). In these cases the rods are situated at the vertices of a rectangular lattice. While the mode overlap of individual waveguides is larger (smaller) for close (far) positioned waveguides, the coupling strength is stronger (weaker). For two identical waveguides, this in turn results in stronger (weaker) mode splitting, $\beta_{\pm} = \beta \pm \kappa$, with respect to the propagation constant β of the un-coupled NPW. Here κ is a coupling coefficient [15]. Note, that the CNPW mode frequencies are still bounded by the position of the projected band structure of the corresponding infinite rectangular PhC (Figure 5.3, dashed lines).

The last parameter which may affect the dispersion of a CNPW is the longitudinal shift between its individual rows. In Figure 5.4 the dispersion diagrams for CNPW with rods placed in the vertices of a triangular lattice are shown for W2, W3, W4 and W5 waveguides. The orientation of the waveguides coincides with the Γ–X direction of the triangular lattice. The mode splitting in a “triangular lattice” W2 waveguide strongly depends on the propagation constant (Figure 5.4, left panel), being large for small β and vanishing near the IBZ boundary. This is in contrast to a “square lattice” W2 waveguide (Figure 5.2), where the mode splitting is approximately constant for all propagation constants. The mode degeneracy near the IBZ boundary leads to regions

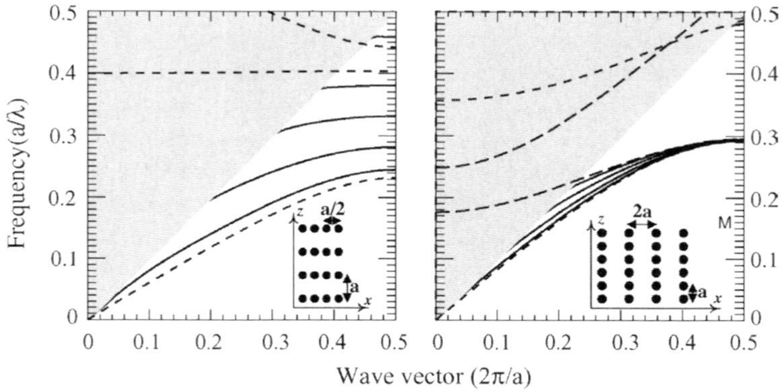

Figure 5.3 Same as in Figure 5.2 for CNPWs with different transverse offsets, $d = 0.5a$ (left) and $d = 2.0a$ (right). The insets show a sketch of waveguides and coordinate system.

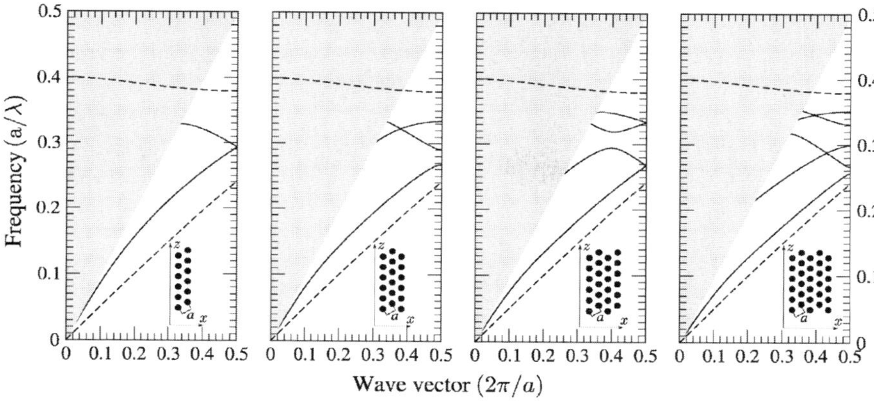

Figure 5.4 Same as in Figure 5.2 for triangular lattices W2, W3, W4 and W5 waveguides. The insets show a sketch of the waveguides and the coordinate system. Here $\varepsilon = 13.0$ and $r = 0.26a$.

with negative dispersion (backward propagating waves) of the second mode. For CNPWs with the number of rods larger than two (Figure 5.4, right panels) it results, furthermore, in the formation of mini-bandgaps and multiple backward waves regions in the dispersion. Note that in spite of the complex nature of the mode splitting, CNPW modes are still bounded by the projected bands of the corresponding triangular lattice PhC.

The longitudinal shift δ can be arbitrarily set to any value between $\delta = 0$ and $\delta = 0.5a$. The concomitant dramatic changes in the CNPW dispersion are illustrated in Figure 5.5 for the case of W2 waveguide. Starting from the simple mode splitting for $\delta = 0$ one can have a very flat second band for $\delta \approx 0.25a$, with negative dispersion regions in the second band for $\delta > 0.25a$ and degenerate first and second bands at the IBZ boundary for $\delta = 0.5a$. By combining such shifted W2 waveguides and appro-

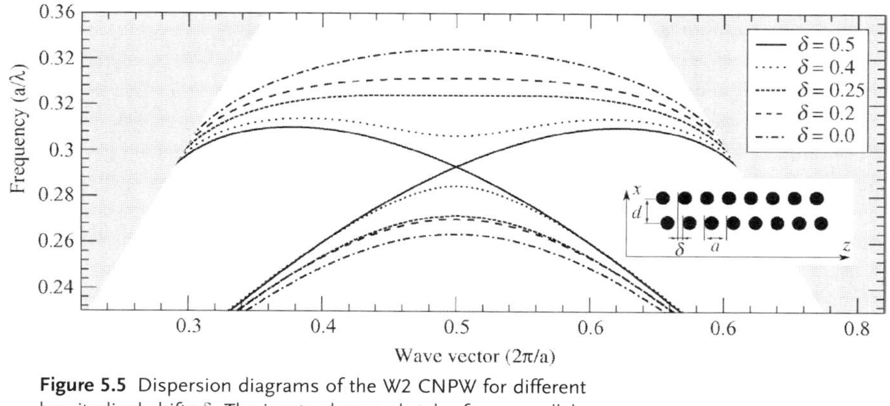

Figure 5.5 Dispersion diagrams of the W2 CNPW for different longitudinal shifts δ. The insets show a sketch of two parallel periodic waveguides with transverse offset d and longitudinal shift δ. Here $\varepsilon = 13.0$, $r = 0.15a$ and $d = a$.

priately choosing the rod radius and transverse offset one obtains large flexibility in designing CNPWs with anomalous dispersion in the frequency range of interest.

5.2.2
Coupled Mode Model

To understand qualitatively the physical mechanism of the anomalous dispersion presented in the last example (Figure 5.5), the coupled mode theory (CMT) can be used [17]. Being an approximate theory, CMT nevertheless manages to combine a simple physical model with accurate qualitative and even quantitative results [18]. In what follows, two identical coupled periodic waveguides a-W1 and b-W1 are arranged in a W2 CNPW. The second waveguide, b-W1, is shifted by δ with respect to the first one (Figure 5.5, inset). We limit ourselves to the scalar CMT, which in our case corresponds to the TM polarization.

The modes of the W2 waveguide are defined as the solutions of the 2D scalar wave equation

$$\left(\frac{\partial^2}{\partial x^2}+\frac{\partial^2}{\partial z^2}\right) E(x,z)+k_0^2\varepsilon\,(x,z)\,E(x,z)=0, \tag{5.1}$$

where the dielectric function of the composite structure is simply the sum of dielectric functions of the two W1 waveguides, $\varepsilon(x,z)=\varepsilon_a(x,z)+\varepsilon_b(x,z)$. Here $k_0=\omega/c$ is a wave number in vacuum. We are looking for a solution of Eq. (5.1) in the form of a linear combination of the propagating modes in two isolated W1 waveguides [19], which allows us to separate spatial variables in the form

$$E(x,z)=\Psi_a(x)\left(f_a(z)e^{-i\beta z}+b_a(z)e^{i\beta z}\right)+\Psi_b(x)e^{-i\beta\delta}\left(f_b(z)e^{-i\beta z}+b_b(z)e^{i\beta z}\right). \tag{5.2}$$

Here $f_m(z)=F_m(z)\exp(i(\beta-\beta_0)z)$ and $b_m(z)=B_m(z)\exp(-i(\beta-\beta_0)z)$ are the slowly varying amplitudes of forward and backward propagating modes near the Bragg resonance condition of a single periodic W1 waveguide with period a and $\beta_0=\pi/a$. The functions $\Psi_a(x)$ and $\Psi_b(x)$ represent the transverse field distributions, and indexes $m=a,b$ refer to a-W1 and b-W1 waveguides, respectively. The spatial shift between the two W1 waveguides is accounted for by the corresponding phase shift $e^{-i\beta\delta}$ of the field of the b-W1 waveguide. Here β is the propagation constant of a homogenized W1 waveguide of width $l=2r$ and dielectric constant $\varepsilon_{\text{eff}}(x)=(1/a)\int_z^{z+a} dz\varepsilon(x,z)$. The dependence of the propagation constant β on frequency is given by the standard planar waveguide dispersion relation [15]

$$\tan^2\left(\frac{l}{2}\sqrt{n_{\text{eff}}^2k_0^2-\beta^2}\right)=(\beta^2-k_0^2)\left(n_{\text{eff}}^2k_0^2-\beta^2\right)^{-1}, \tag{5.3}$$

where we have introduced the effective index of refraction of the homogenized waveguide $n_{\text{eff}}=\sqrt{\varepsilon_{\text{eff}}}$. The transverse field distributions $\Psi_a(x)$ and $\Psi_b(x)$ obey the scalar wave equations

$$\left(\frac{\partial^2}{\partial x^2}-\beta^2\right)\Psi_m(x)+k_0^2\varepsilon_{0m}(x)\Psi_m(x)=0,$$

with $m=a,b$ and the transverse dependent dielectric functions $\varepsilon_{0m}(x)$ being a z-average dielectric constant of the m-th waveguide. Substituting the mode expansion (5.2) into the scalar wave Eq. (5.1) and expanding the dielectric constant $\varepsilon(x,z)$ in a Fourier series with respect to z,

$$\varepsilon(x,z)=\varepsilon_{a0}(x)+\varepsilon_{b0}(x)+\sum_{l\neq 0}(\varepsilon_{al}(x)+\varepsilon_{bl}(x))\mathrm{e}^{-il(2i/a)z}, \tag{5.4}$$

with the Fourier coefficients $\varepsilon_{ml}(x)$, one obtains after some lengthy but straightforward derivations a system of four ordinary differential equations relating slowly varying amplitudes of the forward and backward propagating modes in the two W1 waveguides,

$$\frac{\mathrm{d}}{\mathrm{d}z}\begin{pmatrix}F_a\\F_b\\B_a\\B_b\end{pmatrix}=i\hat{M}\begin{pmatrix}F_a\\F_b\\B_a\\B_b\end{pmatrix}. \tag{5.5}$$

For the propagation constant close to the Bragg point $\beta_0=\pi/a$, the system matrix $\hat{M}$ has the form

$$\hat{M}=\begin{pmatrix}\beta_0-\beta & -\mathrm{e}^{-i\beta\delta}\,\kappa_0 & -\kappa_a & 0\\ -\mathrm{e}^{i\beta\delta}\,\kappa_0 & \beta_0-\beta & 0 & -\mathrm{e}^{2i\beta_0\delta}\,\kappa_a\\ \kappa_a & 0 & \beta-\beta_0 & \mathrm{e}^{-i\beta\delta}\,\kappa_0\\ 0 & \mathrm{e}^{-2i\beta_0\delta}\,\kappa_a & \mathrm{e}^{i\beta\delta}\,\kappa_0 & \beta-\beta_0\end{pmatrix}. \tag{5.6}$$

To simplify the following analysis, we have kept only two coupling constants, namely κ_0, accounting for the coupling between two homogenized waveguides, and κ_a, describing the waveguide's intrinsic periodic structure. These coupling constants are defined in a usual way, as overlap integrals of the transverse field distributions with the corresponding Fourier coefficients of the dielectric function expansion. The resulting propagation constants of the supermodes of the W2 waveguide are given as the eigenvalues of the system matrix $\hat{M}$ (6)

$$\beta_{W2}(\omega)=\beta_0+\Delta\beta(\omega), \tag{5.7}$$

with

$$\Delta\beta(\omega)=\pm\sqrt{(\beta(\omega)-\beta_0)^2+\kappa_0^2-\kappa_a^2\pm\kappa_0\sqrt{4(\beta(\omega)-\beta_0)^2-2\kappa_a^2+2\kappa_a^2\cos(2\pi\delta)}}. \tag{5.8}$$

The implicit dependence of the propagation constant on frequency is given via the dispersion relation (5.3) of a planar homogenized waveguide.

In Figure 5.6 the dispersion diagram of W2 waveguide calculated using Eqs. (5.7), (5.8) is presented for three values of the longitudinal shift $\delta=0.0$ (left), $\delta=0.5$

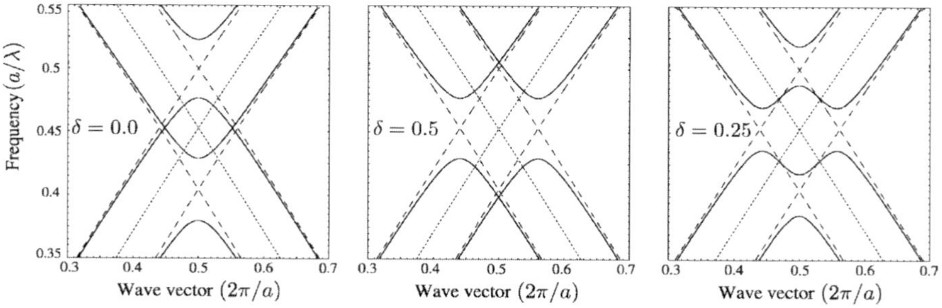

Figure 5.6 Dispersion diagrams of two coupled periodic waveguides (solid lines) within the framework of coupled mode theory for three values of the longitudinal shift $\delta = 0.0$ (left), $\delta = 0.5$ (center) and $\delta = 0.25$ (right). The dotted line is the dispersion of a homogenized waveguide folded into the first Brillouin zone. The dashed lines are the folded dispersions of two coupled homogenized waveguides. Here $n_{eff} = 1.5$, $l = 0.3a$, $\kappa_0 = 0.06$ and $\kappa_a = 0.03$.

(center) and $\delta = 0.25$ (right). See figure caption for further details on the parameters. The dotted line shows the dispersion of a planar homogenized waveguide calculated using the dispersion relation (5.3) and folded back into the first Brillouin zone by the Bragg wave vector corresponding to the periodic W1 waveguide. By setting the self-action coupling constant, κ_a, to zero and choosing some finite value for the inter-row coupling constant, κ_0, one can reproduce the simple band splitting within the CMT model. The split modes are shown as dashed lines (Figure 5.6).

To analyze the influence of the periodic structure and the longitudinal shift on the split band structure, we first consider zero longitudinal shift, $\delta = 0.0$. In this situation the detuning of the propagation constant from the Bragg wave vector, β_0, is given by $\Delta\beta = \pm\sqrt{(\Delta \pm \kappa_0)^2 - \kappa_a^2}$, where $\Delta = (\beta - \beta_0)$ is the detuning of the propagation constant of the homogenized waveguide from the Bragg point. The propagation factor of the supermodes is given by the exponential $e^{\pm i\beta_0 z} e^{\pm i\left(\sqrt{(\Delta \pm \kappa_0)^2 - \kappa_a^2}\right)z}$, which corresponds to propagating modes only if $(\Delta \pm \kappa_0)^2 - \kappa_a^2 > 1$. In the opposite situation, there are two bandgaps at Bragg wave vector β_0 with central frequencies corresponding to $\Delta = \pm\kappa_0$. These bandgaps are due to the destructive interference of the first forward propagating and the first backward propagating supermodes and the second forward propagating and the second backward propagating supermodes, respectively, as can be seen from the left panel of Figure 5.6. In the case of half-period shifted W1 waveguides, $\delta = 0.5$, the detuning of the propagation constant and the supermode propagation factor are given by $\Delta\beta = \pm\left(\sqrt{\Delta^2 - \kappa_a^2} \pm \kappa_0\right)$ and $e^{\pm i(\beta_0 \pm \kappa_0)z}\, e^{\pm i\left(\sqrt{\Delta^2 - \kappa_a^2}\right)z}$. In this case two bandgaps exist at shifted Bragg wave vectors $\beta_0 \pm \kappa_0$ with central frequency at $\Delta = 0.0$. This corresponds to the destructive interference of the first forward propagating and the second backward propagating supermodes and vice versa (Figure 5.6, center). It is important to mention here that at the Bragg condition, β_0, (IBZ boundary) the first forward propagating and the first backward propagating supermodes are in phase, which leads to the degeneracy of the first and second bands at the IBZ boundary (Figure 5.4). The shift of the Bragg condition away from the IBZ boundary, $\beta_0 \pm \kappa_0$, is a reason for the appearance of a region with negative dispersion in the second band (Figure 5.4). In the case of arbitrary shift between the W1 waveguides a destructive interference takes place between all possible pair combinations of forward and backward propagating supermodes leading to the formation of four bandgaps and anomalous dispersion (Figure 5.5). In the right panel of Figure 5.6 an example of a CMT dispersion diagram is shown for the case of quarter-period shifted waveguides, $\delta = 0.25$.

5.3 Transmission Efficiency

An important characteristic of the novel waveguides is their transmission efficiency. To analyze transmission efficiencies of different 2D and 3D CNPW, the finite difference time domain (FDTD) method [20] with perfectly matched layers as absorbing boundary conditions at all sides and a resolution of 16 grid points per lattice constant is used here. The modes are excited by a Gaussian-shaped temporal impulse, the Fourier transform of which is broad enough to cover the frequency range of interest. Fields are monitored by input and output detectors. The transmitted wave intensities are normalized by the ones of the incident waves.

The calculated transmission spectrum of a 20 periods long, straight "square-lattice" W4 CNPW is shown in the top panels of Figure 5.7. There are four modes under the light line as it is shown in the band diagram. In Figure 5.7 the transmission of the fundamental mode is shown together with the dispersion diagram. The W4 waveguide displays high transmission efficiency (close to 100%) over a broad spectral range. The position of the cut-off frequency is clearly seen in the spectrum.

In the bottom panels of Figure 5.7, the band structure and transmission spectra are shown for the W3 "triangular-lattice" CNPW. A 20 period long, straight CNPW is cut in the Γ–X direction of the triangular lattice. A substantial suppression of the transmission is seen in the spectrum, coinciding exactly with the position of the mini bandgap in the band structure. Changing the parity of the signal field distorts the spectrum reflecting the mode symmetries. The even mode displays high transmission efficiency (close to 100%) over a broad spectral range. The odd mode has a lower level of transmission and is mostly transmitted at higher frequencies. Here, by odd and even modes we understand the corresponding first two fundamental modes of a conventional dielectric waveguide. The surprisingly high transmission of the even mode above the cutoff frequency, $\omega \approx 0.34$, can be explained by the resonant behavior of the folded radiation mode with negative group velocity [21]. We found similar behavior above cutoff for other "triangular-lattice" CNPW structures.

An example of 3D calculations for an SOI W4 "triangular-lattice" CNPW is presented in Figure 5.8. In the top panel (left) the dispersion diagram of the structure is presented, while its transmission spectrum for the even mode is plotted in the top panel (right). The dispersion diagram was calculated using 3D supercell PWM. In general, the transmission spectrum is very similar to the corresponding spectrum of 2D structure (Figure 5.7). The transmission band is rather broad with 80% transmission efficiency at maximum and a sizable stopband at the mini bandgap

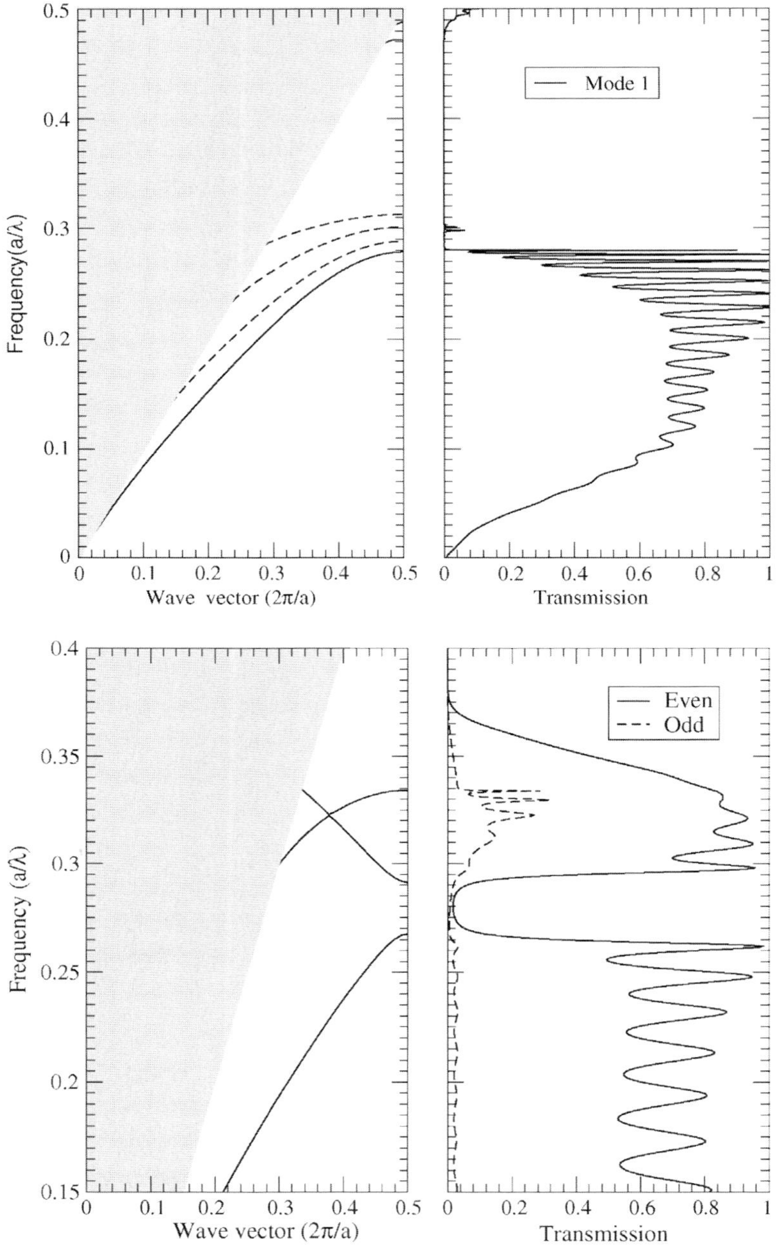

Figure 5.7 Top: Dispersion diagram and transmission spectra of a 2D "square-lattice" W4 CNPW. The fundamental mode was excited. Here $\varepsilon = 13.0$ and $r = 0.15a$ Bottom: Dispersion diagram and transmission spectra of a 2D W3 "triangular-lattice" CNPW. Solid line – even excitation, dashed line – odd excitation. Here $\varepsilon = 13.0$ and $r = 0.26a$.

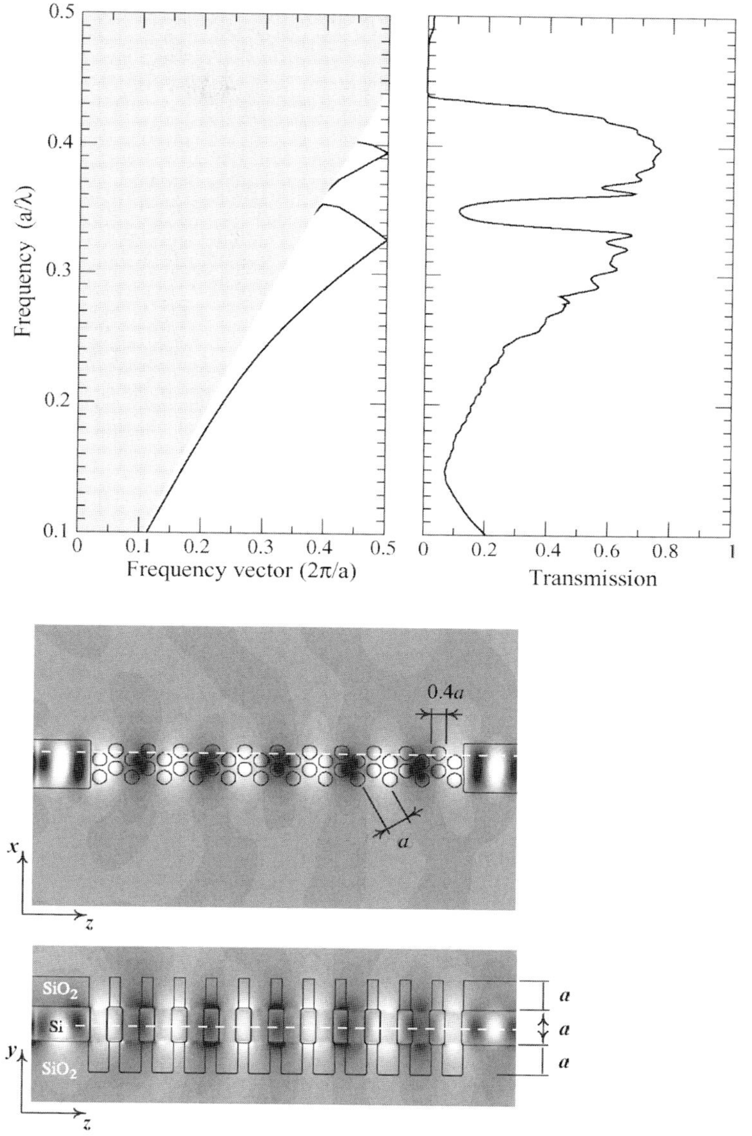

Figure 5.8 Top: Transmission spectra of even mode and dispersion diagram of 3D W4 SOI "triangular lattice" CNPW. Bottom: Field distribution inside W4 SOI CNPW in horizontal and vertical planes. Grey levels mark electric field amplitude. Black contours correspond to waveguide structure. White dashed lines depict positions of the corresponding cuts. Here radius is $r = 0.2a$, the total nanopillar height is $h = 3a$, the thickness of Si layer equals to a. Dielectric constants of Si and SiO_2 were chosen as $\varepsilon = 11.5$ and $\varepsilon = 2.1025$, respectively.

frequencies. We attribute the moderate level of transmission to the impedance mismatch at the conventional waveguide-nanopillar waveguide interface. In the bottom panel of Figure 5.8, the steady-state electric field distribution inside the CNPW is shown for a monochromatic source with normalized frequency $\omega = a/\lambda = 0.3$. The field is well confined within the waveguide core in both the horizontal and vertical planes. There is no evidence for strong energy leakage into the substrate.

5.4 Aperiodic Nanopillar Waveguides

In addition to allowing arbitrary variation of the period and displacement (which is one of the advantages of the nanopillar waveguides as opposed to the PCWs), CNPWs allow arbitrary modification of the longitudinal geometry. A localized change of the properties introduced in one or several nanopillars would create a point defect, which functions as a resonator [2,24]. The design of such micro-resonators on the scale of a few wavelengths is essential for integrated optics applications. Ideally, such resonators should combine the apparently contradictory features of a high Q-factor and of a sufficiently good coupling to a waveguide terminal to inject or extract light into or from the resonator. Due to the absence of a complete bandgap, the breaking of translational symmetry inevitably results in radiation losses of the resonator mode, which raises the need for optimizing the Q-factor of the resonator in 1D nanopillar waveguides. There have been some proposals to decrease the losses based on either mode delocalization [23] or on the effect of multipole cancellation [24]. A delocalized mode typically suffers from a decrease of the Q-factor. On the other hand, the spatial radiation loss profile of a mode described in Ref. [24] has a nodal line along the waveguide axis, which means poor coupling to any components coaxial with the waveguide.

Other than by means of a point defect, a resonant system can also be created by changing the periodic arrangement of nanopillars into a non-periodic one. We show that the use of such aperiodically ordered waveguide leads to improved coupling to the coaxial terminal without considerably sacrificing the Q-factor of the resonant modes. We use fractal Cantor-like NWPs as an example [25]. To construct aperiodic NPW, nanopillars of equal radius are arranged in a 1D chain, where the distances between adjacent pillars are given by the Cantor sequence. If we denote S and L for short and long distance (d_S and d_L), respectively, the Cantor sequence is created by the inflation rule L → LSL, S → SSS and unfolds in the following self-similar fashion, which represents a series of middle third Cantor prefractals L→LSL→LSL SSS LSL→LSLSSSLSL SSSSSSSSS LSLSSSLSL→ . . .

In order to compare the amount of energy gathered by the coaxial terminal and dissipated elsewhere, we excite the system by a dipole source emitting a pulse with a broad spectrum, and use the FDTD method to investigate the process of energy loss into the surroundings. Figure 5.9 shows the results. For the point-defect structure, the radiation of the resonant mode primarily escapes sideways (Figure 5.9(a)), so despite having a high Q-factor (2.1×10^4), the coupling between the resonator and other components cannot be made efficient. The Cantor structure

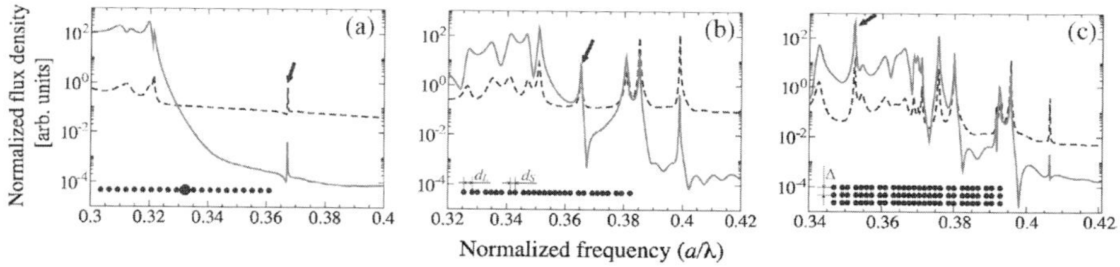

Figure 5.9 Normalized energy flux of electromagnetic radiation escaping from the resonator into the terminal (solid line) and elsewhere (dashed line) for three nanopillar structures shown in the insets: a W1 with a point defect [24] (left); a W1 with Cantor-like longitudinal geometry (center) and a W3 Cantor-like CNPW (right). Here, $d_s = 0.5a$, $d_L = 0.81a$, $r = 0.15a$, $\Delta = 0.75a$ and $\varepsilon = 13.0$. The point defect is created by doubling the radius of a central rod in a periodic waveguide with period d_L. Arrows mark the resonances discussed in the text.

shows a considerably improved coupling (Figure 5.9(b)) accompanied by a drop of the Q-factor down to 2.7×10^3. Using a W3 CNPW with Cantor geometry (Figure 5.9(c)) raises it back to $Q = 1.1 \div 2.4 \times 10^4$ while still providing just as good coupling to the coaxial terminal.

One should notice that the Cantor geometry is only one kind of deterministically aperiodic arrangement. Other kinds, e.g., quasi-periodic Fibonacci-like one, can be used leading to a modification of the mode structure of NPWs as well as the coupling efficiency of resonant mode into coaxial terminal [25]. Engineering the longitudinal geometry of CNPWs appears to be a promising and powerful tool for a further degree of freedom in controlling their dispersion properties.

5.5 Applications

Relatively high transmission efficiency and flexibility in dispersion tuning of CNPWs may initiate their use as components for efficient and compact nanophotonics devices. Here we discuss two possible applications of CNPWs in integrated optics: a coupled nanopillar waveguide directional coupler [22] and a switchable coupled mode laser [31].

5.5.1 Directional Coupler

A pair of CNPWs can be used as an effective directional coupler [22]. An example of such a directional coupler based on two W1 waveguides is shown in Figure 5.10. Analyzing the dispersion diagram of the coupling section, namely W2 CNPW (Figure 5.10, left panel), one can see a pronounced difference in the propagation constants of the even and odd supermodes in the frequency region around $\omega = 0.25$–0.27. It is a result of the strong interaction of coupled waveguides, which now are much closer to each other than in the case of standard line defect

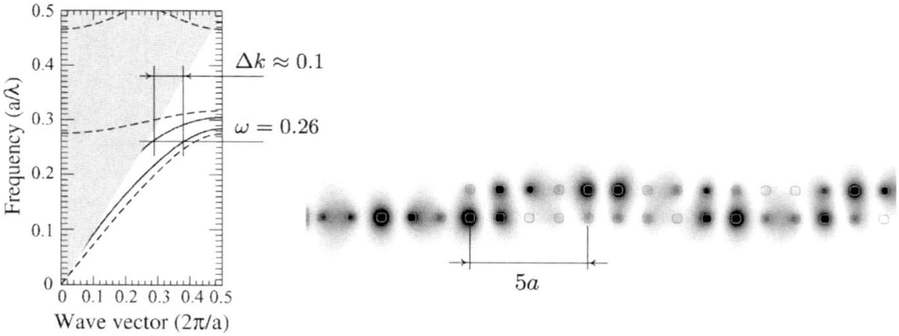

Figure 5.10 Dispersion diagram for the W2 CNPW section of a directional coupler (left). Directional coupler based on two W1 CNPWs (right). Grey levels mark field intensity. Here $\varepsilon = 13.0$ and $r = 0.15a$.

waveguides in a PhC lattice (see, for example, the similar rod structure in [26]). In this frequency range the difference between the even and odd supermode propagation constants is close to $0.1 \cdot 2\pi/a$, which leads to a crude estimate of minimum coupling length [27]: $L = \pi/|k_{\text{even}} - k_{\text{odd}}| = (\pi/0.1)\cdot(a/2\pi) = 5a$. Enhanced interaction leads to a shorter coupling length. This is illustrated in Figure 5.10, right panel, where the time averaged squared electric field pattern is shown for the normalized frequency $\omega = 0.26$. Guided light hops from the bottom W1 waveguide to the top one and back on a distance equal to approximately $5a$, which represents well the estimated value. In contrast to a directional coupler proposed in [28], the CNPW structure does not require a specially adjusted separation layer between coupled waveguides, thus considerably simplifying a directional coupler design and fabrication.

There are several parameters, which can be used to optimize the directional coupler, e.g., length of the coupling region or the number of rows in each of the waveguides. As it has been shown in Section 5.2, longitudinal and transverse offsets between the individual waveguides, as well as variation of the dielectric constant and radius of the rods substantially modify the dispersion of the compound system, thus affecting the coupling efficiency. For example, by shortening the distance between two waveguides one can dramatically increase the propagation constant difference and reduce the coupler size. A similar effect cannot be achieved with standard PhC waveguides without any special design tricks involving intermediate walls, which increases the complexity of the fabrication procedure. An arbitrary longitudinal offset breaks the symmetry of the device with respect to the symmetry plane between the two W1 waveguides, which may further improve the coupling strength similar to the case of an antisymmetric grating coupler [29,30].

5.5.2
Laser Resonators

The periodicity of the coupled nanopillar waveguides ensures the distributed feedback within a finite waveguide section. This can be seen from the flat tails of the nanopillar waveguide modes near the IBZ edge, which correspond to a very low group velocity of CNPW modes. Taking into account that any of the CNPW modes can be efficiently excited in the waveguide using an external seeding signal of the appropriate spatial profile [14], we have proposed the design of a switchable laser resonator [31] with distributed feedback based on the CNPW. The possibility to tune the number of modes, their frequency and separation (Section 5.2) would make such a resonator a promising candidate for a chip-integrated laser source.

The concept of switchable lasing was originally proposed in Ref. [14] and has got further justifications in our recent work [31,32]. In essence, a switchable microlaser comprises a multimode microresonator, where lasing can be switched on demand to any of its eigenmodes by injection seeding [33,34], i.e. by injecting an appropriate pulse before and during the onset of lasing, such that the stimulated emission builds up in a designated mode selected by this seeding field rather than from the random noise present in the system due to quantum fluctuations and spontaneous emission [32].

To provide a basic physical picture of switchable lasing we first consider briefly a simple semi-classical laser model in the case of two identical coupled single-mode cavities [32]. In this case there are two modes, the symmetric and the antisymmetric one, characterized by spatial field distributions $u_{1,2}(\mathbf{r})$ and frequencies $\omega_{1,2} = \omega_0 \mp \Delta\omega$, respectively. Here $\Delta\omega$ is the mode detuning from the frequency of the single-cavity resonance, ω_0. For weak mode overlap the spatial intensity profiles of the two modes nearly coincide, $|u_1(r)|^2 \approx |u_2(r)|^2$. We assume that the cavities contain a laser medium with a homogeneously broadened gain line of width $\Delta\omega_a > \Delta\omega$, centered at frequency $\omega_a = \omega_0 + \delta$. Here δ is the detuning of the gain profile from the cavity frequency ω_0. For this system the semiclassical Maxwell–Bloch equations [33,35], in the rotating-wave and the slowly varying envelope approximation read

$$
\begin{aligned}
\frac{\mathrm{d}E_1(t)}{\mathrm{d}t} &= gR_1\left(\mathcal{L}_1 - \frac{\kappa_1}{gR_1}\right)E_1(t)\\
&\quad - gR_1\eta\mathcal{L}_1(\alpha_{11}^{11}\mathcal{L}_1|E_1|^2 + [\alpha_{22}^{11}\mathcal{L}_2 - \alpha_{21}^{12}\,\mathrm{Re}\,(\chi_1\mathcal{M}_{12})]\,|E_2|^2)\,E_1(t) + F_1(t),\\
\frac{\mathrm{d}E_2(t)}{\mathrm{d}t} &= gR_2\left(\mathcal{L}_2 - \frac{\kappa_2}{gR_2}\right)E_2(t)\\
&\quad - gR_2\eta\mathcal{L}_2([\alpha_{11}^{22}\mathcal{L}_1 - \alpha_{12}^{21}\,\mathrm{Re}\,(\chi_2\mathcal{M}_{21})]\,|E_1|^2 + \alpha_{22}^{22}\mathcal{L}_2|E_1|^2)\,E_2(t) + F_2(t).
\end{aligned} \tag{5.9}
$$

Here all the spatial dependencies of the electric field and atomic polarization were represented in the basis of the two cavity modes, such that $E(\mathbf{r},t) = E_1(t)\,u_1(\mathbf{r})\,\mathrm{e}^{-i\omega_1 t} + E_2(t)\,u_2(\mathbf{r})\,\mathrm{e}^{-i\omega_2 t}$, etc., and the atomic polarization was eliminated adiabatically [36,37]. $E_j(t)$ are slowly varying envelopes of two modes $j = 1, 2$. In Eq. (9) the terms linear in $E_j(t)$ describe stimulated emission driving, where the light–matter coupling constant is denoted by $g \simeq \sqrt{2\pi\omega_0\, d^2/\hbar}$, the pumping rates projected onto the two resonator modes by $R_j = \int_G u_j^*(\mathbf{r})\, u_j(\mathbf{r})\, R(\mathbf{r})\,\mathrm{d}\mathbf{r}$, and the cavity mode decay rates by κ_j. Here d is the dipole moment of the atomic transition. The coefficients $\mathcal{L}_j = Re\,\beta_j^{-1}$, with $\beta_{1,2} = \Delta\omega_a/2 + i(\delta \pm \Delta\omega)$, account for the different mode-to-gain couplings due to asymmetrical detuning of the atomic transition with respect to the resonator frequencies. The terms cubic in $E_j(t)$ describe field saturation above the lasing threshold, where $\eta = d^2/2\gamma_{\|}\hbar^2$ and the overlap integrals $\alpha^{ij}_{kl} = \int_G u^*_i(\mathbf{r})\, u_j(\mathbf{r})\, u^*_k(\mathbf{r})\, u_l(\mathbf{r})\, d\mathbf{r}$ are taken over the regions G containing the gain medium. Here $\gamma_{\|}$ is the non-radiative decay rate. The frequency dependence of the cross-saturation terms is given by $\mathcal{M}_{ij} = \beta_i^{-1} + (\beta_j^*)^{-1}$, $i \neq j$. Since $|u_1(r)|^2 \approx |u_2(r)|^2$ we can further assume that $\alpha^{ii}_{jj} = \alpha^{ij}_{ji} \equiv \alpha$, $R_1 = R_2 = R$ and $\kappa_1 = \kappa_2 = \kappa$.

The inhomogeneous terms $F_j(t)$ originate from the external injection seeding field and from a noise field accounting for spontaneous emission [36]. For vanishing functions $F_j(t)$, Eq. (9) would take the form of the standard two-mode competition equations [33,35], describing bistable lasing [38] and mode hopping in the presence of stochastic noise in the system [39]. If both an external seeding field $\mathcal{E}^s(\mathbf{r},t)$ and a stochastic noise field $\mathcal{E}^n(\mathbf{r},t)$ are present in the cavity, $\mathcal{E}(\mathbf{r},t) = \mathcal{E}^s(\mathbf{r},t) + \mathcal{E}^n(\mathbf{r},t)$, the inhomogeneous terms are given by,

$$F_j(t) \approx \frac{\omega_j \mathcal{L}_j}{\tau} \int\limits_{t-\tau}^{t} dt'\, e^{i\omega_j t'} \int\limits_{G} u_j(\mathbf{r})\, \mathcal{E}(\mathbf{r},t')\, d\mathbf{r} = F^s_j F(t) + F^n_j(t). \tag{5.10}$$

The time integration in Eq. (10) is the averaging over a time interval larger than $1/\Delta\omega$. The function $F(t)$ is determined by the temporal dependence of the seeding signal $\mathcal{E}^s(\mathbf{r},t)$. The coefficients F^s_j and $F^n_j(t)$ are determined by the spatial overlap of each mode with the seeding and noise fields, respectively.

We consider the situation when the seeding prevails over the noise, i.e., $F^s_j F(t) \gg F^n_j(t)$, before and during the onset of lasing. After the onset the E_j become so large that the terms F_j have no effect anymore. During the onset the evolution of the resonator will be determined by the ratio of F^s_1 and F^s_2. In Figure 5.11 (left) the phase trajectories of the temporal resonator state evolution in the $(|E_1|^2, |E_2|^2)$ space is presented for different values of F^s_1 and F^s_2. As seen in Figure 5.11 (left), the lasing

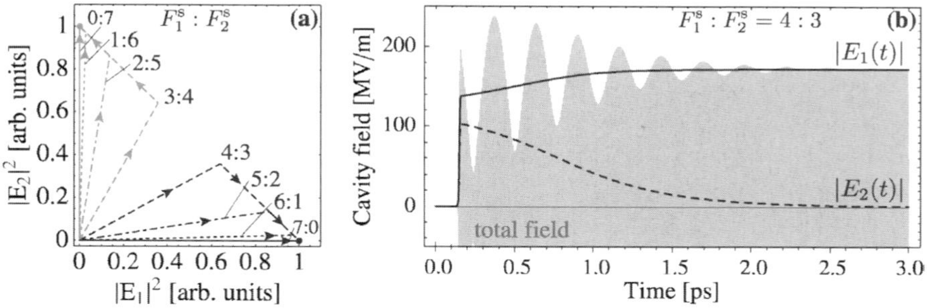

Figure 5.11 (Left) Cavity phase diagrams for a lasing system governed by Eq. (9) for $F^s_j F(t) \gg F^n_j(t)$ in the case of symmetric distribution of mode frequencies with respect to gain. The dots denote the stable cavity states and the curves represent the phase trajectories for their temporal evolution for different ratios $F^s_1 : F^s_2$ in the direction of the arrows. (Right) Time evolution of the total laser field (shown in grey) and the two cavity modes envelops $|E_1|$ (solid line) and $|E_2|$ (dashed line) for the ratio $F^s_1 : F^s_2 = 4:3$. One can see how the first mode wins the competition. Here numerical values of the coefficients in Eq. (9) were calculated for W2 CNPW resonator with the following parameters, period a, $d = 1.21a$, $r = 0.15a$ and $\varepsilon = 13.0$.

state first reaches overall intensity saturation ($|E_1|^2+|E_2|^2 = E_s^2$) and then drifts towards one of the stable fixed points corresponding to single-mode lasing (either $|E_1|^2 = E_s^2$ or $|E_2|^2 = E_s^2$). The drift happens on a longer time scale than the initial overall intensity growth, and the intermode beats decay fast after the lasing onset (Figure 5.11, right). The drift occurs towards the mode whose spatial and temporal overlap with the seeding signal is larger, demonstrating a switchable lasing behavior. It is important to note that even in the case of asymmetric detuning of the cavity modes with respect to the gain frequency ($\delta \neq 0$), single-mode lasing is achieved into the mode whose spatial overlap with the seeding field, F_j^s, is largest, i.e., if one of the following conditions, $F_1^s \gg F_2^s$ or $F_1^s \ll F_2^s$, is satisfied.

To demonstrate the predictions of this simple theory we have modeled the lasing action in four-row CNPW structure with a realistic injection seeding (Figure 5.12) using the FDTD method [31]. The externally pumped laser-active medium is placed in the central 7 pillars of all four rows. This is done to maximize coupling between the active medium and the main localization region of the lasing modes. The population dynamics of an active medium is described at each space point by the rate equations of a four-level laser with an external pumping rate W_p. To achieve population inversion we have chosen the following values for the non-radiative transition times, $\tau_{32} \simeq \tau_{10} \ll \tau_{21}$, with $\tau_{31} = \tau_{10} = 1\times10^{-13}$ s, $\tau_{21} = 3\times10^{-10}$ s, and the total level population is $N_{\text{total}} = 10^{24}$ per unit cell [40]. The Maxwell equations are solved using FDTD scheme supplemented by the usual equation of motion for the polarization density in the medium and by the laser rate equations [40–43]. All calculations were done for TM polarization. The seeding signal is excited by four emitters (linear groups of dipoles) engineered on the regular dielectric waveguide attached to the CNPW structure (see Figure 5.12). Each of the emitters generates a single short Gaussian pulse with carrier frequency ω_a and with half-width duration $\sigma_t = 10^4\,dt$. The relative phase of the fields in these pulses is chosen 0 or π. Technically, the seeding dipoles are realized as point like oscillating current sources in the Maxwell equations [31]. Similarly, the spontaneous emission [42,44,45] can be modeled as an ensemble of point current sources, randomly placed in space, with temporally δ-correlated Langevin noise [45]. The computational domain of size $7a\times22a$ was discretized with a mesh point spacing of $a/16$. The time step is related to the spatial mesh to assure stability and was chosen $dt = 6\times10^{-17}$ s. To simulate an open system, perfectly matched layer (PML) boundary conditions [20] were used.

In Figure 5.12 (top) the lasing spectra in the steady state long after the seeding signal has decayed is shown. The broad shaded area depicts the laser line of width $\Delta\omega_a$ centered at ω_a, which is shifted slightly towards lower frequencies. As a rule the Q-factor is larger for modes with the higher frequency. The shifted laser line compensates this Q-factor difference, so that any of the four CNPW modes can be selected by the appropriate seeding signal with the same symmetry. In Figure 5.12 (bottom) the spatial electric field distribution in the four-row CNPW laser resonator is shown at an instant of time long after the seeding signal has decayed and after the steady state has been reached. The symmetry of the selected lasing modes corresponds to that of the seeding signal (Figure 5.12).

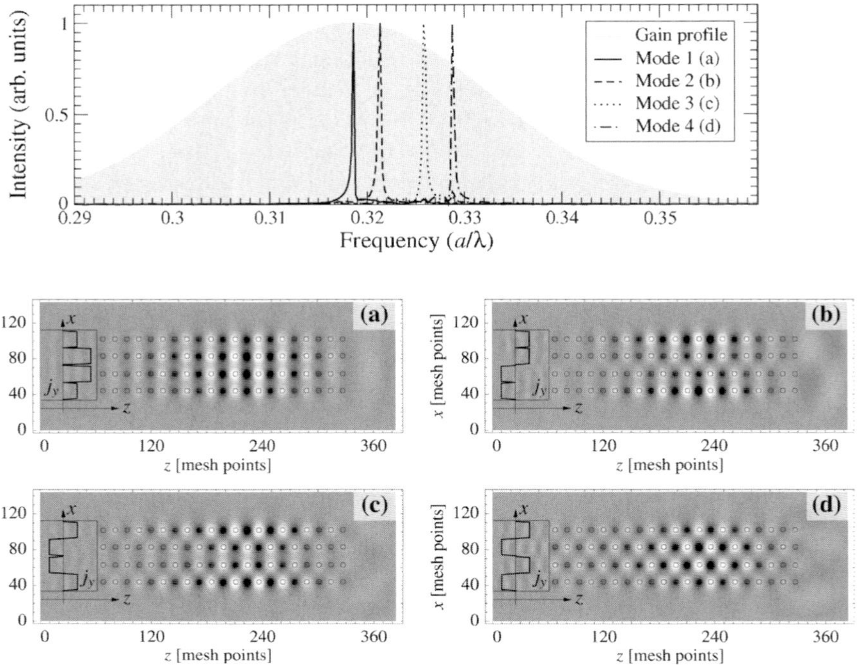

Figure 5.12 Amplitude spectra (top) and laser filed distribution (bottom) for the periodic injection-seeded four-row CNPWs. The lines labeled "Mode-1" to "Mode-4" correspond to the seeding signals (a)–(d) shown in bottom panel. The shaded areas represent the laser amplification line, with its central frequency $\omega_a = 0.3225$. The pumping rate equals $W_p = 1.0 \times 10^{13}\,s^{-1}$. The panels (a)–(d) correspond to different seeding signals, shown schematically as excited in the terminal.

The proposed concept of switchable lasing is not limited to the periodic CNPW structures, but is expected to work in any resonator featuring bi- or multistability. Any coupled cavity based system would be a good candidate for the effects predicted. For example aperiodic CNPW based resonator also show the switchable lasing behavior for resonant modes discussed in Section 5.4 [31].

5.6 Conclusion

We have shown that a novel type of coupled nanopillar waveguides, comprised of several periodic or aperiodic rows of dielectric rods, may have potential applications in compact photonics. The strong coupling regime can be utilized in ultrashort directional couplers or laser cavities, which might possess an additional functionality and flexibility when different longitudinal and transverse offsets among individual waveguides are employed. The factors of major influence upon the mode dispersion have been analyzed. Transmission spectra for 2D and 3D systems prove the possible single mode excitation by imposing specific symmetry conditions onto a field source and high transmission characteristics of coupled nanopillar waveguides.

Acknowledgements

D.N.C., S.V.Z. and J.K. acknowledge partial support by the Deutsche Forschungsgemeinschaft through projects SPP 1113 and FOR 557. A.V.L. acknowledges partial support by Danish Technical Research Council via PIPE project and the EU Commission FP6 via project NewTon (NMP4-CT-2005-017160).

References

1 Joannopoulos, J.D., Meade, R.D. and Winn, J.N. (1995) *Photonic Crystals: Molding the Flow of Light*, Princeton University Press, Princeton, N.J.

2 Fan, S., Winn, J., Devenyi, A., Chen, J.C., Meade, R.D. and Joannopoulos, J.D. (1995) *J. Opt. Soc. Am. B*, **12**, 1267–1272.

3 Johnson, S.G., Villeneuve, P.R., Fan, S. and Joannopoulos, J.D. (2000) *Phys. Rev. B*, **62**, 8212–8222.

4 Bjarklev, A. and Lavrinenko, A. (eds) (2001) Special Issue on Photonic Bandgaps, J. Opt. A **3** (6), S103–S207. De La Rue, R.M. (Ed.) (2002) Special review issue, Opt. Quantum Electron., 34 (1–3), 1–316.

5 Sugimoto, Y., Tanaka, Y., Ikeda, N., Yang, T., Nakamura, H., Asakawa, K. and Inoue, K. (2004) *Opt. Express*, **12**, 1090.

6 Notomi, M., Shinya, A., Mitsugi, S., Kuramochi, E. and Ryu, H.-Y. (2004) *Opt. Express*, **12**, 1551.

7 Chigrin, D.N., Enoch, S., Sotomayor Torres C.M. and Tayeb, G. (2003) *Optics Express*, **11**, 1203–1211.

8 Witzens J. and Scherer, A. (2003) *J. Opt. Soc. Am. A*, **20**, 935.

9 Chen, C., Sharkawy, A., Pustai, D.M., Shi, S. and Prather, D.W. (2003) *Opt. Express*, **11**, 3153–3159.

10 Tokushima, M., Yamada, H. and Arakawa, Y. (2004) *App. Phys. Lett.*, **84**, 4298–4300.

11 Assefa, S., Rakich, P.T., Bienstman, P., Johnson, S.G., Petrich, G.S., Joannopoulos, J.D., Kolodziejski, L.A., Ippen, E.P. and Smith, H.I. (2004) *App. Phys. Lett.*, **85**, 6110–6112.

12 Schonbrun, E., Tinker, M., Park, W. and Lee, J.-B. (2005) *IEEE Photonics Technol. Lett.*, **17**, 1196–1198.

13 Johnson, J.C., Yan, H.Q., Schaller, R.D., Haber, L.H., Saykally, R.J. and Yang, P.D. (2001) *J. Phys. Chem. B*, **105**, 11387.

14 Chigrin, D.N., Lavrinenko, A.V. and Sotomayor Torres, C.M. (2004) *Opt. Express*, **12**, 617–622.

15 Coldren L.A. and Corzine, S.W. (1995) *Diode Lasers and Photonic Integrated Circuits*, John Wiley, New York.

16 Johnson S.G. and Joannopoulos, J.D. (2001) *Opt. Express*, **8**, 173.

17 Lavrinenko A.V. and Chigrin, D.N. unpublished.

18 Huang, W.P. (1994) *J. Opt. Soc. Am. A*, **11**, 963–983.

19 März R. and Nolting, H.P. (1987) *Opt. Quantum Electron.*, **19**, 273–287.

20 Lavrinenko, A., Borel, P.I., Fradsen, L.H., Thorhauge, M., Harpoth, A., Kristensen,

M., Niemi, T. and Chong, H.M.H. (2004) *Opt. Express*, **12**, 234–248.

21 Loncar, M., Nedeljkovic, D., Pearsall, T.P., Vuckovic, J., Scherer, A., Kuchinsky, S. and Allan, D.C. (2002) *Appl. Phys. Lett.*, **80**, 1689–1691.

22 Chigrin, D.N., Lavrinenko, A.V. and Sotomayor Torres, C.M. (2005) *Opt. Quantum Electron.*, **37**, 331.

23 Benisty, H., Labilloy, D., Weisbuch, C., Smith, C.J.M., Krauss, T.F., Kassagne, D., Béraud, A. and Jouanin, C. (2000) *Appl. Phys. Lett.*, **76**, 532–534.

24 Johnson, S.G., Fan, S., Mekis, A. and Joannopoulos, J.D. (2001) *Appl. Phys. Lett.*, **78**, 3388–3390.

25 Zhukovsky, S.V., Chigrin, D.N. and Kroha, J. (2006) *J. Opt. Soc. Am.*, **10**, 2256–2272.

26 Boscolo, S., Midrio, M. and Somedo, C.G. (2002) *IEEE J. Quantum Electron.*, **38**, 47.

27 Zimmermann, J., Kamp, M., Forchel, A. and März, R. (2004) *Opt. Commun.*, **230**, 387.

28 Martinez, A., Cuesta, F. and Marti, J. (2003) *IEEE Photon. Technol. Lett.*, **15**, 694.

29 Perrone, G., Laurenzanno, M. and Montrosset, I. (2001) *J. Lightwave Technol.*, **19**, 1943.

30 Aslund, M., Canning, J., Poladian, L., de Sterke, C.M. and Judge, A. (2003) *Appl. Opt.*, **42**, 6578.

31 Zhukovsky, S.V., Chigrin, D.N., Lavrinenko, A.V. and Kroha, J. (2007) *phys. stat. sol. (b)*, **244**, 1211–1218.

32 Zhukovsky, S.V., Chigrin, D.N., Lavrinenko, A.V. and Kroha, J. (2007) *Phys. Rev. Lett.*, **99**, 073902.

33 Siegman, A. (1986) *Lasers*, University Science Books, Mill Valley, CA.

34 Lee W. and Lempert, W.R. (2003) *Appl. Opt.*, **42**, 4320.

35 Sargent III, M., Scully, M.O. and Lamb, W.E. Jr. (1974) *Laser Physics*, Addison-Wesley, Reading, MA.

36 Hodges, S.E., Munroe, M., Cooper, J. and Raymer, M.G. (1997) *J. Opt. Soc. Am. B*, **14**, 191–197.

37 Florescu, L., Busch, K. and John, S. (2002) *J. Opt. Soc. Am. B*, **19**, 2215–2223.

38 Agrawal G.P. and Dutta, N.K. (1985) *J. Appl. Phys.*, **56**, 664–669.

39 Bang, J., Blotekjar, K. and Ellingsen, R. (1991) *IEEE J. Quantum Electron.*, **27**, 2356–2363.

40 Chang S.-H. and Taflove, A. (2004) *Opt. Express*, **12**, 3827–3833.

41 Nagra A.S. and York, R.A. (1998) *IEEE Trans. Antennas Propag.*, **46**, 334.

42 Jiang X. and Soukoulis, C.M. (2000) *Phys. Rev. Lett.*, **85**, 70–73.

43 Bermel, P., Rodriguez, A., Johnson, S.G., Joannopoulos, J.D. and Soljacic, M. (2006) *Phys. Rev. A*, **74**, 043818.

44 Seo, M.-K., Song, G.H., Hwang, I.-K. and Lee, Y.-H. (2005) *Opt. Express*, **13**, 9645–9651.

45 Slavcheva, G.M., Arnold, J.M. and Ziolkowski, R.W. (2004) *IEEE J. Sel. Top. Quantum Electron.*, **10**, 1052–1062.

6
Investigations on the Generation of Photonic Crystals using Two-Photon Polymerization (2PP) of Inorganic–Organic Hybrid Polymers with Ultra-Short Laser Pulses

R. Houbertz, P. Declerck, S. Passinger, A. Ovsianikov, J. Serbin, and B.N. Chichkov

6.1
Introduction

The interaction of laser light with polymer surfaces and bulk samples is of high technological interest. The adaptation of polymers to laser beam and processing characteristics is very challenging from the scientific as well as from a technological point of view. A method which has recently attracted considerable attention is two-photon absorption (TPA) or two-photon polymerization (2PP) using femtosecond lasers [1–4], where complicated microstructures can be generated in photoresponsive materials with high speed. Among the demonstrated structures are, e.g. photonic crystal structures [3] or mechanical devices [5,6].

The materials used to create these structures were mainly commercially available acrylate- or epoxy-based resins [5,6]. However, in terms of integration, these materials lack of some major requirements: they are often chemically not stable against solvents typically applied in multi-layer processing, their thermal stability is quite low, and they suffer from a low mechanical stability. A material class which has attracted considerable attention for integration and packaging [7–9] is the class of inorganic-organic hybrid polymers such as ORMOCER®s (Trademark of the Fraunhofer-Gesellschaft zur Förderung der Angewandten Forschung e.V., Munich, Germany). Their properties can be tailored towards application and corresponding processing technologies [10–16]. ORMOCER®s can be employed in many devices for a large variety of applications, thus enabling novel properties from micro- down to the nanometer scale with an outstanding chemical, thermal, and mechanical stability. Thus, they overcome the restrictions of purely organic polymers for most applications.

Integrated optical devices with micro- and nanooptical elements using polymers and, particularly nano-scaled organic-inorganic hybrid materials such as ORMOCER®s, will be beyond of the next generation of optical components. For the realization of photonic elements such as, for example photonic crystals (PhC), either high or low refractive index materials are required, dependent on the device

Nanophotonic Materials: Photonic Crystals, Plasmonics, and Metamaterials. Edited by R.B. Wehrspohn, H.-S. Kitzerow, and K. Busch

ISBN: 978-3-527-40858-0

design. For ORMOCER®s, refractive indices range typically between 1.48 and 1.59, whereas new materials can be developed with much lower or much higher refractive indices.

Three dimensional (3D) photonic bandgap materials [17] are expected to be the basis of many devices, the majority of which rely on the incorporation of aperiodic defects to provide functionality. In order to create complete 2- or 3D photonic bandgaps (PBG), materials accounting for high refractive index contrast are needed. We here report on the development of high refractive inorganic-organic hybrid polymers as a novel integral part of the already existing ORMOCER®s in combination with TPA as innovative processing technology for the fast and reliable generation of photonic crystal structures. The materials and the patterning process will be discussed with emphasis on the fabrication of 3D photonic crystal structures. In order to demonstrate the potential of the 2PP method, different classes of materials are investigated.

6.2
High-Refractive Index Inorganic–Organic Hybrid Polymers

In general, ORMOCER®s are synthesized via sol–gel processing [18], where inorganic–oxidic units are connected to organic moieties on a molecular level [19]. This synthesis offers a tremendous flexibility by variation of the catalysts, temperature, and alkoxysilane scaffold. There are several concepts of creating a high-refractive index hybrid polymer material. The most common method for increasing the refractive index is the introduction of nanoparticles either by blending a polymer with particles, or by introducing them upon synthesis into organic/inorganic matrices (see, e.g., [20–24]). The described materials, however, often contain bromide or iodine compounds which are known to increase the refractive index [25,26]. In addition, the syntheses of oxide nanoparticles are generally performed in water or in alcoholic media, resulting in OH groups adsorbed at the surface of the nanoparticles [22,27]. This results in a strong light absorption around 1550 nm. An alternative method is to add metal oxide powders dispersed in a solvent to a polymer matrix which, however, often results in agglomeration [20,21,23]. Another strategy is the binding of an organo-siloxane network to an inorganic matrix by hydrolysis and polycondensation reactions between the organo-siloxane network and a metal precursor. Schmidt *et al.* [28,29] have performed polycondensation reactions between an epoxysilane and Si-, Al-, or Ti-alkoxide, and have investigated the refractive index in dependence of the metal oxide content. The refractive index has increased with increasing metal oxide content up to 1.55, but was found to be surprisingly low with respect to the corresponding inorganic system. By complete substitution of epoxysilane by diphenylsilanediol (DPD), the refractive index increases up to 1.68 which is related to the introduction of phenyl groups. However, this resin did not contain any UV polymerizable groups. Introducing some epoxysilane into the material has yielded refractive indices below 1.6.

In the following, *in situ* syntheses of class II hybrid polymers will be reported, where the organic moieties provide organically polymerizable entities, and the inorganic-oxidic network contributes to an increase of the refractive index of the final material due to the incorporation of titania into the silica network. Class II materials have the advantage that organic and inorganic parts are linked together through chemical bonds (covalent or iono-covalent bonds) on a molecular level [30]. Thus, problems such as phase separation or agglomeration are avoided which result in inhomogeneities and scattering and, consequently, in a poor optical performance. A large variety of syntheses was performed. In order to show the influence of the synthesis conditions on the resulting optical material properties, we will mainly focus on two Ti-containing ORMOCER® modifications with refractive indices of about 1.65 (@ 1035 nm). It has to be mentioned, however, that refractive indices up to 1.84 between 1800 nm and 1900 nm were achieved for resins containing 90 mol% Ti$(OEt)_4$ [31], whereas the material's stability was expectedly poor due to the high content of non-reacted Ti alkoxide.

In class II materials, the molecules used possess at least two distinct functionalities: an alkoxy group (RO-M) which should react in the presence of water upon hydrolysis and condensation reactions, and organometallic bonds (M−C bonds) stable against hydrolysis. The stability of the M−C bonds depend on the nature of the metal used. For most sol–gel conditions, the Si−C bond is stable towards hydrolysis. In contrast, the M−C bond is not stable towards hydrolysis, when M is a transition metal. Organo-alkoxysilanes such as $R'_n\mathrm{Si(OR)}_{4-n}$, where −OR is an alkoxy group and R′ contains organically polymerizable groups such as methacryl, styryl, or epoxy moieties, are typically used to modify the inorganic network. The introduction of metal alkoxides by co-condensation reactions of such organo-alkoxysilanes can also increase the refractive index [30]. Bao-Ling *et al.* [32] studied the dependence of the refractive index and the aging time of the sol on the Ti content in materials containing titanium butoxide [$Ti(OBu)_4$] and glycidoxypropyltrimethoxysilane. The refractive index increases with the Ti content, resulting in $n = 1.5225$ and 1.545 for 20 mol% and 60 mol% $Ti(OBu)_4$, respectively. By aging (17 days) of the sol containing 60 mol% $Ti(OBu)_4$, the refractive index has increased up to 1.57. Very recently, Luo *et al.* [33] synthesized hybrid organic-inorganic titania-silica polymers via an anhydrous sol–-gel process. For a resin based on 40 mol% titanium ethoxide [$Ti(OEt)_4$] and 60 mol% 3-methacryloxy-propyltrimethoxysilane (MEMO), the refractive index was determined to be 1.5685.

In the following, two material examples for novel Ti-modified ORMOCER®s will be discussed which can be processed with conventional technologies such as UV lithography or imprint technology, and by TPA processes. Although the molar composition of the materials is kept equal, their resulting properties are different due to the fact that different solvents have been used for syntheses (c.f., Table 6.1) which result in different inorganic networks. These resins were characterized with multi-nuclei NMR and FT-IR spectroscopy, whereas for the processed layers UV–VIS and μ-Raman spectroscopy as well as ellipsometry were applied [34,35]. The material's processing was performed by patterning them either with UV lithography or by TPA [35].

Table 6.1 Composition of the novel Ti-containing ORMOCER® resins PD92 and PD5.

Composition	PD92	PD5
MEMO	17 mol%	17 mol%
DPD	50 mol%	50 mol%
$Ti(OEt)_4$	33 mol%	33 mol%
solvent	THF	cyclopentanone

The syntheses were carried out with dried tetrahydrofurane (THF) and cyclopentanone. Water and a suitable catalyst were added in order to perform the hydrolysis and condensation reactions. Volatile components were removed under reduced pressure. ^{13}C- and ^{29}Si-NMR spectroscopy measurements were carried out at room temperature with a Bruker Avance DPX 400 NMR spectrometer. The resins were dissolved in deuterated solvents, such as deuterated-chloroform ($CDCl_3$) and deuterated acetone (d_6-acetone). As reference for the chemical shifts, tetramethylsilane (TMS) was used.

A comparison of the FT-IR spectra (Figure 6.1) of both resins reveals that an additional peak at 1745 cm^{-1} was detected for PD5. This is attributed to the stretching vibrational mode of C=O resulting from cyclopentanone. Also by ^{13}C-NMR (Figure 6.2), peaks of cyclopentanone were detected at 23.70, 38.33, and 220.16 ppm, respectively. The occurrence of these peaks might be related to the fact that this solvent was not completely removed under reduced pressure (peak at 220.16 ppm not shown). In resin PD5, more peaks are detected than in resin PD92. In addition to the three peaks of cyclopentanone, seven new peaks with similar intensities were recorded by ^{13}C-NMR spectroscopy. These new peaks are attributed to a new compound formed by an aldol condensation reaction between two cyclopentanone

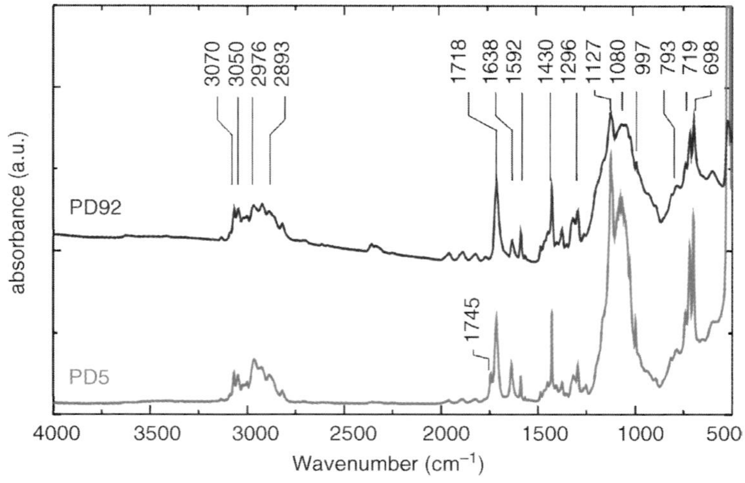

Figure 6.1 FT-IR spectra of the resins PD92 and PD5.

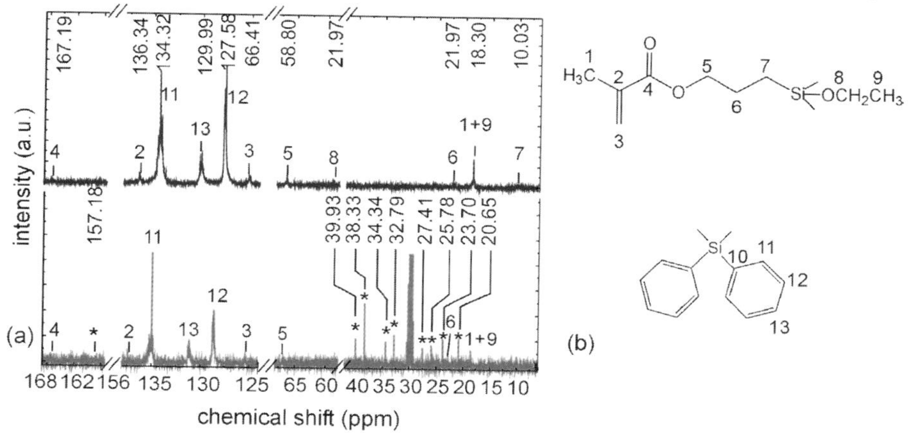

Figure 6.2 (a) Zooms of the ^{13}C-NMR spectra of the resins PD92 (in chloroform-d, top spectrum) and PD5 (in acetone-d_6, bottom spectrum). (b) Structural units of MEMO and DPD with labeled C, corresponding to the peak numbering in (a). The stars correspond to new peaks (see text).

molecules under release of water [34]. This compound has a high molecular weight and cannot be removed under reduced pressure, thus being also detected in the FT-IR spectrum.

The refractive indices of the resins and coatings were determined either with transmission spectroscopy or by ellipsometry for differently treated samples. In Table 6.2, the optical properties of PD5 and PD92 coatings which were characterized by transmission spectroscopy using a UV–VIS spectrometer are summarized, already exhibiting refractive indices to be 1.58 and 1.6 for wavelengths between 950 nm and 1550 nm, respectively [34]. After soft temperature treatment (without using cross-linking initiators), the refractive indices were found to be 1.6 to 1.62 (960 nm to 1460 nm). This clearly demonstrates the potential for high refractive index coatings if suitable initiators and processing methods will be employed. Taking the dispersion relation into account, much higher indices are expected in the VIS spectral

Table 6.2 Refractive indices and optical loss values at 780 nm of the resins PD92, PD5, and, PD92 (n_D^{20}), the cured ($n_{c\ 150°C}$), and non-cured (n_c) coating. The optical loss values are given for 780 nm.

Resin	n_D^{20}	Attenuation (dB/cm)	n_c @ λ (nm)	$n_{c\ 150°C}$ @ λ (nm)
PD92	solid	0.014	1.60 @ 950	1.62 @ 960
			1.59 @ 1120	1.60 @ 1170
PD5	1.59	0.35	1.58 @ 1550	1.60 @ 1460

regime (c.f., Figures 6.4 and 6.5). The optical loss of the materials at 780 nm which is the typical wavelength of the femtosecond laser used for TPA experiments, were found to be very low, thus enabling to focus the laser beam through a large volume of the resin in order to generate first photonic crystal-like structures (c.f., Figure 6.10).

The processing was tested by coating the resins containing a suitable UV initiator such as Irgacure 369 either on Borofloat® glass or on p-Si(100) wafers, following the typical procedures of spin-coating, annealing, UV exposure in a mask-aligner, and development steps (see, e.g., [14]). The UV exposure was carried out with and without a mask. In the standard procedure, a final thermal curing step is typically performed. However, the samples presented here are not finally cured. Results can be found in [36]. In Figure 6.3, optical microscopy images of a PD92 and a PD5 layer are shown, demonstrating the material's ability of being patterned by UV light. Vias of about 10 μm were achieved which can be further improved by changing the processing parameters such as, for example the quantity and/or the kind of the UV initiator introduced into the material, the UV exposure dose, the developing solvent, only to mention some. The ultimate resolution limits are not yet clear.

Wet layers up to 6.5 μm could be coated on the different substrates, while the layer thicknesses have decreased significantly after UV exposure and development. This is related to the fact that upon UV-exposing a photoresponsive polymer under ambient conditions whose organic cross-linking is radically initiated, an inhibition layer is formed since the oxygen from the atmosphere acts as a scavenger [37]. In addition, titania also has a significant influence on the UV absorption of a material [38]. Due to the tita-nia content in the material's inorganic-oxidic network [39], the exposure time needs to be much longer compared to titania-free ORMOCER®s in order to achieve organically highly cross-linked layers. This is attributed to the absorption of titania in the UV regime which takes place in the same wavelength band than the cross-linking by UV light. Since the exposure time is directly correlated to the degree of organic polymerization, it is expected that the material's density is lower. This, however, will

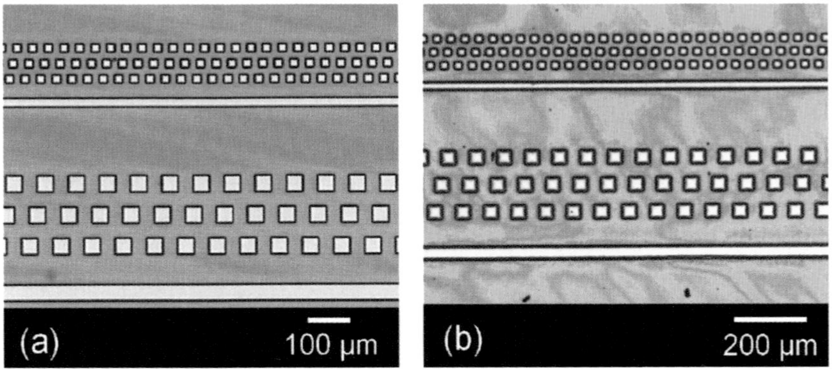

Figure 6.3 Optical microscopy images of a patterned (a) PD5, and a (b) PD92 layer on p-Si(100) wafers (open vias). A via mask was used as test mask.

Table 6.3 Degree of organic polymerization of the methacrylate groups of PD92.

UV exposure time (s)	degree of C = C conversion (%)
120	32
420	37
900	45

also result in lower refractive indices. In Table 6.3, the degree of C = C conversion of UV-exposed coatings determined by μ-Raman spectroscopy is summarized for different UV exposure times for PD92. By further increasing the exposure dose, higher refractive indices are expected, thus resulting in higher index steps.

Figure 6.4 shows ellipsometry data of thin PD92 layers (c.f., Table 6.2), whereas it has to be mentioned, however, that the coatings were not thermally treated. The refractive index increases significantly upon increasing the UV exposure time which is in good agreement with the higher degree of C = C conversion (c.f., Table 6.3). In the visible (e.g., at 600 nm), the refractive index is about 1.62 (for 120 s exposure), and approximately 0.02 higher for 900 s UV exposure. Thus, it is expected that the refractive index will be even higher for completely processed, i.e. developed and thermally cured ORMOCER® layers.

Figure 6.5 shows a dispersion curve of a PD92 layer which was completely processed except for the thermal curing step, measured by ellipsometry. In addition, the refractive indices of the non-processed coating (i.e., without photo-initiator, no thermal treatment), calculated from the transmission spectra of the coating at 950 nm and 1120 nm are shown as well. At 950 nm, the refractive index of non-processed coating is 1.60, whereas 1.65 was measured for the patterned coating. This increase is attributed to the organic cross-linking, which is known to increase the refractive index

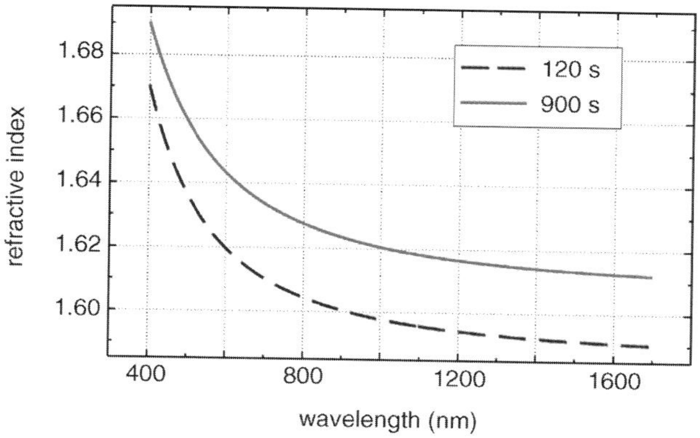

Figure 6.4 Refractive index of Ti-containing ORMOCER®PD92 coatings for different UV exposure durations. The samples were not developed and not thermally cured.

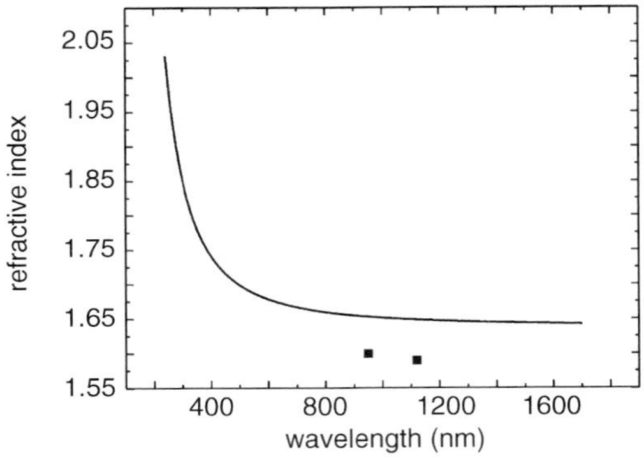

Figure 6.5 Dispersion relation of the refractive index of PD92 coatings after UV exposure and development, determined with an ellipsometer. The dots were determined from transmission spectra on non-processed coatings (without initiator).

due to the densification of the material [40]. Moreover, the addition of the photochemical initiator (Irgacure 369) can also contribute to the increase of the refractive index due to the presence of the aromatic groups which are known to improve the refractive index due to their better electronic polarizability [41]. To our knowledge, it is the first time that patterned structures are obtained on a material containing if such a high titanium content (33 mol%).

6.3 Multi-Photon Fabrication

6.3.1 Experimental Setup

The schematic representation of the experimental setup which was used for the fabrication of structures is shown in Figure 6.6. A 100× immersion-oil microscope objective (Zeiss, NA of 1.4) is used to focus laser pulses in all experiments. Two-photon polymerization (2PP) is initiated by near-IR ultra-short laser pulses from a Ti: Sapphire oscillator. The central emission wavelength, repetition rate, and duration of the laser pulses were 780 nm, 94 MHz, and 120 fs, unless otherwise stated. A wave plate (WP) together with a polarizing beam splitter (BS) is used to attenuate the average power of the transmitted beam. An acousto-optical modulator (AOM) in combination with an aperture is used as a fast shutter. The beam is expanded by a telescope and then coupled into $x-y$ galvo scanner. The sample is mounted on a 3D piezo stage for positioning in all directions. A CCD camera placed behind the dichroic mirror is used for online monitoring of the 2PP process.

Laser pulses, tightly focused into the volume of the photosensitive material, interact with the material via a multi-photon absorption and induce a highly localized

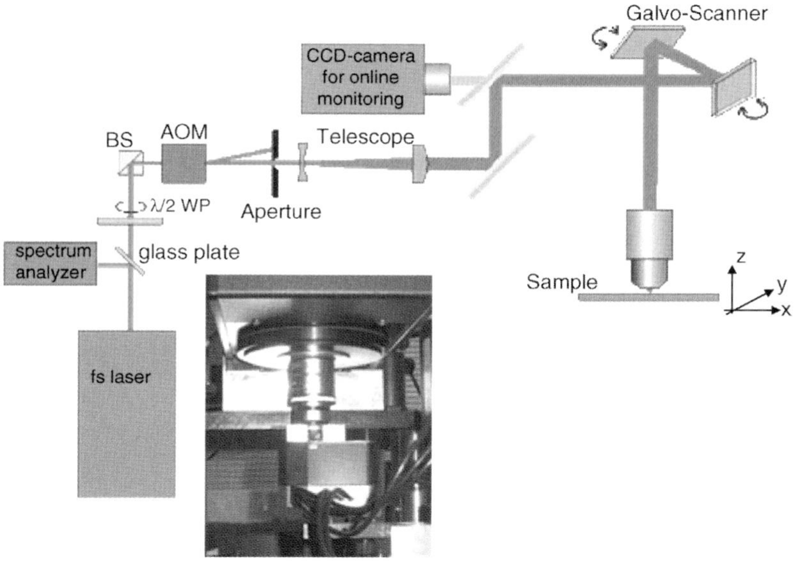

Figure 6.6 Principle setup for the fabrication of 3D structures by 2PP.

chemical reaction. For negative resist materials such as ORMOCER®s, this results in an organic cross-linking, i.e. the material is transformed from the liquid into the solid state. Since the process of multi-photon absorption depends non-linearly on the light intensity, the interaction region is strictly limited to the focal volume, while outside of the focus the material stays unchanged. By moving the focus in x, y, and z-direction, arbitrary 3D structures beyond the diffraction limit can be created.

There are two different illumination strategies which can be used for fabrication of structures: continuous scanning and pinpoint illumination. In the latter case, the positioning system receives a set of coordinates which determines and defines the positions of separate volume pixels (voxels). The voxels then define the structure, whereas each point is illuminated separately. The resolution is changed by varying illumination duration, average laser power, and the overlapping between neighboring voxels. For continuous scanning, the structure is defined by a set of curves. The resolution is changed by adjusting the scanning speed and the average laser power.

6.3.2 Fabrication of PhC in Standard ORMOCER®

First experiments were carried out with an ORMOCER® material originally developed for waveguide applications [14]. The minimal resolution (feature size) of the 2PP technique which could be demonstrated using this material was 100 nm [4]. Figure 6.7 shows SEM images of woodpile structures fabricated in ORMOCER® using the 2PP process under continuous scanning illumination. A major problem

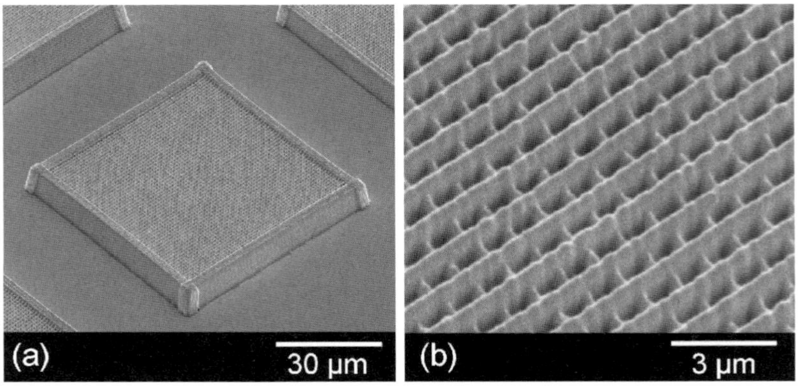

Figure 6.7 SEM images of (a) a woodpile PhC fabricated by 2PP in a standard ORMOCER®, and (b) zoom into the structure.

one always has to cope with when using photosensitive resins in general is that these shrink upon cross-linking processes, leading to distortions within the structures. This can be overcome by either producing a thick frame surrounding the structure (Figure 6.7(a)), or by numerical compensation [42,43]. Photonic crystal structures of different rod sizes and periodicity were written and optically characterized [44] (c.f., Figure 6.8).

Despite the fact that the refractive index of the standard ORMOCER® ($n \approx 1.56$) is too low to expect a complete PBG, a PBG of 8 % in the direction corresponding to perpendicular incidence to the structure is found. Figure 6.8(a) shows the experimentally determined transmission as a function of the wavelength for a crystal having in-layer rod distances of $d = 0.9$ μm to 1.2 μm. The fcc symmetry is preserved for all prepared PhCs. All stop gaps are blue-shifted relative to the theoretically calculated values, since in the calculations shrinkage was not taken into account.

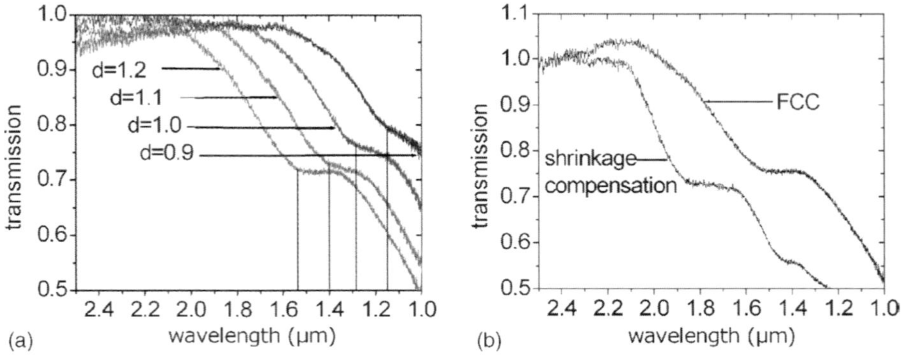

Figure 6.8 Transmission spectra of woodpile structures with (a) various in-layer rod distance d, and (b) various distance between the layers.

In Figure 6.8(b), transmission spectra of both, shrinkage compensated and non-compensated woodpile structures are shown. By an increase of the layer-to-layer distance by a factor of 1.4 compared to the fcc symmetry, the stop-gap is shifted from $\lambda = 1.50\,\mu m$ to 1.86 μm. This behavior is also supported by the fact that the acceptance angle for the band-gap is larger for ideal fcc symmetry than for distorted ones. Since the transmission measurements have been performed using a 15× microscope objective with a semi-aperture angle of ±15°, the band edge is sharper for fcc symmetry, leading to a steeper drop in the transmittance. The transmission spectra indicate that the appearance of band-gaps, especially when using low refractive index materials, is very sensitive to all kinds of disorder such as roughness of the rods, aspect ratio, or rod thickness. Any disorder that leads to deviation from the strictly periodic structure can cause Rayleigh scattering of the incident light. Since Rayleigh scattering scales with $1/\lambda^4$, it becomes more dominant for shorter wavelengths. Hence, scattering occurs particularly at the higher bands leading to a decrease in measured transmission above the stop-gap. As a consequence, the transmission does not regain its value after the stop-band.

6.3.3
2PP of High Refractive Index Materials

TiO_2 is a very promising material for photonic applications due to its high refractive index (from 2.5 to 2.9 depending on the crystalline phase) and high transparency in the visible spectral range. In the following, two different concepts were followed in order to generate PhC structures. One approach was the generation of PhC using a Ti-containing ORMOCER® material, where the patterning is performed analogously to the processing described in chapter 3.2 via cross-linking of the organic moieties upon laser light illumination. The other approach is the breaking of bonds in the TiO_2 containing resist, where the irradiated regions are insoluble for organic solvents such as acetone [45]. Feature sizes down to 400 nm to 500 nm are obtained for both materials.

ORMOCER®PD92 was used in order to investigate the patterning process by means of 2PP. Different parameters such as the writing speed, average power, and also continuous multiple exposure were investigated. The material was spin-coated on glass, and subsequently patterned with the femtosecond laser followed by development step in MIBK. Figure 6.9 shows SEM images of continuously written lines in ORMOCER®PD92. Multiple exposure was incrementally carried out, ranging between 1 to 10 times exposure. The images show that a slow writing speed, a low energy, and multiple exposure are needed in order to obtain well-defined lines.

Based on these results, 3D photonic crystal structures were written in PD92 by means of TPA. In Figure 6.10, the first 3D structures using a Ti-containing high refractive index ORMOCER® are shown. The typical structural dimensions are between approx. 0.4 μm and 1 μm line width with a period of about 2 μm (Figure 6.10(b)). It has to be mentioned, however, that neither the process is optimized so far, nor the quality of the generated structure is good enough to account for optical characterization. Besides, the photo-initiator used for radical initiation (Irgacure 369)

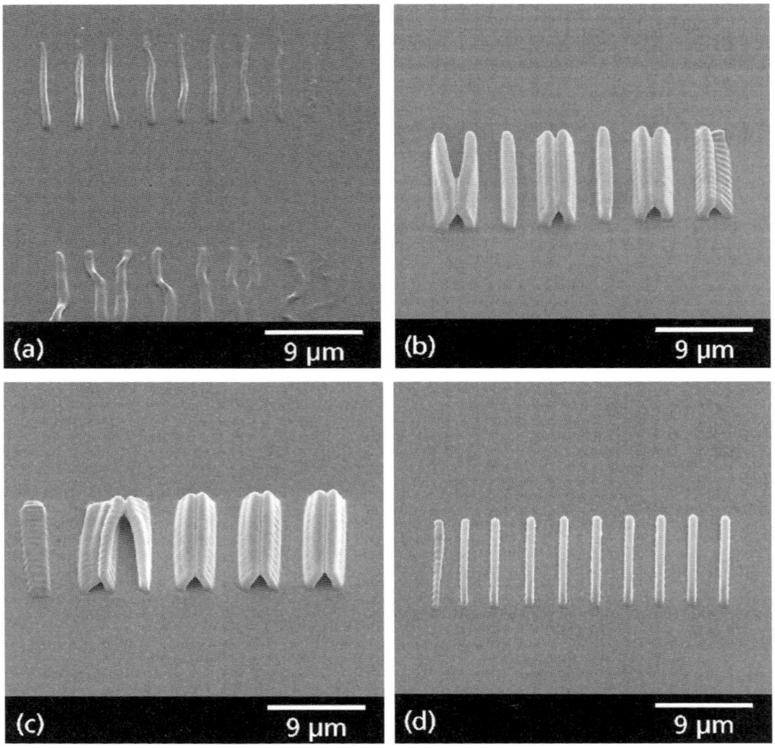

Figure 6.9 SEM images of continuously written lines by TPA in ORMOCER®PD92. (a) Top lines: 40 mW, bottom lines: 35 mW, both at a speed of 200 μm/s, and (b) 40 mW with 20 μm/s. From the right to the left, the exposure was increased with increment 1. (c) 20 mW, and (d) 15 mW, both at a speed of 20 μm/s. From the left to the right, the exposure was increased with increment 1.

has a very low absorption cross-section which limits the possibilities of the 2PP method [4]. Cumpston *et al.* [46] could demonstrate that using π-conjugated components as photo-initiators, the two-photon absorption cross-section can be increased significantly, i.e., absorption cross-sections of up to 1.25×10^{-47} (cm^4 s per photon) were achieved.

In further experiments, the processing will be optimized in order to achieve crystal structures which enable optical characterization. The structure size and precision, the adhesion between the crystal and the glass substrate, the development step, and also the material's shrinkage have to be optimized. However, the latter will be very difficult to be reduced since the Ti-containing resins presented in this paper are significantly densified upon processing. This causes high shrinkage, but will also result in higher refractive indices for completely processed layers and structures. The experiments also have revealed that resins based on titanium compounds require 2PP processing parameters completely different that of the standard ORMOCER®

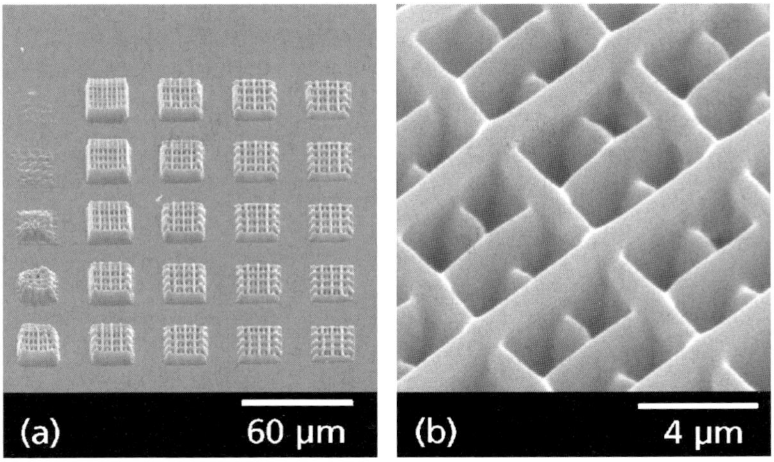

Figure 6.10 (a) 3D photonic crystal structures, produced in ORMOCER®PD92. The average laser power was varied between 7 mW and 32 mW (from the upper left to the lower right), and multiple exposure (five times) was applied. (b) Zoom into a structure.

material as reported in the literature [4]. This is also supported by the low percentage of organic cross-linking determined from μ-Raman spectroscopy (c.f., Table 6.3). Typically, 3D photonic crystal structures are obtained in about 5 min for the entire process. Slower speed, lower energy, and also continuous multiple exposure have to be applied with resin PD92 in order to obtain a similar photonic crystal structure. Comparing the parameters yields that the ORMOCER® material based on Ti approximately requires 40 min processing in order to obtain a photonic crystal structure. This might be related to the high absorption of titania as a competitive process to the organic cross-linking. In addition, the structural feature sizes are much broader than for the standard ORMOCER® [4] which is a result of the non-optimized optics for high refractive index materials within the TPA experiment. Further investigations will be carried out to achieve a more detailed understanding of the underlying processes.

Beside the high-refractive index ORMOCER®PD92, another titania-containing resist which is described elsewhere [47] was used. This spin-coatable TiO_2 resist was prepared by chemical modification of titanium *n*-butoxide, $Ti(OBun)_4$, (Aldrich, >97 %) with benzoylacetone, BzAc. At 1550 nm, its refractive index was determined to be 1.68 before polymerization. Figure 6.11 shows SEM images of woodpile structures fabricated in a 10 μm thick layer of this resist. It was found that in order to produce well-defined homogeneous structures, a multiple scan of each line similar to the findings for ORMOCER®PD92 with a gradually increasing average laser power is necessary. Each line was scanned with a velocity of 2000 μm/s, and average power increasing from 5 mW to 12 mW in 2000 steps. Between each layer, the patterning process was stopped for 10 s, allowing the volatile components formed upon

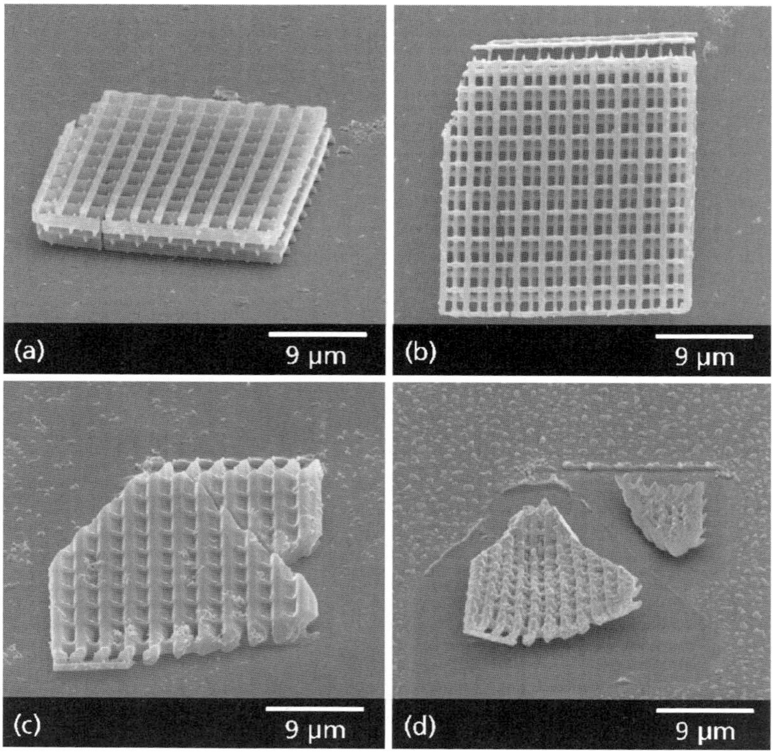

Figure 6.11 SEM images of photonic crystal structures fabricated in a TiO_2 resist. (a), (b) show the structure before, and (c), (d) after thermal treatment at 400 °C for 1 h.

bondbreaking to be removed from the structure. The structure has a periodicity of 2 μm with an interlayer distance of 700 nm.

In comparison to SU-8 photoresist, whose refractive index has been reported to be close to 1.65 nm at 405 nm [48] after a postbake treatment of the TiO_2 resist, it is possible to produce pure metal–oxidic structures from the TiO_2 resist without an organic component. When heated up to more than 500 °C, the TiO_2 resist reaches a refractive index of over 2.1 in the visible and near infrared region [45]. Whereas no improvement of the refractive index of SU-8 may be achieved by a thermal treatment at 500 °C due to its degradation temperature reported to be 380 °C [48]. Thermal treatment of 3D structures is still a challenge, because the TiO_2 resist shrinks about 50 % of its original volume. The particles visible around the structures (Figure 6.11(c), (d)) result from gold that has been sputtered previously for SEM imaging, and conveniently mark the initial size of the structure. It is clear that the thermal treatment results in an increased refractive index of the material, and simultaneously decreased structural size. However, in order to obtain non-distorted PhC structures, more control over the material shrinkage is necessary.

6.3.4 Patterning and PhC Fabrication in Positive Resist Material S1813

Currently, there are no commercially available negative photoresists with a refractive index approaching 2. A popular approach is to fabricate replicas of 3D crystals in higher refractive index materials. Generally, this approach implies the removal original structure. Therefore, due to a relatively high thermal and chemical stability of negative photoresist materials, these materials complicate the fabrication process of replicas. On the other hand, positive resists can be easily chemically or thermally removed without much mechanical stress.

Investigations on the patterning of positive resists were performed with a commercially available positive resist S1813 (Shipley Corp.). This resist was developed for the integrated circuit device fabrication in microelectronic industry. S1813 can be dissolved in propylene glycol monomethyl ether acetate (PGMEA), and is optimized for the G-line (436 nm) absorption of a mercury lamp. This resist is highly transparent down to wavelengths below 450 nm, and hence is well-suited for multi-photon processing.

In case of positive tone resists, light exposure leads to a chain scission. The difference in processing of positive and negative photoresists is illustrated in Figure 6.12. Two-photon scission of polymer chains in the positive photoresist allows one to dissolve the irradiated regions, this way producing 3D hollow structures.

S1813 can be spin-coated in different thicknesses ranging from around 2.5 μm down to a few hundred nanometers. The refractive index of the S1813 is approximately 1.63 at around 800 nm. In thick S1813 droplets, it is possible to write hollow photonic crystal structures [42]. In Figure 6.13 (a) and (b), two examples of woodpile structures written in S1813 are shown. Deep wide ridges are formed on two sides of

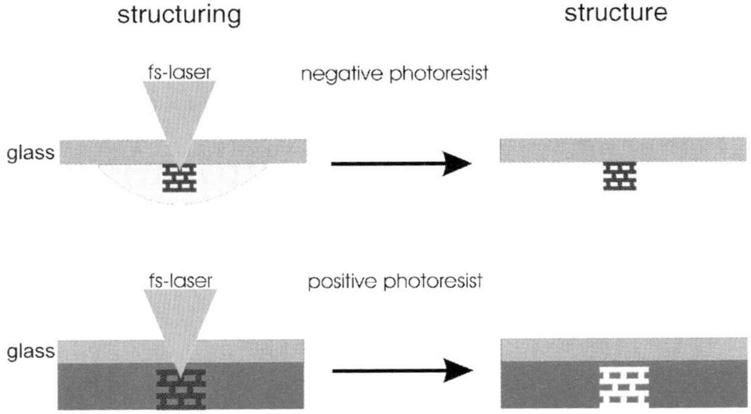

Figure 6.12 Difference in processing of negative photoresists (top) and positive photoresists (bottom).

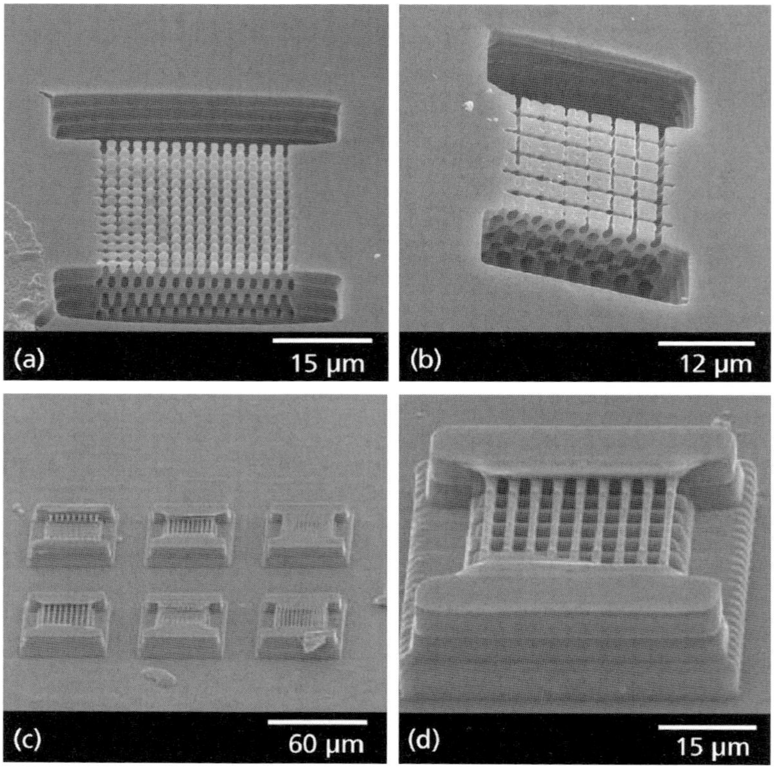

Figure 6.13 Scanning electron microscope images of (a), (b) hollow woodpile structures written in S1813 positive photoresist and (c), (d) replicas fabricated from acrylate monomer.

the PhC in order to provide a better developing of the structure and simplify the infiltration procedure.

For the fabrication of the test replica, a UV-sensitive acrylate monomer (Microresist Technology, Germany) is infiltrated into the structure. Subsequently, the entire sample was exposed to UV light and placed into NaOH for removing the positive resist. This procedure dissolves the S1813 structure completely, while the acrylate monomer remains a solid and stable polymer. The result of this procedure is a free-standing replica as shown in Figure 6.13(c) and (d). Since this is an indirect process, one can produce replicas from a material that is not directly structurable by 2PP.

6.4
Summary and Outlook

Novel high refractive index inorganic-organic hybrid polymers synthesized and characterized by different spectroscopic and optical methods were exemplary discussed for two Ti-containing ORMOCER®s. Based on a standard ORMOCER®, the refractive index could be increased about 7% by substitution of an organo-alkoxysilane with a Ti alkoxide. Refractive indices of 1.67 (at 633 nm) were achieved even without thermal treatment and for low cross-linked samples. Thus, the refractive indices are expected to be much higher for an optimized processing. Patterning was carried out by UV lithography and by TPA using a femtosecond laser. In comparison to other methods currently used for the fabrication of PhC, 2PP is a fast and flexible technology allowing rapid fabrication of PhC with various geometries and structural parameters. A common issue associated with the low refractive index of processable materials can be solved by using specially designed high refractive index materials, or indirect replication methods. We have successfully demonstrated the application of the TPA technology micropatterning. The resolution of this technique allows the realization of structures with PBG in the near-IR region. In order to push the PBG central wavelength values further to the visible part of the spectrum, further improvements in the resolution and the fabrication confidence of this technology have to be made. Direct patterning of high refractive index photosensitive materials will have an impact on the related research areas, such as fabrication of micro-optical components, and surface plasmon guiding structures.

References

1 Maruo, S., Nakamura, O. and Kawata, S. (1997) *Opt. Lett.*, **22**, 132.

2 Kawata, S., Sun, H.-B., Tanaka, T. and Takada, K. (2001) *Nature*, **412**, 697.

3 Kuebler, S.M., Rumi, M., Watanabe, T., Braun, K., Cumpston, B.H., Heikal, A.A., Erksine, L.L., Thayumanavan, S., Barlow, S., Marder, S.R. and Perry, J.W. (2001) *J. Photochem. Sci. Technol.*, **14**, 657.

4 Serbin, J., Egbert, A., Ostendorf, A., Chichkov, B.N., Houbertz, R., Domann, G., Schulz, J., Cronauer, C., Fröhlich, L. and Popall, M. (2003) *Opt. Lett.*, **28**, 301.

5 Sun, H.-B., Takada, K. and Kawata, S. (2001) *Appl. Phys. Lett.*, **79**, 3173.

6 Sun, H.-B., Kawakami, T., Xu, Y., Ye, J.-Y., Matsuo, S., Misawa, H., Miwa, M. and Kaneko, R. (2000) *Opt. Lett.*, **25**, 1110.

7 Houbertz, R., Wolter, H., Dannberg, P., Serbin, J. and Uhlig, S. (2006) *Proc. SPIE*, **6126**, 612605.

8 Schmidt, V., Kuna, L., Satzinger, V., Houbertz, R., Jakopic, G. and Leising, G. (2007) *Proc. SPIE*, **6476**, 64760.

9 Langer, G. and Riester, M. (2007) *Proc. SPIE*, **6475**, 64750.

10 Amberg-Schwab, S., Arpac, E., Glaubitt, W., Rose, K., Schottner, G. and Schubert, U. (1991) in *Mater. Sci. Monographs – High Performance Ceramic Films and Coatings*, Vol. 67 (ed. P. Vincencini), Elsevier, Amsterdam, p. 203.

11 Wolter, H. and Storch, W. and Ott, H. (1994) *Mater. Res. Soc. Symp. Proc.*, **346**, 143.

12 www.voco.com.

13 Streppel, U., Dannberg, P., Wächter, Ch., Bräuer, A., Fröhlich, L., Houbertz, R. and Popall, M. (2002) *Opt. Mater.*, **21**, 475.

14 Houbertz, R., Domann, G., Cronauer, C., Schmitt, A., Martin, H., Park, J.-U., Fröhlich, L., Buestrich, R., Popall, M., Streppel, U., Dannberg, P., Wächter, C. and Bräuer, A. (2003) *Thin Solid Films*, **442**, 194.

15 Fröhlich, L., Houbertz, R., Jacob, S., Popall, M., Mueller-Fiedler, R., Graf, J., Munk, M. and von Zychlinski, H. (2002) *Mater. Res. Soc. Symp. Proc.*, **726**, 349.

16 Haas, U., Haase, A., Satzinger, V., Pichler, H., Leising, G., Jakopic, G., Stadlober, B., Houbertz, R., Domann, G. and Schmidt, A. (2006) *Phys. Rev. B*, **73**, 235339.

17 Toader, O. and John, S. (2001) *Science*, **292**, 1133.

18 Brinker, C.J. and Scherer, W.G. (1990) *Sol–Gel Science*, Academic Press, New York.

19 Sanchez, C., Julián, B., Belleville, P. and Popall, M. (2005) *J. Mater. Chem.*, **15**, 3559.

20 US 6,656,990, Curable high refractive index compositions.

21 Nussbaumer, R.J., Caseri, W.R., Smith, P. and Tervoort, T. (2003) *Macromol. Mater. Eng.*, **288**, 44.

22 Chen, W.C., Lee, S.-J., Lee, L.-H. and Lin, J.-L. (1999) *J. Mater. Chem.*, **9**, 2999.

23 Kim, J.-W., Shim, J.-W., Bae, J.-H., Han, S.-H., Kim, H.-K., Chang, I.-S., Kang, H.-H. and Suh, K.-D. (2002) *Colloid Polym. Sci.*, **280**, 584.

24 Judeinstein, P. and Sanchez, C. (1996) *J. Mater. Chem.*, **6**, 511.

25 Henzi, P., Rabus, D.G., Bade, K., Wallrabe, U. and Mohr, J. (2004) *Proc. SPIE*, **5454**, 64.

26 Yasufuku, S. (1992) *IEEE Electr. Insul. Magx.*, **8**, 7.

27 Dibold, U. (2003) *Surf. Sci. Rep.*, **48**, 53.

28 Schmidt, H. and Seiferling, B. (1986) *Mater. Res. Soc. Symp. Proc.*, **73**, 739.

29 Lintner, B., Arfsten, N., Dislich, H., Schmidt, H., Philipp, G. and Seiferling, B. (1988) *J. Non-Cryst. Solids*, **100**, 378.

30 Sanchez, C. and Ribot, F. (1994) *New J. Chem.*, **18**, 1007.

31 Declerck-Masset, P., Houbertz, R., Domann, G., Reinhardt, C. and Chichkov, B. 3rd European Organosilicon Conference ISOSXIV, Würzburg, Germany, 31 July–5 August 2005.

32 Bao-Ling, W. and Li-Li, H. (2004) *Chin. Phys.*, **13**, 1887.

33 Luo, X. and Zha, C. and Luther-Davies, B. (2005) *J. Non-Cryst. Solids*, **351**, 29.

34 Declerck, P., Houbertz, R. and Jakopic, G. *J. Sol–Gel Sci. Technol.*, (submitted).

35 Declerck, P., Houbertz, R., Passinger, S. and Chichkov, B.N. (2007) *Mater. Res. Soc. Symp.*, Symposium S (Volume 1007E).

36 Declerck, P. Ph.D. thesis, University of Würzburg (in preparation).

37 Croutxé-Barghorn, C., Soppera, O., Simonin, L. and Lougnot, D.J. (2000) *Adv. Mater. Opt. Electron.*, **10**, 25.

38 Kuznetsov, A.I., Kamenava, O., Rozes, L., Sanchez, C., Bityurin, N. and Kanaev, A. (2006) *Chem. Phys. Lett.*, **429**, 523.

39 Chang, C.H., Mar, A., Tiefenthaler, A. and Wostratzky, D. (1992) in *Handbook of Coatings Additives*, Vol. 2 (ed. L.J. Calbo), Marcel Dekker Inc., p. 1.

40 Dorkenoo, K.D., Bulo, H., Leblond, H. and Fort, A. (2004) *J. Phys. IV (France)*, **119**, 173.

41 Ma, H. and Jen, A.K.-Y. and Dalton, L.R. (2002) *Adv. Mater.*, **14**, 1339.

42 Ovsianikov, A., Passinger, S., Houbertz, R. and Chichkov, B.N. (2006) in: Laser Ablation and its Applications, Springer Series in Optical Sciences, Vol. 129, (ed. Claude R. Phipps) p. 123.

43 Houbertz, R. (2005) *Appl. Surf. Sci.*, **247**, 504.

44 Serbin, J., Ovsianikov, A. and Chichkov, B.N. (2004) *Opt. Express*, **12**, 5221.

45 Passinger, S., Saifullah, M.S.M., Reinhardt, C., Subramanian, K.R.V., Chichkov, B.N. and Welland, M.E. (2007) *Adv. Mater.*, submitted.

46 Cumpston, B.H., Ananthavel, S.P., Barlow, S., Dyer, D.L., Ehrlich, J.E., Erskine, L.L., Heikal, A.A., Kuebler, M., Lee, I.-Y.S., McCord-Maughon, D., Qin, J., Röckel, H., Rumi, M., Wu, X.-L., Marder, S.R. and Perry, J.W. (1999) *Nature*, **398**, 51.

47 Saifullah, M.S.M., Subramanian, K.R.V., Tapley, E., Kang, Dae-Joon, Welland, M.E. and Butler, M. (2003) *Nano Lett.*, **3**, 1587.

48 del Campo, A. and Greiner, C. (2007) *J. Micromech. Microeng.*, **17**, R81.

7
Ultra-low Refractive Index Mesoporous Substrates for Waveguide Structures

D. Konjhodzic, S. Schröter, and F. Marlow

7.1
Introduction

Mesoporous materials have pore sizes between 2 and 50 nm and have been studied for potential applications in catalysis, separation, and chemical sensing in many works since their discovery in 1992. The use of these materials as novel optical or electrical materials has also attracted a high interest. They could be a component in one of the next generations of chips [1]. Here, low dielectric constants are required and mesoporous materials can exploit their high porosity for achieving this. It is expected that these materials can substantially help to overcome the problems of cross-talk and propagation delay. The semiconductor industry is currently targeting new dielectric films with dielectric constants $k = 2.5$ to 3.0, and it is anticipated that, as the packing density of metal lines on the semiconductors continues to increase, interlevel dielectric films with ultra-low k ($k < 2.2$) will soon be required [1].

In the field of integrated optics, a similar materials problem is emerging. Waveguides with many functions are designed, but they all need a support with a lower refractive index to operate properly. Many of new interesting waveguides (e.g. 2D PhCs) have unfortunately relatively low effective refractive indices, demanding even lower support indices. Low-n materials have nearly the same requirements as low-k materials. They differ in the relevant frequency range leading to the decisive role of OH groups and water for the low-k materials. Both applications require perfect films of these materials.

In this work we describe an optimized fabrication procedure of mesoporous silica films and analyze the properties of these mesoporous films. The films are used as ultra low-n support for the realization of the polymeric or inorganic waveguides.

The use of such low-n supports could be decisive step towards optical integrated circuits on a highly variable material basis because the index contrast between the guiding and the support layer is then easy to achieve. The refractive-index contrast is very important for dielectric multilayer structures, optical resonators, and photonic

Nanophotonic Materials: Photonic Crystals, Plasmonics, and Metamaterials. Edited by R.B. Wehrspohn, H.-S. Kitzerow, and K. Busch

ISBN: 978-3-527-40858-0

crystals. In integrated optics, low-*n* materials would be very helpful to enable more waveguiding structures.

There have been a number of recent studies on the application of porous xerogel [2] or mesoporous silica films [3] as low-*k* materials. Optically applicable xerogel films for waveguide cladding layers were prepared using an ethylene glycol co-solvent procedure and the core layer was prepared using plasma-enhanced chemical vapor-deposited silicon dioxide or a siloxane epoxy polymer. These polymer-xerogel waveguides had a maximum refractive index contrast of 0.34, whereas the PECVD oxide-xerogel planar waveguides were fabricated with a maximum refractive index contrast of only 0.28. A low dielectric constant of 1.8 to 2.5 was measured for the spin-coated mesoporous silica films [3], but no realization of the waveguides with this system was reported so far.

7.2
Mesoporous Films

7.2.1
Fabrication of Mesoporous Silica Films

7.2.1.1 General Remarks

Ordered mesoporous materials are made with the use of surfactants forming micelles and acting as structure-directing agent (SDA). This kind of templating approach was used in the MCM-41 synthesis for the first time [4]. The ordered mesophase depends on the concentrations of surfactant, the inorganic species, and the processing conditions. The mechanism for the molecular interaction between an inorganic material and a surfactant was first discussed by Beck *et al.* [5] in detail. They proposed two alternative pathways, in which either the liquid-crystal phase is intact before the silica species are added, or the addition of the silica results in the ordering of the silica-encased surfactant micelles. In this case, silica species coat the surface of surfactant micelles, which then self-order to form the phase observed in the final product. The final porous structure is obtained after removal of the organic template upon thermal treatment called calcination.

Thin films of such mesoporous materials can be realized in a dip-coating process [6]. The solution deposited on the supports contains metal-alkoxide as a precursor and surfactant as a SDA. The film thickness is controlled by the evaporation rate of the solvent, the drawing speed, and by the viscosity of the coating solution. The increase in surfactant concentration upon solvent evaporation causes the assembly process of the micelles into a close-packed phase.

A special class of mesoporous materials named SBA-15 has been synthesized by use of amphiphilic triblock-copolymers [7]. The SBA-15 materials are formed in acidic media and show mostly two-dimensional hexagonal (space group p6mm) mesophases consisting of a silica channel framework filled with the block-copolymer. Calcination gives porous silica structures with relatively large lattice constants of 7.5 nm. These materials are highly versatile and can also be used in low-*k* and low-*n* applications.

7.2.1.2 Preparation Details

A solution containing the triblock copolymer poly(ethylene oxide)-block-poly(propylene oxide)-block-poly(ethylene oxide) (P123, $EO_{20}PO_{70}EO_{20}$), ethanol, water, HCl, and Rhodamine 6G was mixed with tetrabutyl orthosilicate (TBOS) as a silica precursor resulting in the final molar composition: TBOS:P123:H_2O:HCl:EtOH:Rh6G=1:0.018:2.83:0.015:5.58:0.0009. A pre-reaction of this solution has been carried out for 2.5 h at 70 °C. The solution was deposited onto carefully cleaned substrates (BK7 glass slides, Si-wafer, mica slides, or epoxy resin blocks) by dip-coating. The porosity of the films was obtained by calcination. The thickness of the resulting transparent films was tuned in the range 300–1100 nm.

According to their visual appearance, two types of films have been observed and denoted as A-type and B-type [8]. A-type films are perfectly clear, whereas B-type films appear slightly milky. Observation with an optical microscope revealed homogeneously distributed, bubble-like defects between 1 µm and 100 µm in size for B-type films, whereas the A-type films are fully non-structured. These two types of films were found in dependence on the processing conditions. A-type films turned out to be well suited as low-*n* supports and were used in Ref. [9], whereas B-type films revealed some academically interesting features.

The most important processing parameter which determines the film type is humidity of the surrounding air. In order to obtain perfect A-type films it is crucial to keep the relative humidity (RH) on a low level during the synthesis. For this, an air conditioner system was used. During the film deposition the relative humidity in the chamber fluctuated about ±4%, due to opening of the chamber for the sample exchange.

The realized film thickness and the relative humidity during the film deposition form a synthesis field depicted in Figure 7.1. Single points in the diagram represent the samples synthesized with the normal dip-coating. The lines in the diagram

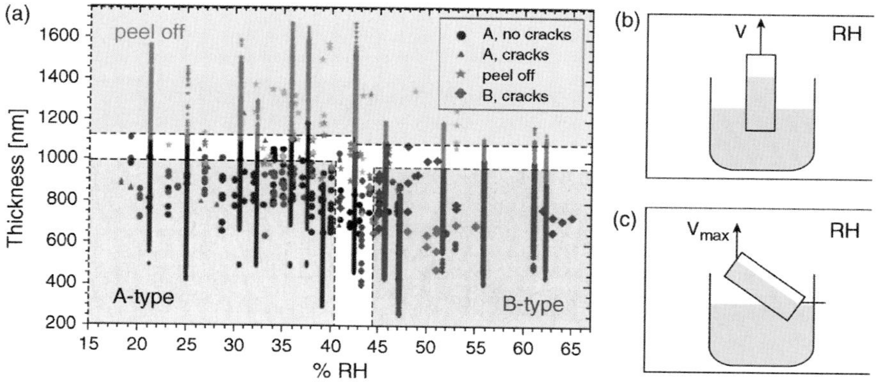

Figure 7.1 (a) Synthesis field. The green region shows mainly the A-type films and the blue region the B-type films. In the red region the films start to peel off. Schemes for (b) normal dip coating and (c) coating with tilted movement.

represent samples with thickness gradient deposited by a tilting movement with the substrate. This process results in the whole range of the thicknesses at a certain value of the relative humidity.

One can distinguish three regions in the diagram. Clear A-type films (black circles) appear predominantly in the region of relative humidities below about 41% RH and thicknesses below about 1000 nm. B-type films appear predominantly in the region above about 44% RH and with thicknesses below about 970 nm. Films with ablations for both film types (red stars) are to be found mainly in the region above 1100 nm (red region). In white regions no one of these regions could be precisely assigned. The region between A-type and B-type of 4% RH width, where both film types occur, can be explained with the typical fluctuations of the relative humidity ($\pm$4%) during the film synthesis.

7.2.2
Characterization and Structure Determination of MSFs

To determine the structures, small angle X-ray scattering (SAXS), transmission electron microscopy (TEM), and atomic force microscopy (AFM) investigations were performed.

As-synthesized A-type films showed a circle-like X-ray diffraction pattern with a radius of $2\Theta = 1.08^\circ$ (8.2 nm) and an isotropic intensity distribution (Figure 7.2a). In brackets we give the deduced lattice constant as a result of application of the Bragg equation. After calcination an ellipse with the half axes of 1.08° (8.2 nm) and 1.95° (4.5 nm) was found (Figure 7.2b). This diffraction pattern can be attributed to a partial ordering of uniform channels with well-defined pore sizes. The structure shrinks in the direction normal to the surface upon calcination. This effect is visible in the longer half-axis in the diffraction pattern and amounts to 55% [8].

The TEM analysis was performed on cross sections of the samples. The resulting micrograph (Figure 7.2c) shows fluctuating pattern typical for the so-called worm-like

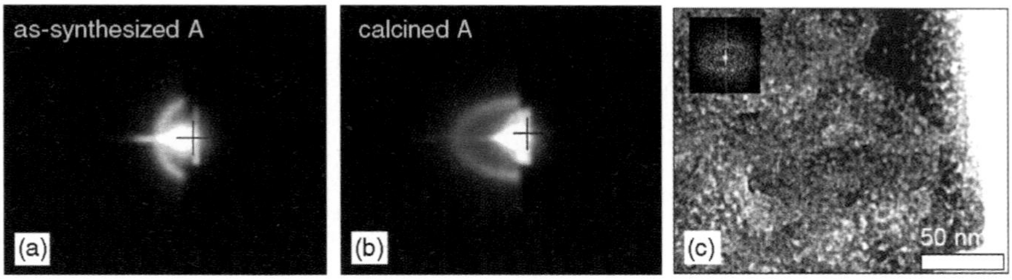

Figure 7.2 Structure determination of A-type films. (a) The SAXS diffraction pattern of as-synthesized and (b) calcined films. The right sides of diffraction patterns are not accessible by the scattered waves. The sample was positioned perpendicular to the image plane in vertical direction. The primary beam position is marked by a cross. (c) TEM micrograph of the cross-section of a calcined A-type film. Scale bar: 50 nm, Inset: FFT. [8]

structure [10]. It can be imagined as nearly dense-packed bundles of long flexible rods (Figure 7.5a). This structure can be considered as a non-equilibrium state of a dense-packed channel array. The equilibrium state would be the hexagonal phase. One may speculate if this partially disordered structure is useful for the application as low-*n* film. Many authors have reported similar films but there are no reliable facts that the synthesized films are really thick enough, i.e. thicker than 1 μm, and defect-free. For a number of syntheses it must be recognized that there are unsolved difficulties if the films becomes thicker than 400 nm. This is, however, too thin for an optical application because the evanescent fields penetrate deeper into the film. Much synthesis work was focused on perfect and, therefore, especially nice structures. However, these nice and highly ordered structures are not so well suited for surviving the stress during the film synthesis because the structure has no freedom to change without destruction. A partially disordered structure allows a more flexible reaction to the stress during drying, condensation, and template removal. A further advantage of partially disordered films is their isotropy. An advantage of the specific channel-type porosity is the good mechanical integrity [8].

As-synthesized B-type films exhibited very pronounced equidistant peaks in the SAXS analysis. The peaks represent X-ray beams diffracted perpendicular to the film plane (Figure 7.3a). This indicates a layered structure with layers ordered parallel to the film surface with a *d*-spacing of 8 nm. In the diffraction pattern of calcined B-type films (Figure 7.3b), one sharp diffraction spot at $2\Theta = 3.1^{\circ}$ was still visible. This means that the layer structure remained stable during calcination. The layer spacing decreased to 3 nm, which means shrinkage to 38%.

In the TEM images of the B-type films, the layers can be made visible (Figure 7.3c and d). Tilting the sample during the investigation did not reveal any additional structure. A calcination-stable layer structure is a surprise, because one expects a collapse of the interlayer spacing after template removal. Such a structure can only be stable if there is a sustaining system among the layers. Although B-type films are not suitable as low-*n* substrates, their internal layer structure is very interesting in itself.

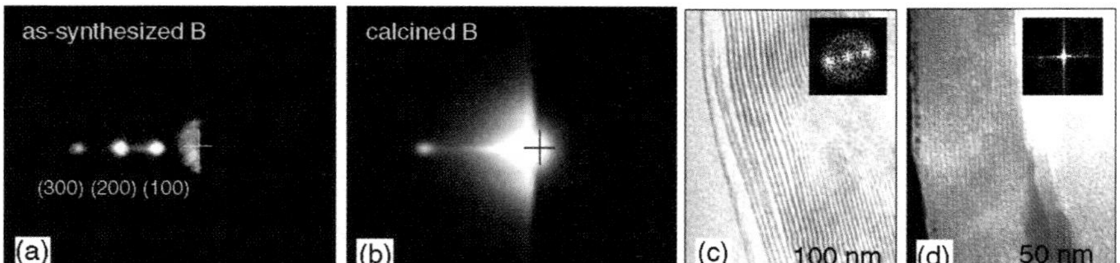

Figure 7.3 Structure determination of B-type films. (a) The SAXS diffraction pattern of as-synthesized and (b) calcined films. Sharp diffraction spots out of film plane indicate layers ordered parallel to the substrate, which remain stable upon calcination. The layers are clearly resolved in the TEM micrographs of the cross-sections of (c) as-synthesized and (d) calcined B-type films. Insets: FFT [8].

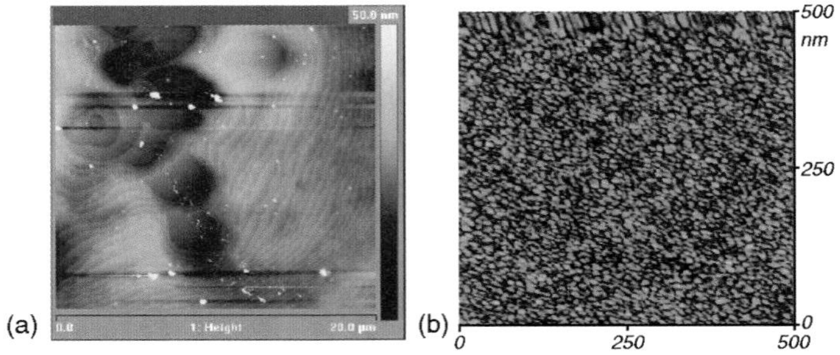

Figure 7.4 (a) AFM height image of a defect of a calcined B-type film. It reveals nano-terraces with a spacing of 3 nm. The image size is $20 \times 20\,\mu m^2$. (b) AFM phase image of the surface with higher lateral resolution. The phase angle variations are in the 20° range. The distance between the dark regions is approximately 20 nm.

Examination of the defects in the B-type films with AFM revealed terraces (Figure 7.4a). They are visible in about 70% of all investigated defects performing hard tapping mode. These nano-terraces have a step height of 3 nm, which is in a very good agreement with the TEM and SAXS results.

In-between the defects and on each terrace step the surface is very flat (RMS <0.5 nm). However, the AFM phase image reveals some pronounced lateral variations. Figure 7.4a shows a dark network which can be assigned to higher energy dissipation. This network might be a picture of the sustainers that support the separated layers. In [8] we describe a model resulting in higher dissipation on the top of the sustainers compared with the dissipation on the bridges between them. Therefore we interpret Figure 7.2b as a picture of the sustaining network between the simple silica layers.

Based on this interpretation a layered structure with novel kind of sustainers is drafted for the B-type films in Figure 7.5b. Flat voids of a typical size of about $20 \times 20 \times 2\,nm^3$ occur among the layers. The special structure of the B-type films is likely responsible for some of their peculiarities. The typical macroscopic defects of these films could be ascribed to shrinkage problems during film condensation. On a

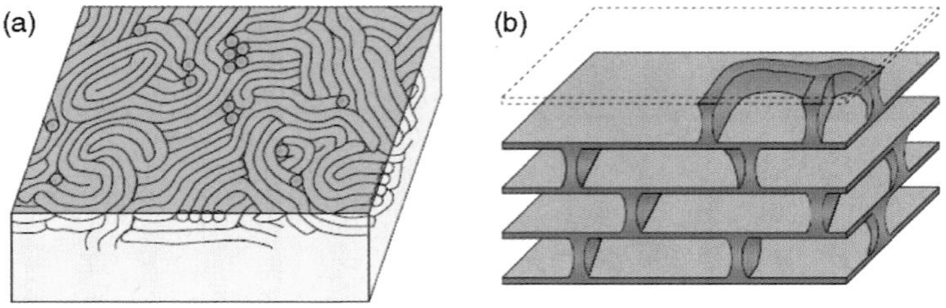

Figure 7.5 Structure models of mesoporous silica films: (a) worm-like structure for the A-type films and (b) layer structure with sustainers for B-type films.

solid support, a parallel lamella has no possibility to relax when it shrinks, except the formation of a defect. Furthermore, we observed in the TEM micrographs also single lamellas produced by the stress during mi-crotoming. This fact points to a reduced integrity between the lamellas, which is consistent with the relatively large typical distance between sustainers of 20 nm.

7.2.3
Optical Properties of MSFs

The refractive index of calcined A-type films was determined from the interferometric measurements using a two-axes goniometer (Figure 7.6a). An example of the measured transmission spectra a special different incidence angle is shown in Figure 7.6b. The interference at the thin layer results in different reflection intensities for different wavelengths of light (Fabry–Pérot oscillations). The interference curves have a cos-like shape with intensity variations between 2 and 10%. The positions of the interference extrema on the wavenumber scale have a linear dependence on the

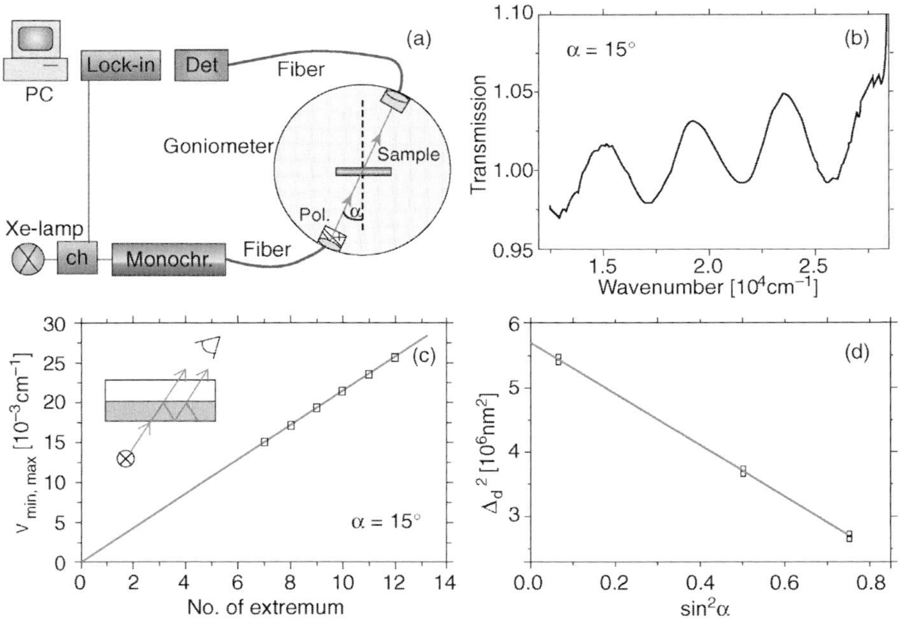

Figure 7.6 (a) Experimental setup of the goniometer for the simultaneous measurement of film thickness and refractive index. The incidence angle α was varied in the range 0°–60°. (b) An example of the measured transmission spectra for the incidence angle α of 15° with s-polarized light. (c) Position of the extremum versus the number of the extremum for $\alpha = 15°$ for s-polarized light. This graph allows the determination of the thickness dependent part Δ_d of the optical path difference. As indicated in the inset, different rays contribute to the transmitted signal. They have an optical path difference Δ. (d) Determination of the refractive index n and film thickness d from the dependence Δ_d on α.[13]

interference order. Therefore, this dependence was fitted with a straight line for each incidence angle (Figure 7.6c). Also the maximal and minimal possible values for the extrema were considered for the error determination. The slope of this line m delivers the thickness-dependent part of the optical path difference $\Delta_d = (2m)^{-1}$. It differs from the total path difference Δ by a reflex pronounced on phase jump: $\Delta = \Delta_d + \lambda/2$. Figure 7.6d shows the dependence of Δ_d on the incidence angle for s-polarized light. Mathematically, it is described by [11]:

$$\Delta_d = 2d\sqrt{n^2 - \sin^2\alpha}. \tag{7.1}$$

Therefore, the axes have been chosen appropriately to allow a linear fit. This fit enables simultaneous determination of the film thickness d and the refractive index n. A refractive index of $n = 1.18 \pm 0.01$ was determined in the visible range of 350–800 nm. This is one of the lowest measured n for a transparent material suited as waveguide support. The air filling fraction, i.e. porosity, of these mesoporous films was calculated using effective medium approximation from Bruggemann [12] to $(59 \pm 2)\%$.

Angle-dependent interferometric measurements enable reliable information on the accuracy of the n-determination and deliver an n-value averaged over the whole film thickness. The difference in the refractive index for parallel polarized light and perpendicular polarized is unfortunately within the experimental error (0.01), which is, therefore, the upper limit for the birefringence of A-type films.

Stability of the refractive index upon humidity changes and ageing is important for the application of calcined A-type films as low-n substrates. For this a sample was first stored in the desiccator at low humidity of about 30% for 107 days and the refractive index was measured with the previously described method to be $n = 1.183 \pm 0.003$. Then, the sample was stored in the desiccator at 57% RH for 17 days, further 6 days at 97% RH and additional 18 days at 97% RH. The resulting refractive indexes were $n = 1.181 \pm 0.006$, $n = 1.190 \pm 0.017$ and $n = 1.185 \pm 0.012$, respectively. It seems that exposition of calcined mesoporous films to higher humidity, as well as the ageing in the period of few months do not influence the refractive index significantly. The measured deviations are within the error bars.

In waveguides, a considerable part of the radiation power is guided in the evanescent field penetrating into low-n support. Therefore, it is very important that the damping due to scattering in the support is low. One possibility to estimate these losses is to measure the diffuse reflectance of these films. Diffuse reflectance spectra were measured using a UV–vis spectrometer with a praying mantis attachment. By this tool the specular reflection was eliminated and only the diffuse reflection was collected. B-type films show a high wavelength-dependent scattering up to $S = 0.2$ (20%) as expected from the visual impression. Contrary to this result the optical scattering S of the calcined A-type films was very low, $S = 0.0026$ (0.26%). However, this value was close to the detection limit of our set-up and may be an overestimation. Having this restriction in mind, we take the value as the basis for the estimation of the scattering coefficient $\alpha_{sc} = -(1/l)\log_{10}(1 - S) = 8.4\,\text{cm}^{-1}$ (in analogy to the absorption coefficient [14]).

Furthermore, one can take this value for the estimation of the waveguide damping with such substrates. Assuming that 10% of the modes are guided in the substrate one obtains $\alpha_{WG} = 0.84\,cm^{-1}$. This allows for sure mm-long applications but also guiding over centimeters can be possible which is agreement with our former waveguide experiments [9]. The possible mistake in this estimation (systematic overestimation) would enable a wider application range.

7.2.4
Synthesis Mechanism

The fabrication procedure used in this work is a special version of the SBA-15 synthesis class [7] for mesoporous materials. In several works [15,16] the synthesis mechanism was discussed with the main focus on the role of the SDA and the surface charge matching possibilities. Although the deeper understanding of the chemical mechanism was not the aim of this work, it gave some new insights in respect to the mechanism.

Experimentally, two surprising facts have been recognized. First, the structure types of the products are more variable than expected. Especially the lamellar structure of the B-type films has not been ob-served before and cannot be considered as a disturbed variant of a “SBA-15 main structure”. Therefore, the abbreviation used for labeling mesopores (here SBA-15) should better be used for the description of the main features of the synthesis and not to assign a structure. A similar non-uniqueness of the structure has been found for the SBA-3 type mesopore synthesis as well [17]. Second, the obtained structure sensitively depends on processing conditions. The same chemical composition of the coating solution leads to different structure types. It seems that the SDA is not able to drive the condensation process efficiently into a unique energetic minimum state.

These two findings are not in a good agreement with the general concept of the SDAs. This concept assumes that the structure-directing agents can control the pore structure of the synthesized material. In our case they form micelles which should determine the film structure. Normally, this concept gives a good guideline for finding new synthesis schemes. However in detail, it cannot be completely right. It can especially not explain strong sensitivities towards processing conditions.

The key for understanding of the synthesis seems to us the fact that the synthesis process goes through a series of non-equilibrium states. The synthesis tries to reach an energetic minimum state determined by the SDA, but because of the simultaneous silica condensation, it is not able to reach this state. Therefore, the synthesis results in a product which is frozen-in somewhere on an assumed reaction path to the ideal structure. The known dependence of the silica condensation on the processing conditions is transferred to the formation of the pore structure. Although our structure-directing agent drives the structure towards one equilibrium state, slightly varying processing conditions lead to the formation of different film types.

The exact description of such processes is difficult. A schematic picture can be given based on the equilibrium phase diagrams. They describe which phase the micelles (separated spherical micelles, hex-agonal arrangements of rod-like micelles etc.) form for certain compositions of the solution. However, they neglect that a certain time is needed to reach these phases. Nevertheless, a reaction path in such a diagram can provide a rough explanation for the observed phenomena. The SBA-15-relevant micelle structures formed by the ternary copolymer-water-oil system were described in phase diagrams, e.g. in Ref. [18]. The chemical recipe for the dip-coating solution determines a starting point in this diagram and the chemical composition of the dried film gives an end point. Inbetween these two points there is the reaction path determined by the evaporation of the solvents. However, the structural forecast at the end point is not relevant because of the freezing-in of the structural transformations somewhere on the reac-tion path. We discussed the possible phenomena of our special system in a former paper [8] and refer to it for the details.

Although the synthesis mechanism is very complex because of the non-equilibrium phenomena, the practical synthesis turned out to be reproducible and controllable. It enables a wider range of products than normal near-equilibrium syntheses, it allows fine-tuning, but it requires careful control of the processing conditions.

7.3 MSFs as Substrates for Waveguide Structures

7.3.1 Polymer Waveguides

Because of the very low refractive index, low optical scattering, sufficient thickness of the films, and the very smooth surface, A-type films are very well suited as ultra-low refractive index substrates, especially for 2D photonic crystal waveguides (PhC WGs). An example of such system is shown in Figure 7.7 representing the result of a common project of the TU Hamburg-Harburg, IPHT Jena and MPI Mülheim. A-type films with a thickness of 1 μm were deposited onto oxidized silicon wafers for the fabrication of this waveguide structure. A slab waveguide was produced by spin coating a polymer poly(methyl methacrylat/disperse red-1) (P(MMA-DR1)) with $n = 1.54$ onto this substrate. Then 2D PhC structures were fabricated by a combination of different etching processes (EBL, RIE, dry etching) [9,19]. They formed a resonator consisting of a line defect (LD) in a 2D PhC. In the shown example only the core layer was etched and the air holes do not penetrate the substrate. The resonator consists of two finite square arrays of holes separated by a non-structured region. A 150 nm hole radius was chosen resulting in an optical stop band around the vacuum wavelength of 1.3 μm. The lattice defect is formed by omitting 4 lines of holes perpendicular to the wave propagation direction. The exact design of the structure was optimized by simulations using a 3D finite integration technique as described in [9]. Two examples of the calculated electrical field distributions for finite 2D PhC are depicted in Figure 7.8a. The light is strongly confined inside the PhC waveguide core when the ultra-low index substrate is used.

The transmission spectra (Figure 7.8b) of this resonator were measured for different polarizations by the prism coupling method. The resonator structure on

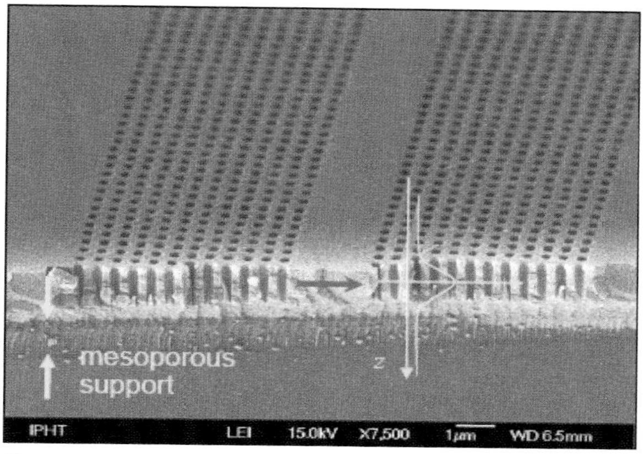

Figure 7.7 SEM micrograph of the realized 2D PhC LD resonator made of P(MMA/DR-1) on an A-type mesoporous silica film as a support [9]. A schematic drawing shows the field distribution in a guided mode (arrow). The evanescent field of the guided mode penetrates into the support up to some 100 nm.

the mesoporous substrate showed a high transmission in the resonance peak of about 60% for TE and 80% for the TM polarization. These values were expected from the simulations and exceed values determined on similar structures on conventional Teflon supports. A further interesting property of this transmission spectrum are the seemingly absent radiation losses at the air band side of the band gap. The transmission has the same height at this band edge as at the dielectric edge. In contrast, the PhC waveguides on Teflon showed strong losses at the air band edge since these states are above the light cone [9].

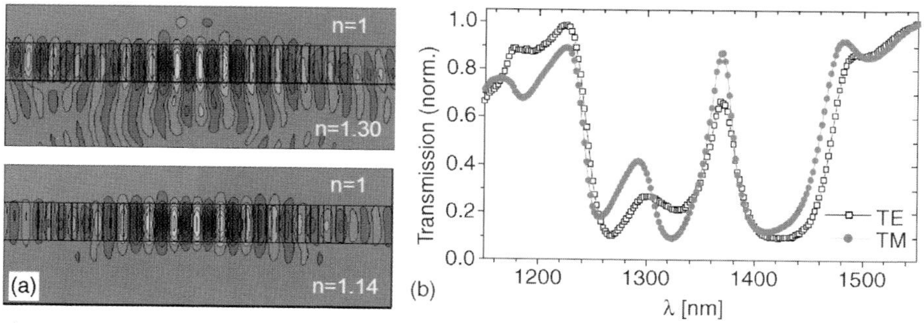

Figure 7.8 Wave propagation in polymer waveguides [9]. (a) Simulations of the electric field distribution inside a finite 2D PhC line defect resonator for two different substrates: Teflon with $n = 1.3$ and MSF with $n = 1.14$. (b) Experimental transmission spectra of a P(MMA/DR-1)/MSF line defect resonator for TE-like and TM-like modes.

Beside the simulations, also simpler considerations can be used to understand the measured result. The decay of the guided modes into the support is schematically depicted in the Figure 7.7. The depth of decay depends on the mode type, waveguide thickness, and significantly on the refractive index of the support. This effect can be estimated by a lower limit of the field decay. The fields of every guided mode of a waveguide decay slower into the support then $exp\{-z\,4\pi(n_0^2-n_{sup}^2)^{1/2}/\lambda_0\}$. Here, z is the distance from the waveguide-support interface, n_0 the effective refractive index of the waveguide, n_{sup} the refractive index of the support, and λ_0 the wavelength of the used light [20]. This estimation delivers the result that a 1 μm thick mesoporous film can be sufficient for the field confinement, but not a Teflon film with $n = 1.3$, since there is still a remarkable field intensity at 1 μm which can be absorbed or irradiated.

7.3.2
Ta_2O_5 Waveguides and 2D PhC Structures

Waveguides made of inorganic materials especially of oxides are regarded to have a wider application range than polymeric ones. They are transparent in larger parts of the spectrum, can resist higher temperatures, and – most importantly – can have larger refractive indices. However, their structuring in the nm-range seems to be generally more difficult. In respect to our low-n supports, it is not clear if such layers are compatible with the support. The inorganic layers will likely create a large stress on the support during their deposition as well as during their structuring. Therefore, we studied some model examples and show the results obtained with Ta_2O_5 top-layers here.

Films of Ta_2O_5 have been sputtered on the MSFs (mso Jena Mikroschichtoptic GmbH). The structuring of these films was performed in a similar manner as for polymers, but with changed processing parameters. It required tempering at 180 °C after the photoresist deposition and heating during structuring (RIE) to 180 °C together with HF heating from the top. Surface temperatures of 200 °C were likely reached in this step.

The deposition of the original Ta_2O_5 layers worked without visible problems. Here, it might have been useful that a surface barrier [21] can protect the porous films against penetration of Ta_2O_5 precursors and that the elasticity of the MSF can help to distribute the stress in the forming Ta_2O_5 layer. Difficulties occurred in some of the structuring steps resulting in delamination of the layer system. We ascribe these difficulties to water with was adsorbed during sample handling. It was possible to avoid these effects and to demonstrate the fabrication of several structures suitable for stripe waveguides, PhC waveguides, PhC WG bends, and for PhC WG splitters.

A planar stripe waveguide is shown in Figure 7.9a. Here, the mesoporous support shows an interesting under-etching effect. Such an effect is not negative for the desired function as waveguide and is likely caused by the very low density of the MSF. Figure 7.9b shows a PhC waveguide. Here, the etching was stopped at the MSF/Ta_2O_5 interface. Alternatively, the etching can be continued deeply into the support as shown in Figure 7.9c. The under-etching effect leads then to very thin ribs under the

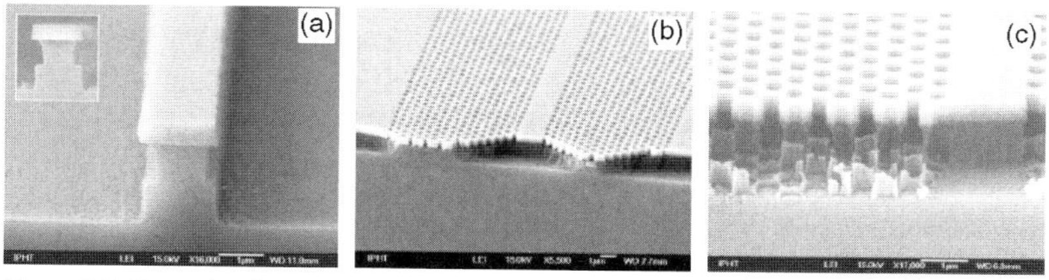

Figure 7.9 Examples for WG structures on MSF supports. (a) Strip waveguide, (b) PhC WG on a homogeneous support, (c) PhC WG on a structured support.

Ta_2O_5. Interestingly, the whole layer arrangement kept stable. It is, up to now, unclear if the remaining rib consists of unchanged mesoporous material or if it is subjected to a densification process during the ion treatment. Such modifications have been observed in the interaction with other ions [22].

The optical characterization of the fabricated structures is still in progress. However, they all fulfilled their principle function in a similar manner as reference structures on glass. There was not observed any additional damping for the fabricated structures originating from the scattering in the mesoporous support. The detailed parameters (bend damping, waveguide damping, and coupling efficiencies) differ of course from the reference structures since the MSF has another refractive index.

These results with the Ta_2O_5 system show that a highly accurate structuring of inorganic top-layers on MSFs is possible. This means that the MSFs can resist the larger processing stress in comparison to the requirements for polymer waveguides. This enables future prospects to use the MSF also in high contrast PhC systems. Especially the optical up-down symmetry can be useful in such systems.

7.3.3
PZT Films

The incorporation of active materials into waveguides and PhCs is a field of strong interest. The materials should be switchable, nonlinear-optical, or amplifying. They can be incorporated as guests or as constituent of the PhC itself. For example, an electro-optic PhC would be very interesting. However, no material with high electro-optic coefficients and sufficient possibilities for structuring in PhCs has been identified up to now. Films of lead-zirconate-titanate (PZT) could be one promising candidate for that. Ferroelectric inorganic films, especially $Pb(Zr_xTi_{1-x})O_3$ (PZT) are known to have large optical nonlinearities and can be prepared by different methods. The sol–gel approach can be very useful, because it is offering a number of tuning possibilities. In addition, this approach seems to be compatible with the MSF support fabrication. Therefore, this approach was used and investigated concerning the fabrication of PZT/MSF layer structures.

The coating solution for the lead zirconate titanate (PZT) films was prepared as described in Ref. [23]. The PZT precursor solution was deposited onto diverse

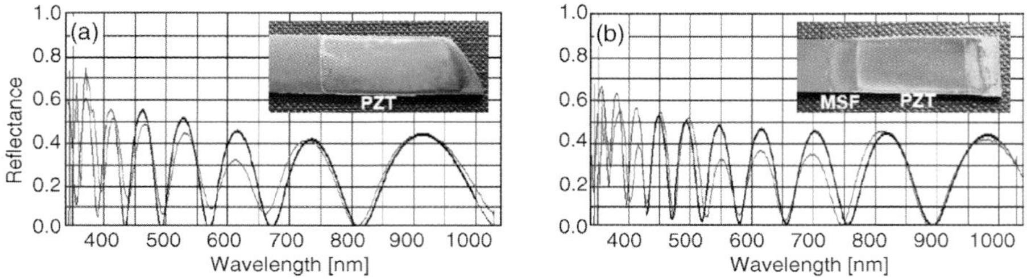

Figure 7.10 Measured and calculated reflection spectra of a PZT film on Si-wafer (a) and of PZT on a mesoporous silica film (b) . Insets are photographs of PZT on Si-wafer and on MSF, respectively. The photographs of the calcined samples were made using a digital camera with diffuse sample illumination.

substrates in the same manner as mesoporous silica films. The temperature in the chamber was 21–23 °C and the humidity varied in the range of 37–50% RH without noticible influence on the optical appearance of the films in normal and microscopical observation.

The calcined films were transparent and clear (Figure 7.10). However, closer observation with optical and electron microscopes reveals a more or less dense crack network on normal supports (Figure 7.11a). The cracks develop as a reaction to the stress in the film on firing as it has been discussed in many publications, e.g. [24]. To avoid this polyvinylpyrrolidone (PVP) was added to the coating solution as a stress-relaxing agent. Although the crack formation was reduced by addition of PVP, it could not be fully eliminated. We assign this fact to the different thermal expansion coefficients of PZT and the support. Therefore, the final tuning of the PZT film properties has to be adjusted to the support system.

PZT films were deposited on the mesoporous silica films in the same manner. They were also transparent and mostly clear. The interference colors, typical for thin films, are visible as shown in the insets in Figure 7.10. The unique color indicates

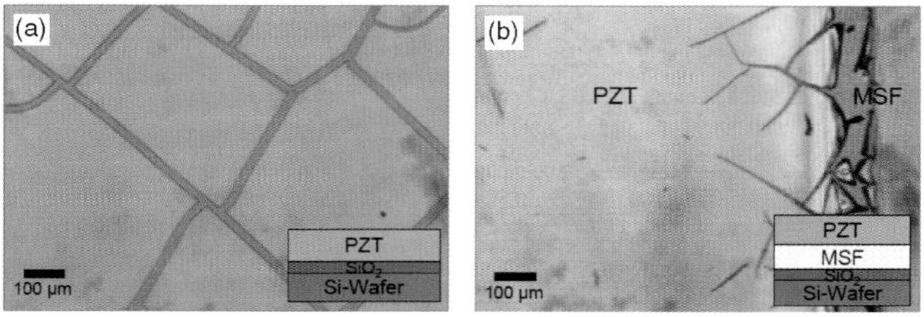

Figure 7.11 Microscope pictures of a PZT film deposited on (a) silicon wafer and (b) mesoporous silica film. The photos were made using an eye-piece camera (MA88, CA Scientific Co) with 640×480 pixels.

good homogeneity of the films. Only the borders, especially the outflow edge, show different colors, due to changes in the film thickness.

Observation with an optical microscope showed that the PZT films deposited on the MSF have much less cracks, mostly only near to the edges (Figure 7.11b). However, the films exhibit slightly blear regions, not visible in the photography and microscope pictures. It is possible that the mesoporous film changed or even crashed during the PZT fabrication. Thickness measurements (Figure 7.10) indicated that the thickness d of the MSF is significantly lower after the deposition of PZT. Such a compression of the film might be caused by infiltration of the PZT precursor. Possibilities to avoid such effects could be the change of surface properties of the MSF or the introduction of surface barriers [21].

The suppression of the regular cracks by deposition on MSF supports is, however, an encouraging result on the way to high-quality PZT films. The flexibility of the porous structure lowers obviously the stress induced in the PZT by the temperature treatment.

7.4 Conclusions

Optically perfect mesoporous films have been synthesized reproducibly. They exhibit an extremely low refractive index, sufficient mechanical and chemical stability, low optical scattering and a thickness up to about 1 μm.

The humidity during film fabrication turned out to be a decisive processing parameter for the fabrication of optically perfect films. In the range between 20 and 40% relative humidity the desired A-type films have been synthesized. The control of the film thickness which is important for many applications is possible via the drawing speed.

Partially ordered mesostructures of worm-like type turned out to have most useful properties (A-type films). They are well suited as low-n supports for waveguides. Low-n supports have three important advantages: one can avoid working with leaky modes, the penetration depth into the substrate is lower, and the system is nearly symmetric (or fully symmetric with a low-n cover layer). This leads to a decoupling of TE and TM modes, which is also very useful for an easier design of functional waveguide structures.

The mesoporous films resist many fabrication procedures for waveguide overlayers. In some cases they may be even advantageous for obtaining high-quality homogeneous overlayers.

Acknowledgements

The authors thank H.L. Li, H. Bretinger, and R. Brinkmann for experimental support and M. Eich for stimulation of these works. Funding by the DPG is gratefully acknowledged.

References

1 Miller, R.D. (1999) *Science*, **286**, 421.

2 Jain, A., Rogojevic, S., Ponoth, S., Agarwal, N., Matthew, I., Gill, W.N., Persans, P., Tomozawa, M., Plawsky, J.L. and Simonyi, E. (2001) *Thin Solid Films*, **398/399**, 513.

3 Baskaran, S., Liu, J., Domansky, K., Kohler, N., Li, X., Coyle, C., Fryxell, G.E., Thevuthasan, S. and Williford, R.E. (2000) *Adv. Mater.*, **12**, 291.

4 Kresge, C.T., Leonowicz, M.E., Roth, W.J., Vartuli, J.C. and Beck, J.S. (1992) *Nature*, **359**, 710.

5 Beck, J.S. *et al.* (1992) *J. Am. Chem. Soc.*, **114**, 10834.

6 Brinker, C.J., Lu, Y., Sellinger, A. and Fan, H. (1999) *Adv. Mater.*, **11**, 579.

7 Zhao, D., Feng, J., Huo, Q., Melosh, N., Frederickson, G.H., Chmelka, B.F. and Stucky, G.D. (1998) *Science*, **279**, 548.

8 Konjhodzic, D., Bretinger, H., Wilczok, U., Dreier, A., Ladenburger, A., Schmidt, M., Eich, M. and Marlow, F. (2005) *Appl. Phys. A*, **81**, 425.

9 Schmidt, M., Boettger, G., Eich, M., Morgenroth, W., Huebner, U., Meyer, H.G., Konjhodzic, D., Bretinger, H. and Marlow, F. (2004) *Appl. Phys. Lett.*, **85**, 16.

10 See e.g.: Grosso, D., Carnol, F., Soler-Illia, G., Crepaldi, E.L., Amenitsch, H., Brunet-Bruneau, A., Bourgeois, A. and Sanchez, C. (2004) *Adv. Funct. Mater.*, **14**, 309.

11 Bergmann, L. and Schäfer, C. (1987) *Lehrbuch der Experimentalphysik Band III*, Walter de Gruyter, Berlin–New York, p. 344.

12 Bruggeman, D.A.G. (1935) *Ann. Phys. (Leipzig)*, **24**, 636.

13 Konjhodzic, D., Bretinger, H. and Marlow, F. (2006) *Thin Solid Films*, **495**, 333.

14 See e.g.: Wöhrle, D., Tausch, M.W. and Stohrer, W.-D. (1998) *Photochemie*, Wiley-VCH, Weinheim, p. 498.

15 Zhao, D., Yang, P., Melosh, N., Feng, J., Chmelka, B.F. and Stucky, G.D. (1998) *Adv. Mater.*, **10**, 1380.

16 Alberius, P.C.A. and Frindell, K.L., Hayward, R.C., Kramer, E.J., Stucky, G.D. and Chmelka, B.F. (2002) *Chem. Mater.*, **14**, 3284.

17 Marlow, F., Khalil, A.S.G. and Stempniewicz, M. (2007) *J. Mater. Chem.*, **17**, 2168.

18 Holmquist, P., Alexandridis, P. and Lindman, B. (1998) *J. Phys. Chem. B*, **102**, 1149.

19 Huebner, U., Boucher, R., Morgenroth, W., Schmidt, M. and Eich, M. (2006) *Microelectron. Eng.*, **83**, 1138.

20 See e.g.: Iizuka, K. (2002) *Elements of Photonics, Vol. II*, Wiley-Interscience.

21 Stempniewicz, M., Rohwerder, M. and Marlow, F. (2007) *Chem. Phys. Chem.*, **8**, 188.

22 Stempniewicz, M., Khalil, A., Rohwerder, M. and Marlow, F. (2007) *J. Am. Chem. Soc.*, **129**, 10561.

23 Takenaka, S. and Kozuka, H. (2001) *Appl. Phys. Lett.*, **79**, 3485.

24 Kozuka, H., Takenaka, S., Tokita, H. and Okubayashi, M. (2004) *J. Eur. Ceram. Soc.*, **24**, 1585.

8
Linear and Nonlinear Effects of Light Propagation in Low-index Photonic Crystal Slabs

R. Iliew, C. Etrich, M. Augustin, E.-B. Kley, S. Nolte, A. Tünnermann, and F. Lederer

8.1
Introduction

The last years have seen a rapid development of the field of photonic crystals and components based thereon. The majority of such components was based on slab geometries which are much easier to realise due to their ease in fabrication. Besides semiconductors [1–11], also materials with a significantly lower refractive index [12–19] were utilised to fabricate such geometries. However, experimental investigations presented in [12–14] were restricted to reflection and transmission spectroscopy in 2D fully periodic photonic crystals.

In the following we investigate amorphous materials with a refractive index in the range of 1.9–2.2 on a silica substrate with an index of 1.43, or free-standing membranes. A 2D photonic crystal (PhC) slab made of such materials can exhibit still an in-plane photonic bandgap for TE-polarised light with a gap-to-midgap ratio of 14–19%. However, due to the smaller index contrasts the light confinement is weaker than in semiconductor PhCs. Therefore, one important goal of our investigations was to determine the limits of these low-index devices and to compare with well-known results from high-index semiconductor PhCs, where different functional elements with 2D (e.g. waveguides) and 3D (e.g. cavities) light confinement where realised experimentally.

The advantages of low-index materials are evident. First, the wavelength range of transparency extends to the visible. Second, the resulting larger structure sizes allow for easier coupling to conventional guiding structures, as fibres and waveguides. Also, the lower index reduces Fresnel losses at air interfaces. Compared to semiconductor heterostructures [3–6] the much higher vertical index contrast reduces losses due to sidewall roughness [20]. The higher vertical index contrast also raises the light cone, allowing for low-loss operation in the first bands. Hence, regarding the vertical confinement, this system is between semiconductor membranes [10,11] and semiconductor-on-insulator (SOI) structures on the one side and semiconductor heterostructures on the other. Regarding the in-plane index contrast, it is between

Nanophotonic Materials: Photonic Crystals, Plasmonics, and Metamaterials. Edited by R.B. Wehrspohn, H.-S. Kitzerow, and K. Busch

ISBN: 978-3-527-40858-0

semiconductor systems and polymer systems with an even lower refractive index of up to 1.6 [21,22]. In these polymer slabs only devices with one-dimensional light confinement were demonstrated experimentally.

This work highlights results obtained in the course of different projects within the framework of the "Photonic crystals" programme of the Deutsche Forschungsgemeinschaft. The main emphasis of these projects was on two-dimensional low-index photonic crystals in slab geometries. In the following we summarise our main results and discuss them in the context of the current state-of-the-art. For more details of the several investigations we refer the interested reader to the original publications cited in the respective context.

The article is organised as follows. In the next section we review briefly the fabrication of substrate-based low-index PhCs. Then theoretical and experimental results of linear effects of light propagation in slab geometries are presented. In Section 8.4 we discuss theoretical results for nonlinear PhCs, before we summarise our work in Section 8.5.

8.2 Fabrication of Photonic Crystal Slabs

The experimental investigations presented here were carried out using samples with a waveguiding layer of niobium pentoxide Nb_2O_5 with a refractive index $n = 2.1$ at a wavelength of 1.55 μm or of silicon nitride SiN_x with $n = 1.91$ at 1.55 μm and $n = 1.95$ at 800 nm. For the first experiments with PhCs of Nb_2O_5 a technology for etching of this very resistant material, employing a multilayer resist, was developed [15–19]. Hereby, after electron beam exposure of the upper resist layer, the photonic crystal pattern is transferred into the chromium layer below. In order to obtain the necessary etching depth, a second etching mask needs to be created. However, due to profound technological benefits (the chemical composition makes it compatible to existing microelectronic processing methods at high etching ratios, resulting in high-aspect ratios, steep sidewalls and very regular structures, only one etching mask is required) the performance of the characterised components realised later in SiN_x turned out to be significantly better. Hence, although the considerably lower refractive index reduces slightly the achievable bandwidth of the photonic bandgap, a major part of PhC structures investigated here is realised in this material. The SiN_x waveguiding layer and a 2000 nm SiO_2 buffer layer are deposited in an ICP-eCVD (inductive coupled plasma enhanced chemical vapour deposition) process on an oxidised silicon substrate. When higher vertical symmetry was required, an additional cladding layer of 300 nm SiO_2 was deposited on top.

The PhC structures designed for operation at infrared wavelengths were fabricated by means of electron beam lithography (Leica ZBA 23 H) using a rectangular-shaped electron beam. Structures for experiments in the visible (see subsection 8.3.3.1) were defined with an electron beam writer (Leica LION LV1) with a spot exposure system (Figure 8.1). Due to its higher resolution the fabrication of these much smaller structures is feasible. For both materials (Nb_2O_5 and SiN_x) the slab system was etched

(a) 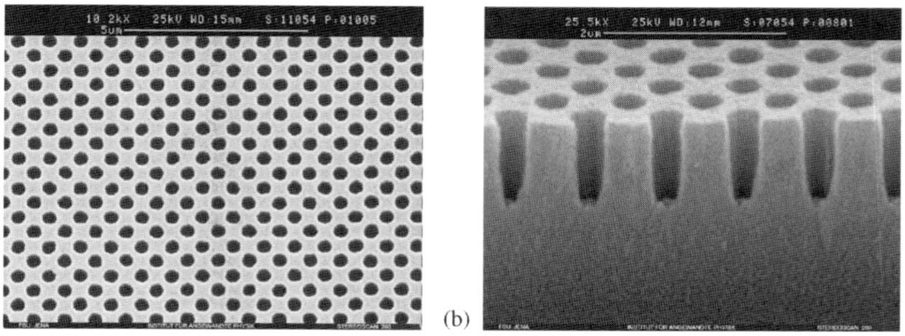 (b)

Figure 8.1 Photonic crystals in SiN_x/SiO_2. After optimisation of the processing parameters very regular structures (a) with deeply etched holes are achieved (b).

in multiple ICP-etching processes in order to account for the different materials (cladding layer, waveguiding layer, substrate) being etched. Finally optical facets were obtained by cleaving and breaking the samples.

A major part of the theoretical investigations is concerned with parametric second order nonlinear processes in photonic-crystal microcavities. The theoretical investigations are based on lithium niobate ($LiNbO_3$), which is available as wafers in optical quality and has a large nonlinear diagonal coefficient. By means of a focused ion beam [23] two-dimensional photonic crystal structures could be realised in this material. A very promising second approach by means of ion beam amorphisation and HF-etching [24] was proposed recently.

8.3
Linear Properties of Photonic Crystal Slabs

Up until now, the majority of theoretical and experimental investigations of PhCs deals still with linear materials. Historically, the photonic bandgap was the first effect being explored. For instance for obtaining light localisation [25] or for modifying the radiation dynamics of emitters [26]. Apart from the bandgap, another aspect of the investigation of periodically modulated dielectrics were the modified properties of light propagation in these media [27–29]. In order to obtain effects as negative refraction or anomalous diffraction a much smaller index modulation than necessary for a photonic bandgap is sufficient.

We utilise the bandgap to realise defect waveguides of different widths and to obtain a microcavity with a high quality factor. The strong light confinement in the waveguides, in conjunction with the periodicity in propagation direction, was shown to give an unusual dispersion [7,30] or to lead to strong suppression of the transmissivity in a certain wavelength range [31].

Here we focus on three different aspects of light control provided by linear PhCs and present the results obtained for low-index PhC slabs. In the first subsection we investigate line defect waveguides realised in low-index PhC slabs regarding their transmission and dispersion behaviour. In the following subsection we present detailed numerical investigations of point-like defects in 2D PhC membranes, where the 2D photonic bandgap leads to light localisation, and hence, a photonic micro-cavity is obtained. We discuss the quality factors obtainable in these low-index materials and compare them to results for high-index membranes. In the third subsection we discuss effects of anomalous light propagation in the low-index SiN_x PhC slab system. We compare the performance of guidance without defects and of anomalous refraction with results for high-index PhC systems.

Bandstructures presented in the following are calculated by preconditioned conjugate-gradient minimisation of the block Rayleigh quotient in a plane wave basis, using a freely available software package [32].

8.3.1 Transmission and High Dispersion of Line-Defect Waveguides

One way of introducing line defect waveguides in a two-dimensional PhC slab is the omission of one or more rows of holes. In a hexagonal lattice the resulting waveguide is referred to as Wn, where n denotes the number of consecutive rows omitted in ΓK-direction (see Figure 8.2). The resulting waveguide modes in high-index (silicon) PhC membranes were shown to exhibit a very large dispersion [7,30]. Signal delays of 1 ns were proposed for a 670 μm long device [33] and a reduction of the velocity of light by a factor of 1000 was experimentally demonstrated in a W3 waveguide in a silicon-on-insulator (SOI) PhC slab [34]. The performance of these devices relies crucially on the waveguide modes lying outside the light cone, because otherwise the associated radiation losses would be detrimental to the

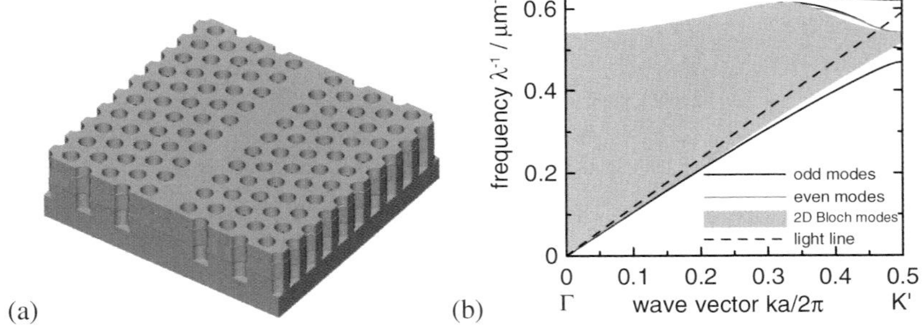

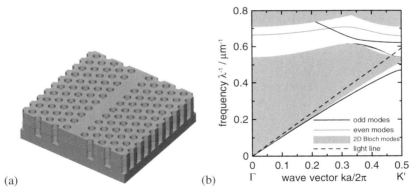

Figure 8.2 (a) Schematic drawing and (b) bandstructure of a W1 waveguide in a photonic crystal slab with a hexagonal lattice (diameter of holes 374 nm, lattice pitch 595 nm). The grey regions indicate frequencies, where light can propagate as 2D-Bloch waves inside the photonic crystal without defect and are obtained by projecting the respective 2D bandstructure in the waveguide direction.

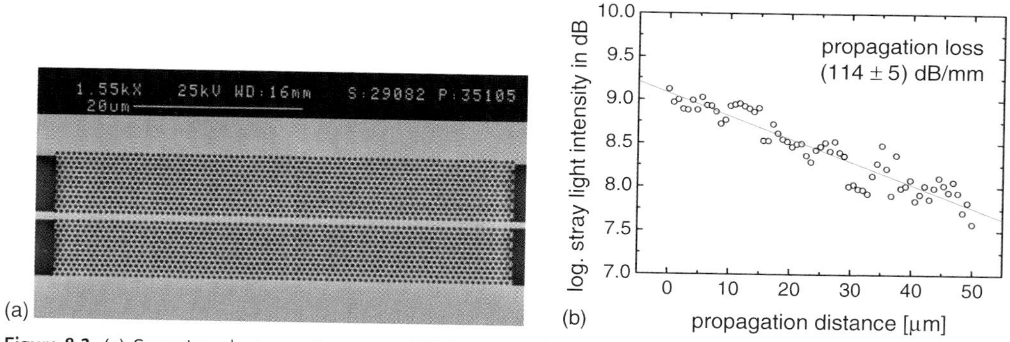

Figure 8.3 (a) Scanning electron microscope (SEM) image of photonic crystal waveguide sample and (b) propagation losses (at $\lambda = 1493$ nm) determined from stray light measurement for TE-polarisation.

transmissivity in the slow-light regime. In high-index membranes and SOI slabs a relatively large fraction of the mode bands in the photonic bandgap is outside of the light cone.

In contrast, normally this cannot be achieved in our substrate-based low-index materials systems, due to an unfavourable location of the light cone. The simplest waveguide structure is a W1 (see Figure 8.2(a)). First a photonic crystal structure comprising a hexagonal lattice of air holes with a diameter of 374 nm, a lattice pitch of 595 nm (Nb_2O_5-system) and a length of 58 μm along the ΓK-direction was realised in a slab system comprising a 500 nm thick Nb_2O_5 guiding and a 300 nm silica cladding layer on a silica substrate. From the bandstructure (see Figure 8.2(b)) it is clear, that the waveguide mode in the photonic bandgap (frequency region without Bloch modes for any vector $\boldsymbol{k}$, here $0.6\ \mu\text{m}^{-1} \leq \lambda^{-1} \leq 0.71\ \mu\text{m}^{-1}$) is always inside the light cone, i.e. couples to radiation, mainly to the substrate. In the following we will investigate experimentally and theoretically the performance of this mode and compare to modes outside the light cone.

Ridge waveguides are used to couple light into the photonic crystal waveguide (cf. Figure 8.3(a)). The corresponding calculated 2D-bandstructure (using an effective index) for TE-like polarisation is displayed in Figure 8.2(b). Here only the fundamental (odd) modes of the defect waveguide are considered, because they are easier to excite and are expected to have lower losses than the even modes. For the odd modes gap-guiding (flat part of the band inside the bandgap) and index-guiding (steeper part of the band inside the bandgap and modes below the light line, as well as modes below all 2D Bloch modes, $\lambda^{-1} < 0.5\ \mu\text{m}^{-1}$) is possible. At the edge of the first Brillouin zone the band corresponding to the gap-guided modes becomes very flat (cf. Figure 8.2(b), black solid curve), indicating that the group velocity approaches zero while the group index,

$$n_g = c\frac{dk}{d\omega} = -\frac{\lambda^2}{2\pi}\frac{dk}{d\lambda},$$

diverges. This leads to a large group velocity dispersion, $\text{GVD} = c^{-1}\text{d}n_g/\text{d}\lambda$, of the gap-guided modes near the edge of the first Brillouin zone. They establish the upper part of a mini-stopband. The lower part (outside of the 2D photonic bandgap) is index-guided. It vanishes into the continuum of 2D Bloch states and does not play a role for our considerations. Depending on the actual parameters, these modes can even be superposed almost completely by the continuum. Note, that the even modes are entirely gap-guided.

First we concentrate on the odd modes within the bandgap. Characterisation of the photonic crystal waveguide yields losses of at least (114 ± 5) dB/mm at a wavelength of 1493 nm (inside the photonic bandgap) in the index-guided region (see Figure 8.3(b)), where the slope of the band is large (see Figure 8.2(b)). The losses are due to the very strong interaction of the mode with the 2D photonic crystal, because a major (although evanescently decaying) part of the modal energy overlaps with the cladding in such a narrow waveguide. Consequently, because the modes are above the light line, efficient out-of-plane scattering induces large propagation losses. This is confirmed by a 3D finite-difference time-domain (FDTD) [35] simulation. On the other hand, the gap-guided part of the modes (flat region) is expected to have even larger losses due to the lower group velocity. Using a wider waveguide the losses in the steep region can be reduced, but in the region of mini-stopbands still prohibit the observation of high dispersion [36].

One possibility to avoid the high losses associated with the light cone in the W1 is to utilise the above mentioned (index-guided) waveguide modes below the light line and below the lowest order Bloch modes of the underlying crystal ($\lambda^{-1} < 0.5\,\mu\text{m}^{-1}$ in Figure 8.2(b)). Here the bands corresponding to the waveguide modes are again very flat. Furthermore a mini-stopband of the waveguide modes can be identified near the edge of the first Brillouin zone. Thus high dispersion can be expected there as well. To access this domain with the available experimental wavelength range (around $\lambda = 1550$ nm) the structure parameters have to be scaled accordingly. For this purpose we used the SiN_x-slab system (500 nm thick SiN_x guiding layer) since it exhibits a better structuring quality for the required photonic crystal parameters (diameter of holes 320 nm and period 500 nm). A 2D-band structure (using an effective index) for these parameters is shown in Figure 8.4(a) which looks very similar to Figure 8.2(b). Note that the large blue ellipse in Figure 8.4(a) marks the mini-stopband mentioned above.

SEM-images of the fabricated photonic crystal waveguides with these parameters reveal somewhat smaller holes of diameter 284 nm with a depth of 700 nm. Again ridge waveguides were used to couple light into the photonic crystal waveguides with the new parameters. Detecting the stray light the propagation losses of the 125 μm long photonic crystal waveguide were determined as (43 ± 16) dB/mm at 1594 nm which is considerable lower as in the case of modes within the bandgap which are located inside the light cone and thus radiate partially into the substrate. For the index-guided modes under consideration this loss does not occur. It should be noted that the smaller structure sizes result in inferior structure quality and thus higher scattering losses due to surface roughness. Hence structures qualitatively equivalent to the larger ones should result in even smaller losses.

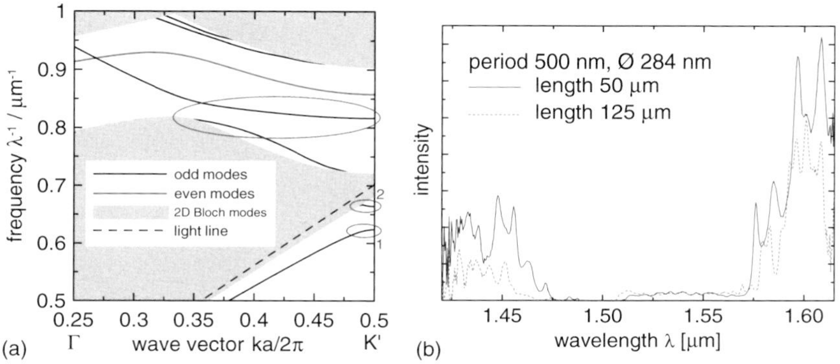

Figure 8.4 (a) 2D bandstructure and (b) experimental transmission spectrum for TE-polarisation of a W1 photonic crystal waveguide (ΓK-direction, diameter of holes 320 nm, period 500 nm) based on the SiN*x* slab system. The index-guided mode lying below the light line is located at 1550 nm. The ellipses in (a) mark the regions of small group velocity and high dispersion, where 1 and 2 are below the light line.

Comparing the transmission spectra with the location of the waveguide bands in the band structure shows that the region of low transmission corresponds to the mini-stopband (1450–1530 nm). Since the waveguide bands are relatively flat, an increase of the group index and GVD can be expected there. The upper band yields anomalous (GVD > 0) and the lower band normal dispersion (GVD < 0). In contrast, for the gap-guided modes investigated before only anomalous dispersion (GVD > 0) is possible.

The reflectivity due to the interface between the ridge and photonic crystal waveguides increases with the frequency approaching the mini-stop-band. This can be exploited to determine the group index experimentally using the formula of a Fabry–Pérot resonator:

$$n_g = \frac{\lambda_1 \lambda_2}{2d(\lambda_2 - \lambda_1)},$$

where λ_1 and λ_2 are two consecutive oscillations in the transmission spectra and d is the length of the photonic crystal waveguide. In Figure 8.5(b) the derived group index based on the spacing of the oscillations is shown. To recheck these values the measurement was repeated with a tunable laser. Although here different coupling conditions were used and a different height of the oscillation peaks was observed, the spacing in the transmission spectra was the same.

The values obtained for the group index were fitted by means of an appropriate function (see Figure 8.5(a)). The maximum value here is around 6, for the gap-guided modes in a silicon membrane system a group index as high as $n_g = 100$ was found [7]. Calculating the derivative of these functions with respect to the wavelength the group velocity dispersion is obtained (see also Figure 8.5). The largest dispersions were found to be +1,000,000 ps/nm/km and −500,000 ps/nm/km. For comparison, the

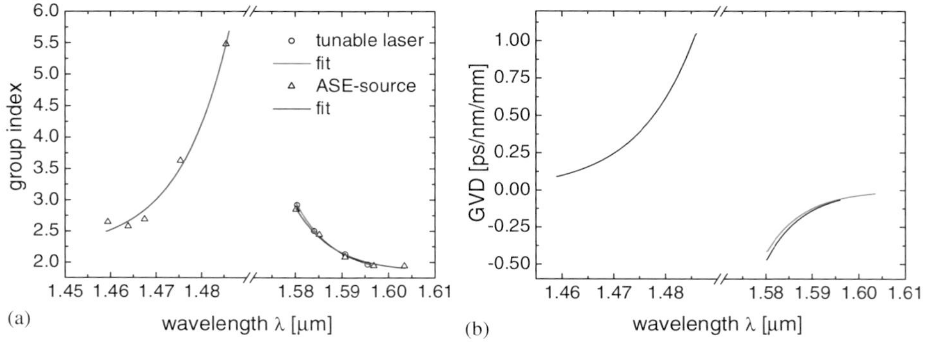

Figure 8.5 (a) Wavelength dependence of the group index determined from the experimental transmission spectra in Figure 8.4(a). (b) Group velocity dispersion (determined calculating the first derivative of the function fitted in (a)).

material dispersion of a silica fibre at a wavelength of 1.55 μm amounts to 17 ps/nm/km. With such a high dispersion only a length of 34 μm would be necessary to compensate for the dispersion in a 1 m long silica fibre.

In summary, the excitation of the index-guided mode (below the light line) of a photonic crystal waveguide is a possibility to achieve high dispersion with moderate losses without the need of a complete photonic band gap. However, at a wavelength of 1.55 μm very small structure sizes are needed in order to use these index-guided modes. On the other side, for using the gap-guided modes of a photonic crystal waveguide the propagation losses need to be reduced.

8.3.2
High-Quality Factor Microcavities in a Low-Index Photonic Crystal Membrane

Many modern applications in optics or quantum electrodynamics require optical structures, where light can be confined with a very long lifetime (high quality factor Q) to a very small modal volume, e.g., high quality semiconductor lasers [37,38]. But also for on-chip optical integration of resonators a small footprint is essential. Quality factors of up to 1,000,000 have been demonstrated in microcavities realised by placing a carefully designed defect in a 2D PhC in a high-index semiconductor membrane [39]. However, often not only the linear but also the nonlinear material properties play an important role for applications. For example, quadratic nonlinear interactions cannot easily be exploited in all-semiconductor geometries, but they can in materials such as $LiNbO_3$, with a substantially lower refractive index. A second motivation for the following investigations is to explore the limits of materials, where still a usable microcavity can be realised.

Again, as in the case of line defect waveguides, the out-of-plane radiation is the main obstacle which limits Q, because in-plane radiation can be made arbitrarily small using larger PhC reflectors around the defect, as long as the crystal provides a

2D photonic bandgap for at least one polarisation. Quantitatively, the radiated power can be expressed in terms of the energy density of spatial Fourier components $\boldsymbol{k}$ of the mode which are located inside a circle with radius $\omega_0 n_{subst}/c$ defined by the light cone of the substrate [40]. Here ω_0 is the resonance frequency and n_{subst} is the refractive index of the substrate.

Following the approach of a modal gap [9], where a "gentle confinement" of the mode strongly reduces the spectral fraction of wavevectors inside the light cone, we propose a novel structure with a high quality factor. We are interested in cavities obtained by omitting a certain number of consecutive holes in ΓM-direction, resulting in a short W1-waveguide [41]. The Q factor of such cavities was dramatically increased by means of the concept of gentle confinement [8], which was improved later using a modal gap of waveguide modes [9]. In this way a complete W1-waveguide is created in a hexagonal PhC. To obtain localisation in direction of the waveguide, this waveguide is locally modified by slightly stretching the lattice constant leading to a tetragonal lattice. This modification leads to a slight shift of the waveguide dispersion relation, and hence, for a frequency close to the band edge of one waveguide the same frequency may already be outside the dispersion relation of the respective mode in the unmodified waveguide. This mechanism gives rise to a very gentle reflection of the light at the boundaries and finally leads to very high Q factors of the resulting defect mode.

Because in our low-index material the influence of the light cone is much more severe than for semiconductors, a careful design for avoiding spectral components with small in-plane wavevectors is crucial even to obtain much lower Q values. Therefore we follow the approach of gentle confinement at a modal gap. However, in order to keep the translation symmetries of the underlying hexagonal crystal, which would be required in a photonic chip based on this lattice, we do not stretch the lattice but rather change the radius of the innermost holes of the waveguide. In order to raise the light cone as much as possible also a membrane is used here. A substrate with $n = 1.43$ is detrimental to the quality factor, as we shall see later. In the following a membrane with $n = 2.21$, which corresponds to $LiNbO_3$ neglecting the material anisotropy is used. Accounting for the anisotropy is not a problem and can be done afterwards by slightly adjusting the parameters.

In a first step the parameters of the 2D PhC membrane (thickness, lattice constant, radius) are optimised to obtain a large photonic bandgap for TE polarisation in the wavelength region of interest (1.55 μm) and a light cone as far as possible from the bandgap. A bandgap with a gap-to-midgap ratio of 18% is obtained (lattice constant $a = 600$ nm, hole radius $r = 190$ nm, membrane thickness $h = 500$ nm, see Figure 8.6). Then a W1-waveguide is introduced omitting one row of holes. In order to obtain a modal gap, we vary the radius R of the holes of the two rows adjacent to the waveguide to a maximum value of 220 nm. This leads to an increase of the frequency of the bands of this modified W1 (see Figure 8.6(b)). Intermediate values of the radius increase the frequency by an accordingly smaller amount and will be used for fine-tuning of the mode. One important idea behind our design is the easy coupling to a regular (unmodified) W1-waveguide. This means, for $R = 190$ nm, the waveguide should be transmitting at the resonance wavelength λ_0 of the defect. On the other

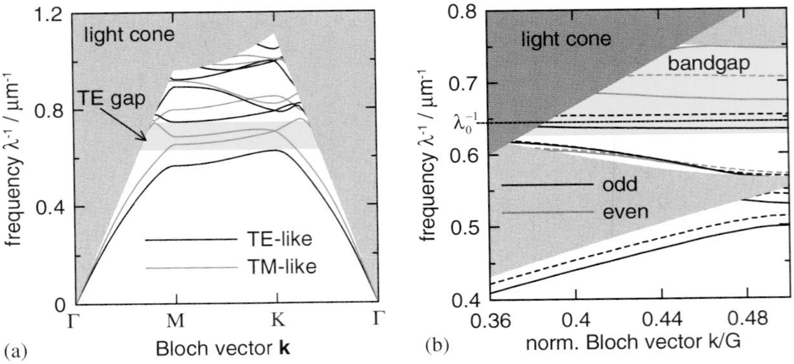

Figure 8.6 (a) Bandstructure of a 500 nm thick PhC membrane ($n = 2.211$) with triangular lattice of holes with $r = 190$ nm and $a = 600$ nm. (b) Dispersion relations of a W1 waveguide in this PhC membrane (solid lines) and of a modified W1 waveguide (radius R of innermost holes 220 nm, dashed lines). $G = 2\pi/a$ is the lattice vector, λ_0 the resonance wavelength.

hand, the modified waveguide should not be transmitting in order to obtain a model gap which confines the light. This can be achieved using values of R larger than 210 nm (see Figure 8.6(b)).

We performed a systematic study of different defects employing the 3D FDTD method where the radii of the two rows adjacent to the W1-waveguide were changed near the defect centre. A first result was that the Q factor could only be increased substantially using a defect mode spanning over three periods. For smaller modes (stronger confinement) the spatial Fourier components inside the light cone are too strong and the associated radiation limits Q [40], for our low-index membrane to values around 6,000. On the other hand, we need to ensure that the waveguide mode of the defect is outside the light cone, because only then a larger defect increases Q. Second, a low group velocity improves Q [42]. Hence, the best choice for the waveguide Bloch vector is $\mathbf{k}/G = 0.5$, where the lattice vector $G = 2\pi/a$ was introduced.

Furthermore, using holes with $R = 200$ nm at the defect centre is advantageous because it slightly increases the resonance frequency shifting it more into the bandgap. As a compromise between Q and modal volume we use two such holes on every side of the waveguide. Second, in order to have a high reflectivity at the mirrors based on the modal gap ("Bragg" mirrors), the radius of the next-nearest neighbour holes was gradually changed. This resulted in the following values: on both sides (left and right) of the defect centre (and on both sides of the W1, respectively) we first use one hole with $R = 210$ nm, then 4 holes with $R = 220$ nm, followed by one hole with $R = 210$ nm, one with $R = 200$ nm and a regular W1-waveguide with $R = r = 190$ nm. In the resulting structure we obtained from 3D FDTD calculations a cavity mode with $Q = 14{,}000$ shown in Figure 8.7. We also confirmed the possibility of exciting the mode via the W1 waveguide.

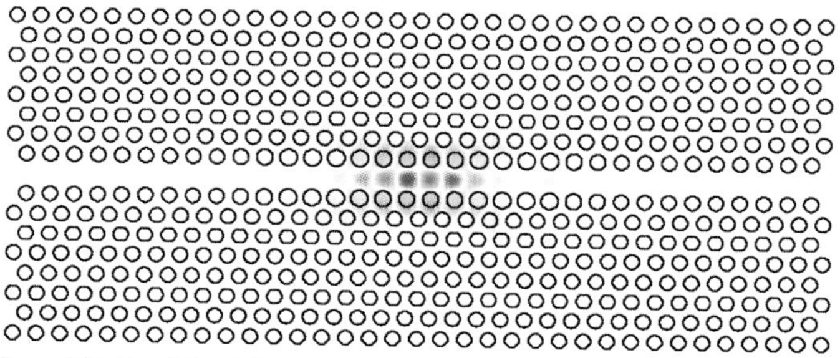

Figure 8.7 *Hz* of the defect mode with $Q = 14{,}000$ (parameters see text).

This Q factor is substantially smaller than for a semiconductor membrane, what can be explained with the different location of the light cone. Assuming a Gaussian spatial distribution of the fields in direction of the waveguide with a similar normalised width of $\sigma = \sqrt{2}a$ for both systems, we obtain a Gaussian normalised spatial spectral distribution $A(\boldsymbol{k}/G)$ of the individual fields, centred at $k_x/G = \pm 0.5$ and with a width $1/\sqrt{2}a$ [40]. From this value we can estimate to which magnitude the energy density has dropped at the boundary of the light cone, defined by $|\boldsymbol{k}/G| = n_{subst}a/\lambda_0$. The important difference between low-index and high-index membranes is the large difference in the normalised resonance frequency $\Omega_0 = a/\lambda_0$. In the semiconductor we have $\Omega_{semi} \approx 0.25$, whereas in our system $\Omega_0 = 0.39$. This results in an energy density larger by a factor of 50 for the low-index system, explaining the much smaller Q. Hence, to strongly increase Q one would have to use a much longer cavity or to try to decrease the frequency Ω_0 of the defect mode. However, in a low-index membrane, it is virtually impossible to obtain a photonic bandgap at much lower normalised frequencies. Still it can be expected that a slight increase of Q is possible by introducing more degrees of freedom, as shifting also holes and stretching the lattice, to further improve the modal gap reflectivity [42]. On the other hand, we also can see that using a substrate with $n_{subst} = 1.43$ is detrimental to Q because the light cone region is greatly enlarged.

To sum up, we presented a cavity mode design with a high Q factor at small modal volume, which intrinsically includes the waveguides for coupling. This cavity can be realised in a low-index membrane, and hence, can be used for quadratic nonlinear effects in $LiNbO_3$, as will be shown later.

8.3.3
Unusual Diffraction and Refraction Phenomena in Photonic Crystal Slabs

Recently a large deal of interest in photonic crystals shifted from the investigation of light localisation within the photonic bandgap to the utilisation of the unusual properties of light propagation in PhCs without defects [43–50]. These effects rely on the well-known fact, that the direction of beam propagation in homogeneous and periodic media, comprising nonabsorbing and nondispersive dielectrics, always points normal to the so-called isofrequency curves (IFC) of the dispersion relation [51]. In PhCs these IFCs can be tailored to attain various shapes. In the following two different cases are considered. First, if the IFC is flat over a wide spatial spectral range, the light can propagate only in the corresponding directions, their number given by the symmetry of the crystal. Self-guidance occurs in these directions of low IFC curvature, i.e., there is almost no diffraction of a finite beam. The second case of interest is a concave IFC, leading to negative refraction.

8.3.3.1 **Self-Collimated Light at Infrared and Visible Wavelengths**

Depending on the vertical index contrast, the TE- or TM-like bands where self-guiding occurs can be located inside the light cone, which results in huge scattering losses compared to defect waveguides operated inside the light cone, because here the entire field interacts with the lattice. Designs in semiconductor heterostructures with a low vertical index contrast utilise a square lattice with the first TM-like band completely inside the light cone [6]. Using a higher vertical index contrast this problem can be lifted, even with a lower in-plane index contrast [52]. Alternatively, the rectangular shaped IFCs of the second TE-like band below the light cone of a silicon membrane were used for low-loss guidance [53].

To date the operation domain of PhC films is almost exclusively restricted to the infrared spectral region. For various applications it might be appealing to extent this towards visible light. In the following it is demonstrated that almost diffractionless propagation for both TE- and TM-polarised light with wavelengths in the visible is feasible in homogeneous 2D PhC slabs. Related results of the performance of mirrors and splitters in the near infrared are presented elsewhere [54].

To allow for low-loss light propagation at visible wavelengths we use a SiN_x slab waveguide sandwiched between SiO_2 cladding and substrate. The thickness of the waveguiding layer amounts to 250 nm. Flat IFCs were obtained for $a = 320$ nm and $d = 210$ nm and a wavelength range between 790 nm and 860 nm (see Figure 8.8(b)).

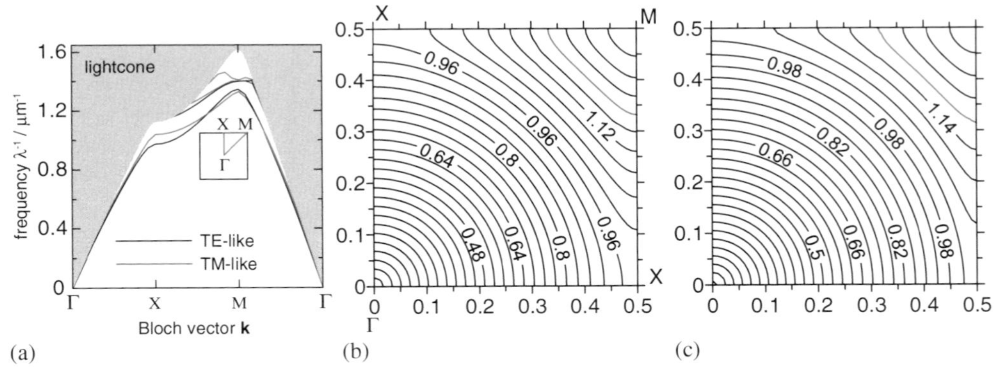

Figure 8.8 (a) Bandstructure diagram and iso-frequency curves for the 2D PhC slab with parameters $a = 320$ nm and $d = 210$ nm for (b) TE and (c) TM polarisation. The curves of good self-guidance at $\lambda = 830$ nm for TE and $\lambda = 845$ nm for TM are highlighted.

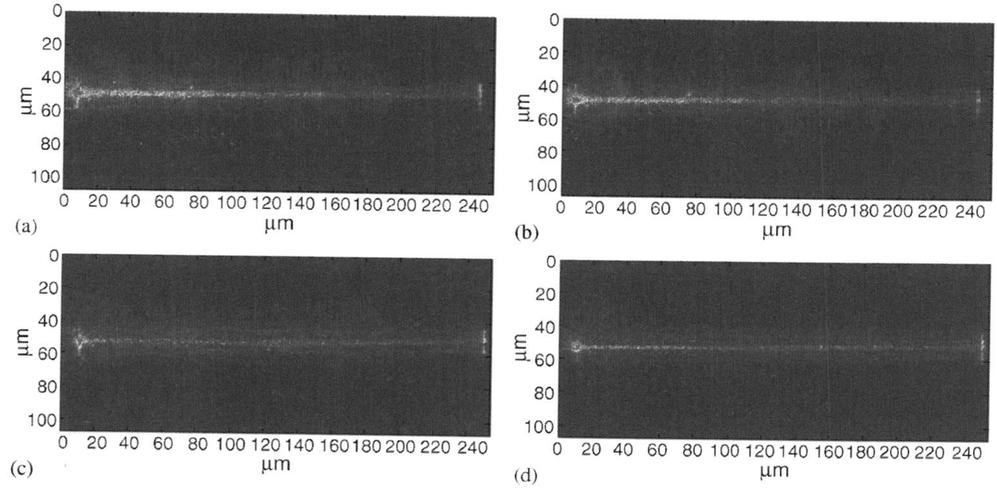

Figure 8.9 Stray light images of self-guiding of visible light in a PhC (period 320 nm, Ø 210 nm) for TE polarisation at 790 nm (a) and 800 nm (b) and for TM polarisation at 800 nm (c) and 810 nm (d).

Self-guiding in the same wavelength region can be expected for TM-polarised light (see Figure 8.8(c)). However, the wavelength for optimal guidance is about 10 nm larger.

Based on these calculations PhCs with a width of 25 μm and lengths of 60 μm, 120 μm and 240 μm were fabricated (see Section 8.2). The overall hole depth amounts to 650 nm. In order to couple beams of a defined width into the structure ridge waveguide tapers down to 1.8 μm were used. The self-guiding behaviour was monitored recording the top view stray light for different fixed wavelengths via a microscope objective (125×, NA = 0.8) and a CCD camera.

The stray light images revealed that for wavelengths around 800 nm self-guiding of light over a distance of 240 μm occurs (see Figure 8.9), with the best self-guidance at 790 nm for TE-polarisation and at 800 nm for TM-polarisation. Again, based on Gaussian optics, in an unstructured medium a beam with a spot size of 1.8 μm would spread up to a diameter of 75 μm. For a wavelength of 800 nm both TE- and TM-polarised light is simultaneously self-guided. Since the CCD-camera used for detecting the stray light did not show the necessary linearity regarding the detected power no assertion with respect to propagation losses is possible.

To sum up, we theoretically and experimentally demonstrated diffractionless propagation for both polarisations at visible wavelengths. The experimentally obtained wavelengths for optimal guidance are in excellent agreement with theory.

8.3.3.2 Negative Refraction of Light

For frequencies slightly larger than those required for diffractionless propagation the isofrequency curve becomes concave, leading to negative refraction (see, e.g., [45] for all-angle negative refraction in the first TE-band of a 2D square lattice or [55], where the fourth TE band of a 2D hexagonal lattice is used). The concave shape of the IFC leads also to negative diffraction due to the negative Gaussian curvature. A negative index of refraction at optical wavelengths was confirmed experimentally [56]. We found negative refraction for our substrate-based slab system in the first TE-like band for a hexagonal arrangement of air holes, where the IFC at the wavelengths of interest is a triangle with round corners. This again allows for operation below the light cone, and hence for a reduced loss. For a beam coming from an unstructured part and impinging on the interface to this PhC cut along ΓM we expect from the IFCs negative refraction for small angles of incidence from -9° to 9°. This relatively small angular range is mainly attributed to two facts. First, the (circular) IFC of the unstructured layer is larger than for air (compare with [45]). Second, the lower index of the dielectric reduces the size of the triangle. The construction of the refraction of waves is shown for an angle of incidence of 5° in Figure 8.10(c).

The negative refraction in our system is validated via full 3D FDTD calculations in a large computing window of $60\,\mu m \times 62\,\mu m \times 2.2\,\mu m$ (see Figure 8.10(a) and (c)). In addition, we find that the diffractive spreading of the beam increases with the angle of incidence due to the larger curvature of the IFC near the corner, leading to second and higher order diffraction. At an angle of 10° no light enters the crystal anymore. From the IFCs of the first TM-like band we expect a similar behaviour for slightly smaller wavelengths.

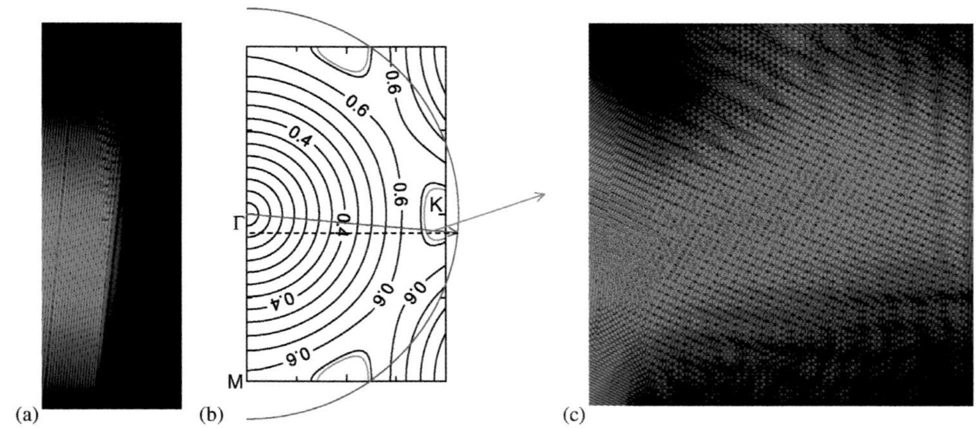

Figure 8.10 (a) TE polarised beam impinging on the interface from a homogeneous region at an angle of 5° with $\lambda = 1.52\,\mu m$ and (b) construction of the refraction at the interface (red curve: $\lambda = 1.52\,\mu m$, blue curve: unstructured layer at $\lambda = 1.52\,\mu m$, blue and red arrow: pointing vector of incoming and refracted waves) for the hexagonal crystal with parameters $a = 620$ nm, $r = 210$ nm. The refracted and diffracted beams in the crystal are shown in (c). The frequencies λ^{-1} are given in μm^{-1}.

First estimates from the isofrequency diagram give a minimum width of the incident beam of 12 μm in order to increase the diffraction length and to allow for a propagation of the beam inside the crystal over at least a few tens of lattice constants.

8.4 Light Propagation in Nonlinear Photonic Crystals

The combination of photonic crystals with nonlinearity is a very promising novel direction of investigation of growing interest, due to advances in microstructuring of nonlinear materials. The strong linear light control provided by photonic crystals can be used to realise nonlinear or active devices with a very small footprint [38,57]. In the following two different systems are investigated. First the feedback of a cavity is used to realise a microsized optical parametric oscillator, second one- and two-dimensional arrangements of microcavities are used to obtain strongly localised conservative solitons.

8.4.1 An Optical Parametric Oscillator in a Photonic Crystal Microcavity

One prominent parametric effect in a medium with quadratic nonlinearity is difference frequency generation (see, e.g., the textbook by Yariv [58]). Here, a strong pump at frequency ω_P interacts with a (usually) weak signal input at $\omega_S < \omega_P$ generating an idler wave at the difference frequency $\omega_I = \omega_P - \omega_S$ and amplifying the signal at the expense of the pump wave. Still, without signal input, the vacuum fluctuations of the optical field are amplified, an effect termed parametric fluorescence. When a resonator is added for providing positive feedback self-sustained oscillations can be obtained in a so-called optical parametric oscillator (OPO) above a certain threshold [58]. The frequencies ω_S and ω_I adjust themselves according to the lowest losses in the system. These types of OPOs were studied in the early days of nonlinear optics in conventional cavities [58–60]. Later the focus shifted to the investigation of the transverse dynamics in driven wide aperture resonators with quadratic media, including pattern formation [61–63] and cavity solitons [64,65]. Recently the spectral response of a four-wave mixing OPO realised in a finite semiconductor Kerr PhC without defect was computed by means of FDTD simulations [66]. Here we investigate theoretically a three-wave mixing OPO realised in a high-Q microcavity in a $LiNbO_3$ photonic crystal. First, we restrict ourselves to a 2D model. We simulate the transient behaviour of the oscillator for a stationary pump till we obtain a stationary solution. First results for a 3D extension are presented also.

In order to achieve a TE bandgap in $LiNbO_3$ we use a hexagonal lattice of cylindrical air holes and introduce a point defect by removing just one hole. In an isotropic material this kind of defect leads to the formation of two degenerate dipole defect modes [67] reducing the performance of the nonlinear interaction. However, due to the anisotropy of $LiNbO_3$ (uniaxial crystal with $n_e = 2.146$, $n_o = 2.220$) the degeneracy is lifted and we obtain two dipole modes with normalised frequencies $a/\lambda = 0.36200$ (fundamental, oriented along ΓM) and 0.36588 (higher order mode, oriented in ΓK, with the major component E_y, i.e., polarised in ΓM-direction) for a normalised hole radius $r/a = 0.3$ and an orientation of the optical axis of $LiNbO_3$ in ΓM-direction.

We choose to model the system with a nonlinear version of the FDTD method, where Maxwell's equations are discretised in time and space without further approximation. This treatment ensures that all effects of light propagation, e.g., field discontinuity at dielectric interfaces, any radiation losses, and local focusing due to nonlinear interactions, are properly accounted for. Moreover, the limits of frequently used approximations, e.g., the modal or mean-field approach, can be identified. To mimic radiation losses of a PhC film, which will be used to implement a real PhC OPO, and to obtain realistic Q-factors, we model a finite system, alternating 15 and 14 air holes in ΓK-direction (x-direction) and 13 such rows in the perpendicular ΓM-direction (y-direction). This gives $Q = 806$ for the fundamental mode and $Q = 5{,}150$ for the higher order mode of this finite system. Due to the mode symmetry and the orientation of the largest tensor component (d_{33}) in y, a linearly polarised pump couples best to the higher order mode when it is polarised in this direction [68]. In the following we use a continuous wave pump with an appropriate Gaussian spatial distribution, linearly polarised in y-direction and centred in the defect.

Because there is only one resonance we expect degenerate operation ($\omega_S = \omega_I$) for pump frequencies not too far from twice the resonance frequency ω_0. Hence, we use pump frequencies around $a/\lambda = 0.73176$. To quantify the detuning from resonance the normalised detuning $\delta = Q(\omega_P/\omega_0 - 2)$ is introduced. The temporal evolution of the system is calculated for three different detunings and for different pump fields. Above a certain pump threshold and after a certain time of propagation, we see a sudden onset of parametric oscillations. By inspecting the Fourier transform of the temporal evolution we confirm operation in the degenerate regime.

In a mean field approach a parametrically driven complex Ginzburg-Landau equation without transverse terms can be obtained [68], giving a parabolic shape of the respective bifurcation curves. Therefore we compare the FDTD results with a parabolic fit. The agreement for small pumps is excellent, as can be seen from Figure 8.11(a). However, for stronger pumps (Figure 8.11(b)) due to concurrent nonlinear effects as second harmonic generation, three-wave mixing, or optical rectification a growing deviation is found.

To check the feasibility of a microcavity OPO in a realistic system we use a defect based on a modal gap (see subsection 8.3.3.1) in a PhC membrane, where in-plane-coupling to the defect is intrinsically included. In this way more complex designs, as used for second harmonic generation in a semiconductor microcavity [69] can be avoided. However, in a first step we are interested in coupling from the top of the structure. Hence, a Gaussian beam polarised in ΓM-direction (y-direction) is excited above the defect and passes the membrane, part of it being reflected and another part interacting with the nonlinear membrane. Again, the largest diagonal nonlinear coefficient is oriented in y-direction. The defect mode is similar to the mode shown in Figure 8.7, but the number of altered hole radii in the reflection layer has been modified to allow for generated light leaving the structure via the W1 waveguides. Consequently, Q is reduced to 5,000. In Figure 8.12 the temporal evolution of the envelope of the pump and signal fields obtained for $\delta = 0$ from a 3D FDTD

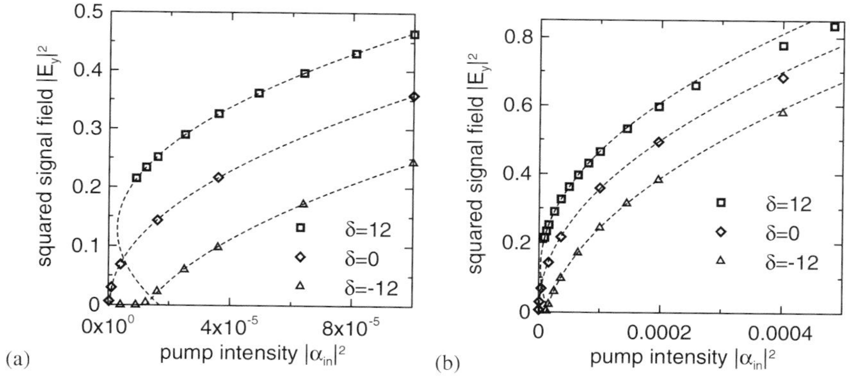

Figure 8.11 Data points obtained from FDTD calculations (symbols) for the signal field in dependence on the injected pump for different detunings. (a) is a zoom of (b) for pump intensities below 9×10^{-5}.

calculation is shown. One clearly sees the exponential growth of the signal and the oscillations before the stead-state is reached, where the pump field is weaker than before the oscillation started.

8.4.2 Discrete Solitons in Coupled Defects in Photonic Crystals

The investigation of discrete optical systems has been a subject of scientific interest for many years. Particular interest was on the propagation of monochromatic light in

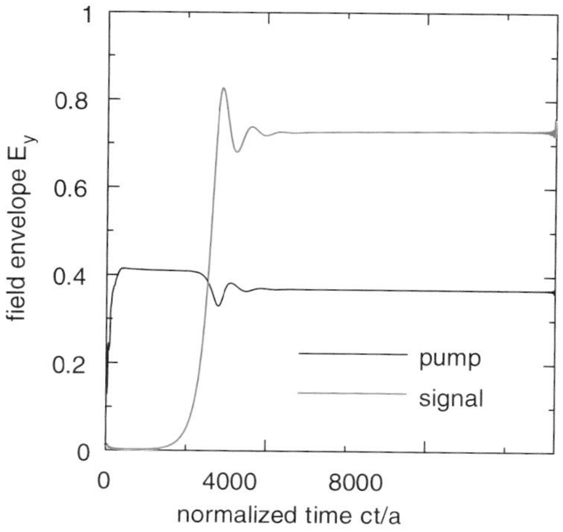

Figure 8.12 Temporal evolution of pump and signal field envelopes for $\delta = 0$ and well above threshold in the $LiNbO_3$ membrane.

linear and nonlinear coupled waveguides [70–75]. With the advance in experimental technology also 2D optical lattices could be realised [76–79]. In both 1D and 2D systems one subject of interest are discrete solitons. Recently, the concept of coupled optical resonators (microcavities) was introduced [80], where instead of the dynamics in propagation direction the temporal evolution is of interest.

Solitons are stable intrinsically localised solutions of the governing equations in a medium with a nonlinear response. This can be for instance a quadratic or saturable nonlinearity. We shall focus on the simplest case of a Kerr nonlinearity, where the refractive index changes linearly with the light intensity. In a homogenous medium these solutions depend on one or more continuous coordinates. In a waveguide array, within the framework of a modal approach, however, one assigns to every of the discrete waveguide sites one (or a few) amplitudes with a discrete coordinate corresponding to a linear mode of the waveguide (so-called tight-binding approximation). The resulting system of discrete equations can have localised solutions, termed *discrete solitons*. These solutions in one- [81,82] and two-dimensional [82] arrays of coupled nonlinear waveguides were studied in detail in the past. Recently, the coupled-mode approach (CMA) was also used to describe arrays of coupled defects (microcavities) in PhCs, and discrete soliton solutions were predicted [83].

A different description of soliton formation in these periodic media is possible with the nonlinear paraxial wave (Schrödinger) equation [74]. Here the amplitudes depend on two continuous spatial coordinates and the dielectric properties are periodic functions (lattice). Local effects within the waveguide, linear as well as nonlinear interaction with all neighbours and spatial inhomogeneities of the nonlinear response, are intrinsically included. Localised solutions of this equation are called *lattice solitons* [71,76]. However, the derivation of the Schrödinger equation relies on small refractive index changes. This model can be applied without problems to optically induced waveguide structures [84], because the induced index change is very small. In PhCs with a high index contrast or for high nonlinearly induced refractive index changes this continuous approach is inadequate.

In the following we investigate solitons in coupled defects in photonic crystals. Here the transverse light localisation at the defects is provided by a photonic bandgap. In order to achieve a 2D in-plane photonic bandgap for TE-polarised light a relatively large refractive index contrast of the order of $\Delta n \approx 0.5$ is required. In contrast, when the propagation is mainly perpendicular to the plane of periodicity, as in coupled waveguides, a transverse bandgap appears already for very weak index modulations [79]. Because we are interested in the temporal evolution of light in defects without components propagating out-of-plane, we use photonic crystals with a high-index contrast. In order to take all lattice effects as well as arbitrary large discontinuities into account we discretised Maxwell's equations directly in space and time without further approximations. Thus, the arising localised solutions may be termed Maxwell lattice solitons being a generalisation of both discrete and lattice solitons. With a 2D nonlinear FDTD method we identified one- and two-dimensional Maxwell lattice solitons of different topologies in defect arrays embedded in PhCs. We scanned the dispersion relation, studied the stability behaviour and compared the results to those obtained from the CMA-equations.

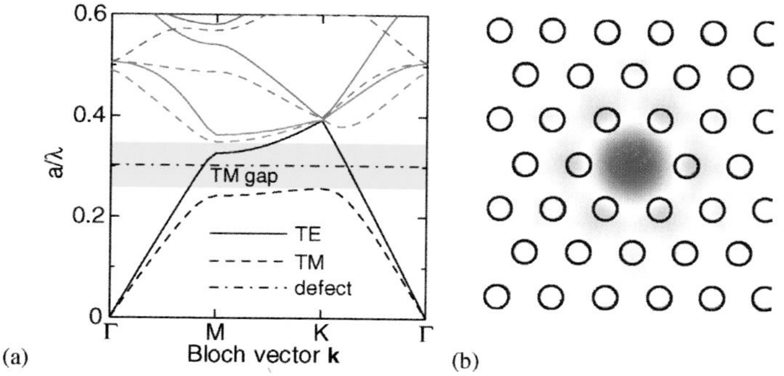

Figure 8.13 Bandstructure diagram (a) of the PhC and field (E_z) distribution of the defect mode (b).

For exploring Maxwell solitons we designed a 2D PhC model structure with nondegenerate single-moded defect cavities where the light localisation is provided by a photonic bandgap. Degenerate defect modes introduce more degrees of freedom and could possibly lead to vector solitons or cause instabilities. The model system comprises a hexagonal arrangement of high-index ($n = 3.4$) dielectric rods embedded in a low-index ($n = 1.45$) background (ratio radius/pitch $r/a = 0.22$) and has a bandgap for TM-polarisation (see Figure 8.13(a)).

We assume a homogeneous focusing Kerr nonlinearity. The microcavities are formed by introducing defects by removing rods. First we are interested in an approximate description of the system in tight-binding approximation. Using the reciprocity theorem in Fourier domain we obtain the governing discrete equations in nearest neighbour approximation in time domain

$$i\frac{\partial b_l}{\partial t} = -\frac{\omega_0}{2}\alpha(b_{l-1} + b_{l+1}) - \omega_0 d|b_l|^2 b_l, \tag{8.1}$$

where $b_l(t)$ are the time dependent modal amplitudes. For stationary soliton solutions we get the normalised system of equations

$$\Omega B_l + \mathrm{sgn}(\alpha)\frac{1}{2}(B_{l-1} + B_{l+1}) + \mathrm{sgn}(d)|B_l|^2 B_l = 0 \tag{8.2}$$

with the normalised amplitudes $b_l(t) = B_l\sqrt{|\alpha/\delta|}\exp(-i\,\Delta\omega t)$ and frequency $\Omega = \omega/|\alpha|\omega_0$. Bright solitons with evanescently decaying tails require a certain phase relation between neighbouring defects. Bright soliton solutions are either staggered, i.e., the phase difference between adjacent defects is π, for $\alpha > 0, \delta > 0$ or unstaggered (phase difference 0) for $\alpha > 0, \delta > 0$ [81]. Hence, Eq. (8.1) can be transformed to a real one by separating a trivial phase from all amplitudes.

The coefficients α and δ are obtained from the first- and third-order overlap integrals of the defect field. For a first check of the validity of the discrete equations we compared the obtained linear dispersion relation $\Delta\omega(K) = -\omega_0\alpha\cos(K\Lambda)$ of an infinite defect chain (Λ... defect spacing) with the results from rigorous plane wave supercell calculations. Depending on the defect spacing the coincidence varies indicating the influence of next-nearest neighbour interactions.

One feature not investigated so far is the possibility of a negative coupling constant for defect arrays in photonic crystals. Due to the Bragg reflection guidance mechanism the field components change their sign with the lattice period (see Figure 8.13(b)) and hence, also the overlap integral α alters its sign [85]. Thus simply by changing the defect spacing, the sign of the coupling constant can be adjusted. In a conventional waveguide this is usually not possible due to nonoscillatory tails. From (18.1) we see that a solution B_l for a given set of parameters (Ω,α,δ) is a solution for the parameter set $(-\Omega,-\alpha,-\delta)$ as well. Thus, for positive sign of the nonlinearity δ we can also find staggered solitons when $\alpha < 0$ because we have the same solutions as for $\alpha > 0$ and $\delta < 0$.

To investigate the Maxwell solitons, corresponding to the solutions obtained from the discrete equations, we calculated the temporal dynamics of appropriate pulsed excitations in finite linear defect chains with a nonlinear 2D FDTD method. To track the dispersion relation of the solitons, relating a typical amplitude B_0 to the soliton frequency, one field component was sampled at a given point. Due to the finite crystal size the soliton slowly loses energy upon propagation in time, and transforms adiabatically and continuously into a soliton with lower energy following the dispersion curve. From the sampled time series the dispersion curve $\Omega = \Omega(|B_0|^2)$ was extracted by means of spectral methods. In agreement with the discrete approximation we found staggered solitons for a defect spacing of two lattice constants ($\Lambda = 2a$, $\alpha = -0.02$) and unstaggered ones for $\Lambda = 3a$ ($\alpha = 0.003$).

As expected from CMA, for negative coupling constant the soliton is staggered, i.e. the phase difference between adjacent defects is π. From Figure 8.14 two remarkable differences between discrete and Maxwell lattice solitons can be observed. First, for small soliton centre power $|B_0|^2$, the dispersion relation departs from the CMA dispersion relation and the soliton is finally cut off. This is an effect of the finite PhC size because the soliton gets wider and hits the PhC boundaries. For high soliton power the excitation will be ultimately located at a single defect. Thus the dispersion relation of the defect (green line) and the soliton should almost coincide. This is only true for the Maxwell soliton (red line) but not for the discrete soliton (black line). The reason is that the CMA does not account for defect mode shape variations, which come obviously into play for high power.

We have also verified the existence of Maxwell solitons in a two-dimensional defect lattice. Here the most stable solution is the fundamental on-site soliton (see Figure 8.15). Again the expected differences between Maxwell and discrete solitons appear for very wide and narrow high-power solitons. However, in contrast to the 1D case, here the soliton disappears before it hits the boundaries. This can be seen from a stability analysis of the CMA equations which give instability for $\Omega > -1.7$ (see Figure 8.15(b)). Introducing the total power (energy) as sum over all defects, $W = \sum_i |B_i|^2$, we find the minimum at this point, suggesting that the Vakhitov–Kolokolov criterion [86] applies.

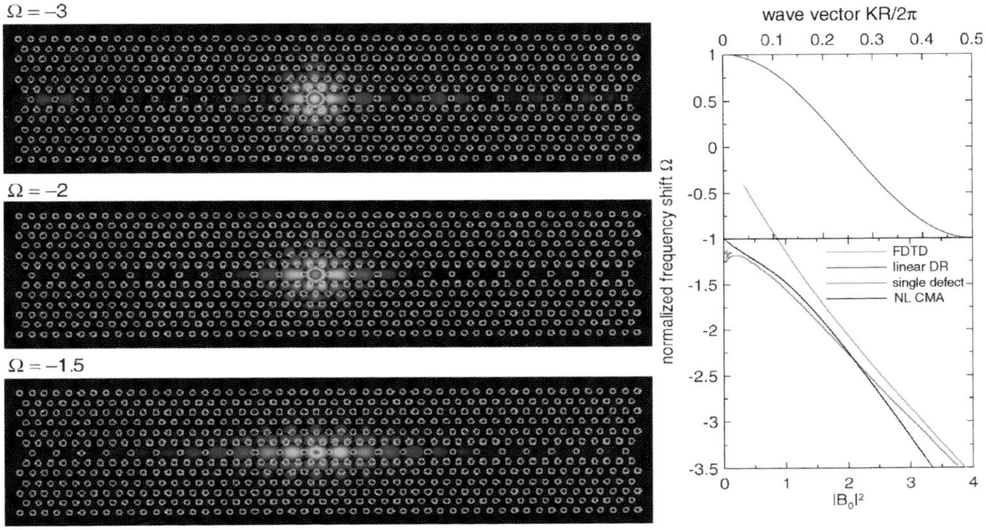

Figure 8.14 Magnetic field of the fundamental Maxwell soliton in the defect chain with $\Lambda = 2a$ at different stages of the propagation, i.e., for different soliton parameters (left, with the calculated normalised frequency Ω) and corresponding full dispersion relation (right), both obtained from 2D FDTD calculations.

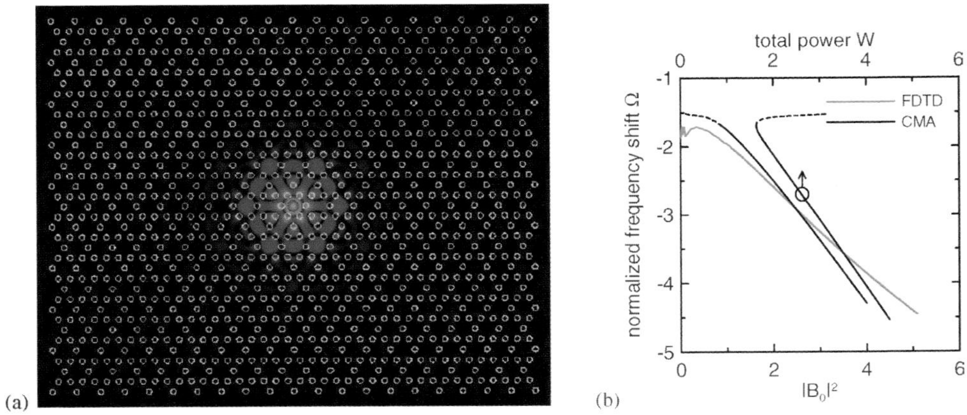

Figure 8.15 (a) Snapshot of a FDTD propagation of the fundamental two-dimensional soliton and (b) soliton dispersion relation.

8.5 Conclusion

We presented theoretical and experimental results for photonic crystals made of different low-index materials. Furthermore, we investigated theoretically spatial solitons in 1D and 2D defect lattices in photonic crystals.

The transmission in the highly dispersive regions of a W1-waveguide is significantly increased by utilising a mini-stopband below all 2D Bloch modes of the underlying photonic crystal, and hence, below the light cone. We proposed a high-Q (14,000) microcavity in a low-index photonic crystal membrane and investigated the limiting parameters compared to semiconductors. We proposed theoretically and verified experimentally self-collimation of TE- and TM-polarised visible light at the same wavelength. The feasibility of negative refraction in the low-index material was demonstrated theoretically and the minimum width and maximum inclination of the incident beam was estimated. By means of nonlinear FDTD calculations we obtained the bifurcation diagrams for a degenerate, doubly resonant OPO realised in a photonic crystal microcavity in lithium niobate. A first proof of principles was given for an OPO in a 3D membrane system. We investigated strongly localised solitons in 1D and 2D lattices of point defects in nonlinear photonic crystals with high-index inclusions. Staggered as well as unstaggered solitons for a focusing Kerr nonlinearity due to the unique Bragg localisation mechanism were obtained.

Acknowledgements

This work was supported by the Deutsche Forschungsgemeinschaft within the framework of the priority programme 1113 "Photonic Crystals" (Grants Nos. Le 755/8-1,2,3).

References

1 Krauss, T.F., De La Rue, R.M. and Brand, S. (1996) *Nature*, **383**, 699.

2 Smith, C.J.M., Benisty, H., Olivier, S., Rattier, M., Weisbuch, C., Krauss, T.F., De La Rue, R.M., Houdré, R. and Oesterle, U. (2000) *Appl. Phys. Lett.*, **77**, 2813.

3 Zimmermann, J., Kamp, M., Schwertberger, R., Reithmaier, J.P., Forchel, A. and März, R. (2002) *Electron. Lett.*, **38**, 178.

4 Olivier, S., Benisty, H., Weisbuch, C., Smith, C.J.M., Krauss, T.F., Houdré, R. and Oesterle, U. (2002) *J. Lightwave Technol.*, **20**, 1198.

5 Mulot, M., Qiu, M., Swillo, M., Jaskorzynska, B., Anand, S. and Talneau, A. (2003) *Appl. Phys. Lett.*, **83**, 1095.

6 Wu, L., Mazilu, M., Gallet, J.-F. and Krauss, T.F. (2003) *Phot. Nanostruct. Fund. Appl.*, **1**, 31.

7 Notomi, M., Yamada, K., Shinya, A., Takahashi, J., Takahashi, C. and Yokohama, I. (2001) *Phys. Rev. Lett.*, **87**, 253902.

8 Yoshihiro Akahane, Takashi Asano, Bong-Shik Song and Susumu Noda (2003) *Nature*, **425**, 944.

9 Song, B.-S., Noda, S., Asano, T. and Akahane, Y. (2005) *Nature Mater*, **4**, 207.

10 Assefa, S., McNab, S.J. and Vlasov, Y.A. (2005) *Opt. Lett.*, **31**, 745.

11 Dulkeith, E., McNab, S.J. and Vlasov, Y.A. (2005) *Phys. Rev. B*, **72**, 115102.

12 Baumberg, J.J., Perney, N.M.B., Netti, M.C., Charlton, M.D.C., Zoorob, M. and Parker, G.J. (2004) *Appl. Phys. Lett.*, **85**, 354.

13 Netti, M.C., Charlton, M.D.B., Parker, G.J. and Baumberg, J.J. (2000) *Appl. Phys. Lett.*, **76**, 991.

14 Crozier, K.B., Lousse, V., Kilic, O., Kim, S., Fan, S. and Solgaard, O. (2006) *Phys. Rev. B*, **73**, 115126.

15 Augustin, M., Böttger, G., Eich, M., Etrich, C., Fuchs, H.-J., Iliew, R., Hübner, U., Kessler, M., Kley, E.-B., Lederer, F., Liguda, C., Nolte, S., Meyer, H.G., Morgenroth, W., Peschel, U., Petrov, A., Schelle, D., Schmidt, M., Tünnermann, A. and Wischmann, W. (2004) in *Photonic Crystal Optical Circuits in Moderate Index Materials*, (eds Kurt Busch, Stefan Lölkes, Ralf B. Wehrspohn and Helmut Föll), *Photonic Crystals – Advances in Design, Fabrication, and Characterization*, Wiley-VCH, Weinheim, p. 289.

16 Augustin, M., Fuchs, H.-J., Schelle, D., Kley, E.-B., Nolte, S., Tünnermann, A., Iliew, R., Etrich, C., Peschel, U. and Lederer, F. (2003) *Opt. Express*, **11**, 3284.

17 Augustin, M., Fuchs, H.-J., Schelle, D., Kley, E.-B., Nolte, S., Tünnermann, A., Iliew, R., Etrich, C., Peschel, U. and Lederer, F. (2004) *Appl. Phys. Lett.*, **84**, 663.

18 Tünnermann, A., Schreiber, T., Augustin, M., Limpert, J., Will, M., Nolte, S., Zellmer, H., Iliew, R., Peschel, U. and Lederer, F. (2004) *Adv. Solid State Phys.*, **44**, 117.

19 Augustin, M., Iliew, R., Fuchs, H.-J., Peschel, U., Kley, E.-B., Nolte, S., Lederer, F. and Tünnermann, A. (2004) *Jpn. J. Appl. Phys. Part 1*, **43** (8B), 5805.

20 Bogaerts, W., Bienstman, P. and Baets, R. (2003) *Opt. Lett.*, **28**, 689.

21 Schmidt, M., Böttger, G., Eich, M., Morgenroth, W., Hübner, U., Boucher, R., Meyer, H.G., Konjhodzic, D., Bretinger, H. and Marlow, F. (2004) *Appl. Pys. Lett.*, **85**, 16.

22 Böttger, G., Schmidt, M., Eich, M., Boucher, R. and Hübner, U. (2005) *J. Appl. Phys.*, **98**, 103101.

23 Roussey, M., Bernal, M.-P., Courjal, N. and Baida, F.I. (2005) *Appl. Phys. Lett.*, **87**, 241101.

24 Schrempel, F., Gischkat, T., Hartung, H., Kley, E.-B. and Wesch, W. (2006) *Nucl. Instrum. Methods Phys. Res. B*, **250**, 164.

25 John, S. (1987) *Phys. Rev. Lett.*, **58**, 2486.

26 Yablonovitch, E. (1987) *Phys. Rev. Lett.*, **58**, 2059.

27 Russell, P.St.J. (1984) *Electron Lett.*, **20**, 72.

28 Russell, P.St.J. (1986) *Appl. Phys. B*, **39**, 231.

29 Zengerle, R. (1987) *J. Mod. Opt.*, **34**, 1589.

30 Notomi, M., Shinya, A., Mitsugi, S., Kuramochi, E. and Ryu, H.-Y. (2004) *Opt. Express*, **12**, 1551.

31 Olivier, S., Rattier, M., Benisty, H., Weisbuch, C., Smith, C.J.M., De La Rue, R.M., Krauss, T.F., Oesterle, U. and Houdré, R. (2001) *Phys. Rev. B*, **63**, 113311.

32 Johnson, S.G. and Joannopoulos, J.D. (2001) *Opt. Express*, **8**, 173.

33 Mori, D. and Baba, T. (2005) *Opt. Express*, **13**, 9398.

34 Gersen, H., Karle, T.J., Engelen, R.J.P., Bogaerts, W., Korterik, J.P., van Hulst, N.F., Krauss, T.F. and Kuipers, L. (2005) *Phys. Rev. Lett.*, **94**, 073903.

35 Taflove, A. and Hagness, S.C. (2000) *Computational Electrodynamics: The Finite-Difference Time-Domain Method*, Artech House, Boston, MA.

36 Augustin, M., Iliew, R., Etrich, C., Setzpfandt, F., Fuchs, H.-J., Kley, E.-B., Nolte, S., Pertsch, T., Lederer, F. and Tünnermann, A. (2006) *New J. Phys.*, **8**, 210.

37 Yamamoto, Y. and Machida, S. and Björk, G. (1991) *Phys. Rev. A*, **44**, 657.

38 Altug, H., Englund, D. and Vučković, J. (2006) *Nature Physics*, **2**, 484.

39 Asano, T., Song, Bong-Shik, Akahane, Y. and Noda, S. (2006) *IEEE J. Sel. Top. Quantum Electron.*, **12**, 1123.

40 Englund, D., Fushman, I. and Vučković, J. (2005) *Opt. Express*, **13**, 5961.

41 Akahane, Y., Asano, T., Song, Bong-Shik and Noda, S. (2003) *Appl. Phys. Lett.*, **83**, 1512.

42 Sauvan, C., Lalanne, P. and Hugonin, J.P. (2005) *Phys. Rev. B*, **71**, 165118.

43 Notomi, M. (2000) *Phys. Rev. B*, **62**, 10696.

44 Gralak, B., Enoch, S. and Tayeb, G. (2000) *J. Opt. Soc. Am. A*, **17**, 1012.

45 Luo, Chiyan Johnson, S.G., Joannopoulos, J.D. and Pendry, J.B. (2002) *Phys. Rev. B*, **65**, 201104.

46 Luo, C., Johnson, S.G. and Joannopoulos, J.D. (2002) *Appl. Phys. Lett.*, **81**, 2352.

47 Cubukcu, E., Aydin, K., Ozbay, E., Foteinopoulou, S. and Soukoulis, C.M. (2003) *Nature*, **423**, 604.

48 Luo, C., Johnson, S.G., Joannopoulos, J.D. and Pendry, J.B. (2003) *Phys. Rev. B*, **68**, 045115.

49 Foteinopoulou, S. and Soukoulis, C.M. (2005) *Phys. Rev. B*, **72**, 165112.

50 Iliew, R., Etrich, C. and Lederer, F., (2005) *Opt. Express*, **13**, 7076.

51 Yeh, P. (1979) *J. Opt. Soc. Am.*, **69**, 742.

52 Iliew, R., Etrich, C., Peschel, U., Lederer, F., Augustin, M., Fuchs, H.-J., Schelle, D., Kley, E.-B., Nolte, S. and Tünnermann, A. (2004) *Appl. Phys. Lett.*, **85**, 5854.

53 Prather, D.W., Shi, S., Pustai, D.M., Chen, C., Venkataraman, S., Sharkawy, A., Schneider, G.J. and Murakowski, J. (2004) *Opt. Lett.*, **29**, 50.

54 Augustin, M., Schelle, D., Fuchs, H.-J., Nolte, S., Kley, E.-B., Tünnermann, A., Iliew, R., Etrich, C., Peschel, U. and Lederer, F. (2005) *Appl. Phys. B, Lasers Optics*, **81**, 313.

55 Foteinopoulou, S. and Soukoulis, C.M. (2003) *Phys. Rev. B*, **67**, 235107.

56 Berrier, A., Mulot, M., Swillo, M., Qiu, M., Thylén, L., Talneau, A. and Anand, S. (2004) *Phys. Rev. Lett.*, **93**, 073902.

57 Yoshie, T., Scherer, A., Hendrickson, J., Khitrova, G., Gibbs, H.M., Rupper, G., Ell, C., Shchekin, O.B. and Deppe, D.G. (2004) *Nature*, **432**, 200.

58 Yariv, A. (1989) *Quantum Electron*. John Wiley & Sons, Inc.

59 Harris, S.E. (1969) *Proc. IEEE*, **57** (12), 2096.

60 Smith, R. (1973) *IEEE J. Quantum Electron.*, **9**, 530.

61 Oppo, G.-L., Brambilla, M. and Lugiato, L.A. (1994) *Phys. Rev. E*, **49**, 2028.

62 Staliunas, K. (1995) *J. Mod. Opt.*, **42**, 1261.

63 Etrich, C., Michaelis, D. and Lederer, F. (2001) *J. Opt. Soc. Am. B*, **19**, 792.

64 Staliunas, K. and Sánchez-Morcillo, V.J. (1997) *Opt. Commun.*, **139**, 306.

65 Etrich, C., Peschel, U. and Lederer, F. (1997) *Phys. Rev. Lett.*, **79**, 2454.

66 Conti, C., Falco, A.D. and Assanto, G. (2004) *Opt. Express*, **12**, 823.

67 Vučković, J., Painter, O., Xu, Yong Yariv, A. and Scherer, A. (1999) *IEEE J. Quantum Electron.*, **35**, 1168.

68 Iliew, R., Etrich, C., Peschel, U. and Lederer, F. (2006) *IEEE J. Sel. Top. Quantum Electron.*, **12**, 377.

69 Di Falco, A., Conti, C. and Assanto, G. (2006) *Opt. Lett.*, **31**, 250.

70 Christodoulides, D.N. and Joseph, R.I. (1988) *Opt. Lett.*, **13**, 794.

71 Eisenberg, H.S., Silberberg, Y., Morandotti, R., Boyd, A.R. and Aitchison, J.S. (1998) *Phys. Rev. Lett.*, **81**, 3383.

72 Pertsch, T., Dannberg, P., Elflein, W., Bräuer, A. and Lederer, F. (1999) *Phys. Rev. Lett.*, **83**, 4752.

73 Kobyakov, A., Darmanyan, S., Pertsch, T. and Lederer, F. (1999) *J. Opt. Soc. Am. B*, **16**, 1737.

74 Christodoulides, D.N. and Eugenieva, E.D. (2001) *Phys. Rev. Lett.*, **87**, 233901.

75 Iwanow, R., Schiek, R., Stegeman, G.I., Pertsch, T., Lederer, F., Min, Y. and Sohler, W. (2004) *Phys. Rev. Lett.*, **93**, 113902.

76 Fleischer, J.W., Carmon, T., Segev, M., Efremidis, N.K. and Christodoulides, D.N. (2003) *Phys. Rev. Lett.*, **90**, 023902.

77 Pertsch, T., Peschel, U., Kobelke, J., Schuster, K., Bartelt, H., Nolte, S., Tünnermann, A. and Lederer, F. (2004) *Phys. Rev. Lett.*, **93**, 053901.

78 Trompeter, H., Krolikowski, W., Neshev, D.N., Desyatnikov, A.S., Sukhorukov, A.A., Kivshar, Y.S., Pertsch, T., Peschel, U. and Lederer, F. (2006) *Phys. Rev. Lett.*, **96**, 053903.

79 Fischer, R., Träger, D., Neshev, D.N., Sukhorukov, A.A., Krolikowski, W., Denz, C. and Kivshar, Y.S. (2006) *Phys. Rev. Lett.*, **96**, 023905.

80 Yariv, A., Xu, Y. and Scherer, A. (1999) *Opt. Lett.*, **24**, 711.

81 Lederer, F. Kobyakov, A. Darmanyan, S. (2001) in *Discrete Solitons, Spatial Solitons, Springer Series in Optical Science*, (eds S. Trillo and W.E. Torruellas), Vol. 82, Springer-Verlag, Berlin, p. 269–292.

82 Christodoulides, D.N., Lederer, F. and Silberberg, Y. (2003) *Nature*, **424**, 817.

83 Christodoulides, D.N. and Efremidis, N.K. (2002) *Opt. Lett.*, **27**, 568.

84 Efremidis, N.K., Sears, S., Christodoulides, D.N., Fleischer, J.W. and Segev, M. (2002) *Phys. Rev. E*, **66**, 046602.

85 Iliew, R., Etrich, C., Peschel, U. and Lederer, F. (2004) in *Nonlinear Guided Waves and Their Applications on CD-ROM*, The Optical Society of America, Washington, DC. TuC44.

86 Vakhitov, M.G. and Kolokolov, A.A. (1973) *Radiophys. Quantum Electron.*, **16**, 783.

9
Linear and Non-linear Optical Experiments Based on Macroporous Silicon Photonic Crystals

Ralf B. Wehrspohn, Stefan L. Schweizer, and Vahid Sandoghdar

9.1
Introduction

From the beginning of research on photonic crystals, one of the major areas of investigation concerned two-dimensional (2D) structures [1]. This was mainly caused by experimental reasons as the fabrication of 3D photonic crystals is more difficult and cumbersome than that of 2D photonic crystals. However an ideal 2D photonic crystal consists of a periodic array of infinitely long pores or rods so that a structure which approximates this theoretical model has to exhibit very high aspect ratios (ratio between pore/rod length to pore/rod diameter). Using conventional dry etching techniques only structures with aspect ratios up to 10–30 are possible. To avoid scattering of light out of the plane of periodicity and to reduce the corresponding loss the so-called slab structures were developed and thoroughly investigated [2,3]. In such low-aspect structures, one relies on guiding of light by total internal reflection in the third dimension and, consequently, deals with a full 3D problem. As an alternative approach, Lehmann and Gruning [4,5], as well as Lau and Parker [6] proposed macroporous silicon as a model system for 2D photonic crystals. This system consists of a periodic array of air pores in silicon. The fundamental bandgap appears in general for wavelengths which are approximately twice the lattice constant, the pores are 50–250 times longer than the wavelengths of the corresponding 2D fundamental bandgap. Therefore, macroporous silicon represents an excellent system to study ideal 2D photonic crystal properties. Typically, high-quality photonic crystals with lattice constant of $a = 500$ nm to 8000 nm and a depth up to 1 mm can be produced with this process. These structures exhibit photonic bandgaps from the near infrared to the far infrared. Possible applications include miniaturized sensors or selective thermal emitter [7]. In the following, we present our improved fabrication technologies for macroporous silicon and novel optical experiments on beaming, near-field optical and non-linear optics.

Nanophotonic Materials: Photonic Crystals, Plasmonics, and Metamaterials. Edited by R. B. Wehrspohn, H.-S. Kitzerow, and K. Busch

ISBN: 978-3-527-40858-0

9.2
Fabrication of 2D Photonic Crystals

9.2.1
Macroporous Silicon Growth Model

The etching method used in this work to produce trenches and ordered arrays of macropores in Si is based on photo-electrochemical etching (PECE) of silicon. This process is revisited briefly below following the model for macropore formation by Lehmann [8,9].

While Si is easily etched in aqueous alkaline solutions it is quite stable in most aqueous acids. However, hydrofluoric acid (HF) is an exception to this general observation. Figure 9.1a shows the current density j across the HF/n-Si interface versus the applied voltage V for different illumination conditions. In the regime of cathodic currents (I in Figure 9.1a) the Schottky-like HF/n-Si contact is forward biased. The current is determined by the majority charge carriers, i.e., the electrons – independent of the illumination state – and leads to the reduction of the H^+ ions in the acidic solution followed by formation of molecular hydrogen (H_2).

The regime of anodic currents (II and III in Figure 9.1a) is the more interesting one. Figure 9.1b shows the arrangement of the valence and conduction bands, respectively, within n-Si in contact with HF. When the semiconductor Si is brought into contact with the electrolyte HF the situation resembles a Schottky contact in which the rather conductive electrolyte represents the metal. The different chemical potentials of the aqueous HF and the Si will adapt. This leads to the formation of a Helmholtz double layer in the electrolyte and a surface charge resulting from the ionized donor atoms in the Si from which a depletion of majority charge carriers (electrons in n-Si) at the HF/Si interface follows. Due to the mobile ions in the electrolyte the width of the Helmholtz double layer is only a few nm while due to the

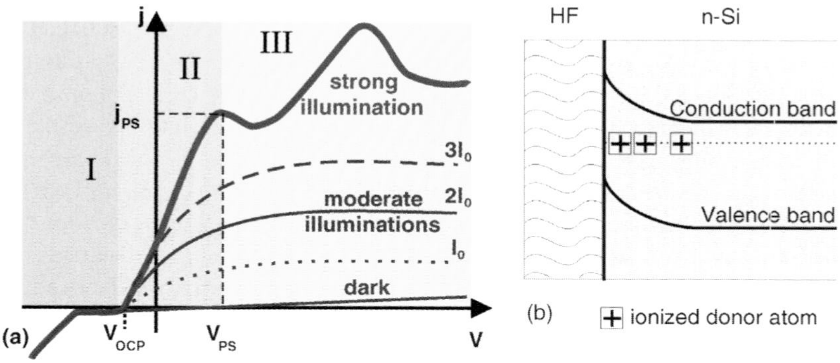

Figure 9.1 (a) Schematic plot of current density across the HF/Si interface for n-Si under no (violet), medium (blue) and strong (red) illumination. (b) Unbiased HF/n-Si Schottky-like contact.

stationary nature of the donors in Si the depletion region in n-Si is on the order of a few μm wide. If no illumination is applied to the n-Si, a small (anodic) dark current resulting from thermally generated holes is observed (violet line Figure 9.1a). If the Si is strongly illuminated the HF/n-Si contact behaves like an HF/p-Si contact (red line in Figure 9.1a). An increase in applied voltage V leads to an increase in current across the HF/Si interface. For anodic currents below the critical current density j_{PS} and $V < V_{PS}$ (II in Figure 9.1a) divalent[1] dissolution of Si occurs along with the formation of hydrogen. Here the etching current is limited by charge carrier supply from the Si electrode and porous Si is formed. A suggested reaction is

$$Si + 4HF_2^- + h^+ \rightarrow SiF_6^{2-} + 2HF + H_2 + e^- \tag{9.1}$$

with e^- and h^+ denoting an electron and a hole in the Si. In aqueous HF electrolytes the critical current density j_{PS} was experimentally found to only depend on electrolyte concentration c_{HF} (in wt%) and electrolyte temperature T_{HF} and can be described by

$$j_{PS} = C c_{HF}^{3/2}\, e^{-E_a/(k_B T_{HF})} \tag{9.2}$$

with C being a constant of 3300 A/cm^2, $E_a = 0.345$ eV and k_B being Boltzmann's constant [9]. For anodic currents and $V > V_{PS}$ (III in Figure 9.1a) tetravalent dissolution of Si is observed. In a first step, under consumption of 4 holes, an anodic oxide is formed on the Si electrode:

$$Si + 2H_2O + 4h^+ \rightarrow SiO_2 + 4H^+. \tag{9.3}$$

In a second step this oxide is then chemically etched by the fluorine species HF, $(HF)_2$ or HF_2^- in the electrolyte [9]

$$SiO_2 + 2HF_2^- + 2HF \rightarrow SiF_6^{2-} + 2H_2O. \tag{9.4}$$

Here the current is limited by the chemical reaction rate during the removal of the SiO_2. As a consequence the Si electrode is electropolished, i.e., all Si surface atoms are removed uniformly. For medium illumination intensities the IV-curve of the HF/n-Si is similar to the blue curves in Figure 9.1a. Here the current density j across the interface is below j_{PS}. j is limited by charge supply from the Si electrode and therefore porous Si is formed. The electronic holes necessary for the dissolution of Si at the HF/Si interface are created by illuminating the sample with light energy $E_\nu = h\nu \geq E_{g,Si} = 1.1$ eV. Due to the high absorption of Si for the IR light used ($\lambda \approx 880$ nm, $a \approx 10^2$ cm^{-1} [9]) electron hole pairs are produced within the first few μm from the air/Si interface. For the formation of porous Si it is necessary that the current density across the HF/Si interface is smaller than j_{PS}. According to the diameter d_{pore} of the pores three regimes are distinguished: microporous Si with $0\,nm \leq d_{pore} \leq 2\,nm$, mesoporous Si with $2\,nm < d_{pore} \leq 50\,nm$ and macroporous Si with $50\,nm < d_{pore}$. For p-Si $j < j_{PS}$ can only be fulfilled for potentials $V_{OCP} < V < V_{PS}$. For the n-Si used in this work $j < j_{PS}$ can be achieved for potentials $V_{OCP} < V$ by appropriate adjustment of the illumination. Stable macropore growth is possible for

[1] Divalent (tetravalent) means that in the external electrical circuit 2 (4) electrons are necessary for the removal of one Si atom from the electrode.

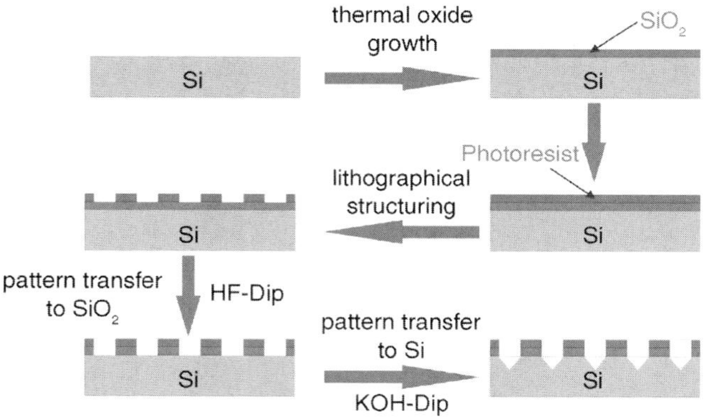

Figure 9.2 Process steps for lithographical prestructuring of a Si-wafer for subsequent photo-electrochemical etching.

$j < j_{PS}$ and $V_{PS} < V$. The ratio j/j_{PS} only controls the average porosity p of the sample. The x-y-positions as well as the diameters of individual pores show a random distribution under the constraint of the average porosity being $p = j/j_{PS}$. For ordered arrays of pores in the hexagonal or square lattice, a periodic pattern on top of the Si wafer is defined lithographically and subsequently transferred into the Si using a silicon oxide mask as shown in Figure 9.2. The silicon oxide thickness is adjusted between 10 nm to 200 nm depending on the interpore spacing. By this procedure, so-called *etch-pits* in the form of inverse pyramids are generated which serve as starting points for the subsequent pore growth. The porosity of such an ordered macropore array is given by

$$p = \frac{j}{j_{PS}} = \frac{A_{Pores}}{A_{Sample}}, \tag{9.5}$$

with A_{Pores} being the total pore area and A_{Sample} the total HF/Si interface area. Figure 9.3 schematically shows the principle of photo-electrochemically etching ordered macropore arrays. Electron–hole pairs are generated by appropriate illumination of the back of the n-Si wafer. Due to the anodic potential the electrons are sucked away into the voltage source while the holes diffuse through the wafer towards the HF/Si interface where a space charge region (SCR) has formed. To ensure that the holes can reach the HF/Si interface high quality float-zone Si has to be used in which the diffusion length of the holes is on the order of the thickness of the Si wafer. The shape of the SCR follows the physical shape at the interface and is therefore curved. Because the electric field lines are perpendicular to the HF/Si interface the electronic holes that come into the vicinity of the pore tips are focused onto the pore tips where they promote the dissolution of Si. The width x_{SCR} of the SCR depends on the applied anodic voltage and can be described by [9]

$$x_{SCR} = \sqrt{\frac{2\varepsilon_0 \varepsilon_{Si} V_{eff}}{q N_D}}, \tag{9.6}$$

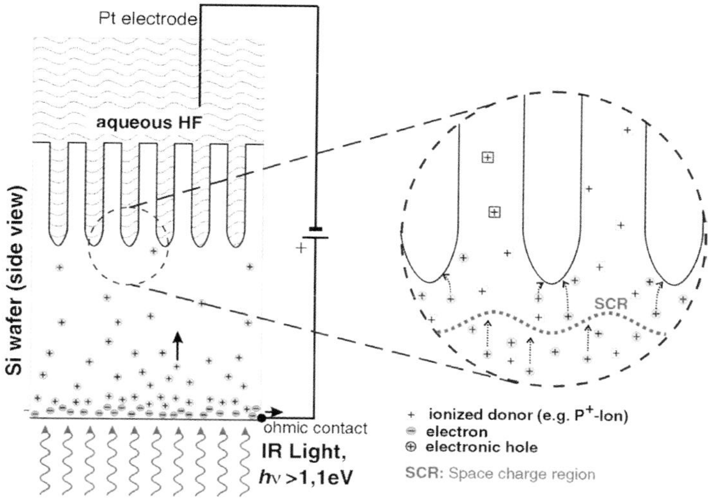

Figure 9.3 Schematic diagram explaining the formation of ordered macropores by photo-electrochemical etching of n-Si under backside illumination using an anodic potential.

where ε_0 is the free space permittivity, ε_{Si} is the dielectric constant of Si, N_D the doping density of Si and $V_{eff} = V_{bi} - V - k_B T/e$ is the effective potential difference between the electrolyte and the Si anode. $V_{bi} \approx 0.5$ V represents the built-in potential of the contact, V the applied external potential ($k_B T \approx 25$ meV at room temperature). The applied anodic bias has to be chosen high enough such that all of the incoming electronic holes are focused onto the pore tips and none of them penetrate into the Si wall remaining between two pores. If this condition is fulfilled the pore walls are passivated against dissolution. The porosity of such an ordered array of macropores with radius r_{Pore}, e.g., arranged in a hexagonal lattice with lattice constant a, can be expressed as

$$p_{hex} = \frac{2\pi}{\sqrt{3}} \left(\frac{r_{Pore}}{a}\right)^2. \tag{9.7}$$

The growth speed of the pores along the [100] direction in the model of Lehmann only depends on the temperature T_{HF} and the concentration c_{HF} of the electrolyte and can be described by [9]

$$v_{P100} = \frac{j_{PS}}{n_{Val} q N_{Si}}, \tag{9.8}$$

with N_{Si} being the particle density of Si ($5 \times 10^{22}\,\mathrm{cm}^{-3}$) and $n_{Val} \approx 2.6$ the dissolution valence for the dissolution process, i.e., the number of electrons supplied by the external circuit needed for the dissolution of 1 Si atom and $q = 1.6 \times 10^{-19}$ C the elementary charge [9].

9.2.2
Extension of the Pore Formation Model to Trench Formation

For the realization of the trenches during PECE, it was necessary to go a step further [10]. A trench has to be etched in close proximity to an ordered array of macropores to prevent the formation of non-lithographically defined pores. Having in mind the design rule of constant porosity within a unit cell, the thin silicon layer between the last row of pores and the trench, called in the following ARL, can be realized by lithographically defining a trench to be etched next to the last row of pores (Figure 9.4a). The thickness t_{ARL} of the remaining ARL is given by the distance of the edge of the trench and the center of the adjacent pores as shown in Figure 9.4b. By lithography the x-y-positions of the pores and the trench are fixed. But the width of the etched trench depends on the r/a ratio chosen during PECE according to

$$t_{\mathrm{trench}} = \frac{\pi\left(\frac{r}{a}\right)^2 (2t_{\mathrm{ARL}} - 0.5)}{\sqrt{3} - 2\pi\left(\frac{r}{a}\right)^2}, \qquad (9.9)$$

with the symbols used in Figure 9.4a. As a consequence the intended r/a ratio has to be taken into account when defining the position of the trench on the lithography mask.

9.2.3
Fabrication of Trenches and More Complex Geometries

Figure 9.5 depicts successfully etched, 450 μm deep trenches next to arrays of hexagonally arranged macropores. Both, the macropores as well as the trenches

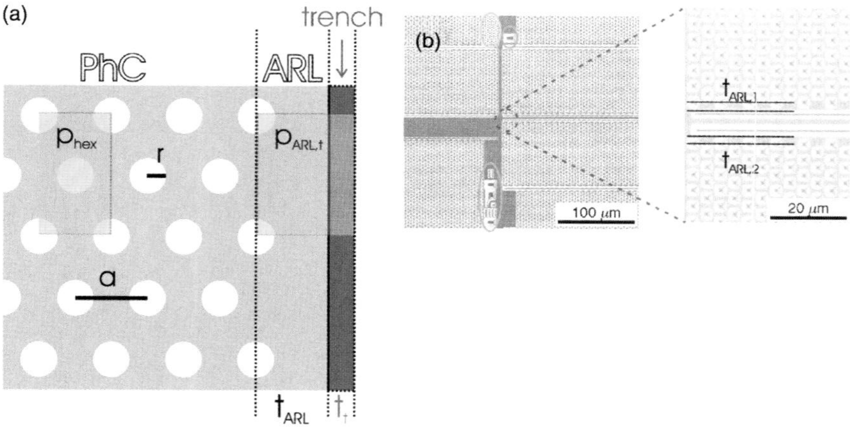

Figure 9.4 Realization of the ARL during PECE. (a) Design principle, (b) SEM micrographs of the etch pits for the macropores and the trenches in a Si wafer. Left: overview. Right: zoom revealing ARLs with different thickness [9].

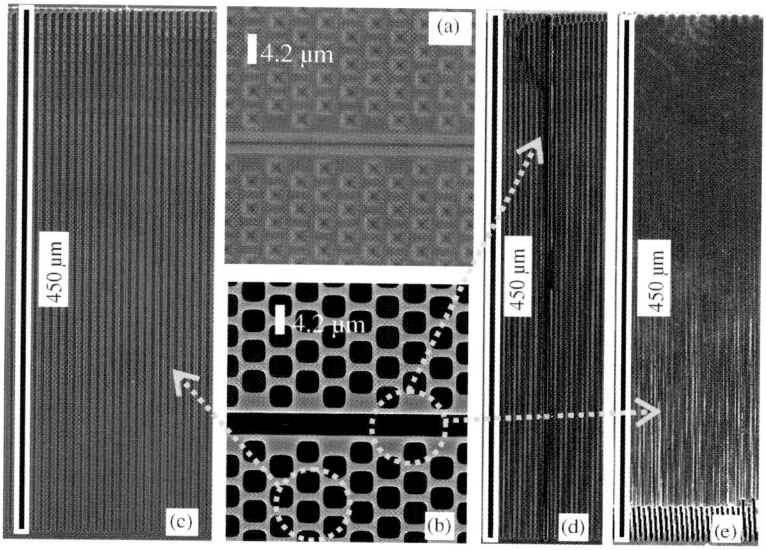

Figure 9.5 Deep trenches next to macropores realized by PECE (SEM micrographs). (a) Lithographically prestructured n-Si wafer (top view). (b) Trench and adjacent macropore array after PECE. (c) Side view of the cleaving edge through the macropore array. (d) Side view of the cleaving edge perpendicular to the trench. (trench highlighted by the dotted line). (e) Side view of the cleaving edge parallel to the trench. (The broken trench surface at the bottom is a consequence of cleaving) [9].

grow stable with depth. Due to the observance of the before stated design rules, stable trench-next-to-macropores-growth could be achieved by using PECE parameters comparable to the parameter set used for the fabrication of pure macropore arrays. However, it is clear that constant porosity in the bulk PhC and the ARL/trench regions is a necessary but not sufficient prerequisite to achieve stable macropore/trench growth. If the system is disturbed too much by, e.g., creating too thick ARLs with $t_{\mathrm{ARL}} \gtrsim 1a$ the PECE process becomes unstable.

Trench etching is expected to work in principle independent of the lattice constant of the macropore array. Thin stripes are generated during the PECE process without any post processing (except from cleaving using a pair of tweezers). By standard procedures, i.e., inscribing the Si with a diamond scribe and mechanical cleaving, such thin stripes cannot be realized. The thin macropore stripes can, e.g., be used to determine the transmission through macroporous Si waveguide structures as a function of the sample length Figure 9.6c. However, effects resulting from the presence of the ARL have to be accounted for. The importance of the surface termination not only of bulk PhCs but also of PhC waveguide structures will be discussed later in this review. The now available macroporous Si structures with ARL terminations of varying thickness allow verification of the theoretically predicted effects. Figure 9.6b shows that PECE of trenches allows the realization of well defined hole structures in macropore arrays. The macropore array in the center of the hole can easily be removed after membrane fabrication by pushing with a small tip or blowing

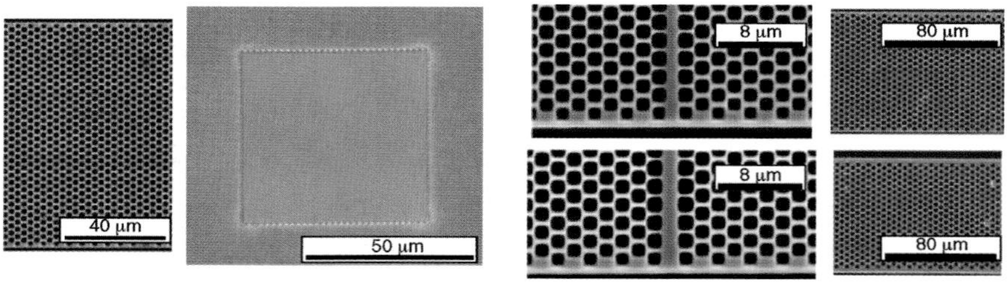

(a) (b) (c) (d)

Figure 9.6 PECE trenches: (a) two trenches used to define a thin stripe of 61 pore rows in ΓK direction ($a = 2\,\mu m$). (b) Trenches along ΓK and ΓM directions to realize a $100 \times 100\,\mu m^2$ hole within a macropore array ($a = 2\,\mu m$). (c) W0.7 (top) and W1 (bottom) PhC waveguides, terminated with an ARL ($a = 2\,\mu m$). (d) Two trenches used to define thin stripes of 35 (bottom) and 39 (top) pore rows in ΓK direction ($a = 4.2\,\mu m$). (SEM pictures: a, c, d; optical microscope pictures: b.

with pressurized air. These holes could either be used to host light sources or for mechanically mounting macroporous Si structures onto device substrates or as alignment markers. Later in this article we will discuss modified thermal emission form 2D macroporous silicon photonic crystals.

9.2.4
Current Limits of Silicon Macropore Etching

Electrochemical etching of macropores in silicon can fabricate ordered pore arrays with pore diameters between 350 and approx. 8 µm. The pores are straight and have no variations in the pore diameter if the current is adjusted properly during the etching process. For photonic crystals applications where coherent light propagation is needed, three structural materials parameter are important. The roughness of the pore walls, the pore diameter variations and the interpore spacing variations. The roughness of the pore walls can be reduced significantly by oxidation and chemical etching steps below the resolution of current scanning electron microscopes, thus below a few nanometers in this case. More difficult is the pore diameter variation and the variation of the interpore spacing. These variations are typically in the range of 1–2% due to doping variations (striations) in the silicon starting material. Due to the local doping density variations, also the space charge region and thus the electrochemical etching is influenced [see Eq. (9.6)].

9.3
Defects in 2D Macroporous Silicon Photonic Crystals

Since the beginning of the study of photonic crystals special attention was paid to intentionally incorporated defects in these crystals. Point defects in photonic crystals lead to microresonators, line defect result in waveguides. However, both functional optical elements have a significantly different optical behavior compared to ridge or strip waveguides and microring or microdisk resonators.

9.3.1
Waveguides

To demonstrate waveguiding through a W1-waveguide, a 27 μm long line defect was incorporated along the Γ–K direction into a triangular 2D photonic crystal with a r/a-ratio of 0.43 ($r = 0.64$ μm) [11].

The transmission through the line defect was measured with a pulsed laser source which was tunable over the whole width of the H-stopband in Γ–K direction ($3.1 < \lambda < 5.5$ μm). The measured spectrum (Figure 9.7) exhibits pronounced Fabry–Perot-resonances over a large spectral range which are caused by multiple

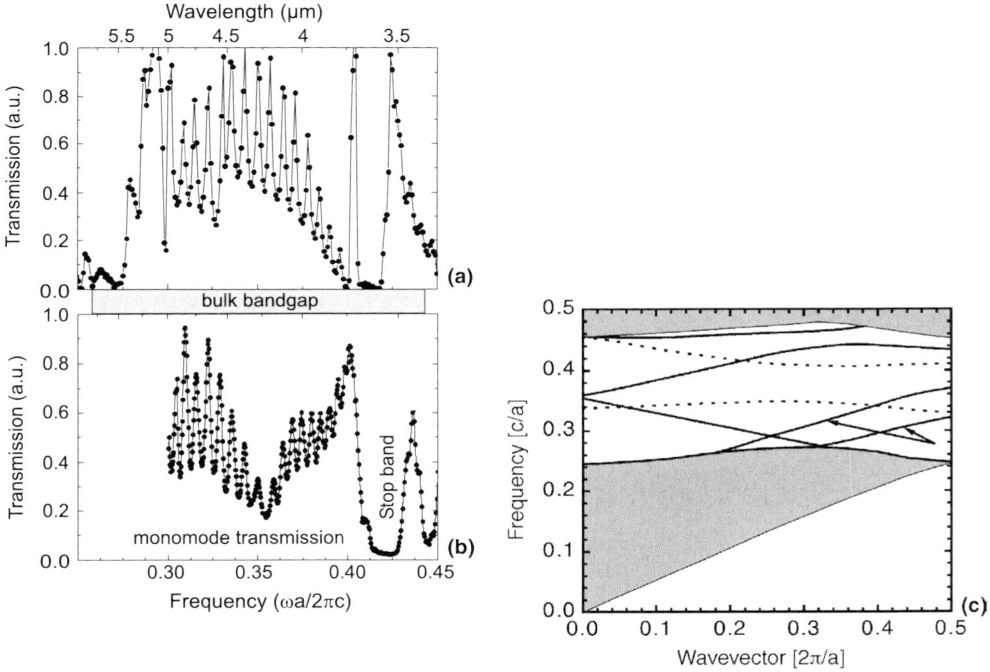

Figure 9.7 (a) Measured and (b) calculated H-polarized transmission spectrum of a 27 μm long waveguide directed along Γ–K covering the spectral range of the H-bandgap of the surrounding perfect photonic crystal. The transmission is in %. Only the even waveguide modes contribute to the transmission as the incoming plane wave cannot couple to the odd waveguide modes. The small stopgap at a frequency of 0.45 c/a is caused by the anticrossing of 2 even waveguide modes [10]. (c) Computed H-polarized band structure of the waveguide oriented along Γ–K. Solid and dotted curves correspond to even and odd modes, respectively. The two bands which are labelled with arrows appear due to the overetched pores on either side of the waveguide. The shaded areas correspond to the modes available in the adjacent perfect crystal regions [10].

reflections at the waveguide facets. Comparing the spectrum with an 2D-FDTD-transmission calculation reveals very good agreement and the comparable finesse of the measured and calculated resonances indicate small losses inside the sample. A band structure calculation for H-polarization along Γ–K including waveguide modes is depicted in Figure 9.7. Here, the 2D band structure has been projected onto the new 1D Brillouin zone in Γ–K direction, since the line waveguide reduces the symmetry. The grey shaded regions represent all possible modes inside the perfect crystal areas adjacent to the line defect. Defect modes bound to the line defect, therefore, occur only in the bandgap, i.e., in the range $0.27 < f < 0.46$. They split into even and odd modes with respect to the mirror plane which is spanned by the waveguide direction and the direction of the pore axis. As the incoming wave can be approximated by a plane wave, the incident radiation can only couple to the even modes of the waveguide. The odd modes do not contribute to the transmission through the waveguide and, therefore, in this experiment transmission is solely connected with the even modes. The small stop band between the even modes around a frequency of 0.45 is reproduced as a region of vanishing transmission in Figure 9.7 due to anticrossing of the waveguide modes. Furthermore, from the band structure it can be concluded that for $0.37 < f < 0.41$ c/a only a single even mode exists. Its bandwidth amounts to 10%.

9.3.2
Beaming

Efficient coupling directly into and out of a waveguide that is less than a wavelength wide is in general considered to be at odds with the diffraction limit. As a result, several solutions including coupling via out-of-plane gratings, combinations of ridge waveguides and tapers, or evanescent coupling have been investigated, recently [12,13]. We have shown that proper structuring and truncating of the output facet of a PhC waveguides offers a convenient way to obtain a beam with a very low divergence.

Figure 9.8(b) shows the core of the experimental arrangement. Light from a continuous wave optical parametric oscillator is coupled into the first waveguide of

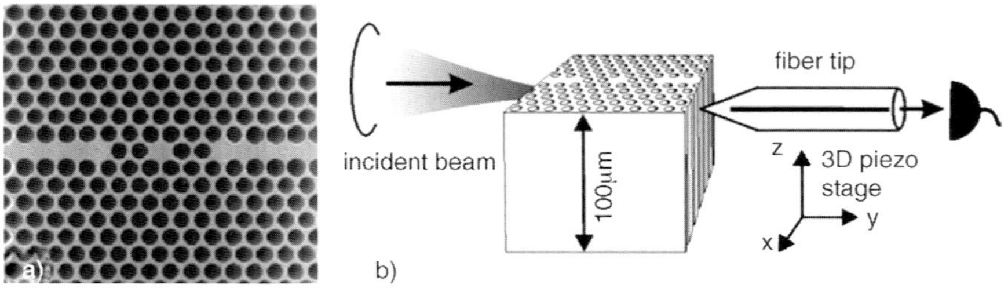

Figure 9.8 (a) An electron microscope image of the photonic crystal structure studied experimentally. (b) The schematics of the setup. The laser beam is focused on the entrance of the first waveguide and a fiber tip is used to detect the light locally at the output side [11].

the photonic crystal. A fluoride glass fiber with a core diameter of 9 μm is etched to form a tip with a radius of curvature around 1 μm, serving as a local detector for the optical intensity. The fiber tip is mounted in a SNOM device with a sample-probe distance control mechanism. This allows us to regulate the gap between the tip and the PhC facet to better than a few tens of nanometers, which in this experiment corresponds to distances smaller than $\lambda/100$. By using a calibrated piezoelectric element, we can also retract and place the tip at well-defined distances away from the PhC exit. At each y location the tip is scanned in the x,z-plane so as to map the lateral intensity distribution in the output beam. Figure 9.9A (a) displays the intensity distribution right at the exit of the waveguide while Figure 9.9A(b) to 9A(i) show the same at successively increasing y distances up to about 24 μm away from the PhC's exit facet. In Figure 9.9A(j) the blue, black, and red curves display the beam profiles along the z direction from Figure 9.9A(a), A(b), and A(i), respectively. Comparison of these plots reveals that the beam does not undergo a notable spread in this direction upon exiting the PhC waveguide. This is not surprising because the light has not experienced any confinement in the PhC along this direction. The central issue of interest in our work concerns the beam divergence along the x direction. Therefore,

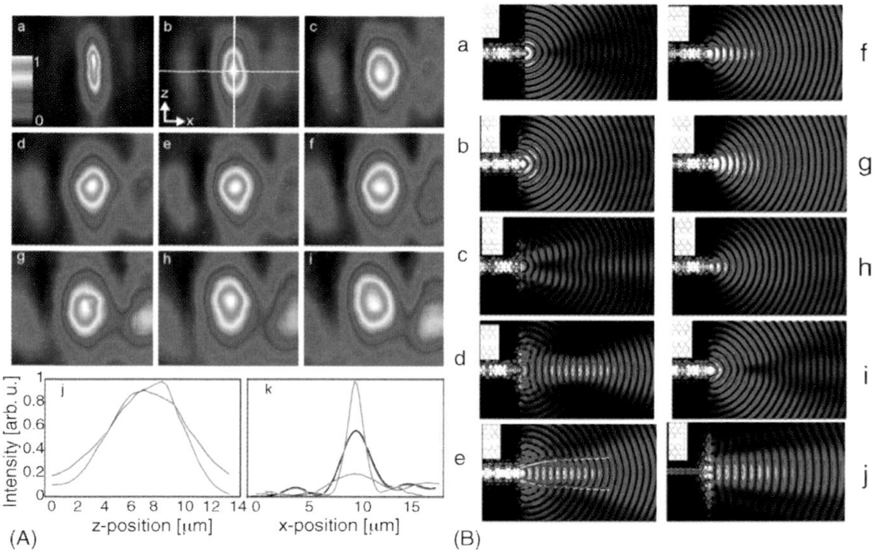

Figure 9.9 (**A**) (a) Experimental lateral intensity distribution as seen by the tip only a few nanometers from the crystal exit. (b)–(i) same as in (a) but for tip–sample separations of 3.5, 6.5, 9.4, 12.4, 15.3, 18.3, 21.2, and 24.2 micrometers, respectively. The scale of the color code is adapted for each image individually to show the full contrast. (j) Vertical cross sections of figures (a) and (i) plotted by the blue and red curves, respectively. Here, the curves were rescaled to facilitate the comparison. (k) Horizontal cross sections of figures (a), (b) and (i) plotted by the blue, black and red curves, respectively. [11]. (**B**) (a)–(h) Calculated intensity distribution of the light at $\omega a/2\pi c = 0.39$, exiting a photonic crystal waveguide for 9 different structure terminations (shown in the insets). (i) The intensity distribution for the structure in (b) but at $\omega a/2\pi c = 0.33$ [11].

in Figure 9.9A(k) we plot the x profiles of images A(a), A(b), and A(i). The blue curve displays the beam cross section when the tip is nearly in contact with the PhC surface. One finds that the great majority of the power is contained in a spot with the full width at half maximum of less than 2 μm, corresponding to the initial confinement of light in a subwavelength region about the waveguide. The black and red curves show that as the tip-sample distance is increased, this spot broadens and becomes weaker. The remarkable fact is, however, that its width does not grow nearly as fast as one would have expected for a beam that emerges out of a subwavelength waveguide. At first sight this might appear to violate the laws of diffraction. However, the side lobes of the blue curve in Figure 9.9A(k) very clearly indicate that, in fact, light is not confined to a subwavelength region. Interestingly, Martin-Moreno et al. discuss a similar effect in their theoretical study of light emission out of a nanoscopic aperture in a corrugated metallic film [14] and also later confirmed our interpretation for dielectric materials numerically [15]. Although there is no equivalent to plasmon polaritons in photonic crystals, there exist surface modes that can be excited at the PhC-air interface, therefore mediating the extension of light to the sides of the waveguide exit.

In order to investigate the angular spread of the emerging beam, we have performed two-dimensional finite-difference time-domain (FD-TD) calculations. We consider a PhC that contains a single straight waveguide but has otherwise the same crystal parameters as the experimentally examined sample in Figure 9.8(a). We set the wavelength to the experimental value of 3.84 μm corresponding to $\omega a/2\pi c$ 0.39. Figure 9.9B(a) to B(i) display the snap shots of the intensity distribution and wave fronts of the outgoing beam in the xy plane for nine different terminations of the PC structure (see the insets). These images let us verify that the spread of the beam depends on the termination in an extremely sensitive manner, therefore supporting the hypothesis that surface modes might be involved. For example, Figure 9.9B(b) and B(h) display very large beam divergence while Figure 9.9B(d) and B(e) show output beams that contain more than 70% of their total radiated power within a small full angle of 20 degree, representing the lowest numerical aperture in these series. In order to facilitate the comparison between the results of measurements and simulations, the symbols in Figure 9.9B(e) mark the locations where the central spots of images A(a) to A(j) reach their $1/e^2$ values in the x direction. The very good agreement between the FDTD outcome and the experimental data is clear. Scanning electron microscope images as well as topography images taken with our fiber tip indicate that the termination of the PhC used in this work, indeed, corresponds to that in Figure 9.9B(e).

9.3.3 Microcavities

Besides line defects also point defects consisting only of 1 missing pore are of special interest. Such a micro-resonator-type defect also causes photonic states whose spectral positions lie within the bandgap of the surrounding perfect photonic crystal. The light fields belonging to these defect states are therefore confined to the very

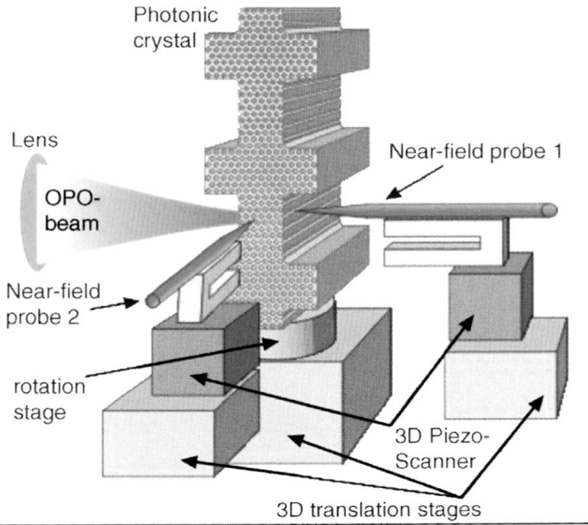

Figure 9.10 Left: Top view of the photonic crystal region containing the waveguide-microresonator-waveguide structure. The r/a-ratio of the pores amounts to 0.433. The waveguides on the left and on the right serve to couple the light into the point defect (microresonator) [14]. Right: Setup of the optical measurement.

small volume of the point defect resulting in very high energy densities inside the defect volume. As the point defect can be considered as a microcavity surrounded by perfect reflecting walls, resonance peaks with very high Q-values are expected in the transmission spectra. Since the symmetry is broken in both high-symmetry direction, a band structure cannot be used anymore to describe point defect. To study this experimentally, a sample was fabricated including a point defect which was placed between 2 line defects serving as waveguides for coupling light in and out [16]. Figure 9.10 shows an SEM-image of the described sample with $r/a = 0.433$.

Measuring transmission through this waveguide-microresonator-waveguide structure demands an optical source with a very narrow linewidth. Therefore, a continuous wave optical parametric oscillator (OPO) has been used which is tunable between 3.6 and 4 μm and delivers a laser beam of 100 kHz line width. For spatially resolved detection an uncoated tapered fluoride glass fibre mounted to a SNOM-head was applied and positioned precisely to the exit facet of the outcoupling photonic crystal waveguide (Figure 9.10). In the transmission spectrum 2 point defect resonances at 3.616 μm and 3.843 μm could be observed (Figure 9.11). Their spectral positions are in excellent agreement with the calculated values of 3.625 μm and 3.834 μm predicted by 2D-FDTD calculations taking into account the slightly widened pores surrounding the point defect. The measured point defect resonances exhibited Q values of 640 and 190 respectively. The differences to the theoretical predicted values of 1700, 750 originate from the finite depth not considered in

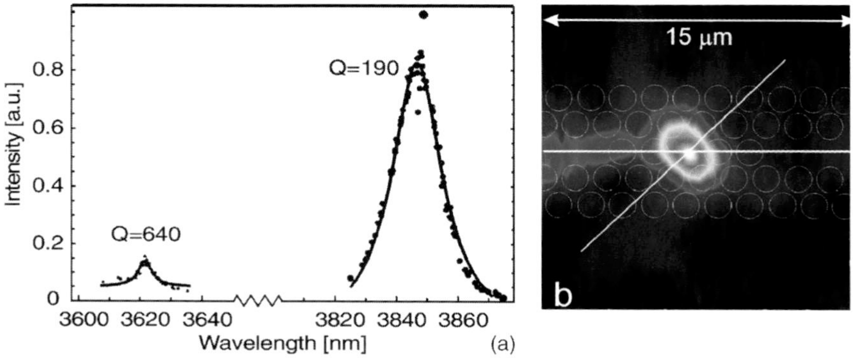

Figure 9.11 (a) Measured monopole ($Q = 647$) and decapole resonances ($Q = 191$) of the point defect at wavelengths 3.616 μm and 3.843 μm, respectively. (b) SNOM signal recorded for the resonance at 3.843 μm [15].

2D-calculations and the exact pore shape near the cavity. Recent 3D-FDTD calculations show that for high Q-values, the finite depth as well as the shape of the pores near the cavity play an important role in the determination of the Q-value [17]. Therefore, the 2D-limit breaks for high-Q cavities under realistic conditions. Intuitively, this can be explained as follows. Any out-of-plane com-ponent of the incoming light will result in a spreading of the mode with depth and to a reduction of the Q-value. Therefore an improved design needs to take care for a 3D confinement of the resonator mode. Nevertheless, already the reported high Q-values of this 2D microresonator might already be sufficient for studying the modification of radiation properties of an emitter placed in such a point defect.

9.4 Internal Emitter

9.4.1 Internal Emitter in Bulk 2D Silicon Photonic Crystals

Silicon itself is a poor light emitter. Recent electroluminscent devices based on silicon interband transitions are in the region of external quantum efficiencies of about a few percent. Therefore, one route to optically active systems around 1.3 to 1.5 μm are the use hybrid devices such as HgTe quantum dots incorporated into a silicon photonic crystals. To incorporate the QDs into the macroporous Si, we prepared QD/polymer composite tubes within the pores. The polymer matrix embeds and fixates the QDs. If melts or solutions of a homopolymer are brought into contact with a macroporous material having pore walls with a high surface energy such as macroporous Si, a mesoscopic wetting film of about 20 nm covering the pore walls will develop. A complete filling of the pore volume with the liquid, however, does not occur for high

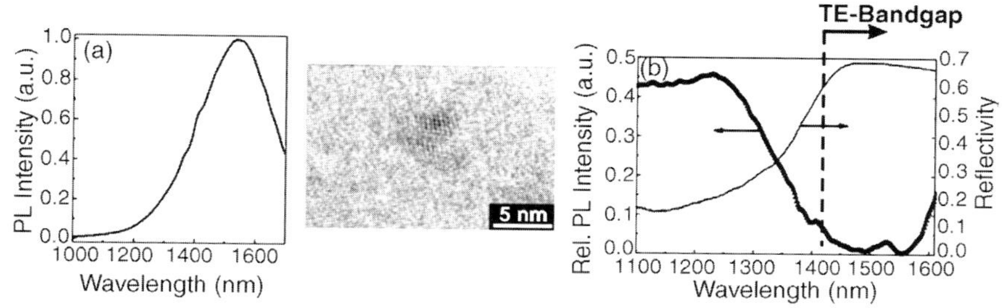

Figure 9.12 (a) Free space PL spectrum of HgTe/PS tubes with transmission electron high resolution image of a QD cluster within a polymer tube wall; (b) PL spectra of HgTe QDs in a hexagonal 2D photonic crystal of macroporous silicon (lattice constant 700 nm, TE polarization, $r/a = 0.45$), normalized to the free space emission spectrum, compared to the corresponding reflection measurement [19].

molecular weight polymers. Quenching at this stage by solidifying the polymer results in the conservation of tubules [18]. HgTe QDs were synthesized in water as previously reported [19] and were transferred to toluene by stabilizer exchange. Quantum confinement increases their electronic band gap, which amounts to -0.15 eV for bulk HgTe, to the average value of 0.83 eV. The size distribution of the particles causes a broadening of the luminescence spectrum from 1300 nm to 1700 nm, which is of use to study the emission modification under the influence of the photonic band gap. The lifetime of the luminescence is about 10 ns, nearly exponential and varies with the concentration of nanocrystals [20]. The suspensions were mixed with 1 wt% solutions of poly(methyl methacrylate) (PMMA, Mw = 120,000 g/mol) and polystyrene (PS, Mw = 250,000 g/mol) in the same solvent. To wet the templates, the polymeric solutions containing the dispersed QDs were dropped on the macroporous Si at ambient conditions. Transmission electron microscopy verified that QDs are along the pore walls (Figure 9.12a). Since the QDs are even found in the vicinity of the blind ends of the template pores, they had been moved over a distance of several tens of microns in the course of the wetting procedure.

The QD/polymer tubes in solution generally exhibit a strong photoluminescence (PL) signal corresponding to that of colloidal HgTe QDs suspended in toluene (Figure 9.12a). In order to investigate the modification of the emission spectrum of the QDs in a photonic crystal, a hexagonal 2D photonic crystal of macroporous silicon (lattice constant 700 nm, $r/a = 0.45$) was prepared and infiltrated with HgTe/PS composite (weight ratio PS:HgTe of 10:1). The polymer film inside the pores had a thickness of about 20 nm, hardly effecting the photonic crystals bandgap. The QDs were excited by an argon-laser (488 nm) with incidence parallel to the pores axes. The partial spectral overlap of the HgTe luminescence and the fundamental bandgap allows only the investigation of the upper band edge. Comparing the photoluminescence (PL) spectra to reflection measurements a coincidence of the decrease in PL intensity and the high reflectivity for wavelengths above 1300 nm for TE occurs, as shown in Figure 9.12b. Here, the PL spectrum is normalized to the free space emission of the HgTe/PS tubes. We believe that the photonic bandgap prevents the existence and propagation of light in the plane of the 2D crystal [21].

9.4.2
Internal Emitter in Microcavities of 2D Silicon Photonic Crystals

To analyze the change of the emission properties of internal emitters in 2D PhC, we developed a coupled cavity mask sets. These 2D photonic coupled cavity structures (see Figure 9.13a) possessed a band gap that overlapped with the emission of HgTe-colloidal quantum dots ($\lambda_{emit} = 1300$–1600 nm). The infiltration was carried out by a novel wetting process described above, allowing well defined positioning of QDs inside the pores of the photonic crystal cavities [21]. This is of particular importance since theoretical studies predict that the local density of states varies within the pore [22]. Since only the central pore of the cavity should be infiltrated with the emitters, a lithographic process to open only the central cavity hole was developed. We then succeeded in the local infiltration of the emitters just in every central cavity hole. Optical characterization by FT-IR spectroscopy of the non-infiltrated samples shows that spectral dips inside the photonic band gap occur (Figure 9.13b) [22]. Since the cavities are very large, there are about 18 modes inside the photonic bandgap. Group theoretical analysis determined that about 10 modes couple to a plane wave. After the local infiltration of the emitters, the broad emission of the quantum dots was modulated strongly by the resonator modes of the cavities (Figure 9.13c). We expect from LDOS calculations a factor of 10 in the suppression of the 3D-LDOS for our configuration. This should be in principle detected by time-resolved spectroscopy.

We have imaged these infiltrated point-defect structures using confocal microscopy in reflection mode [23]. The green excitation ($\lambda = 532$ nm) was focused onto the sample using a NA = 0.95 objective that was also used to collect the infrared

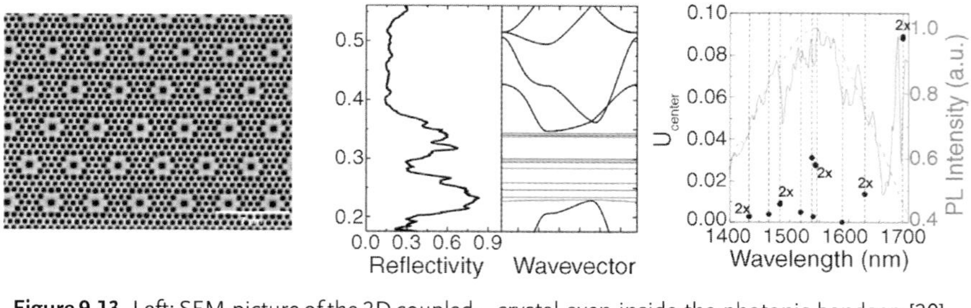

Figure 9.13 Left: SEM-picture of the 2D coupled-cavity photonic crystal (interpore distance $a = 700$ nm). (centre) Reflectivity of the non-infiltrated structure. Inside the photonic bandgap there are more than 10 cavity modes which result in a coupling of light to the photonic crystal even inside the photonic bandgap [20]. (Right) Solid line: Emission spectrum of HgTe quantum dots which are only infiltrated into the central pore of the cavity. Dashed line: Emission of HgTe quantum dots outside the photonic crystals [20].

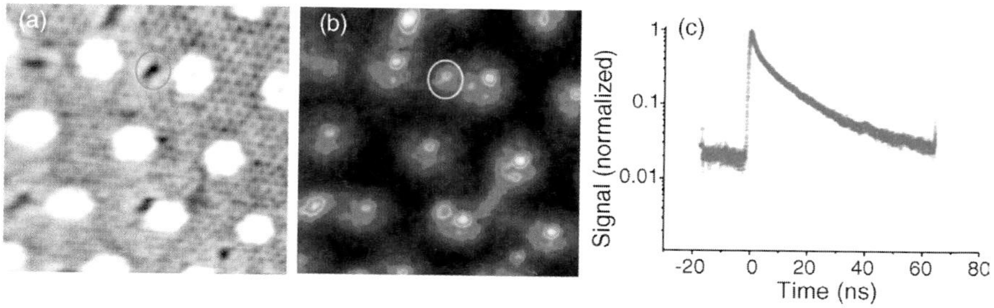

Figure 9.14 (a) A reflection confocal image of the sample presented in Figure 9.2. (b) A luminescence confocal image of the sample. The bright regions indicate luminescence from quantum dots. Note that in addition to the hexagonal pattern of the cavities, one sees other spots from dots that have ended up on the surface of the sample during infiltration (see e.g. the region circled white (red in a)). (c) Luminescence decay of quantum dots from a single cavity [21].

luminescence. As the objective was optimized for infrared, the divergence of the incident beam had to be carefully tuned in order for the excitation and detection focus to overlap on the sample surface. Raster scanning the sample not only provides an image of the infrared luminescence, as the reflected pump beam provides the topography of the sample. As shown in Figure 9.14(a), the defect cavities as well as the underlying lattice are clearly resolved in the scattered light image. This is not unexpected since the lattice spacing far exceeds the diffraction limit at 532 nm. A remarkable advantage of the set up, however, is that the resolution at infrared wavelengths is much smaller than the diffraction limit for the detected wavelength, as it is set by the focus size of the pump beam. In the scattered light image (Figure 9.14(a)), the cavities show as bright six-lobed rings, with a central dark spot. Here it should be noted that the collected signal is light reflected off the sample. The bright ring is hence explained by the stronger Fresnel reflection of unetched silicon, while the dark spot is due to the lack of reflection off the central air hole. The topography further shows a few dark patches apparently uncorrelated to the designed lattice. As further discussed below, we interpret the reduced reflectivity at those positions as due to 16 islands of (nanocrystal) material with surfaces not aligned along the sample surface, which therefore do not reflect light towards the detector. Figure 9.14(b) shows the infrared luminescence collected in parallel with the rasterscan shown in Figure 9.14(a). The infrared luminescence that was collected by the objective was detected by a fiber-coupled InGaAs avalanche photodiode (Id-Quantique) without dispersing it in its spectral components. As we do not detect single dots, the images represent luminescence over a wide spectrum from 1.3 μm to 1.7 μm, as set by the size distribution of the HgTe dots. The luminescence image demonstrates that quantum dots have indeed been deposited preferentially inside the defect cavities, which show as bright dots on a background that is at the dark count level. The few isolated patches that appear dark in the topography also luminesce brightly, which validates their interpretation as clusters of HgTe nanocrystals on the surface of the sample. Close inspection of the cavities in the luminescence images shows a remarkable feature: the bright spots are not simply gaussian in cross sectional profile. Instead it appears that the cavities also luminesce when the excitation spot is located on the ring of unetched holes enclosing the central defect hole. Preliminary fluorescence lifetime measurements demonstrate near single exponential decay, with a lifetime around 15 ns (Figure 9.14(c)). This shows that the quantum dots do not suffer quenching and are hence promising light sources for probing emission lifetime modifications in silicon photonic crystals. Our first results, which are integrated over a wide spectrum do not conclusively confirm that the defect cavities cause a modified spontaneous emission rate [23].

9.4.3
Modified Thermal Emission

We also measured the spectral modification of the thermal emission of 2D macroporous silicon photonic crystals by either heating the photonic crystal directly up to about 1000 K or by introducing inside a 2D photonic crystal a heated tungsten microwire [24]. To avoid direct emission from the crystal side-surface, the in-plane emission measurements were carried out using an integrated heat source. A 50 μm diameter tungsten wire was used as local heat source. To guide the wire through the crystal plane, we etched an area of 100×100 μm out of the 2D photonic crystal membrane. The cutted-out region can be fabricated during the same photoelectrochemical process as the pore structure itself using appropriate mask geometries [10]. Figure 9.15 shows the complete structure and a zoom of the cutout-region. A reduced emission of about 20% can be observed for the in-plane emission spectrum (Figure 9.15) for frequencies in the range from 1250 cm^{-1} to 2000 cm^{-1} (marked as region B). A more complicated structure for the out-of plane emission is found. In particular, in the area of low group velocities in-plane, an enhanced out of plane emission is observed [24]. This technique can lead to selective thermal emitters for integrated infrared sensors [25].

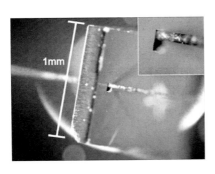

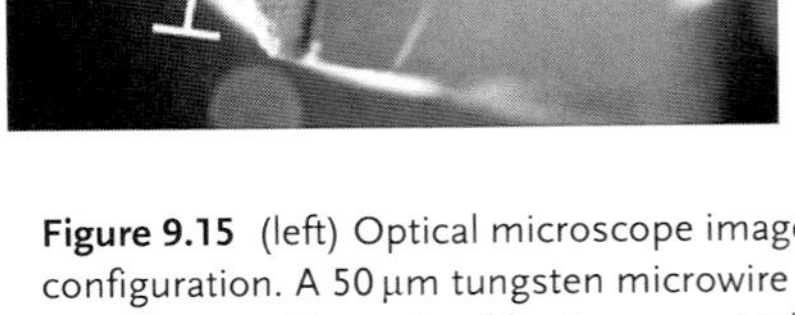

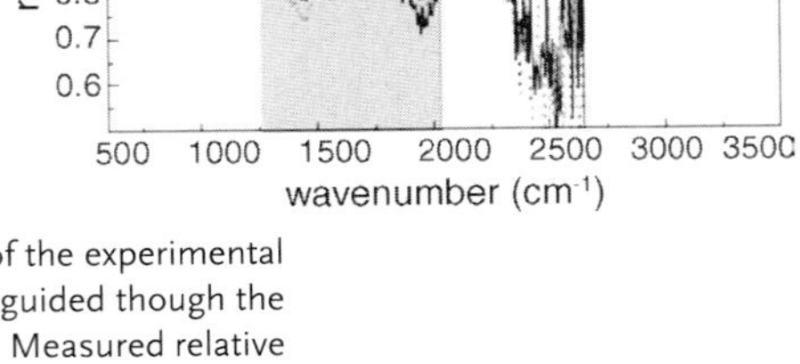

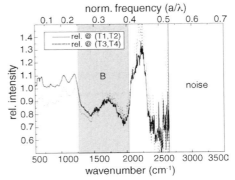

Figure 9.15 (left) Optical microscope image of the experimental configuration. A 50 μm tungsten microwire is guided though the cutted-out section as local heat source. (right) Measured relative in-plane emission intensity of a homogeneously heated silicon 2D-photonic-crystal perpendicular to the crystal pores [21].

9.5 Tunability of Silicon Photonic Crystals

Small deviations of the fabricated experimental structures from designed ones have serious influence on their optical properties. In particular, the design of a microresonator (point defect) with a well defined resonance frequency in the near IR allows only fabrication tolerances in the sub-nanometer regime, a demand which currently can not be fulfilled reproducibly. Additionally, for many applications e.g. optical switches one would like to shift the band gap during operation. Therefore, tuning the optical properties during operation is a major point of interest. In the following we present our recent results on liquid crystal, free carrier and Kerr tuning of silicon photonic crystals.

9.5.1 Liquid Crystals Tuning

One way to achieve this behavior, is to change the refractive index of at least one material inside the photonic crystal. This can be achieved by controlling the orientation of the optical anisotropy of one material incorporated in the photonic crystal [26]. As proof of principle of the latter, a liquid crystal was infiltrated into a 2D triangular pore array with a pitch of 1.58 μm and the shift of a band edge depending on the temperature was observed [27]. The liquid crystal E7 is in its nematic phase at room temperature but becomes isotropic at $T > 59\,^{\circ}\mathrm{C}$.

Transmission for H-polarized light was measured along the Γ–K direction through a 200 μm thick bar first without and then with the infiltrated liquid crystal in the isotropic regime (Figure 9.16). In the case of room temperature the first stop band of the H-polarization is observable in the range between 4.4–6 μm. Although a large bandgap

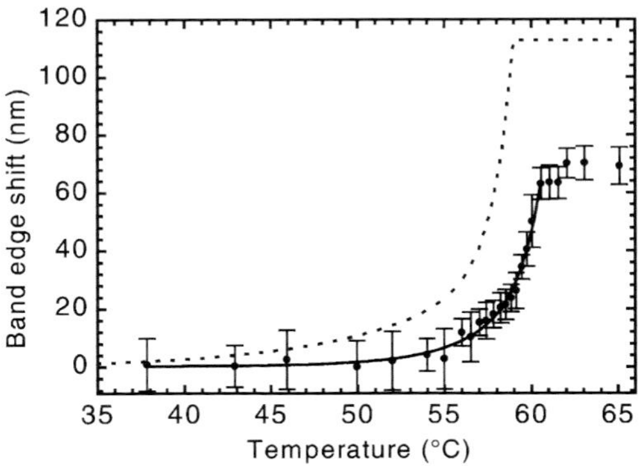

Figure 9.16 Temperature dependence of the band edge shift caused by temperature induced phase transition of the infiltrated liquid crystal. Solid line: Fit to experimental data points, Dashed line: Calculation assuming a simple axial alignment of the liquid crystal in the pores [24].

for the H-polarization still exists, the complete band gap, which is characteristic for the unfilled structure, is lost due to the lowered refractive index contrast within the infiltrated crystal. Therefore the investigations were only carried out for H-polarization. When the structure is heated up, the upper band edge at 4.4 μm is red shifted while the lower band edge exhibits no noticeable shift. At a temperature of 62 degree the red shift saturates and the total shift amounts to $\Delta\lambda = 70$ nm as shown in Figure 9.16. This corresponds to 3% of the band gap width. The shift is caused by the change in orientation of the liquid crystal molecules inside the pores. In a simplified model one can assume that all liquid crystal molecule directors line up parallel to the pore axis when the liquid crystal is in its nematic phase at room temperature. Then the H-polarized light (E-field in plane) sees the lower refractive index n_0 inside the pores. If the temperature is increased above 59°C a phase transition occurs and the liquid crystal molecule directors are randomly oriented. The H-polarized light sees now a refractive index inside the pores which is an average over all these orientations. According to this model a red shift of $\Delta\lambda = 113$ nm is expected which is slightly larger than the measured one. The difference in the observed and calculated shift has been investigated by Kitzerow *et al.* (see this volume) for 2D and 3D macroporous silicon photonic crystals and is also discussed for other semiconductor materials by Ferrini *et al.* [29].

9.5.2
Free-carrier Tuning

Fast tuning of the band edge of a 2D macroporous silicon photonic crystals can be obtained by free carrier injection electrically or optically. In contrast to LQ switching,

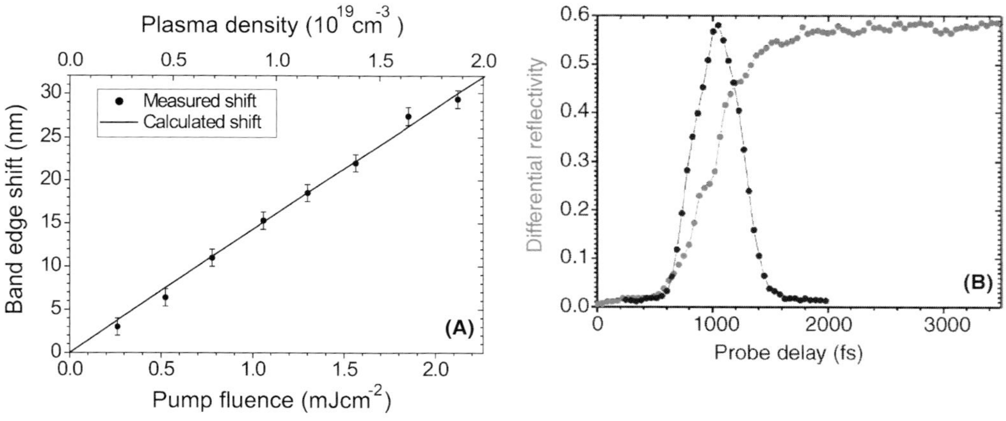

Figure 9.17 (A) Band edge shift as a function of the pump fluence, i.e., the plasma density. A maximum shift of 29 nm at 1.9 μm has been observed in good agreement with numerical calculations. (B) Transient behavior of differential reflectivity at $\lambda = 1900$ nm for a pump beam at $\lambda = 800$ nm and a fluence of 1.3 mJ/cm^2. The band edge shift within 400 fs with a dynamic of 25 dB [25].

the refractive index of the silicon matrix was tuned by optical injection of free carriers by us [30]. The photonic crystal was illuminated by laser pulse at $\lambda = 800$ nm, so well in the absorbtion region of silicon, with a pulse duration of 300 fs. The rise time of the change in the refractive index and thus the shift of the band edge was about 400 fs, slightly slower than the pulse due to the themalization of the excited carrier (Figure 9.17). The band edge shift observed goes linearly with the pulse intensity as expected from Drude theory. For example, for a pump fluence of 2 m J cm^{-2}, a band shift of 29 nm was observable. A drawback of the free carrier injection is the long lifetime of the carriers, here in the ns range due to the high surface area.

9.5.3
Nonlinear Optical Tuning

Another way to change the refractive index dynamically in silicon [31] or even III–V compound semiconductors [32] is the Kerr effect. Figure 9.18A shows the time dependent change in reflectivity at 1.3 μm for a 2.0 μm pump pulse and the cross-correlation trace of both pulses. The pump and probe intensities are 17.6 and 0.5 GW/cm^2, respectively. The decrease in reflectivity is consistent with a redshift of the band edge due to a positive nondegenerate Kerr index. The FWHM of the reflectivity trace is 365 ± 10 fs which is 1.83 times larger than the pump-probe cross-correlation width as measured by sum frequency generation in a BBO crystal. This difference can be explained in terms of pump and probe beam transit time effects in the PhC. From the pump group velocity and probe spot size, one can deduce that the reflected probe pulse is delayed by 110 fs within the PhC sample. The intrinsic interaction times are therefore pulse width limited, consistent with the Kerr effect.

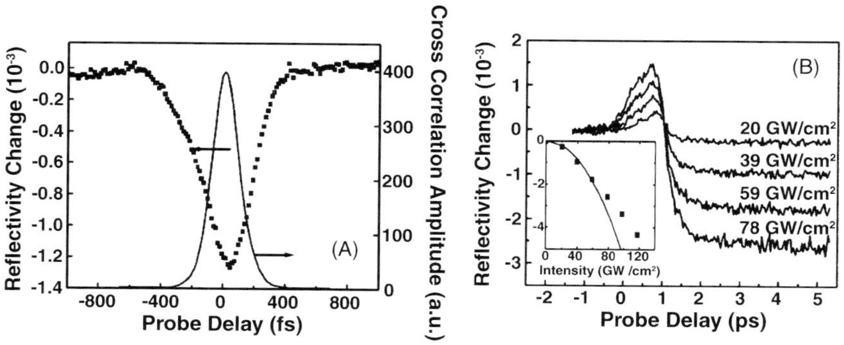

Figure 9.18 (A) Temporal response of the reflectivity change at the 1.3 μm band edge generated by a pump wavelength of 2.0 μm with 17.6 GW/cm^2. The cross correlation trace of the pump and probe pulses is also shown. (B) Temporal response of the reflectivity change at the 1.6 μm band edge for different pump intensities at 1.76 μm. The inset shows the dependence of the carrier-induced reflectivity change on pump intensity for low pump powers [31].

One can estimate a value for the nondegenerate Kerr coefficient n_2 in the silicon PhC from

$$\Delta R = \frac{dR}{d\lambda}\frac{d\lambda}{dn} n_2 \frac{1-R_u}{f} I_0, \tag{9.10}$$

with the steepness of the reflectivity at the band edge $dR/d\lambda$, the differential change in band edge $d\lambda/dn$, the measured pump reflectivity ($R_u = 0.30$) and the effectively pumped fraction of the sample ($f = 0.13$) due to the non-uniform pump field. Because of the relatively large values of $dR/d\lambda = 0.04\,\text{nm}^{-1}$ and $d\lambda/dn = 174\,\text{nm}$, induced reflectivity changes in the vicinity of the 1.3 μm band edge are found to be 70 times more sensitive than that in bulk materials for the same refractive index change. Indeed, when the PhC is replaced by bulk crystalline silicon, no change in reflectivity is observed for our range of pump intensity.

We found good correlation between the change in probe reflectivity and the steepness of the band-edge reflectivity (measured separately) at different wavelengths and for a range of pump intensities. The linear dependence is consistent with the Kerr effect and the nondegenerate Kerr index is estimated to be $n_2 = 5.2 \times 10^{15}\,\text{cm}^2/\text{W}$. This is within an order of magnitude of the degenerate Kerr index reported in literature at 1.27 μm and 1.54 μm and represents reasonable agreement considering uncertainty in the lateral position of the pump pulse in and its intensity at the probe location. It should also be noted that linear scattering losses as the pump pulse propagates through the PhC along the pore axis have not been taken into account.

Results from experiments used to probe the 1.6 μm band edge when the sample is pumped with 1.76 μm pulses are illustrated in Figure 9.18B which shows the temporal response of the change in probe reflectivity at different pump intensities for a probe intensity of 0.13 GW/cm^2. There is an initial increase and decrease in probe reflectivity on a sub-picosecond time scale followed by a response that decays on a time scale of 900 ps and partially masks the Kerr effect near zero delay. At this band edge, the sub-picosecond behavior is consistent with a Kerr effect similar to the previous experiments. The long time response could possibly be due to thermal or Drude contributions to the dielectric constant due to the generation of free carriers. Using a peak pump intensity of 120 GW/cm^2 and a two photon absorption coefficient of 0.8 cm/GW, for 1.55 μm as an upper limit, one can estimate the surface peak carrier density to be smaller than $10^{19}\,\text{cm}^{-3}$ and the maximum change in temperature to be less than 0.15 K. From the thermo-optic coefficient $\partial n/\partial T \approx 10^{-4}\,\text{K}^{-1}$ at the probe wavelength, the change in silicon refractive index is in the order of $\Delta n = 10^{-5}$ and the (positive) induced change in reflectivity is expected to be about the same. From free carrier (Drude) contributions to the refractive index at our probe wavelength, changes to the imaginary part of the dielectric constant are about 2 orders of magnitude smaller than that of the real part which is $\Delta n = 10^{-3}$. Hence free carrier absorption of the probe pulse as well as thermally induced changes can be neglected in what follows and the change in reflectivity is ascribed to changes in the real part of the dielectric constant due to Drude effects. At low pump powers the change in probe reflectivity scales quadratically with pump intensity, consistent with free carrier generation by two photon absorption. However, at higher pump intensities, there is an apparent deviation from this quadratic dependence due to pump saturation effects as the inset in Figure 9.18 shows. With increasing intensity in a two-photon-absorption process, an increasing fraction of the carriers are created closer to the surface where the pump pulse enters and the probe region develops a reduced and increasingly nonuniform carrier density. We estimate that at a depth of 60 μm, the expected saturation pump intensity is about an order of magnitude larger than the maximum pump intensity used here. The carrier lifetime of 900 ps is most likely associated with surface recombination within the PhC sample with its large internal surface area. Details on the effect of the excited modes within the photonic crystal are discussed in more details in [33,34].

9.6 Summary

In summary we have reviewed that macroporous silicon is a model system to fabricate almost perfect 2D photonic crystals for the infrared spectral range. Due to the high refractive index contrast between silicon and air the bandgaps are large and for a triangular array of pores a complete bandgap for the light propagating in the plane of periodicity appears.

In the last few years, we have improved the material technology so that the range for designs has significantly increased. In particular, the fabrication technology has focused on the surface truncation of photonic crystals. Based on this material, we have characterized the linear optical properties in waveguides and microresonators. In particular, we are able to show the beaming of light exiting W1 waveguides based on excited surface modes. We also analysed 2D microresonators optically. We have shown in this case the 2D approximation fails and full 3D simulations are necessary to interpret the experimental data.

We then infiltrated quantum dots inside the silicon PhC, first in 2D bulk PhC, then in specially designed cavities that allow the infiltration with QD. We have seen a strong spectral change of the emission of the quantum dots inside the microresonators and first lifetime measurement show a slight change in the lifetime in comparison to quantum dots in solution.

Finally, we studied in detail three different kinds of tuning of silicon photonic crystals: by liquid crystals, free-carriers and by non-linear effects.

Acknowledgements

We are grateful to our Master, Diploma and PhD-students as well as postdoctoral fellows that made this work possible over the last six years. We are also grateful to our collaborators Andrew Rogach, Henry von Driel, Heinz Kitzerow, Kurt Busch, Armin Lambrecht, Mario Agio, Costas Soukoulis, Maria Kafesaki, Martin Steinhart, Ulrich Gösele and Volker Lehmann. Finally, we like to thank the German Science Foundation for their support within the Focused Program SPP1113.

References

1 Joannopoulos, J.D., Meade, R.D. and Winn, J.N. (1995) *Photonic Crystals*, Princeton University Press, New Jersey.

2 Johnson, S.G., Fan, S., Villeneuve, P.R., Joannopoulos, J.D. and Kolodziejski, L.A. (1999) *Phys. Rev. B*, **60**, 5751.

3 Labilloy, D., Benisty, H., Weisbuch, C., Krauss, T.F., De La Rue, R.M., Bardinal, V., Houdré, R., Oesterle, U., Cassagne, D. and Jouanin, C. (1997) *Phys. Rev. Lett.*, **79**, 4147.

4 Grüning, U., Lehmann, V. and Engelhardt, C.M. (1995) *Appl. Phys. Lett.*, **66**, 3254.

5 Grüning, U., Lehmann, V., Ottow, S. and Busch, K. (1996) *Appl. Phys. Lett.*, **68**, 747.

6 Lau, H.W., Parker, G.J., Greef, R. and Hölling, M. (1995) *Appl. Phys. Lett.*, **67**, 1877.

7 Lambrecht, A., Hartwig, S., Schweizer, S.L. and Wehrspohn, R.B. (2007) *Proc. SPIE*, **6480**, 64800D.

8 Lehmann, V. and Föll, H. (1990) *J. Eletrochem. Soc.*, **137**, 653.

9 Lehmann, V. (1993) *J. Electrochem. Soc.*, **140**, 2836.

10 Geppert, T., Schweizer, S.L., Gösele, U. and Wehrspohn, R.B. (2006) *Appl. Phys. A*, **84**, 237.

11 Leonard, S.W., van Driel, H.M., Birner, A., Gösele, U. and Villeneuve, P.R. (2000) *Opt. Lett.*, **25**, 1550.

12 Kramper, P., Agio, M., Soukoulis, C.M., Birner, A., Müller, F., Wehrspohn, R.B., Gösele, U. and Sandoghdar, V. (2004) *Phys. Rev. Lett.*, **92**, 113903.

13 Vlasov, Y.A. and McNab, S.J. (2006) *Opt. Lett.*, **31**, 50–52.

14 Martin-Moreno, L., Garcia-Vidal, F.J., Lezec, J.J., Degiron, A. and Ebbesen, T.W. (2003) *Phys. Rev. Lett.*, **90**, 167401.

15 Moreno, E., Garcia-Vidal, F. and Martin-Moreno, L. (2004) *Phys. Rev. B*, **69**, 121402.

16 Kramper, P., Birner, A., Agio, M., Soukoulis, C., Gösele, U., Mlynek, J. and Sandoghdar, V. (2001) *Phys. Rev. B*, **64**, 233102.

17 Kramper, P., Kafesaki, M., Birner, A., Müller, F., Gösele, U., Wehrspohn, R.B., Mlynek, J. and Sandoghdar, V. (2004) *Opt. Lett.*, **29**, 174.

18 Steinhart, M., Wendorff, J.H., Greiner, A., Wehrspohn, R.B., Nielsch, K., Schilling, J., Choi, J. and Gösele, U. (1997) *Science*, **296**.

19 Rogach, A., Kershaw, S., Burt, M., Harrison, M., Kornowski, A., Eychmüller, A. and Weller, H. (1999) *Adv. Mater.*, **11**, 552.

20 Olk, P., Buchler, B.C., Sandoghdar, V., Gaponik, N., Eychmüller, A. and Rogach, A.L. (2004) *Appl. Phys. Lett.*, **84**, 4732.

21 Richter, S., Steinhart, M., Hofmeister, H., Zacharias, M., Gösele, U., Schweizer, S.L., Rhein, A. v., Wehrspohn, R.B., Gaponik, N., Eychmüller, A. and Rogach, A. (2005) *Appl. Phys. Lett.*, **87**, 142107.

22 Richter, S., Hillebrand, R., Jamois, C., Zacharias, M., Güsele, U., Schweizer, S.L. and Wehrspohn, R.B. (2004) *Phys. Rev. B*, **70**, 193302.

23 Koenderink, A.F., Wüest, R., Buchler, B.C., Richter, S., Strasser, P., Kafesaki, M., Rogach, A., Wehrspohn, R.B., Soukoulis, C.M., Erni, D., Robin, F., Jäckel, H. and Sandoghdar, V. (2005) *Photon. Nanostruct.*, **3**, 63.

24 Gesemann, B., Rhein, A.v., Schweizer, S.L. and Wehrspohn, R.B. (2007) *Appl. Phys. Lett.*, submitted.

25 Wehrspohn, R.B., Rhein, A.V. and Lambrecht, A. (2007) Patent DE 102005008077.

26 Busch, K. and John, S. (1999) *Phys. Rev. Lett.*, **83**, 967.

27 Leonard, S.W., Mondia, J.P., van Driel, H.M., Toader, O., John, S., Busch, K., Birner, A., Gösele, U. and Lehmann, V. (2000) *Phys. Rev. B*, **61**, R2389.

28 Leonard, S.W. (2001) Ph.D. thesis, University of Toronto.

29 Ferrini, R., Martz, J., Zuppiroli, L., Wild, B., Zabelin, V., Dunbar, L.A., Houdre, R., Mulot, M. and Anand, S. (2006) *Opt. Lett.*, **31**, 1238.

30 Leonard, S.W., van Driel, H.M., Schilling, J. and Wehrspohn, R.B. (2002) *Phys. Rev. B*, **66**, 161102.

31 Tan, H.W., van Driel, H.M., Schweizer, S.L. and Wehrspohn, R.B. (2004) *Phys. Rev. B*, **70**, 205110.

32 Raineri, F., Cojocaru, C., Monnier, P., Levenson, A., Raj, R., Seassal, C., Letartre, X. and Viktorovitch, P. (2004) *Appl. Phys. Lett.*, **85**, 1880.

33 Tan, H.W., van Driel, H.M., Schweizer, S.L. and Wehrspohn, R.B. (2005) *Phys. Rev. B*, **72**, 165115.

34 Tan, H.W., van Driel, H.M., Schweizer, S.L. and Wehrspohn, R.B. (2006) *Phys. Rev. B*, **74**, 035116.

10
Dispersive Properties of Photonic Crystal Waveguide Resonators

T. Sünner, M. Gellner, M. Scholz, A. Löffler, M. Kamp, and A. Forchel

10.1
Introduction

Since the initial proposal of photonic crystals (PhCs) as a way to modify the optical properties of dielectrics by a (periodic) modulation of the refractive index [1,2], a wide variety of physical phenomena, structures and devices based on these fascinating materials has been reported. Most of the devices and functional elements realized in photonic crystals use two-dimensional geometries, where a periodic array of air holes is etched into either a membrane or a solid semiconductor slab waveguide which provides optical confinement in the vertical direction. The photonic bandgap that arises from the high contrast of the refractive indices allows the fabrication of very compact planar waveguide devices [3]. A major part of the experimental work on photonic crystals has focussed on the intensity response of photonic crystals and photonic crystal devices. This includes e.g. the demonstration of low loss PhC waveguides [4], bent waveguides [5], waveguide junctions [6] and the optimization of quality factor of PhC cavities [7]. However, photonic crystals also exhibit unusual dispersive properties. A prominent example is the superprism, where the propagation direction of light in the PhC undergoes large changes if the wavelength of the incident light is changed by only a small amount [8]. Other types of dispersive PhCs are e.g. waveguides with small group velocities or with a strong dependence of the group velocity on the wavelength of the incident light.

Waveguides with a small group velocity are interesting for applications as delay lines, but also for the enhancement of various light–matter interactions, such as optical amplification or absorption, electro-optic and non-linear effects. Group velocities two to three orders of magnitude smaller than those of conventional waveguides have already been observed in PhC waveguides [3,9,10]. Another type of structures with interesting dispersion properties are coupled resonator optical waveguides (CROWs), which allow a wide control over the dispersion properties by a change of the waveguide geometry [11,12]. Modes with low group velocities have already been utilized in PhC based laser devices, where the increased gain of these modes was used to define the laser mode [13].

Nanophotonic Materials: Photonic Crystals, Plasmonics, and Metamaterials. Edited by R. B. Wehrspohn, H.-S. Kitzerow, and K. Busch

ISBN: 978-3-527-40858-0

In this paper, we report on the dispersive properties of photonic crystal waveguide resonators. It is organized as follows: Section 10.2 gives a description of the resonator design and the fabrication technology. Section 10.3 provides an overview of the basic optical properties of the waveguide resonators. The setup used for the dispersion measurements and the dispersion characteristic of the resonators will be described in Section 10.4. An analysis of the results following two different approaches (Hilbert transformation and Fabry–Perot resonator model) will be presented in Section 10.5. Section 10.6 discusses the possibility of post-fabrication tuning, which allows an adjustment of the dispersive properties of the resonators after fabrication and an initial characterization. A conclusion is given in Section 10.7.

10.2 Design and Fabrication

10.2.1 Resonator Design

The last years have seen an increased understanding of the loss mechanisms in 2D membrane based PhC resonators. It was found that the radiative losses can be considerable reduced if the envelope of the localized field distribution is as close as possible to a Gaussian. If the resonant mode is decomposed into its Fourier components, a mode whose envelope closely follows a Gaussian has very little Fourier components above the light line of the membrane [14,15]. This leads to an elimination of loss channels and a corresponding increase of the quality factor. Based on these principles, a number of new cavity designs were reported, which deviate remarkably from the early 'remove holes from a PhC lattice' type [7,15]. The design used for this work is based on a heterostructure cavity [15]. The name originates from the fact that it is composed of several sections of a W1 waveguide (one missing line of holes in a PhC lattice) with different lattice periods.

A scanning electron microscopy (SEM) image of a heterostructure cavity is shown in Figure 10.1. The sections which act as mirrors for the cavity (highlighted in Figure 10.1) have a lattice constant of 400 nm along the direction of the waveguide; the cavity and the access sections have a lattice constant of 410 nm along the waveguide. The lattice constant perpendicular to the waveguide remains unchanged in order to maintain matching lattices. The ratio of hole radius and lattice constant (r/a) was varied between 0.2 and 0.25.

The W1 waveguide has a mode gap below the lower edge of the guided fundamental mode [7]. Any change of the lattice constant of the waveguide leads to a shift of this mode gap. In our case, the smaller lattice constant in the mirror sections results in a shift of the mode gap to higher frequencies. The resulting position of the mode gap along the waveguide is shown in Figure 10.2. The gray area above the mode gap corresponds to the guided modes in the W1 waveguide, the gray area below to either guided modes or extended modes of the PhC lattice. The difference of the spectral

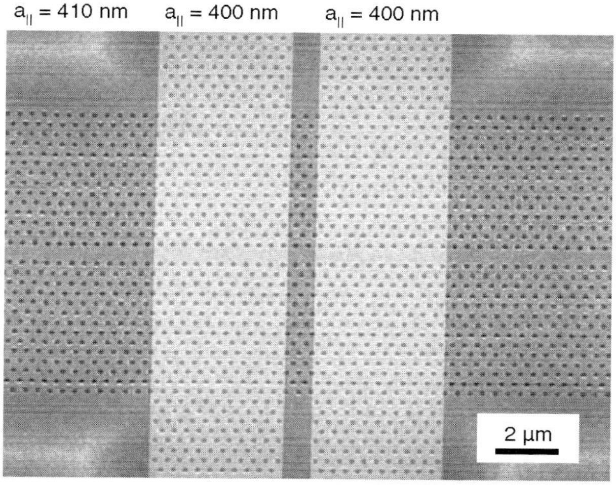

Figure 10.1 Scanning electron microscope image of a PhC heterostructure cavity. The lattice period of the PhC outside the marked areas is 410 nm. The ratio of hole radius and lattice constant (r/a) is 0.2.

positions of the mode gap leads to the formation of an 'optical well', which can localize light in the cavity (solid green line in Figure 10.2). Light with a frequency above the mode gap of the mirror section can propagate through the structure (yellow line in Figure 10.2). The envelope of the confined mode can by tuned to an almost

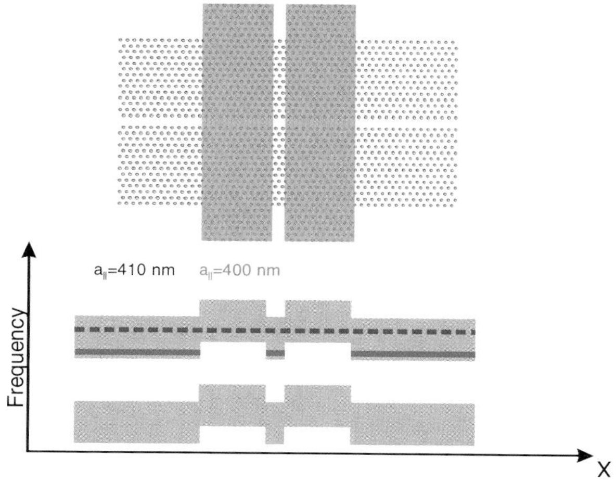

Figure 10.2 Upper part: Schematic of waveguide cavity. The areas with smaller lattice constant are marked. Lower part: Spectral position of the mode gap versus position. The structure is transparent for light with a frequency above the mode edge of the PhC waveguide with the smaller lattice constant (blue line). Light with a frequency just above the mode edge of the PhC waveguide with the larger lattice constant is localized in the cavity (green line).

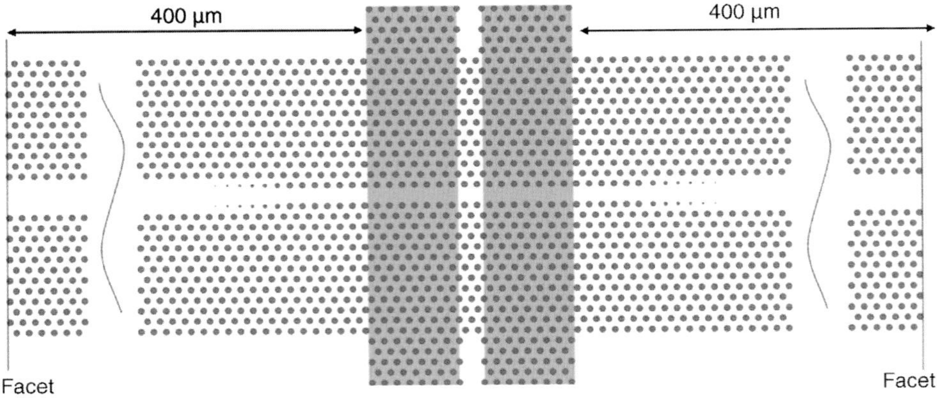

Figure 10.3 Schematic of the complete device. The lattice period of the photonic crystal is 400 nm in the marked regions and 410 nm everywhere else. Access guides with a length of around 400 μm and a width of three missing lines of holes (W3 guides) guide the light from the facets to the cavity. The taper to the W1 waveguide is realized by a gradual change of the hole radius.

perfect Gaussian shape be a careful adjustment of the lattice constants, resulting in very large quality factors.

Using this design, quality factors of over 2×10^5 have been reported for GaAs based cavities [16,17], cavities in Si have reached quality factors of up to 10^6 [18]. There are several possibilites to couple light in and out of such a cavity, e.g. by coupling from a neighbouring waveguide into the cavity and detecting the light emitted perpendicular to the surface [7]. We have used the W1 waveguides before and after the mirror sections as access guides. The coupling to the access guides (and therefore the quality factor) can be adjusted over a wide range by changing the length of the mirror regions. On the samples discussed in this paper, the length was varied between 5 to 15 lattice periods in order to realize resonators with different quality factors. At some distance from the cavity, the W1 is tapered to a W3 in order to increase the coupling efficiency to the lensed fiber which couples the probe light into the sample. Figure 10.3 shows a schematic of the entire device including the tapers and the access waveguides.

10.2.2 Fabrication

The PhCs used in this work are based on etched air holes in a membrane, which is fabricated from a GaAs/AlGaAs heterostructure by a combination of dry and wet etching [19]. The samples consists of a 240 nm thick GaAs layer on top of a 2 μm thick sacrificial $Al_{0.7}Ga_{0.3}As$ layer grown on a GaAs substrate by molecular beam epitaxy. The aluminium content of the sacrificial layer was chosen to be around 70%, in order to ensure a high selectivity during its wet chemical removal. A 100 nm thick SiO_2 layer was sputtered onto the sample, followed by spin coating of a 500 nm thick polymethymethacrylate (PMMA) resist layer. The PhC patterns were defined by electron

Figure 10.4 SEM image of a PhC waveguide facet. The access waveguide is realized by leaving out three lines of holes from the PhC lattice (W3 waveguide) and is tapered down to a W1 just before the cavity. The AlGaAs layer below the membrane has been removed by selective wet chemical etching.

beam lithography in the PMMA and then transferred into the SiO_2 hard mask by a reactive ion etch step. Subsequently, the holes were etched through the GaAs layer by electron-cyclotron-resonance reactive ion etching (ECR/RIE). This etch step was optimized with respect to vertical and smooth sidewalls of the holes. The optimized process uses a flow rate of 3.5 sccm Cl_2 and 27 sccm Ar at a pressure of 3.6×10^{-3} mbar, the RF power was 70 W and the ECR power 250 W. The etching time was 200 s, resulting in an etch depth well beyond the thickness of the waveguide layer. Residual PMMA was removed by soaking the sample in pyrrolidon and acetone. The sacrificial $Al_{0.7}Ga_{0.3}As$ layer was then removed by immersing the sample in diluted HF solution, leading to the formation of freestanding membranes. This step also removes the residual SiO_2 from the surface. Figure 10.4 shows an electron microscopy image of a finished PhC waveguide sample.

10.3 Transmission Measurements

The first characterization was performed in a setup for transmission measurements. This setup is schematically shown in Figure 10.5. The signal light is provided by a tunable semiconductor laser with a tuning range between 1460 nm and 1580 nm. The laser can be tuned in steps of 0.001 nm, which allows a precise measurement of fine spectral features. The polarization of the launched laser light was adjusted to be parallel to the membrane (TE polarization) by a fiber polarizer and coupled into the PhC waveguide by a lensed fiber. Behind the PhC, the light was collected by a microscope objective (NA = 0.35). Since the resonator interacts only with TE light and is non-existent for TM light, residual TM components of the transmitted light have to be blocked by a linear polarizer. An adjustable aperture was used to

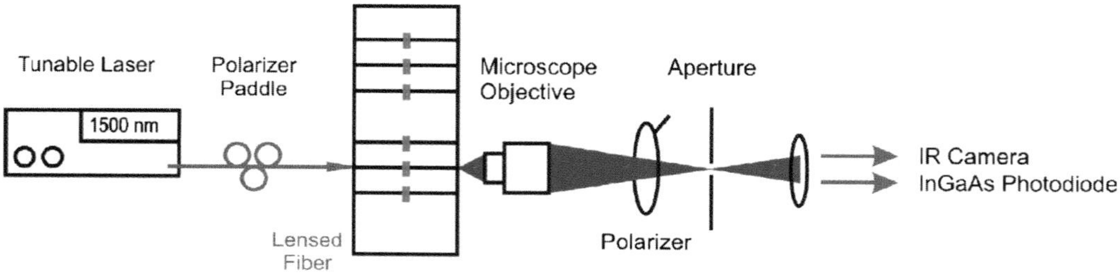

Figure 10.5 Setup for transmission measurements. Light from a tunable laser source is coupled to the PhC waveguides by a lensed fiber. The polarizer paddle is used to adjust the polarization of the light parallel the membrane. A microscope objective collects the transmitted light and creates an intermediate image of the waveguide facet at the opening of an adjustable aperture, thereby filtering out stray light. The polarizer blocks residual light with TM polarization. The IR camera is used to align the lensed fiber to the waveguide, the signal intensity is measured by the InGaAs photodiode.

eliminate stray light. An InGaAs photodiode connected to a lock-in amplifier measured the light intensity.

An example transmission measurement is depicted in Figure 10.6a. The upper plot shows the transmission of a W1 waveguide without resonator and a constant lattice period of 410 nm. For this geometry, the mode edge is located around 1532 nm. For light with a larger wavelength (smaller frequency), the transmission of the waveguide goes to zero. The lower part of Figure 10.6a shows the transmission of an actual cavity. The W1 mode edge in this spectrum is now determined by the regions with the smaller lattice constant of 400 nm, resulting in a shift of the mode edge to 1496 nm. The cavity resonance is located between those two spectral positions and

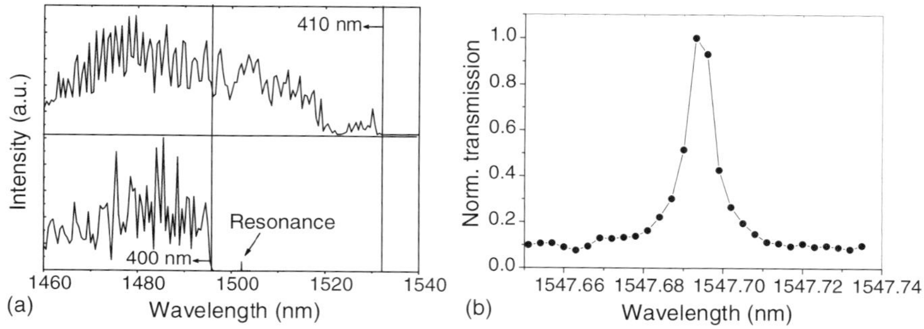

Figure 10.6 (a) Upper curve: transmission of a uniform W1 waveguide with 410 nm lattice period. Light with a wavelength larger than 1532 nm is within the mode gap of the waveguide and not transmitted. Lower curve: transmission measurement of a cavity. The additional waveguide segments with 400 nm lattice period have their mode edge at 1496 nm. The resonance appears a few nm above that mode edge (here at 1502 nm). (b) High resolution scan of the cavity transmission. The quality factor of this cavity is 220 000 and was determined using a Lorentzian fit to the data.

can indeed be seen around 1503 nm. This transmission characteristic allows a straighforward alignment of the setup, since the lensed fiber and the optics can first be adjusted at a wavelength below the mode edge (e.g. at $\lambda = 1480$ nm). A low resolution scan with a stepsize of typically 1 nm is then used to locate the precise position of the mode edge. Starting at the mode edge, a high resolution scan is now used to detect the cavity resonance. Figure 10.6b shows a high resolution scan of a cavity resonance. The quality factor was determined by a Lorentzian fit to the data and was found to be 220 000.

10.4 Dispersion Measurements

Measurements of the group delay and the dispersion are not as straighforward as measurements of the transmission or reflection, since they require either a phase sensitive detection or time resolved measurement of the transmitted signal. A number of approaches for measuring group delay and dispersion have been reported in the literature. One possibility is to use interferometric techniques. In combination with a near-field scanning optical microscopy, this has been used to observe the local phase and group velocities of modes propagating in a PhC waveguide [10]. It is also possible to use an 'on-chip' interferometer, e.g. a Mach–Zehnder, where one arm contains the device under test and the other provides a reference signal [20]. Another interferometric technique relies on measuring the spacing of Fabry–Perot fringes, which is inversely proportional to the group velocity [9,21]. A direct way to determine the group delay is a measurement of the time delay of short light pulses propagating through the PhC structure [22,23]. Finally, the phase shift technique can be used, which is commonly employed to characterize the dispersion of optoelectronic components and optical fibers [24]. This technique has been used by a number of groups to investigate the dispersive properties of PhC waveguides [3,25], resonators [26] and coupled resonator optical waveguides [12].

In this case, the probe light is modulated with a microwave signal, with the modulation frequency f_{mod} being typically around 1 GHz. The modulation creates sidebands with a frequency spacing of f_{mod}. The dispersion of the device under test leads to a phase shift of the modulation sidebands with respect to the carrier wave. The light signal is detected by a high speed photodiode and converted back to a microwave signal. The group delay can be calculated from the phase shift $\varphi(\lambda)$ of the microwave signal by the following equation:

$$\tau_g(\lambda) = -\frac{\varphi(\lambda)}{360^\circ \cdot f_{\mathrm{mod}}}. \tag{10.1}$$

The dispersion is given by the derivative of the group delay with respect to the wavelength: $D = \mathrm{d}\tau_g/\mathrm{d}\lambda$.

Figure 10.7 shows the setup for the dispersion measurement. A tunable laser source is again used as the optical source. The laser light is modulated using a $LiNbO_3$

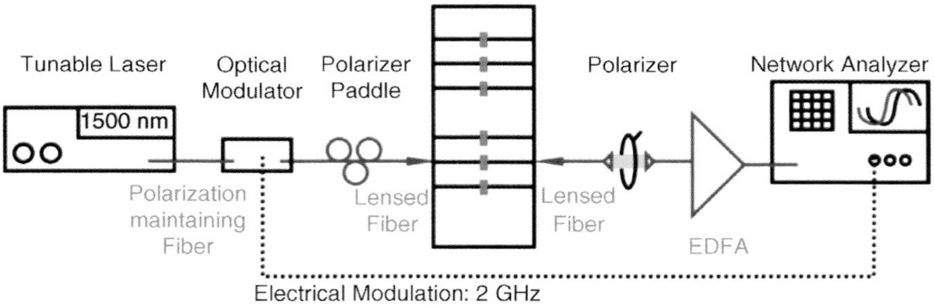

Figure 10.7 Setup for group delay measurements. The left part of the setup is similar to the one used for the transmission measurements, the only extra component is an optical modulator which is driven by a network analyzer. The transmitted light is picked up by a second lensed fiber. Residual TM light is again filtered out by a linear polarizer. In order to compensate the coupling losses to the PhC waveguides, an erbium doped fiber amplifier (EDFA) amplifies the light before it is fed into the optical port of the network analyzer.

Mach–Zehnder modulator, which is driven by a network analyser. The polarization was aligned parallel to the membrane (TE polarization) by fiber polarizer paddles and injected into the PhC waveguide by a lensed fiber. Behind the sample, the light was picked up by a second lensed fiber and residual TM polarized light was removed by a linear polarizer. An erbium doped fiber amplifier is inserted into the optical path in order to compensate the coupling losses, the losses of the modulator and the losses of the PhC devices. The amplifier is inserted into the optical signal path after the sample, since probing the resonators with a high input power can lead to nonlinear effects (predominantly two photon absorption), which will degrade the quality factor [27]. We estimate that the power coupled to the resonators is on the order of 10 μW, which is low enough to avoid nonlinear effects. A high speed optical receiver converts the optical signal back to a microwave signal, which is then analysed by the electrical network analyser (NWA). The phase sensitive detection of the microwave signal by the NWA allows a simultaneous measurement of the resonator transmission and the phase shift.

The choice of the modulation frequency f_{mod} is determined by two competing factors. On one hand a larger frequency enlarges the phase shift of the signal, thereby increasing the accuracy of the phase measurements. On the other hand, the spacing of the sidebands created by the modulation should by smaller than the spectral features of the device under test. For most of our measurements, a modulation frequency of 2 GHz was chosen. At a signal wavelength of 1.5 μm, this leads to sidebands with a spacing of 16 pm from the carrier wavelength. Figure 10.8 shows a measurement of the resonator transmission and the phase shift of the microwave signal.

This measurement was performed with a modulation frequency of 2 GHz. The transmission curve shows a resonance at 1558.9 nm with a quality factor of 48000. At the position of the resonance, the phase measurement shows a dip. The propagation of light is delayed on resonance, which results in a phase lag of the microwave signal.

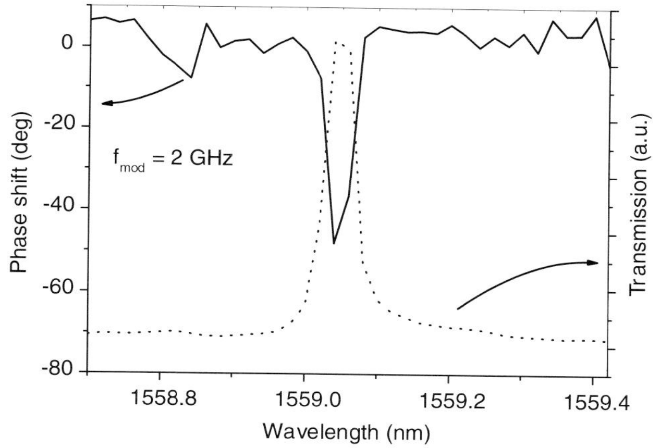

Figure 10.8 Tansmission (dotted line) and phase shift (solid line) measurement of a heterostructure cavity. The quality factor is $Q = 48000$ and the phase shift on resonance 51°, which translates into a group delay of 71 ps.

Using Eq. (10.1), the phase difference of 51° can be translated into a group delay of 71 ps. The scatter of the phase shift for wavelengths outside the resonance is caused by the small transmission in these regions. By taking the derivative of the group delay, one obtains the chromatic dispersion, which is shown in Figure 10.9. The minimum and maximum values of the dispersion are –1.635 ns/nm and 1.766 ns/nm, respectively. A typical single mode optical fiber has a dispersion of 17 ps/nm/km. With the measured anomalous dispersion of –1.635 ns/nm in this cavity, it is possible to compensate for the normal dispersion of around 100 km of optical fiber.

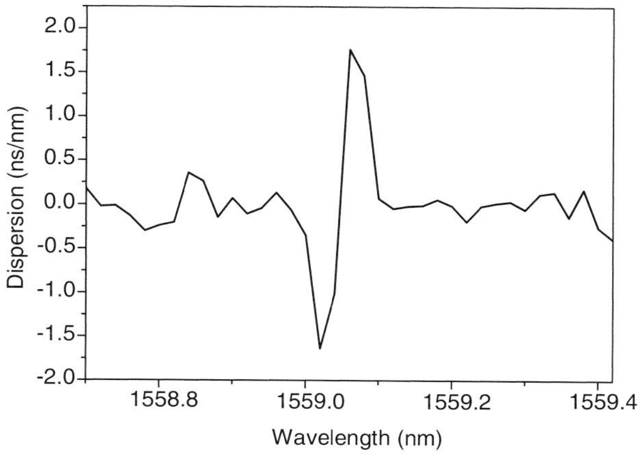

Figure 10.9 Dispersion corresponding to the measurement of the group delay plotted in Figure 10.7. The dispersion has a maximum of 1.77 ns/nm and a minimum of −1.64 ns/nm.

The highest group delay observed was 132 ps in a cavity with quality factor 81600. In this case, the mirror regions had a length of 12 lattice periods. This results in an effective length of the resonator (distance between the two outer edges of the mirrors) of 10.4 μm. The propagation speed of resonant light through the cavitiy is therefore 7.9×10^4 m/s, which corresponds to c/3800. For this cavity, the dispersion reaches values of up to ±2.9 ns/nm.

10.5 Analysis

10.5.1 Hilbert Transformation

The operation of any linear optical device on an optical signal can by described by its impulse response function $h(t)$, which satifies the causality relation $h(t)=0$ for $t<0$, where it is assumed that the impulse is applied at $t=0$. Alternatively, the filter can by described by the frequency transfer function $H(\omega)$, which is the Fourier transform of the impulse response. If the impulse response function $h(t)$ is real, satifies the causality relation and has no singularities at $t=0$, then the real and imaginay parts of the corresponding transfer function $H(\omega)=R(\omega)+I(\omega)$ are related by a pair of Hilbert transforms:

$$I(\omega)=\frac{1}{\pi}P\int_{-\infty}^{\infty}\frac{R(\omega')}{\omega-\omega'}\mathrm{d}\omega', \tag{10.2}$$

$$R(\omega)=-\frac{1}{\pi}P\int_{-\infty}^{\infty}\frac{I(\omega')}{\omega-\omega'}\mathrm{d}\omega', \tag{10.3}$$

where P denotes the principal value of the integral. These relations can be established because the Fourier transform of a real and causal function is analytic in the upper half of the complex plane. A well known example is the Kramers–Kronig relation, which connects the real and imaginary part of the dielectric constant. In case of optical filters however, one does not usually have access to the real and imaginary part of the transfer function, since a transmission measurement determines only the magnitude of the transfer function. However, it is still possible to determine the phase [28]. The transfer function has to be rewritten such that:

$$H(\omega)=\exp\left(-\beta(\omega)\right)\exp\left(i\phi(\omega)\right). \tag{10.4}$$

Now $|H(\omega)|=\exp\left(-\beta(\omega)\right)$ is the magnitude of the transfer function and $\phi(\omega)$ its phase. Taking the logarithm of Eq. (10.4), we obtain:

$$\log\left(H(\omega)\right)=-\beta(\omega)+i\phi(\omega). \tag{10.5}$$

The logarithm of $H(\omega)$ has to be analytic in the upper half of the complex plane. In particular $H(\omega)$, must not have any zeros in this region. This is not always the case for optical filters, a prominent example is the Gires-Tournois interferometer, which has a reflection of 100% for all wavelengths, but a quite complex phase response [24]. The condition is however met for Fabry–Perot resonators and Eqs. (10.2) and (10.3) can be applied. Although we cannot strictly show that the condition is met for the PhC resonators presented in manuscript, the similarity to Fabry–Perot resonators gives us resonably confidence that the Hilbert transform can be applied in this case. $\beta(\omega)$ can be calculated from a transmission experiment and Eq. (10.2) can be used to determine the phase $\phi(\omega)$. The group delay can then be obtained by taking the derivative of the phase with respect to the wavelength:

$$\tau_g = \frac{\partial \phi(\omega)}{\partial \omega}. \tag{10.6}$$

Figure 10.10 shows the measured transmission of a resonator and the group delay calculated using the Hilbert transform oulined above. The measured and calculated values agree very well. As already discussed above, the small signal intensity in the regions outside the resonance leads to an increase of the noise. The ripple on the calculated curve is a result of the derivative (Eq. (10.6)), which is used to calculate the group delay from the recovered phase information. The small peaks on the left and right of the resonance are most likely Fabry–Perot resonances of the entire sample. The smaller quality factors of these resonances result in a smaller group delays, visible in both the measures and calculated values.

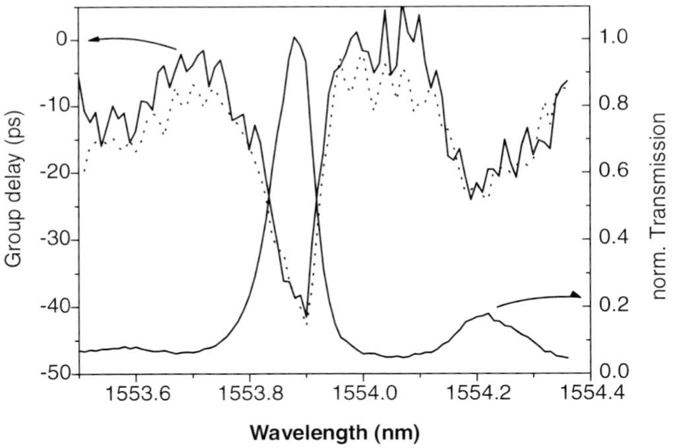

Figure 10.10 Tansmission and group delay of a PhC resonator. The solid line corresponds to the measured group delay, the dotted line is the group delay calculated from the transmission by using a Hilbert transformation.

10.5.2
Fabry–Perot Model

The heterostructure cavity consists of a cavity region sandwiched between a pair of mirrors, resembling the design of a Fabry–Perot (FP) cavity. For the equivalent FP resonator, the mirror waveguides can be replaced by effective mirrors with an effective reflectivity R and spacing L. Adding up the amplitudes of transmitted and reflected beams leads to the complex transfer function of the FP resonator:

$$H(\omega) = \frac{T\mathrm{e}^{-i\delta/2}}{1 - R\,\mathrm{e}^{-i\delta}}. \tag{10.7}$$

T denotes the (intensity) transmission coefficient and R the (intensity) reflection coefficient of the mirrors. The phase shift of the light after one round trip in the resonator is δ. For a lossless resonator, we have $T + R = 1$. With this function, the phase delay of the transmitted light $\phi(\omega)$ can be calculated by taking the phase of the complex transfer function.

$$\phi(\omega) = -\arctan\left(\frac{1+R}{1-R}\tan\frac{\pi\omega}{\omega_{\mathrm{FSR}}}\right). \tag{10.8}$$

In the latter equation, we have introduced the free spectral range ω_{FSR} of the Fabry–Perot resonator, which is the spacing of the resonances. Taking the negative derivative with respect to the frequency gives the group delay as a function of frequency:

$$\tau_g(\omega) = \frac{\pi}{\omega_{\mathrm{FSR}}}\frac{1-R^2}{1+R^2-2R\cos\left(\frac{2\pi\omega}{\omega_{\mathrm{FSR}}}\right)}. \tag{10.9}$$

For high quality factors, the mirror reflectivity is very close to unity. Taking this into account and including the quality factor of the FP resonator given by:

$$Q = \frac{\omega_{\mathrm{res}}}{\Delta\omega} = \frac{\omega_{\mathrm{res}} \cdot R^{1/2} \cdot \pi}{\omega_{\mathrm{FSR}}(1-R)} \tag{10.10}$$

one ends up with a linear relation between the group delay and the quality factor:

$$\tau_g = \frac{2Q}{\omega_{\mathrm{res}}}. \tag{10.11}$$

Figure 10.11 shows a series of measurements of the group delay for cavities with different mirror sections, and therefore different quality factors.

As expected, cavities with larger quality factors have also larger group delay. The scatter of the phase for wavelengths out of resonance leads to an uncertainty of the group delay (error bars in Figure 10.11), which was calculated from the standard deviation of the phase out of resonance. The straight line was calculated using Eq. (10.11) for an average resonance wavelength of 1530 nm.

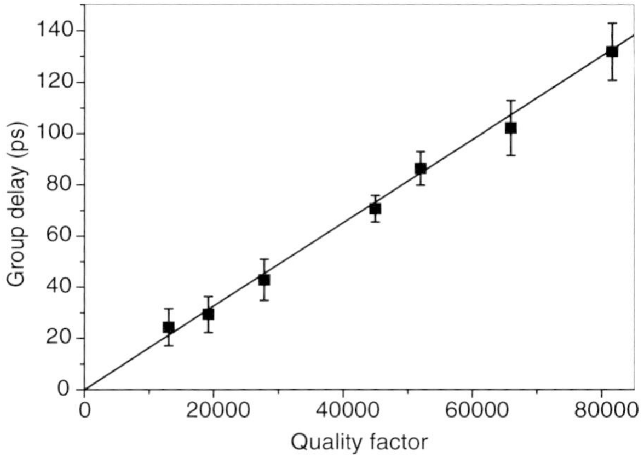

Figure 10.11 Group delay versus quality factors for different cavities. The error bars are estimated from the scatter of the phase for wavelengths out of resonance, which leads to an uncertainty of the group delay. The solid line depicts the theoretical group delay for a resonance wavelength of 1530 nm.

10.6
Postfabrication Tuning

The ability to adjust the resonance wavelength of the cavities after fabrication and an initial characterization is highly desirable. Due to the high index contrast of the PhC, even small, nm-size deviations from the target geometry result in significant changes of the resonance wavelength. This leads to a scatter of the resonance wavelength of several nm for nominally identical structures. In order to tune the resonance to a given wavelength, a precise postfabrication tuning method is therefore required. Several techniques for tuning the resonant wavelength have been investigated. The first one is a chemical etching technique refered to as digital etching [17,29]. In a traditional etching method, the etching time defines the etching depth. In this method, the etch depth is defined by the number of etching cycles and not the etching time, allowing a precise control of the etch depth. One etching cycle consists of two steps. In the first step an oxide is grown on the surface of the sample. This is done either by a wet chemical oxidation or by exposure to an oxidizing gas. In the next step, the oxide (and only the oxide) is removed by soaking the sample in a selective etchant, which in our case is HCl. This enlarges the hole radius and thins the membrane, leading to a blueshift of the resonance. Part a) of Figure 10.12 shows the wavelength shift of two samples. The oxidation was performed by a simple exposure to ambient conditions. After 10 min (30 min) of air exposure, the position of the resonance was measured and the oxide layer was removed by another soaking in HCl solution. Tuning steps as small as 1.9 nm (2.6 nm) per etch cycle were achieved.

A different tuning method uses the shift of the resonance towards longer wavelengths with increasing temperature. This is caused by the temperature depen-

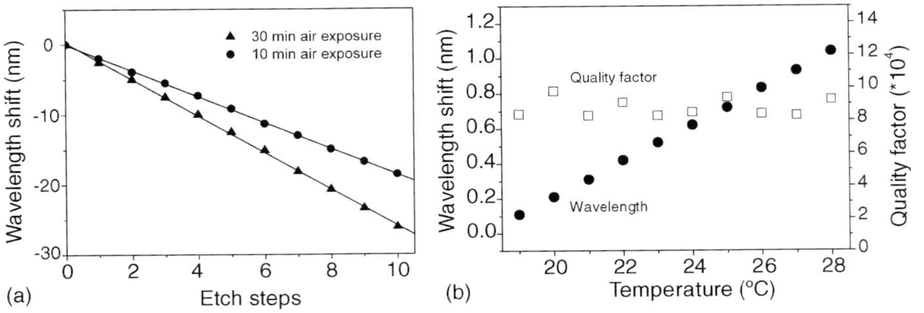

Figure 10.12 Wavelength shift for different tuning methods. (a) Wavelength shift after several cycles of digital etching. Oxidation was performed by 10 min (circles) and 30 min (triangles) of air exposure, resulting in wavelength shifts of 1.9 nm and 2.6 nm per etch cycle. (b) Wavelength (solid circles) and quality factor (open rectangles) versus sample temperature. The change of the refractive index with temperature leads to a wavelength shift of 0.1 nm per K. The quality factor remains constant.

dence of the refractive index, which in case of a semiconductor increases with temperature. Part b) of Figure 10.12 shows the wavelength shift of the resonance with increasing temperature. The linear shift has a slope of 0.1 nm per degree Celsius. There is no significant change of the quality factor around its average value of 86500 as the temperature is changed.

10.7 Conclusion

We have fabricated PhC waveguide resonators in GaAs membranes with quality factors of up to 220 000. The group delay of light propagating through the resonators was measured using a phase shift technique. A maximum group delay of 132 ps for a cavity with a quality factor of 82 000 was observed. The group delay was found to be proportional to the quality factor of the cavity. These results are in line with values published by other groups. Photon lifetimes of 9 ps (which correspond to a group delay of 18 ps) were observed for cavities with quality factors of 12 000 [22]. Cavities with extremely high quality factors of 1.2×10^6 have been reported to store photons for as long as 1 ns [23].

The cavities discussed in this paper have an overall length of 10.4 μm, yielding a propagation speed of 7.88×10^4 m/s (c/3800) for the light transmitted through the structure. This is smaller than the effective propagation speed observed in PhC waveguides, which is on the order of c/1000 – c/100 [9,10]. However, this comes at the expense of a smaller bandwidth, which is on the order of a few pm. The dispersion of resonators with group delays of 71 ps is around 1.7 ns/nm, which is equivalent to 100 km of standard optical fiber. This is much larger than the dispersion realized with PhC cavities on slab waveguides, which had quality factors of only 12 000 and dispersions of up to $\pm$250 ps/nm [26].

Two different models were used to analyse the measurements, one where the group velocity of the transmitted light was reconstructed from the transmission amplitude using a Hilbert transform, and a second one based on replacing the PhC cavity with an equivalent Fabry–Perot resonator. The group delay determined by the Hilbert transform and the measured values were found to be in good agreement. The Fabry–Perot model predicts a linear increase of the group delay with increasing quality factor, which was also observed in the experiments. Two postfabrication techniques, which allow an adjustment of the resonance position (and hence the dispersive properties) after an initial characterization of the resonator, were investigated. Digital etching allows a discrete tuining of the resonance position by a controlled removal of material, whereas a change of the temperature leads to a continuous change of the resonace position cause be the temperature dependence of the refractive index.

References

1 Yablonovitch, E. (1987) *Phys. Rev. Lett.*, **58**, 2059.

2 John, S. (1987) *Phys. Rev. Lett.*, **58**, 2486.

3 Notomi, M., Shinya, A., Mitsugi, S., Kuramochi, E. and Ryu, H. (2004) *Opt. Express*, **12**, 1551.

4 Sugimoto, Y., Tanaka, Y., Ikeda, N., Nakamura, Y., Asakawa, K. and Inoue, K. (2004) *Opt. Express*, **12**, 1090.

5 Têtu, A., Kristensen, M., Frandsen, L.H., Harpøth, A., Borel, P.I., Jensen, J.S. and Sigmund, O. (2005) *Opt. Express*, **13**, 8606.

6 Boscolo, S., Midrio, M. and Krauss, T.F. (2002) *Opt. Lett.*, **27**, 1001.

7 Song, B.-S., Noda, S., Asano, T. and Akahane, Y. (2005) *Nature Mater.*, **4**, 207.

8 Kosaka, H., Kawashima, T., Tomita, A., Notomi, M., Tamamura, T., Sato, T. and Kawakami, S. (1999) *J. Lightwave Technol.*, **17**, 2032.

9 Notomi, M., Yamada, K., Shinya, A., Takahashi, J., Takahashi, C. and Yokohama, I. (2001) *Phys. Rev. Lett.*, **87**, 253902.

10 Gersen, H., Karle, T.J., Engelen, R.J.P., Bogaerts, W., Korterik, J.P., van Hulst, N.F., Krauss, T.F. and Kuipers, L. (2005) *Phys. Rev. Lett.*, **94**, 073903.

11 Hosomi, K. and Katsuyama, T. (2002) *IEEE J. Quantum Electron.*, **38**, 825.

12 Talneau, A., Aubin, G., Uddhammar, A., Berrier, A., Mulot, M. and Anand, S. (2006) *Appl. Phys. Lett.*, **88**, 201106.

13 Happ, T.D., Kamp, M., Forchel, A., Gentner, J.-L. and Goldstein, L. (2003) *Appl. Phys. Lett.*, **82**, 4.

14 Akahane, Y., Asano, T., Song, B.S. and Noda, S. (2003) *Nature*, **425**, 944.

15 Srinivasan, K. and Painter, O. (2003) *Opt. Express*, **11**, 579.

16 Weidner, E., Combrié S., Tran, N.V.Q., De Rossi, A., Nagle, J., Cassette, S., Talneau, A. and Benisty, H. (2006) *Appl. Phys. Lett.*, **89**, 221104.

17 Sünner, T., Herrmann, R., Löffler, A., Kamp, M. and Forchel, A. (2007) *Microelectron. Eng.*, **84**, 1405.

18 Kuramochi, E., Notomi, M., Mitsugi, S., Shinya, A., Tanabe, T. and Watanabe, T. (2006) *Appl. Phys. Lett.*, **88**, 041112.

19 Herrmann, R., Sünner, T., Hein, T., Löffler, A., Kamp, M. and Forchel, A. (2006) *Opt. Lett.*, **31**, 1229.

20 Vlasov, Y.A., O'Boyle, M., Hamann, H.F. and McNab, S.J. (2005) *Nature*, **438**, 65.

21 Letartre, X., Seassal, C., Grillet, C., Rojo-Romero, P., Viktorovich, P., Le Vassor

d'Yerville, M., Cassagne, D. and Jouanin, C. (2001) *Appl. Phys. Lett.*, **79**, 2312.

22 Asano, T., Kunishi, W., Song, B.-S. and Noda, S. (2006) *Appl. Phys. Lett.*, **88**, 151102.

23 Tanabe, T., Notomi, M., Kuramochi, E., Shinya, A. and Taniyama, H. (2007) *Nature Photonics*, **1**, 49.

24 Costa, B., Mazzoni, D., Puelo, M. and Vezzoni, E. (1982) *IEEE J. Quantum Electron.*, **18**, 1509.

25 Jacobsen, R., Lavrinenko, A., Frandsen, L., Peucheret, C., Zsigri, B., Moulin, G., Fage-Pedersen, J. and Borel, P. (2005) *Opt. Express*, **13**, 7861.

26 Zimmermann, J., Saravanan, B.K., März, R., Kamp, M., Forchel, A. and Anand, S. (2005) *Electron. Lett.*, **41**, 45.

27 Barclay, P.E., Srinivasan, K. and Painter, O. (2005) *Opt. Express*, **13**, 801.

28 Beck, M., Wamsley, I.A. and Kafka, J.D. (1991) *IEEE J. Quantum Electron.*, **27**, 2074.

29 DeSalvo, G.C., Bozada, C.A., Ebel, J.L., Look, D.C., Barette, J.P., Cerny, C.L.A., Dettmer, R.W., Gillespie, J.K., Havasy, C.K., Jenkins, T.J., Nakano, K., Pettiford, C.I., Quach, T.K., Sewell, J.S. and Via, G.D. (1996) *J. Electrochem. Soc.*, **143**, 3652.

II
Tuneable Photonic Crystals

Nanophotonic Materials: Photonic Crystals, Plasmonics, and Metamaterials. Edited by R. B. Wehrspohn,
H.-S. Kitzerow, and K. Busch

ISBN: 978-3-527-40858-0

11
Polymer Based Tuneable Photonic Crystals

J.H. Wülbern, M. Schmidt, U. Hübner, R. Boucher, W. Volksen, Y. Lu, R. Zentel, and M. Eich

11.1
Introduction

A long standing challenge in photonics is the implementation of a nanophotonic circuit into a non-centrosymmetric medium which exhibits a second-order nonlinear optical susceptibility based on electronic displacement polarization. The inherent quasi-instantaneous response of the nonlinear polarization in such media generates the potential of ultra fast electro-optical sub micrometer photonic devices with switching bandwidths well beyond 100 GHz. Such functionalities will play a vital role in next-generation computer technologies which will employ optical chip-to-chip and even on-chip optical communications on micron dimensions.

Optically transparent dielectrics which are periodically structured in one, two, or three dimensions generally are called photonic crystals (PhCs) [1,2]. The structuring typically results in frequency gaps which inhibit propagation of waves at such frequencies. Discontinuities or defects implanted on purpose into the otherwise regular lattice lead to localized states which provide the option for micron-sized waveguides [3,4] and optical resonators of high quality-factors [5–8]. We concentrate on photonic crystal slabs from optical slab waveguides into which vertical holes are drilled and which typically exhibit gaps for transverse electric (TE) polarization.

The concept proposed here utilizes the ultra fast response times of the Pockels effect, which is a second-order nonlinear optical effect based on electronic displacement polarization. This effect can only be observed in noncentrosymmetric media, such as single-crystalline GaAs, poled LiNbO3, and in special poled organic polymers which carry nonlinear optically active groups. Covalently functionalized organic nonlinear optical polymers [9] poled using high-field strengths have been shown to exhibit very high electro-optical susceptibilities from several pm/V to well above 100 pm/V [10,11]. These materials can be structured on submicron scales to form photonic crystal slabs [12]. Whenever direction independent stop gaps are not

Nanophotonic Materials: Photonic Crystals, Plasmonics, and Metamaterials. Edited by R. B. Wehrspohn, H.-S. Kitzerow, and K. Busch

ISBN: 978-3-527-40858-0

needed, such as in ultra refractive dispersive photonic crystals and in one dimensional resonators the relatively small air-polymer refractive index contrast does not impose a particular disadvantage.

11.2
Preparation of Photonic Crystal Structures in Polymer Waveguide Material

11.2.1
Materials

PMMA/DR-1 [12], APC (amorphous polycarbonate) and TOPAS® (a copolymer of ethylene and a cycloolefin) which can be doped and covalently functionalized with molecules that possess strong second-order hyperpolarisability, were used as the waveguide cores. These materials are advantageous for photonic crystal structures because they show very low optical waveguide losses in the near infrared regime around 1300 nm (<1.0 dB/cm) due to their amorphous structures and the low absorption of their constituents. Typically, these polymers have a refractive index of $n = 1.54$ at a wavelength $\lambda = 1300$ nm.

PMMA/DR-1 is a poly(methyl methacrylate) polymer in which 10 mol% of its monomeric units are covalently functionalized with the nonlinear chromophore "disperse red" as a side group [12]. PMMA/DR-1 has a glass transition temperature of $T_g = 113\,°C$. It is a second-order nonlinear optical medium, which can be used for the realization of electro-optical photonic crystals. APC (amorphous polycarbonate) and TOPAS® have much higher glass transition temperatures ($T_g = 180\,°C$ to $203\,°C$) than PMMA/DR-1. Higher T_g-polymers are advantageous because they hinder the reorientation of the chromophores and prevent the relaxation of their non-linear optical properties (NLO) at ambient conditions after poling, which is used to break the centrosymmetry of the polymer and create a bulk second-order nonlinear polarisability in the material.

The polymers were spin coated and dried using a baking procedure. The whole waveguide stack, consisting of a 1.5 μm thick waveguide core layer and an optical waveguide substrate (described below), were deposited on n-doped polished or oxidized 3″-[100]Si-wafers.

The fabrication of the PhC structures involved two different waveguide substrate materials: amorphous poly (tetrafluoro-ethylene) (a-PTFE, Teflon, DuPont) with a refractive index of $n = 1.30$ at $\lambda = 1300$ nm (Figure 11.1a) and an ultra low refractive index "air like" mesoporous silica substrate with $n = 1.14$ at $\lambda = 1300$ nm, (Figure 11.1b). This "air-like" material has an air-filling fraction of 70%. It provides a higher vertical index contrast than the Teflon layer, and a nearly symmetric vertical index profile, which reduces the mode mismatch in the perforated region of the waveguide between core and substrate. Consequently, the PhC structures made on low index substrates are easier to fabricate because only a small over etch into the substrate is needed to achieve total internal reflection at the core/substrate boundary [13] and reduce optical losses.

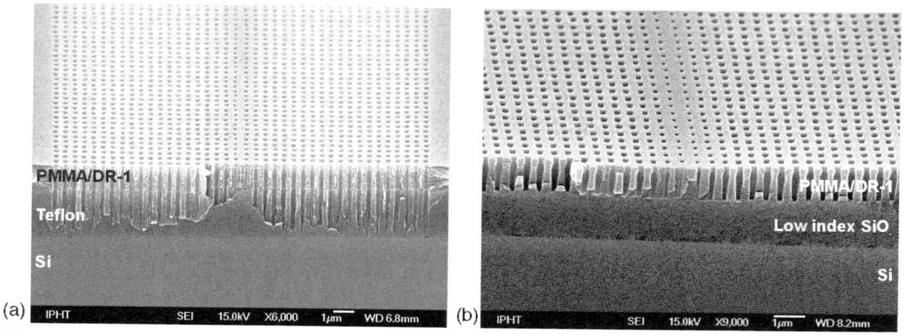

Figure 11.1 SEM-images show cross-sections of "tapered" line defect resonators (hole widths from 160 nm to 300 nm, period 485 nm, a top view is shown in Figure 11.4); (a) PMMA/DR-1 on Teflon, etch depth: 4 μm, aspect ratio (hole width: hole depth) = 1:13, (b) PMMA/DR-1 on mesoporous silica, etch depth: 1.5 μm, aspect ratio = 1:5.

11.2.2
Fabrication

The fabrication of the PhC pattern starts with an electron beam lithography step in which the etch mask is created. On top of the polymer waveguide, an e-beam resist and a thin NiCr-film are deposited. The patterns of the resist mask are transferred into the NiCr-film by Ar ion beam etching. The NiCr-hard mask serves as the mask in the subsequent deep etch into the waveguide and the substrate layer (Figure 11.2).

The fabricated two-dimensional PhCs consist of a slab waveguide perforated by a periodic array of air holes. The introduction of defects into the PhC structure leads to states in the band gap, which can be used to form high Q cavities or PhC defect waveguides. Designed for an optical wavelength of about $\lambda = 1300$ nm the lattice of the finite two dimensional photonic crystal consists of two sections of square hole

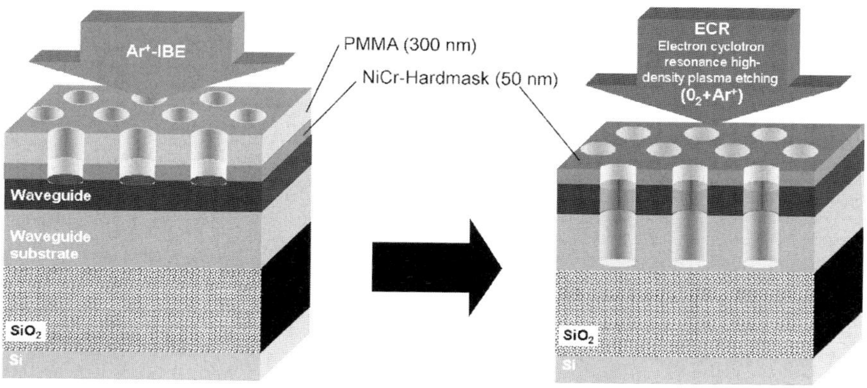

Figure 11.2 Sketch of the overall fabrication process.

lattices with hole diameters of around 300 nm and a lattice constant (period) of 485 nm. Typically, both sections were 8000 lattice constants wide and 10 lattice constants long and were positioned with a separation of 3 or 4 lattice constants, thereby creating a W3 or a W4 linear defect (10_3_10, see also Figure 11.10b), respectively. In addition, a sophisticated version of the waveguide slab, the so called radius tapered line defect resonator, was fabricated in both material combinations (Figure 11.4b). In the tapered resonator the region between the two sections was filled by a hole array with variable hole diameters from 160 nm to 300 nm and a period of 485 nm. Simulations show that such structures should enhance the Q-factor of the mirror by a factor of 10 compared to the PhCs without a taper section [5–8].

Foresi-like resonators [14] are planar ridge waveguides with hole arrangements in the waveguide (see Figure 11.3). The fabrication of the ridge waveguides starts with the realisation of the Au-electrodes and the alignment marks. On the 3″ wafer the 100 nm thick Au-structures were deposited by using a lift-off process. Afterwards, the polymer waveguide material was spun, baked out, either poled or not poled and the above mentioned NiCr- and PMMA-film were deposited. The etching procedure then continued in the same manner as established for the line-defect resonators. The planar waveguides have a tapered width, beginning at the wafer edge with a width of about 4 mm and reducing to a width of 700 nm near the electrodes. By using the alignment marks the large scale pattern (waveguide taper) was exposed by means of the shaped e-beam machine and the inner and high quality part of the structure was exposed using the Gaussian e-beam tool in the same run.

For the e-beam writing step two shaped beam machines (ZBA23H – 40 keV, SB350OS – 50 keV) and in special cases a Gaussian beam writer LION LV-1 (20 keV) were used. The ZBA23H is an older shaped beam writer with a resolution of down to 200 nm, with the capability to handle up to 6″ substrates. The SB350OS is a state-of-the-art shaped beam machine, which can fulfil the 65 nm node requirements of the semiconductor industry, deal with 300 mm substrates and can be used to undertake

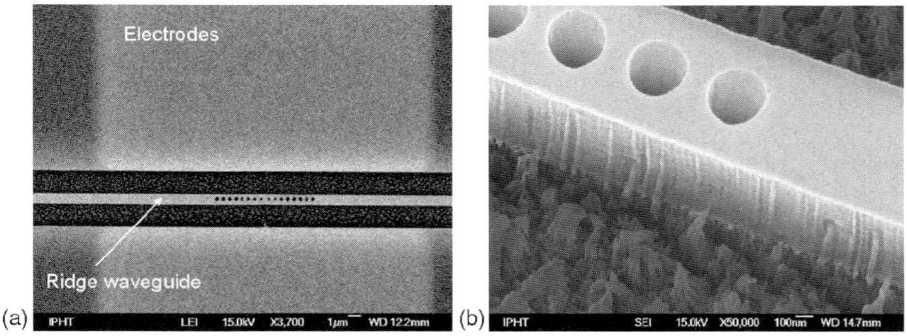

Figure 11.3 SEM-images show a planar ridge waveguide with Foresi-like resonator pattern etched in PMMA/DR-1 on mesoporous silica (e-beam: LION-LV1). The planar waveguide has a width of 700 nm and the holes in the waveguide are between 160 nm and 300 nm in diameter.

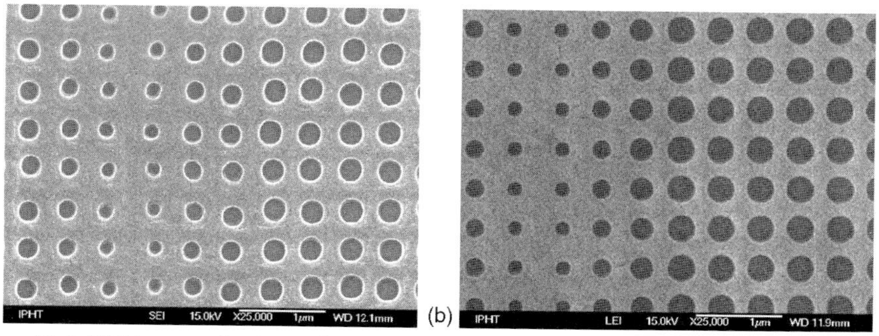

Figure 11.4 SEM-images of resist masks written with the different e-beam tools; (a) ZBA23H, an older e-beam writer (error of image placement <150 nm over a 6″ wafer) and (b) the "65 nm node" e-beam writer SB350 OS (error of image placement <18 nm over a 6″ wafer).

very accurately placed and relatively fast e-beam exposures down to sub-100 nm structures over large areas. Because of the change from the ZBA23H to the SB350OS strong improvements of the pattern shape quality (i.e. the circularity of the holes) and the pattern placement (grating arrangement) were achieved (Figure 11.4). The e-beam system LION-LV1 provides minimal Gaussian beam diameters from 2 nm at 20 keV to 6 nm at 2.5 keV and has a high precision 6″-x,y-stage. All these e-beam tools are made by Vistec Electron Beam GmbH (formerly Leica Microsystems Lithography GmbH). These tools are compatible and allow *Mix&Match*-procedures based on the use of alignment marks.

For the e-beam lithography resist ARP671 (a PMMA from Allresist GmbH) was used. The PMMA combines high resolution with high process reproducibility and lifetime, but it has only low e-beam sensitivity, and also low ion beam etch resistance. By using a resist thickness of 300 nm the ARP671 allows the reproducible realisation of micro holes with diameters of down to 160 nm in the 50 nm thick NiCr-hard mask. The resist film was prepared by the spin-coating technique and baked for 1 h at 100 °C on a hotplate. This relatively low temperature (normally used: 1 h at 180 °C) was chosen because of the T_g of the waveguide and substrate polymers. In order to avoid charging effects during the e-beam exposure the resist is covered with a 10 nm Au-film. After e-beam exposure the Au-film is removed and the development is carried out for 60 s at 21 °C in a 1:1 mixture of methylisobutylketone (MIBK) and isopropanol (IPA). More details about the e-beam exposure of the PhC structures are described in [15].

In order to achieve high quality submicron holes with diameters of down to 160 nm and an etch depth of 4 μm (into the PMMA/DR-1/Teflon-system) a hard mask of 50 nm NiCr was used. The NiCr-layer was deposited by Ar ion beam sputtering. During the patterning, the e-beam written resist mask is transferred by Ar ion beam etching (IBE) into the NiCr-layer. The NiCr-mask is needed as the etch mask for the subsequent electron cyclotron resonance (ECR) etch process. The results of the complete mask making process are holes with an average deviation from the design diameter of about +10%, i.e. 30 nm for a 300 nm diameter.

For the pattern transfer of the NiCr-mask into the waveguide core and the waveguide substrate an ECR high-density plasma system with a radio frequency (RF) biased substrate was used [16]. A big issue for the ECR etching step is the realisation of cylindrical shaped holes with the designed hole diameter and hole depth. The etch result depends on several process parameters like the RF- and the ECR-power, and the choice of and the mixing ratio of the etch gases.

The holes are etched through the core into the substrate in order to reduce the effective refractive index of the substrate. Consequently, total internal reflection at the core/substrate boundary is achieved [17]. For PMMA/DR-1 on mesoporous silica little etching is needed into the underlayer, because the already high index contrast demands only a small over etch into the waveguide substrate. For PMMA/DR-1 on Teflon the opposite situation is at play, the low index contrast needs a large over etch into the waveguide substrate. The PMMA/DR-1, etched with a mixture of O_2 and Ar, has a very high etch rate, 800 nm per minute for 250 nm diameter holes. The Teflon etches with a large rate of 5.04 μm per minute. Large aspect ratios can be achieved, 16 nm for 250 nm diameter holes [18], see also Figure 11.1a. Good mask to hole diameter integrity and cylindrical holes are achieved when using high biases on the substrate.

The applied substrate bias, along with the gas mixture have been found to have the strongest influence on the etch rate and hole shape. A reduction in the amount of O_2 in the O_2/Ar gas mixture causes the etch rate to reduce. This is especially noticeable when there is a >2 times higher Ar than O_2 flow rate. For pure Ar plasmas the etch rate drops down to a few tens of nm per minute. Conversely, for pure O_2 plasmas large etch rates are achieved, but with the penalty of an increased amount of undercut. Therefore, a compromise is needed in order to obtain a balance between the hole shape and etch rate.

The RF power influences the bias on the substrate and as a consequence both the etch rate and the amount of under cut seen, where this decreases with increasing bias. Therefore, a higher bias is desirable, especially as it also increases the etch rate. Both the ECR power and substrate-resonance region separation have an influence on this bias, in that for both a higher ECR power and for a smaller separation lower biases are generated across the substrate. Therefore, a low ECR power and large separation is desirable. Added to this the polymers are often mechanically and thermally sensitive and the heating power of the plasma can damage them. Consequently, good thermal contact of the sample to the substrate is a pre-requisite.

For some systems such as BCB (Benzocyclobutene), Ta_2O_5 and SiO_2 it is also possible to find a regime where a small bias on the substrate can be used and still straight walls can be achieved (Figure 11.5). This occurs because there is some polymerisation on the surface, in the case of SiO_2 and Ta_2O_5 due to the etch gas (typically CHF_3 and/or CF_4).

The polymerisation rate on the sidewalls of the holes then finds a balance with the wall etch rate at some low bias such that no undercut is caused. However, at the bottom of the holes the etch rate is still higher than the polymerisation rate and so

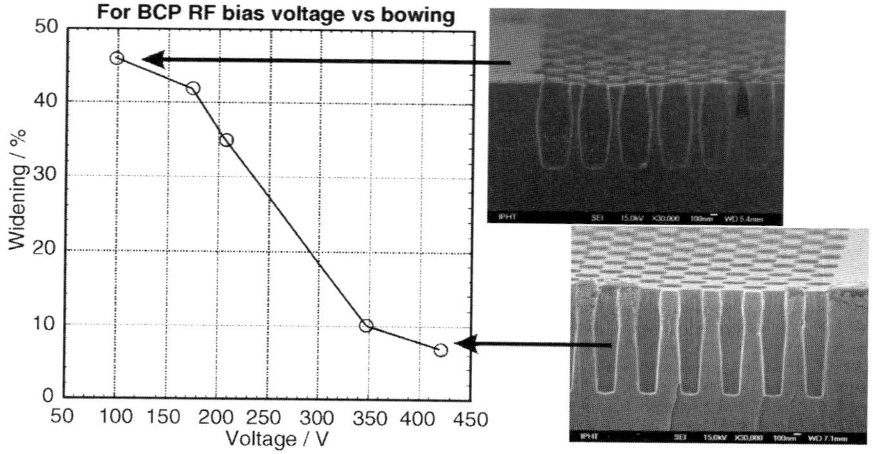

Figure 11.5 Dependence of the hole width on applied substrate bias. The widening is defined as the difference between the hole entrance width and the wider region divided by the hole entrance width.

overall etching occurs. BCB contains Si and a surface coating is formed with O_2 so that this material can also be etched with a small bias without any undercut. However, this is not the case for most polymers when etched in O_2. For these material etch gas combinations no surface protection layers are formed and so reducing the bias just has the effect of continually widening the holes. With the addition of some CHF_3 on the other hand a small bias regime can be established where there is little or no undercut because of its polymerisating nature. Some etching results using the regime are shown in Figure 11.6 for APC and TOPAS®. For the high bias regime APC and TOPAS® etch with a rate of 500–700 nm min^{-1}. However, for the low bias regime this reduces ten fold.

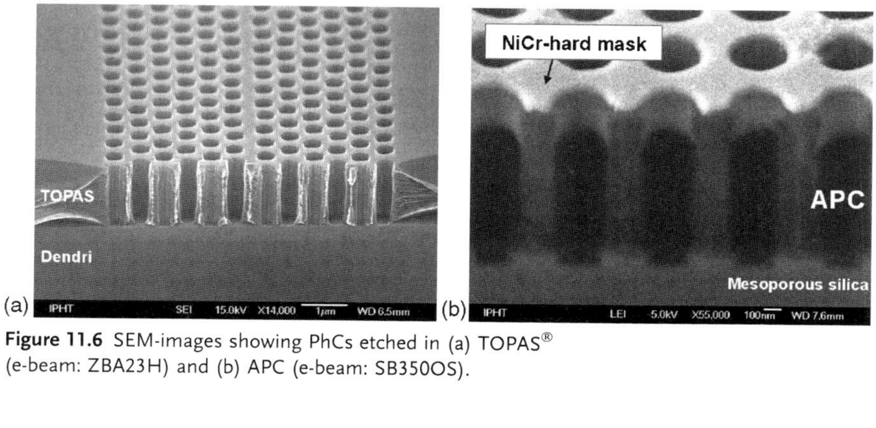

Figure 11.6 SEM-images showing PhCs etched in (a) TOPAS® (e-beam: ZBA23H) and (b) APC (e-beam: SB350OS).

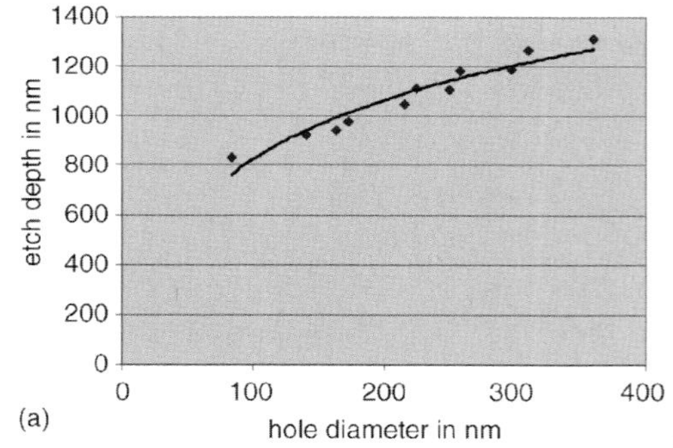

(b)

Figure 11.7 (a) Dependence of the etch depth inside the hole on the hole diameter for a fixed etch time (TOPAS®, Ar^+/O-ECR, e-beam: ZBA23H), and (b) a visual example of this dependence.

In order to reduce the optical losses of the resonators it is important to etch the holes completely through the waveguide layer [17]. However, the achieved etch depth during the etching of very small holes ($<1\,\mu m$) is strongly dependent on the hole diameters, as shown in Figure 11.7a and demonstrated in the cross-sectional SEM-image of Figure 11.7b. This behaviour was found for PMMA/DR-1 and TOPAS®. Etch tests in APC using in addition to O_2 and Ar^+ some CHF_3 have shown that also very small holes can be etched completely through the waveguide layer, i.e. the etch rate difference between larger and smaller holes has been reduced.

11.3
Realization and Characterization of Electro-Optically Tuneable Photonic Crystals

11.3.1
Characterization

A schematic representation of the employed EO-modulation apparatus, which was used to characterize the transmission behaviour of the PhC line defect resonators with applied electric field, is shown in Figure 11.8. The tuneable laser sources provide light in the wavelength range from 1260–1640 nm with a line width of 0.01 nm which allows spectral transmission measurements with very high resolution. To couple light into to the slab waveguide the prism coupling technique was used. Prism coupling allows mode selective coupling to the waveguide [18]. The coupling angle depends on the selected wavelength and has to be adjusted accordingly. For this reason, the sample together with the prism is mounted on a rotation stage, which is driven by a stepper motor. The transmitted light is collected from a cleaved edge of the wafer with a germanium photo diode. The electric signal of the detector is fed to a lock-in amplifier, a multimeter and finally into the recording computer.

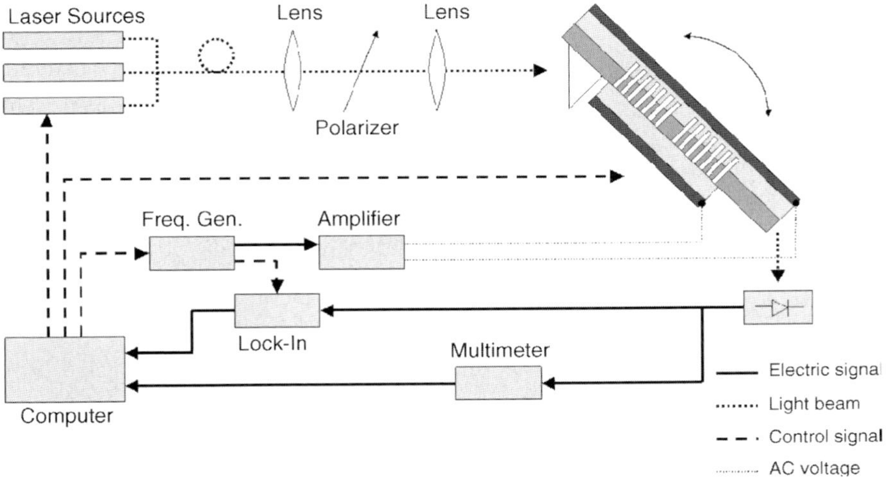

Figure 11.8 Schematic diagram of the EO-modulation setup.

The EO modulation is performed by applying amplified AC voltages provided by a frequency generator to the sample electrodes. The lock-in amplifier, which is locked to the frequency generator, is capable to detect the very small changes in transmission. In contrast, the multimeter signal remains unaffected, since the modulated signal is too small to be detected with the multimeter. It has been verified experimentally, that applying voltages of up to 400 V or disconnecting the lock-in do not induce any changes in the multimeter signal output. Consequently the transmittance, which is recorded by the multimeter, is separated from the modulation response via the lock-in technique. In addition, higher order modulation harmonics are also measurable, because the lock-in amplifier performs generally a Fourier transformation of the injected signal. This also imposes a strong requisite on the modulation signal quality. Since any deviations of the AC voltage from purely sinusoidal oscillations would induce parasitic contributions in the higher harmonics, the single frequency characteristics of the signal has been checked up to 20 kHz and 400 V by performing a Fourier analysis of the electric modulation signal. The phase correlation between the stimulating voltage and the EO-response is also accessible, since the lock-in technique provides a phase sensitive measurement method.

The presented setup can be used to investigate the fundamental properties of the EO-modulated PhC line defect resonators in three different ways. Firstly, it can be used for basic transmission measurements, without an applied modulation voltage, to determine the passive transmission spectrum of the PhC structure. Secondly, it allows varying amplitude measurements to find the relation between modulation voltage amplitude and the modulated optical signal. The laser is fixed at the desired wavelength and the detected optical signal strength is recorded by the computer as a function of applied voltage to the electrodes. Lastly, the setup can be employed for varying wavelength measurements where the modulation response as a function of wavelength is recorded. For this purpose the modulated optical signal is recorded while changing the emission wavelength and keeping the modulated signal constant.

Light at vacuum wavelengths between 1260 nm and 1360 nm was coupled into the slab waveguides. The photonic crystal resonators were first characterized with respect to their transmission spectra in the absence of an external modulation voltage. After that, the wavelength was selected at which the first derivative in the Lorentzian fit to the resonator transmission spectrum achieved its maximum ($\lambda_r = 1340$ nm). Sinusoidal AC voltages with amplitudes between 0.5 V and 100 V and at frequencies between 200 Hz and 10^4 Hz (a high radio frequency setup was not available, which limited the possible bandwidth for characterization) were then applied at the gold electrodes for modulation. Transmission signals were guided into a lock-in amplifier where the magnitude of the relative transmission modulation was recorded both at fundamental and second-harmonic frequencies. In order to separate electromechanical effects, such as a piezo-electrical response and electrostriction from the Pockels effect response, a piezo-stabilized symmetrical Mach–Zehnder interferometer (MZI) was set up into which the samples were placed as one of the two mirrors. With this MZI operating at a wavelength of 632.8 nm a field-induced thickness change of our samples could be detected with a resolution of $\approx$0.1 nm.

11.3.2 Experimental Results

In order to create the necessary noncentrosymmetric order of the DR-1 nonlinear side groups, a subset of the structured samples was corona poled at elevated temperatures under high electric field ($E_{pol} \geq 1$ MV/cm). While the poling field was applied, the samples were cooled down to room temperature. A second 2.5 μm thick layer of Teflon carrying a top gold electrode was placed on top of the PhC via decal technique [17].

The structures are excited in the ΓX direction, from the band diagram it can be seen, that in this propagation direction frequencies within the stop gap cannot propagate (Figure 11.9). By leaving out lines transversal to the direction of propagation a Fabry–Pérot resonator structure is formed. For the in-plane propagation direction ΓX, a distinct Lorentzian-shaped transmission maximum was observed (Figure 11.10). The wavelength of maximum transmission (1333 nm) comes close to the centre of the stop gap (which extends from 1200 nm to 1410 nm) of our photonic crystal slab at the X-point [18].

The laser source was tuned to the wavelength of the maximum of the first derivative of the Lorentz curve λ_r. Applying a modulation field across the slab causes a refractive index change Δn in its electro-optically active PMMA/DR–1 core of $\Delta n = -0.5 r_{33} n^3 E_{mod}$. All optical characterization was carried out on waveguides using PMMA/DR–1 as core material, even though it has only a moderate electro-optic response. However, it is available with the required chemical quality and quantity. An outlook for newer materials with potentially better nonlinear performance is given in Section 11.4.

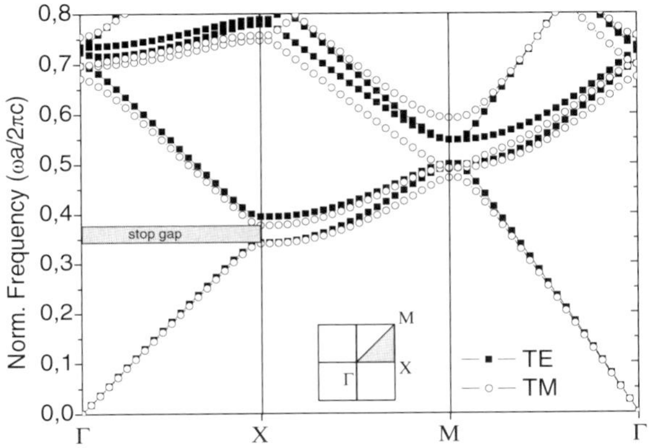

Figure 11.9 Band diagram of a 2D square lattice ($r/a = 0.3$) PhC in a low index material ($n = 1.54$) for both TE and TM polarization. The low index contrast does not allow an omni directional band gap for both polarizations. A range of forbidden frequencies only exists in the ΓX direction (stop gap).

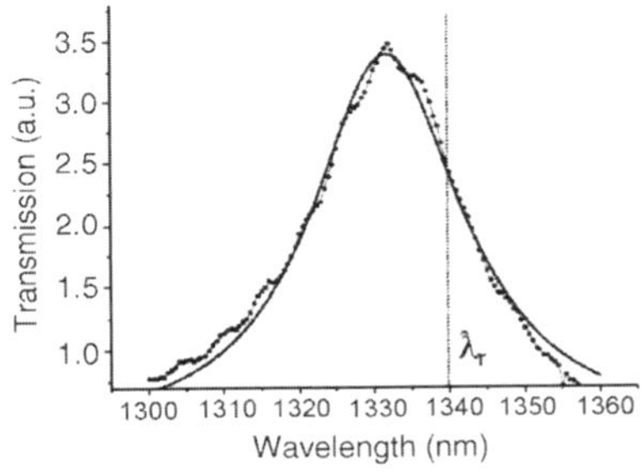

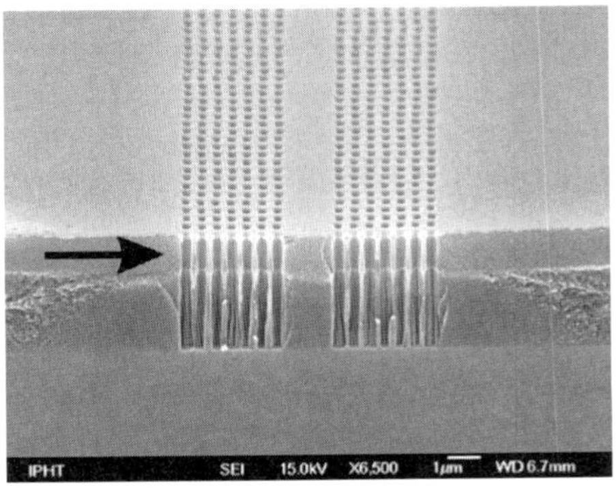

Figure 11.10 (a) Transmission of a 7_4_7 PhC line defect resonator in TM polarization (linked solid black dots). The experimental data are fitted by a Lorentzian shape function ($Q = 60$, solid curve). (b) SEM-image of a 7_3_7 PhC line defect resonator without cladding layer (the arrow indicates the direction of light propagation).

The modulation field strength E_{mod} was calculated from the applied voltage taking into account the thicknesses and dielectric constants of the different layer media. The electro-optical coefficient r_{33} was measured separately on indium-tin-oxide coated glass substrates by an ellipsometric technique [19] (8 pm/V at 632.8 nm) and proved to be stable over ten weeks at ambient temperatures. In the photonic crystal slabs, we observed linear refractive index alterations between 5.4×10^{-7} and 1.1×10^{-4} for AC modulation amplitudes between 0.5 V and 100 V corresponding to comparable transverse magnetic (TM) mode effective index changes. A time periodical transverse

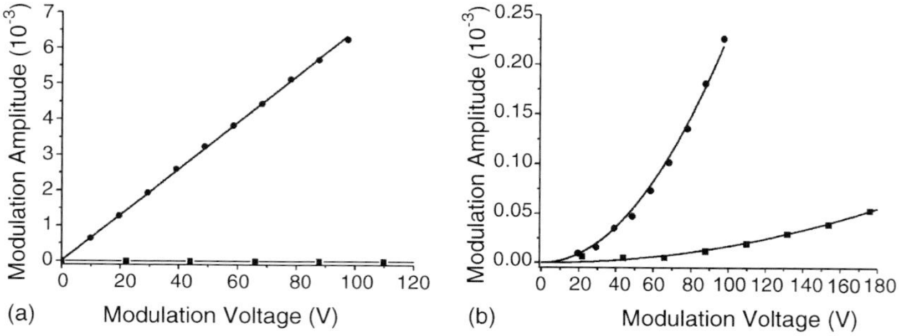

Figure 11.11 Modulation amplitude as a function of modulation voltage for PhC line defect resonators (circles: poled 7_4_7, $\lambda_r = 1340$ nm and squares: unpoled 7_3_7, $\lambda_r = 1296$ nm). Lines represent linear and square fits. (a) Modulation response at the fundamental frequency ($\nu = 200$ Hz). The slope corresponds to an electro-optic coefficient of 5 pm/V. No response in the unpoled PhC. (b) Modulation response at second harmonic frequency ($\nu = 400$ Hz) of the lock-in modulation frequency. A stronger Kerr response is seen in the poled sample ($K_K = 1.8 \times 10{-}20\,\text{m}^2/\text{V}^2$) as compared to the unpoled resonator sample ($K_K = 14 \times 10\ 22\text{m}^2/\text{V}^2$).

shift of the transmission spectra on the wavelength scale and therefore a transmission modulation at a fixed wavelength are observed. Figure 11.11 shows the relative transmission modulation detected at the fundamental and second harmonic frequency as a function of the amplitude of the applied electric modulation field both for a poled and for an unpoled sample. Clearly observable is a linear dependence of the transmission modulation in the case of the poled noncentrosymmetric waveguide core at fundamental frequency, whereas the unpoled sample does not show any response up to modulation amplitudes of 400 V (Figure 11.11a).

Using a Fabry–Pérot model, we could determine the electro-optical susceptibility that effectively caused this shift. The value of $r_{33} = 5$ pm/V compares well with the r_{33} values measured directly and with those from Ref. [16]. We have analyzed the sensitivity of our setup by stepwise reduction of the modulation amplitude to 0 V. A sub-1 V modulation amplitude (0.5 V) applied over the total polymer layer of thickness 1.4 μm induces a refractive index change of only 5.4×10^{-7} and still leads to a resolvable change of 3.4×10^{-5} in the relative transmission modulation signal. We believe this is a remarkable result given the fact that we have neither used a highly effective electro-optical polymer nor an optimized high-Q photonic crystal resonator design [5–8], thus leaving room for substantial improvement. The relative transmission modulation

$$\frac{T_{\text{mod}}}{T_{\text{trans}}} = \frac{3}{8}\frac{\Delta n}{n}(2\sqrt{3}Q-1) \approx \frac{3\sqrt{3}}{8} r_{33} n^2 E_{\text{mod}} \cdot Q, \tag{11.1}$$

profits from a large refractive index and electro-optical susceptibility of the electro-optically active core and from a high-quality factor. A gradual shift of positions of holes adjacent to the cavity in a k-spaced engineered design can suppress Fourier components above the light line and thus reduces vertical radiation losses and enhances Q dramatically [5–8].

The Kerr electro-optical response as observed at the second harmonic of the modulation frequency scales with the square of the applied modulation field and occurs in both noncentrosymmetric and centrosymmetric media. Figure 11.11b shows the relative transmission modulation as a function of the external modulation amplitude for both a poled and an unpoled sample. From the square fit to the experimental data, we first calculated the Kerr constant [20]

$$K_K = \frac{\Delta n}{n} \cdot \frac{1}{E_{mod}^2}, \tag{11.2}$$

for the unpoled PMMA/DR-1 sample. We achieved a value of $14 \times 10^{-22}\,m^2/V^2$ which is comparable to the value of $13 \times 10^{-22}\,m^2/V^2$ reported for unsubstituted PMMA [21].

We also had to address the issue of possible other contributions to the electro-optical response in addition to electronic displacement polarization. At room temperature, well below the glass transition, pure PMMA shows a β-relaxation at ν_β around 5 Hz to 10 Hz [22]. The δ-relaxation in side chain polymers substituted with rod like chromophores, however, occurs at ambient with extremely slow rates $\nu_\delta \approx 10^{-8}$ Hz [23]. At room temperature, therefore, no significant contributions from orientational polarization to the electro-optical response are expected if electric modulation frequencies from 200 Hz to 10 kHz or above are applied. This conclusion is supported by the fact that indeed no linear electro-optical response is observed for the unpoled centrosymmetric sample.

Second, a possible contribution of the electro-mechanical piezo-effect was addressed. We measured the field-induced thickness change by placing a poled sample into the MZI. The thickness variation linear in the modulation frequency yielded a piezo-coefficient of $d_{33} = 3$ pm/V. This value compares well with what was found for other comparable poled electro-optic polymers [24]. Neglecting the transverse contraction (Poisson ratio), we calculated an upper estimate for the change of the waveguide mode effective refractive index of 1.3×10^{-8} for 0.5 V and 2.7×10^{-6} for 100 V AC modulation voltage. This d_{33} translates into a "masked," hence, virtual contribution to the electro-optical susceptibility r_{33} of less than 2×10^{-13} m/V. Therefore, the piezo-contribution to the linear electro-optic modulation is less than 2.5%.

11.4 Synthesis of Electro-Optically Active Polymers

As active component for the preparation of "Electro-optically Tuneable Photonic Crystals" electro-optically active polymers (EO-polymers) with high glass transition temperatures (T_g) are preferable. Organic EO-materials have, the advantage compared to their inorganic counterparts e.g. $LiNbO_3$, that the non-linear optical response is of purely electronic origin and does not include deformations of the crystal lattice. Thus the response is extremely fast. In addition, thin homogenous films of low loss are accessible from amorphous polymers (no grain boundaries).

Polymeric materials for nonlinear optical applications like frequency doubling or the linear electro-optic effect (the Pockels effect) have been the topic of intensive research since the end of the eighties [25,26]. More recent reviews are given in [27,28]. Generally this work was divided into a search for efficient nonlinear optical moieties (the so called NLO-chromophores) and a search for a matrix to stabilize the polar order obtained by the poling process. This matrix includes guest-host polymers, polymers with covalently bound NLO-chromophores [25,26], glasses from low molar mass materials [25–28] and hybrid inorganic–organic materials obtained by sol–gel processes [29,30].

For some time progress was limited on the $\mu\beta$-values of the individual NLO-chromophores, and on their thermal stability needed for proper processing in high T_g materials. An additional obstacle is the formation of aggregates with an antiparallel dipolar orientation in the polymer matrix. Thus the macroscopic $\chi^{(2)}$-properties determined at small concentrations in guest-host systems could not be extrapolated linearly to the higher NLO-chromophore concentration needed for sufficient macroscopic NLO properties. This problem got solved by a proper combination of newly designed electron donor and acceptor groups in elongated π-systems [31–36] and the introduction of sterically hindered NLO-chromophores [34,37–41]. With these NLO-chromophores it is claimed to be possible to prepare guest-host systems with r_{33}-values well above 60 m/V [37,39,41,42] and a suitable thermal stability.

For an application within tuneable nanophotonic devices an EO-polymer should fulfil three criteria: (i) it has to form a low loss (low absorption and low scattering) film, (ii) because of the patterning process, which involves elongated processing above 100 °C, the T_g-value should be well above 140 °C and (iii) the EO-coefficient should be as high as possible (r_{33} above 40 pm/V at $\lambda = 1.5\,\mu m$) to allow a strong variation of the transmission characteristics of the photonic structure. This requires a high concentration of about 20 wt% (good solubility) of optimized NLO-chromophores.

As matrix for the NLO-chromophores we selected two transparent, amorphous high T_g-polymers. Amorphous high T_g polycarbonate APC is a copolymer of classical polycarbonate with a sterically demanding biphenyl-unit (Figure 11.12). It has a T_g-value of 183 °C and a refractive index of 1.54. TOPAS® is a copolymer of

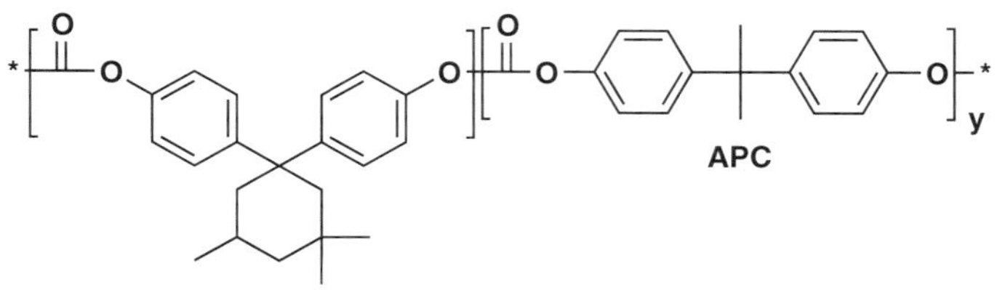

Figure 11.12 Molecular structure of amorphous high T_g polycarbonate APC.

ethylene and a cycloolefin and posses a T_g-value of 190 °C. It is a pure hydrocarbon compound and thus very hydrophobic. The preparation of high quality films with a thickness around 1.5 μm is challenging because of limited solubility. With chlorobenzene at elevated temperatures or cyclohexanone good solvents for TOPAS® or APC respectively were found. From 14 wt% solutions of APC in cyclohexanone, filtered through a 0.2 μm filter and spin-coated at 3.000 rpm high quality APC films with 1.5 μm thickness can be obtained on glass, silicon or mesoporous silica. They were used for the patterning experiments. Because of the better solubility of APC and its more polar structure, which makes it the better solvent for the NLO-chromophores, APC was investigated in more detail.

As NLO-chromophores the chromophores 1–4 presented in Figure 11.13 are under investigation. They can be synthesized in analogy to the synthetic routes described in [39–41] and possess an amino group as electron donor and dicyano- (2, 4) or tricyano-moieties (1, 3) as acceptor. Chromophores 1 and 2 employ a short π-conjugation and are more easily accessible. Chromophore 3 should have a rather high molecular hyperpolarizibility according to [39]. As donor group a triarylstructure was used to increase the thermal stability in comparison to the alkylated derivatives of [39]. Its synthesis was performed according to Figure 11.14. Chromophore 4 possesses a long π-conjugation and a hydroxyl group for covalent linkage to reactive high T_g-polymers as the ones described in [43].

From these chromophores guest-host systems in APC can be produced; e.g. homogenous films with 20 wt% chromophore 3 can be obtained by spin-coating. In these films the T_g-value is reduced to 140 °C (183 °C for the pure APC), which is still just enough for the processing scheme. To test the long term stability of these films we investigated the solubility of 3 (20 wt%) in neat APC in more detail. Thereby it turned out that chromophore 3 crystallizes from APC in drop-casted films (thick

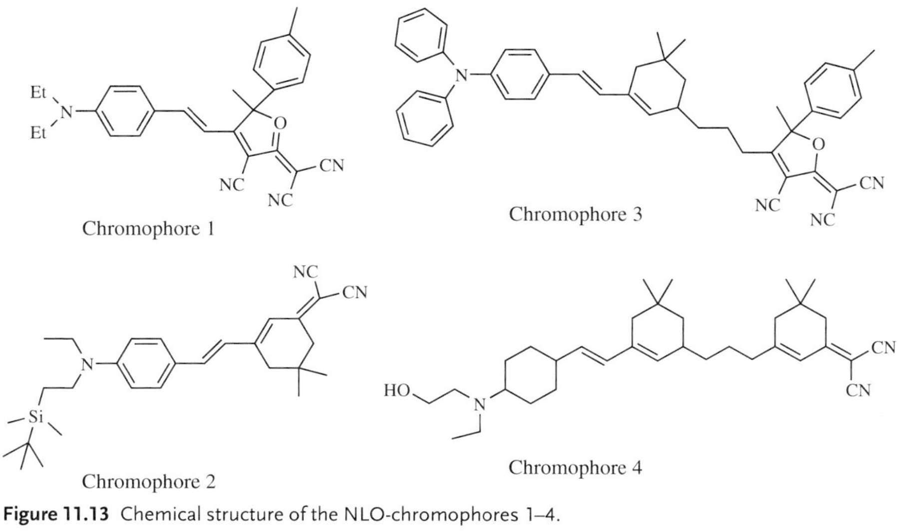

Figure 11.13 Chemical structure of the NLO-chromophores 1–4.

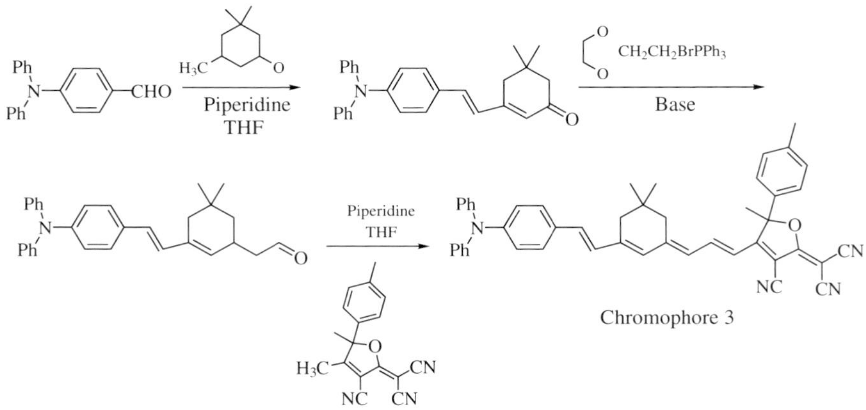

Figure 11.14 Synthesis route to chromophore 3.

films cast from cyclohexanone). In these systems the solvent evaporates more slowly and the system has more time for crystallization before the rising T_g-value (due to loss of solvent) stops crystal growth. Annealing of such films around T_g leads to a further increase of the crystal size. Thus single phase homogenous films are accessible only by spin-coating. Such films, however, are not thermodynamically stable and crystallization may start during long periods at elevated temperatures, as they are required for the complex poling and patterning processes within this project.

In addition, the thermal stability of chromophore 3 against decomposition was measured for long time spans, as they are necessary for drying, poling and especially patterning of these high T_g EO-polymers. Thermo-gravimetry shows no weight loss of chromophore 3 up to 300 °C. This is in agreement with the observations described in [39] for chromophores of similar structure. UV-measurements performed on samples annealed at 150 °C over days are presented in Figure 11.15. After 30 minutes at 150°C (a time span sufficient for poling) only minor changes of the spectrum are detectable. If poling is performed under inert gas conditions even this effect could possibly be prevented. It is not clear, if these changes result from starting decomposition or just from aggregation or an orientation of the chromophores. After 2 days at 150 °C the absorbance at 650 nm is, however, strongly reduced and the absorbance at 400 nm has increased. Thus chromophore 3 with the elongated π-system is not temperature stable at high temperatures for elongated time spans. This had not been expected based on the claims of [39] for similar structures (only thermo-gravimetric measurements). Chromophores 1 and 2 are long term temperature stable at 150°C, but they give only insufficient r_{33} values below 5 pm/V at 1318 nm due to their short π-conjugation.

Thus new chromophores for guest-host systems have been reported in the literature. They are, however, not yet generally useful for complex processing schemes as presented here, because the solubility in high T_g-polymers is obviously poorer than in PMMA or classical polycarbonate. In addition, comprehensive data of their long-term stability at high temperatures is not available, while tasks like the use

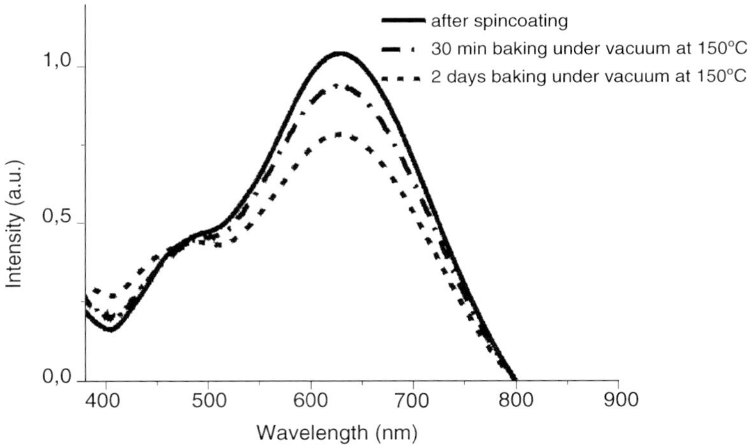

Figure 11.15 Variation of the UV-spectrum chromophore 3 (20 wt%) in an APC film during annealing at 140 °C (substrate ITO covered glass).

of EO-polymers for photonic crystal resonator based modulators requires stability for days at temperatures above 140°C. The problem of demixing can obviously be solved by integrating the chromophores into the polymer system; the thermal stability requires more work on the selection of the chromophores.

11.5 Conclusions and Outlook

In the present paper we have shown that sub micrometer structures can be transferred into polymer thin film waveguides with high quality via state of the art electron beam lithography and high density plasma reactive etching techniques. Such films can be made from high glass transition polymers which are either doped or covalently functionalized with nonlinear optically active chromophore molecules. The design concepts we investigated consisted of line defect structures in square lattice photonic crystal slabs, thereby forming Fabry–Pérot-type optical nanophotonic resonators of which we have investigated the optical transmission spectra. For the standard EO-polymer DR-1 we could realize an electro-optically tuneable photonic crystal by shifting the resonance frequency of the resonator via the Pockels effect with an externally applied electric modulation field. We were able to detect a sub-1-Volt sensitivity for the modulation effect which was proven to stem from electro-optical activity rather than from Kerr-effect, electrostriction or piezo-responses. The next steps to go will include the optimization of the designs in the direction of lower losses and higher quality factors as well as the successful application of our concept to novel polymers with high electro-optical coefficients.

Acknowledgements

The authors would like to thank K. Kandera and K. Pippardt for their work and advice, and acknowledge the financial support of the DFG (HU 1309/1-1, ZE 230/13-1, EI 391/10-2). We also thank CST, Darmstadt for their support with CST Microwave software products.

References

1 Yablonovitch, E. (1987) *Phys. Rev. Lett.*, **58**, 2059.
2 John, S. (1987) *Phys. Rev. Lett.*, **58**, 2486.
3 Busch, K., Lölkes, S., Wehrspohn, R.B. and Föll, H. (2003) *Advances in Design, Fabrication, and Characterization*, Wiley-VCH, Berlin.
4 Baba, T. *et al.* (2002) *IEEE J. Quantum Electron.*, **38**, 743.
5 Böttger, G. *et al.* (2002) *Appl. Phys. Lett.*, **81** (14), 2517.
6 Srinivasan, K. and Painter, O. (2002) *Opt. Express*, **10**, 670.
7 Akahane, Y. *et al.* (2003) *Nature*, **425**, 944.
8 Ryu, H.Y., Notomi, M., Lee, Y.H. (2003) *Appl. Phys. Lett.*, **83**, 294.
9 Eich, M. *et al.* (1989) *J. Appl. Phys.*, **66**, 2559.
10 Shi, Y. *et al.* (2000) *Science*, **288**, 119.
11 Jen, A. *et al.* (2002) *Mater. Res. Soc. Symp. Proc.*, **708**, 153.
12 Liguda, C. *et al.* (2001) *Appl. Phys. Lett.*, **78**, 2434.
13 Kim, Tae-Dong and Jen, A.K.-Y. (2006) *Adv. Mater.*, **18**, 3038.
14 Foresi, J.S. *et al.* (1997) *Nature*, **390**, 143.
15 Huebner, U. *et al.* (2005) *Microelectron. Eng.*, **83**, 1138.
16 Boucher, R. *et al.* (2004)MNE2003 73–74C, P. 330.
17 Schmidt, M. *et al.* (2005) *CFN Lecture of Functional Nanostructures, 658 Springer Series Lecture Notes in Physics*, Springer, Berlin, p. 71.
18 Schmidt, M. *et al.* (2004) *Appl. Phys. Lett.*, **85**, 16.
19 Teng, C.C. *et al.* (1990) *Appl. Phys. Lett.*, **56**, 1734.
20 Krause, S. (1981) *Molecular Electro-Optics*, Plenum, New York.
21 Jungnickel, B.J. (1981) *Polymer*, **22**, 720.
22 Hartig, C. *et al.* (1995) *Polymer*, **36**, 4553.
23 Eckl, D.M. *et al.* (1995) *Proc. SPIE*, **2527**, 92.
24 Winkelhahn, H.-J. *et al.* (1994) *Appl. Phys. Lett.*, **64**, 1347.
25 Eich, M. *et al.* (1989) *J. Appl. Phys.*, **66**, 6.
26 Eich, M. *et al.* (1990) *Polym. Adv. Technol.*, **1**, 189.
27 Samyn, C. *et al.* (2000) *Macromol. Rapid Commun.*, **21**, 1.
28 Kajzar, F. *et al.* (2003)in *Polymers for Photonics Application I*, Nonlinear optical and electroluminescence polymers, (ed. Lee K-.S.), Advances in Polymer Science, **161** 1.
29 Lacroix, P.G. (2001) *Chem. Mater.*, **13**, 3495.
30 Sanchez, C. *et al.* (2003) *Adv. Mater.*, **15**, 1969.
31 Grahn, W. *et al.* (2001) *Proc. SPIE*, **4461**, 33.
32 Pond, S.J.K. *et al.* (2002) *J. Phys. Chem. A*, **106**, 11470.
33 Lebeau, B. *et al.* (1997) *Chem. Mater.*, **9**, 1012.
34 Ostroverkhova, O. *et al.* (2002) *Adv. Funct. Mater.*, **12**, 621.
35 Staub, K. *et al.* (2003) *J. Mater. Chem.*, **13**, 825.
36 Shi, Y.Q. *et al.* (2000) *Science*, **288**, 119; Dalton, L.R. (2000) *Opt. Eng.*, **39**, 589.
37 Zhang, C. *et al.* (2000) *Proc. SPIE*, **4114**, 77.

38 Reyes-Esqueda, J. *et al.* (2001) *Opt. Commun.*, **198**, 2007.

39 He, M. *et al.* (2002) *Chem. Mater.*, **14**, 4669.

40 Luo, J. and Jen, A.K.-Y. (2006) *Org. Lett.*, **8**, 1387.

41 Zhang, C. *et al.* (2001) *Chem. Mater.*, **13**, 3043.

42 Ermer, S. *et al.* (2002) *Adv. Funct. Mater.*, **12**, 605.

43 Dörr, M. *et al.* (1998) *Macromolecules*, **31**, 1454.

12
Tuneable Photonic Crystals obtained by Liquid Crystal Infiltration

H.-S. Kitzerow, A. Lorenz, and H. Matthias

12.1
Introduction

Liquid crystals [1–4] combine fluidity and anisotropy in a unique way. Anisotropy, i.e. the dependence of bulk material properties on the direction, indicates non-spherical symmetry of the respective material. This phenomenon is traditionally attributed to solid crystals. However, organic compounds that consist of anisometric (for example rod-like or disk-like) molecules can show an orientational order of these molecules even in a fluid state. If so, this fluid state is referred to as being liquid crystalline. The least complicated liquid crystalline structure, the nematic phase, is uniaxial, i.e. one preferred axis is sufficient to describe its local anisotropy. The director $\boldsymbol{n}$ (a pseudo-vector) can be used to indicate the local molecular alignment (see Figure 12.1) and a scalar order parameter S is sufficient to describe the degree of orientational order. The temperature dependence of the order parameter and the influence of external electric or magnetic fields on the orientation of the director can lead to a strong dependence of the effective refractive index of the liquid crystal on temperature and external fields. Chiral liquid crystals can show a helical superstructure of the local alignment, thereby leading to a spatially periodic director field $\boldsymbol{n}(\boldsymbol{r})$. In the chiral nematic (cholesteric) phase, the director is twisted along a pitch axis. Blue phases [5–9], which appear in the temperature range between the cholesteric and the isotropic liquid state, are characterized by double twist and a cubic superstructure. Beautiful colours of cholesteric and blue phases arise from Bragg scattering in the visible wavelength range. They led to the discovery of liquid crystals [10] and even to their first commercial applications (as temperature sensors) [11]. Moreover, distributed feedback lasing of dye-doped periodic liquid crystals was suggested many years ago [12] and extensively studied during the last few years [13–15]. Like photonic crystals [16–24], these liquid crystalline helical structures show a periodic modulation of the refractive indices with lattice constants comparable to the wavelength of light and thus give rise to Bragg scattering. Even more interesting is the selectivity of certain states of polarization. A right handed helix shows Bragg reflection of right

Nanophotonic Materials: Photonic Crystals, Plasmonics, and Metamaterials. Edited by R.B. Wehrspohn, H.-S. Kitzerow, and K. Busch

ISBN: 978-3-527-40858-0

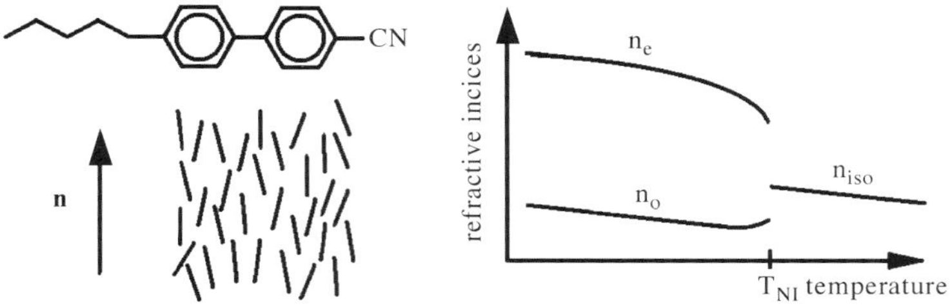

Figure 12.1 Chemical structure of 4-cyano-4′-pentyl-biphenyl, arrangement of rod-like molecules in the (non-chiral) nematic phase (N), and temperature dependence of the ordinary refractive index n_o (effective for light that is linearly polarized with its electric field perpendicular to the director $\boldsymbol{n}$), the extraordinary refractive index n_e (effective for light that is linearly polarized with its electric field parallel to the director $\boldsymbol{n}$), and the isotropic refractive index n_{iso}.

circularly polarized light but is transparent for left circularly polarized light at the same wavelength and vice versa. The variation of the local optical axis in a homogeneous material is sufficient to cause these effects. As a consequence, however, the refractive index contrast cannot exceed the birefringence $\Delta n = n_e - n_o$ of the liquid crystal. This difference between the extraordinary refractive index n_e and the ordinary refractive index n_o is not larger than 0.3. The cubic blue phase structures are fascinating, but their local birefringence is too small for the appearance of a photonic band gap. However, only a few years after the first pioneering works on photonic crystals [25,26], Hornreich *et al.* [27] predicted a photonic band gap in artificial cubic structures that are composed of dielectric ($\varepsilon \geq 10$) or conducting cylinders and exhibit the same space group as the blue phase modification BP1 (O^8, $I4_132$), see Figure 12.2. Fabrication of such structures is still a challenge.

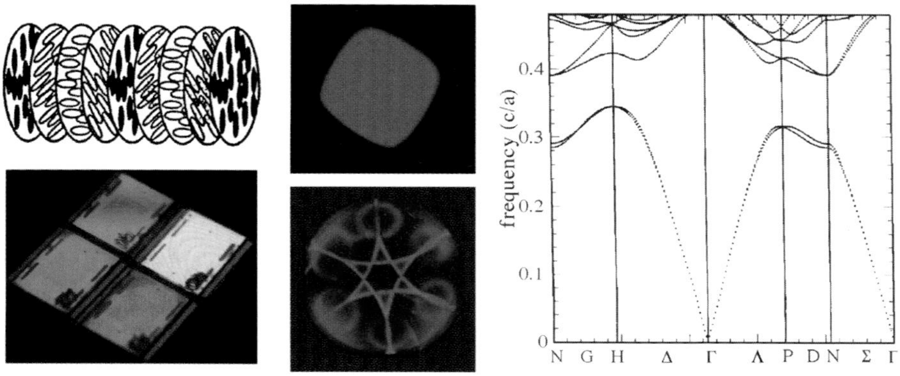

Figure 12.2 To the left: Structure of the chiral nematic (cholesteric) phase (N*) and liquid crystal cells showing a bright selective reflection due to Bragg scattering of the N* phase. Centre: Liquid single crystal and optical diffraction pattern (Kossel diagram) of blue phases. To the right: Dispersion relation of an artificial structure that is composed of dielectric ($\varepsilon \geq 10$) cylinders and shows the same space group (O^8) as BP1 (after Hornreich *et al.* [27]).

Overviews about the variety of possibilities to adjust or switch the optical properties of photonic crystals were given, previously [28–30]. The present article is focussed on one particular approach, namely the infiltration of solid photonic crystals with liquid crystals. Busch and John [31–33] suggested to fill the voids of a colloidal crystal with a (non-chiral) nematic liquid crystal and to make use of the temperature- and field-dependence of its refractive indices in order to achieve a tuneable photonic band gap.

Independently, Yoshino *et al.* [34–36] did the first respective experiments. Kang *et al.* [37] demonstrated field-induced switching of a photonic stop band in liquid crystal-doped colloidal crystals. Leonhard *et al.* [38] infiltrated macroporous silicon with a liquid crystal and observed a thermally induced shift of the photonic band edge. However, until 2001 there was little knowledge about the liquid crystal alignment and the precise explanation of the photonic effects that were observed. Subsequent investigations at the University of Paderborn and the analysis of director fields by means of ^{2}H-NMR spectroscopy and fluorescence confocal polarizing microscopy are reviewed in the following section. The current state and perspectives of liquid crystal-infiltrated photonic crystals are then summarized in Section 12.3.

12.2 Experimental Results

12.2.1 Colloidal Crystals

Opal and corresponding artificial structures consist of spherical colloidal particles, which are arranged in a cubic closed packed, i.e. face centred cubic (fcc), lattice. Unlike the diamond structure, such fcc structures are not expected to show a photonic band gap [39], but inverted fcc structures have been predicted to show a photonic band gap [40]. In particular, Busch and John [31,32] calculated that an inverse opal structure where an fcc lattice of air spheres is embedded in a silicon matrix ($\varepsilon_b \approx 11.5 \ldots 11.9$) should show a vanishing density of states in a frequency range around $\omega = 1.6\ \pi c/a$. The same authors suggested infiltrating such a structure with liquid crystals.

The results reported in this section correspond to normal and inverted colloidal crystals that are filled with a nematic liquid crystal. Monodisperse spheres made of poly-(methyl-methacrylate) [PMMA] were used to form a face centred cubic (fcc) colloidal crystal, which in turn served as a template to manufacture an inverse structure. It is well known that suspensions of colloidal spheres can be used to generate self-organized colloidal crystals, either by dipping a solid substrate into the suspension and slow removal of this substrate or by spreading the suspension on a substrate and subsequent gentle drying. If the sample is flat, the spheres form a close packed hexagonal monolayer at the surface. On this first layer (A), further hexagonal layers (B, C, . . .) grow which are phase shifted with respect to the first layer, thereby filling the voids of the existing layer(s) closely. Assembling of the layers in the sequence ABCABC ... results in an fcc lattice that is oriented with its threefold axis (1, 1, 1) perpendicular to the substrate plane. The resulting colloidal crystals show Bragg reflection in the visible wavelength range if the sphere diameter and thus the layer spacings d_{hkl} have the appropriate size (a few 100 nm). The Bragg wavelength is given by

$$l_{hkl} = 2n_{\mathrm{eff}}\, d_{hkl} \quad \text{and} \quad d_{hkl} = \mathrm{a}(h^2+k^2+l^2)^{-1/2}, \tag{12.1}$$

where d_{hkl} is the layer spacing of a set of planes with Miller indices h, k, l within a cubic structure with lattice constant a. If a heterogeneous material is composed of two components with the dielectric constants ε_1 and ε_2, the average dielectric constant $\varepsilon_{\mathrm{av}}$ can approximately be calculated using the Maxwell-Garnett relation [41]

$$(\varepsilon_{\mathrm{av}}-\varepsilon_1)(\varepsilon_{\mathrm{av}}+2\,\varepsilon_1)^{-1} = f_2\,(\varepsilon_2-\varepsilon_1)(\varepsilon_2+2\,\varepsilon_1)^{-1}, \tag{12.2}$$

where f_2 is the volume fraction of component 2. Thus, the effective refractive index of the colloidal crystal is approximately given by [37]

$$n_{\mathrm{eff}} = (\Sigma f_i\, n_i^2)^{1/2}, \tag{12.3}$$

where f_i is the volume fraction of component i of the heterogeneous system. The close packed structure of opal is characterized by the filling fractions $f_{\mathrm{spheres}} = 0.74$ and $f_{\mathrm{air}} = 0.26$. Here, colloidal crystals were made of monodisperse poly-(methyl-methacrylate) (PMMA) spheres with a diameter D of about 200 nm. Spreading of a suspension on a flat glass plate and subsequent drying results in the expected fcc arrangement, oriented with the threefold axis (1, 1, 1) perpendicular to the substrate plane (Figure 12.3, to the left). The colloidal crystal composed of PMMA spheres ($n_{\mathrm{PMMA}} \approx 1.49, f_{\mathrm{PMMA}} = 0.74$) and air ($n_{\mathrm{air}} \approx 1, f_{\mathrm{air}} = 0.26$) shows a reflection peak at $\lambda_{111} = 448$ nm for normal light incidence. According to Eq. (12.2), an average refractive index of $n_{\mathrm{eff}} \approx 1.38$ is expected. With $a = D\sqrt{2}$ and $(h, k, l) = (1, 1, 1)$, the

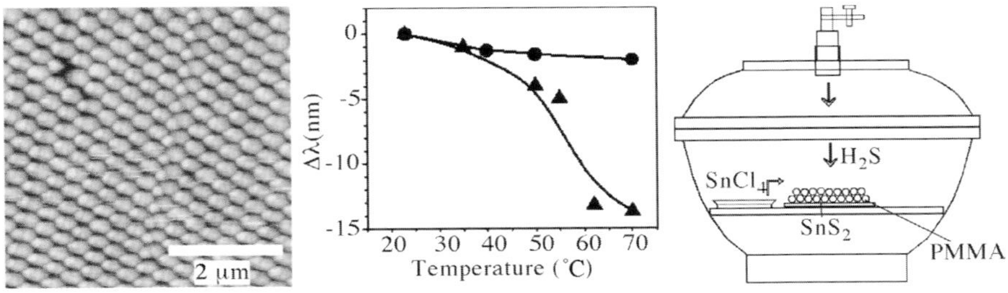

Figure 12.3 Left: AFM picture of a colloidal crystal made of PMMA spheres. Centre: Spectral shift of the stop band as a function of temperature for (•) a colloidal crystal made of PMMA spheres with a liquid crystal (E7) in the tetrahedral and octahedral gaps and (▲) an inverse opal made of SnS_2 where the spherical cavities are filled with the same liquid crystal. Right: Set-up to fill the voids of the colloidal crystal with SnS_2. For details, see Ref. [43].

relation $\lambda_{111} = \sqrt{8/3}n_{\mathrm{eff}}D$ is obtained from Eqs. (12.1) and (12.3). Thus, the measured reflection peak indicates an average sphere diameter of 199 nm, which was confirmed by electron microscopy and light scattering experiments.

In order to achieve tuneable properties, the colloidal crystal was infiltrated with the nematic liquid crystal mixture E7 (Merck), which consists of 4-n-alkyl-4′-cyano-biphenyls and respective terphenyls. Filling the voids with the liquid crystal results in a red-shift of the (1, 1, 1) reflection peak. At 23 °C, the reflection peak was observed at $\lambda_{111} = 493.4$ nm. From the average refractive index of E7, $n_{\mathrm{E7}} = 1.612$, an effective refractive index of $n_{\mathrm{eff}} = 1.523$ is expected, and thus a slightly larger reflection wavelength, $\lambda_{111} = 494.3$ nm. In spite of the transition of E7 from the nematic to the isotropic state (at $T_{\mathrm{NI}} = 60.5$ °C), increasing the temperature from 23 °C to 70 °C results in a very small blue shift of the reflection peak by only 2 nm (Figure 12.3, centre, circles). The direction and the amount of the shift are in agreement with the expectation that the average refractive index of the liquid crystal decreases with temperature.

In the second experiment, the PMMA colloidal crystal was used as a template to generate an inverted fcc crystal made of tin sulphide (SnS_2), following the procedure by Müller *et al.* [42]. The inverted structure was formed by chemical vapour deposition. The template was exposed to $SnCl_4$ and H_2S (Figure 12.3, to the right). Subsequently, the PMMA template was removed by an organic solvent and the remaining inverted SnS_2 colloidal crystal was filled with the nematic liquid crystal using capillary forces. For this experiment, larger PMMA spheres were used. The reflection peak ($\lambda_{111} = 632$ nm) of the template (PMMA spheres and air) and light scattering data indicated sphere diameters slightly larger than 280 nm. After filling the voids with SnS_2 and removing the template, an inverted colloidal crystal was obtained with a lattice constant slightly smaller than what would be expected from the bead diameter. Electron microscopy indicated that the distance of the centres of neighbouring gaps was only 270 nm. From the reflection peak ($\lambda_{111} = 536$ nm, indicating $n_{\mathrm{eff}} \approx 1.22$) and from $n_{\mathrm{SnS_2}} = 3.32$, it was concluded that the volume fraction of SnS_2 is less than expected. Nevertheless, the infiltration of the inverted colloidal crystal with the liquid crystal mixture E7 resulted in stable samples. The filled structure showed a reflection peak at 736 nm at room temperature. This study [43,44] confirmed the theoretical expectation that the temperature-induced shift of the reflection peak of the filled inverted colloidal crystal is much larger than the temperature-induced shift of the normal colloidal crystal filled with a liquid crystal. This behaviour is mainly due to the different volume fractions. In the inverted structure (SnS_2/E7), a change of the temperature from 23 °C to 70 °C resulted in the large wavelength shift of 14 nm (Figure 12.3, centre, triangles), where the transition from the nematic to the isotropic phase (at $T_{\mathrm{NI}} = 60.5$ °C) can be clearly seen.

In further experiments [45,46], monodisperse colloidal spheres with functionalised surfaces were synthesized. Either charged functional groups [45] or chromophores [46] were covalently attached to the surfaces of monodisperse spheres. Investigations on the colloidal crystals made thereof are still in progress and will be reported elsewhere.

12.2.2 Photonic Crystals Made of Macroporous Silicon

The experiments described in this section are focused on three-dimensional (3D) structures consisting of macroporous silicon that are filled with a liquid crystal. Two-dimensional hexagonal or rectangular arrays of pores with an extremely high aspect ratio (diameter $\geq 1\,\mu m$, depth $\leq 100\,\mu m$) can be fabricated by a light-assisted electrochemical etching process using HF [47]. The pore diameter varies periodically, thereby forming a three-dimensional photonic crystal (PhC) [48]. The macroporous structure was evacuated and filled with a liquid crystal. The photonic properties for light propagation along the pore axis were studied by Fourier transform infrared (FTIR) spectroscopy [49–51]. Deuterium-nuclear magnetic resonance (^{2}H-NMR) [49,50] and fluorescence confocal polarizing microscopy (FCPM) [52,53] were used in order to analyse the director field of the liquid crystal inside the pores.

For example, Figure 12.4 shows the infrared transmission of samples that show a two-dimensional hexagonal array of pores with a lattice constant $a = 1.5\,\mu m$. Along the pore axis, the diameter of each pore varies periodically between D_{min} $(0.76 \pm 0.10)\,\mu m$ and $D_{max} = (1.26 \pm 0.10)\,\mu m$ with a lattice constant $b = 2.6\,\mu m$. The pores were filled with the nematic liquid crystal 4-cyano-4′-pentyl-biphenyl (5CB, Figure 12.1) which exhibits a clearing temperature of $T_{NI} = 34\,°C$. For light propagation along the pore axes, the FTIR transmission spectrum of the silicon-air structure shows a stop band centred at $\lambda = (10.5 \pm 0.5)\,\mu m$. Filling the pores with 5CB decreases the dielectric contrast to silicon and results in a shift of the stop band to $\lambda \approx 12\,\mu m$. The band edge was found to be sensitive to the state of polarization of the incident light. For linearly polarized light, rotation of the sample with respect to the plane of polarization was found to cause a shift of the liquid crystal band edge by $\Delta\lambda \approx 152\,nm$ (1.61 meV). This effect can quantitatively be explained by the square shape of the pore cross section, which brakes the threefold symmetry of the hexagonal lattice. Due to the presence of the liquid crystal, the band edge at lower wavelengths ("liquid crystal band" edge) can be tuned by more than 140 nm (1.23 meV) by heating the liquid crystal from 24 °C (nematic phase) to 40 °C (isotropic liquid phase).

The shift of the photonic band edge towards larger wavelengths indicates an increase of the effective refractive index with increasing temperature. This effect can be explained by a predominantly parallel alignment of the optical axis (director) of the nematic liquid crystal along the pore axis. For a uniform parallel alignment, the effective refractive index of the nematic component corresponds to the ordinary refractive index n_o of 5CB. Increasing the temperature above the clearing point causes an increase to the isotropic value $n_{iso} \approx (1/3n_e{}^2 + 2/3n_o^2)^{1/2}$, where n_e is the extraordinary refractive index of the liquid crystal ($n_e > n_o$). From the respective dielectric constants [$\varepsilon_{LC}(24\,°C) = n_o{}^2$ and $\varepsilon_{LC}(40\,°C) = n_{iso}^2$], the average dielectric constant ε_{av} of the heterogeneous structure can approximately be calculated using the Maxwell-Garnett relation (Eq. (12.2)) [41]. The relative shift of the stop band edge towards larger wavelengths corresponds approximately to the relative increase of the average refractive index by $\approx$0.65%.

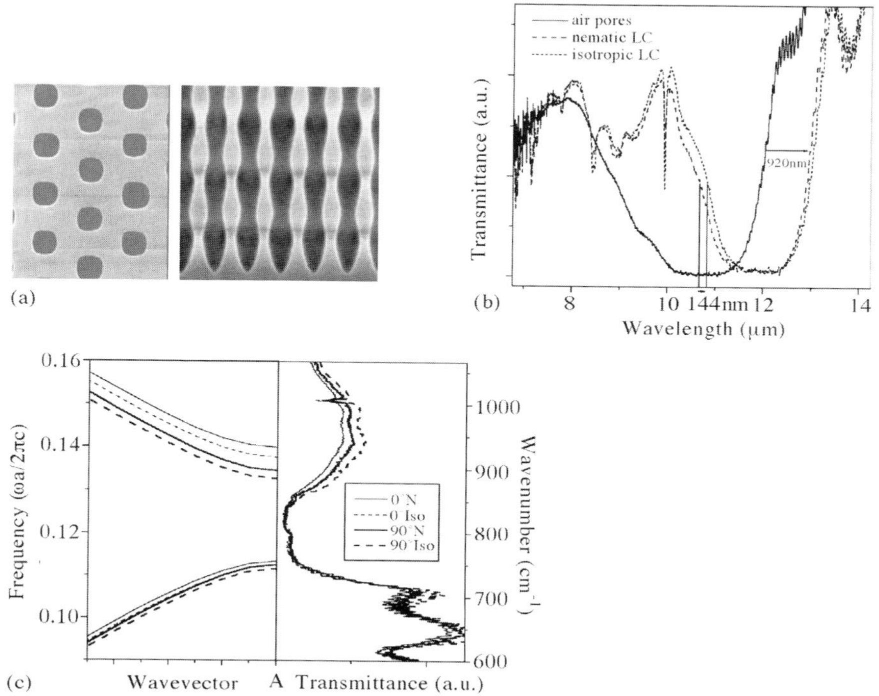

Figure 12.4 (a) SEM top-view and side-view of a photonic crystal made of macroporous silicon containing a two-dimensional hexagonal array of pores with periodically modulated diameter. (b) Transmission spectra of the same photonic crystal for light propagation along the pore axes if the sample is filled with (—) air, (---) 5CB in its nematic phase, and (. . .) 5CB in its isotropic liquid state, respectively. (c) Comparison of the calculated dispersion relation using the plane wave approximation and the experimental spectra. For details, see Ref. [49].

Planar microcavities inside a 3D photonic crystal appear when the pore diameter is periodically modulated along the pore axis, stays constant within a defect layer and is continued to vary periodically. Figure 12.5 shows a structure where a defect layer is embedded between five periodic modulations of the pore diameter. The pores are arranged in a 2D square lattice with a lattice constant of $a = 2\,\mu m$. The pore width varies along the pore axis between $D_{min} = 0.92\,\mu m$ and $D_{max} = 1.55\,\mu m$. The length of a modulation is $b = 2.58\,\mu m$. The defect has a length of $1 = 2.65\,\mu m$ and pore diameters $D_{def} = 0.82\,\mu m$. Within the defect layer, the filling fraction of the liquid crystal is $\xi_{def} = 0.17$. For infrared radiation propagating along the pore axes, a fundamental stop band at around 13 μm and a second stop band at around 7 μm are expected from calculations using the plane wave approximation [54]. The experiment shows a transmission peak at $\lambda = 7.184\,\mu m$ in the centre of the second stop band, which can be attributed to a localized defect mode. Filling the structure with the liquid crystal 4-cyano-4′-pentyl-biphenyl (5CB, Merck) at 24 °C causes a spectral red-shift of the stop band. Together with the stop band, the wavelength of the defect state is shifted by 191 nm to $\lambda = 7.375\,\mu m$. An additional shift of $\Delta\lambda = 20$ nm to

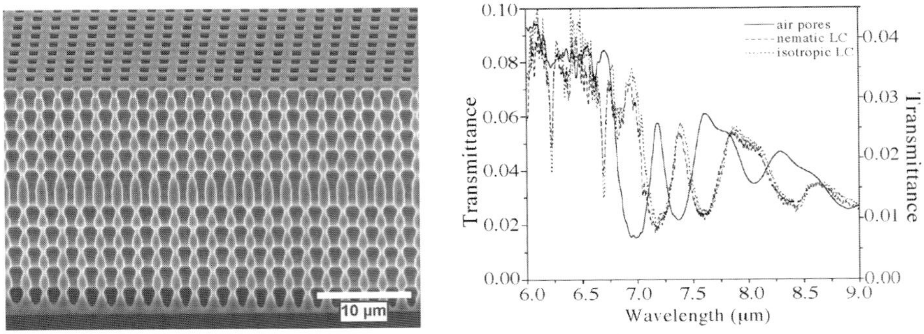

Figure 12.5 Left: SEM image (bird eye's view) of a silicon structure with modulated pores including a planar defect layer without modulation. Right: Transmission spectra of this structure if the sample is filled with (—) air, (---) 5CB in its nematic phase, and (. . .) 5CB in its isotropic liquid state, respectively. For details, see Ref. [51].

$\lambda = 7.395\,\mu m$ is observed when the liquid crystal is heated from 24 °C (nematic phase) to 40 °C (isotropic liquid phase). Again, the shift toward larger wavelengths indicates an increase of the effective refractive index n_{eff} of the liquid crystal with increasing temperature and can be attributed to the transition from an initially parallel aligned nematic phase ($n_{LC,eff} = n_o$) to the isotropic state ($n_{LC,eff} = n_{iso}$). During continuous variation of the temperature, a distinct step by 20 nm is observed at the phase transition from the nematic to the isotropic phase. The quality factor Q of the investigated structure, $Q = \lambda/\delta\lambda = 52$, is rather small and thus the shift by 20 nm appears to be small compared to the spectral width of the defect mode. However, the same order of magnitude of the temperature-induced wavelength shift can be expected for structures with a much higher quality factor and might be quite large compared to the band width of the defect mode.

Analysis of ^{2}H-NMR line shapes is a very sensitive tool to measure the orientational distribution of liquid crystals in non-transparent samples. Corresponding measurements were performed using two-dimensional structures of pores with constant radius ($R = 0.45\,\mu m$ and $R = 1.00\,\mu m$, respectively), filled with 5CB that is deuterated in the α-position of the alkyl chain. The quadrupolar splitting of the ^{2}H-NMR signal is given by

$$\Delta n = \frac{1}{2}\Delta n_0\,(3\cos^2\vartheta - 1), \tag{12.4}$$

where ϑ is the angle between the local director and the magnetic field [3]. Thus, the intensity distribution of the ^{2}H-NMR signal indicates the orientational distribution $f(\vartheta)$, averaged over the sample volume, see Figure 12.6. Comparison between experimental NMR results and calculated spectra confirms a parallel (P) alignment of the director along the pore axis for substrates that were treated like the samples described above. However, also an anchoring of the director perpendicular to the silicon surfaces ("homeotropic" anchoring) can be achieved if the silicon wafer is

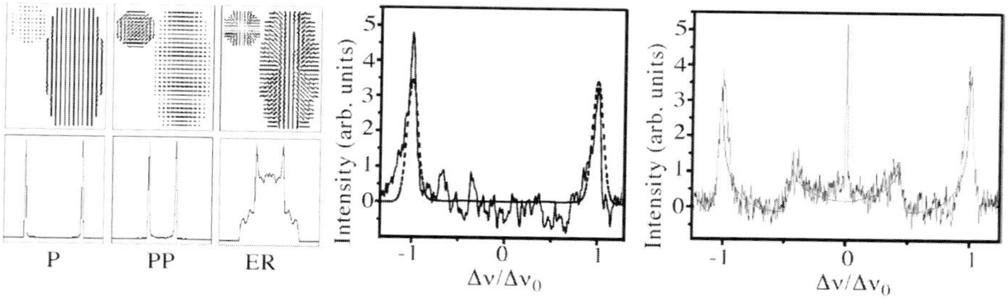

Figure 12.6 Left: Schematic relation between parallel (P), planar polar (PP) or escaped radial (ER) director fields and the respective ^{2}H-NMR-lineshapes. Centre: ^{2}H-NMR spectrum of α-deuterated 5CB in cylindrical pores with parallel anchoring [dotted line: spectrum expected for a parallel (P) structure]. Right: ^{2}H-NMR spectrum of α-deuterated 5CB in cylindrical pores with perpendicular anchoring [thin line: spectrum expected for an escaped radial (ER) structure with weak anchoring]. For details, see Refs. [49,50].

cleaned with an ultrasonic bath and a plasma-cleaner and subsequently pre-treated with N,N-dimethyl-n-octadecyl-3-aminopropyl-trimethoxysilyl chloride (DMOAP). NMR data indicate the appearance of an escaped radial (ER) director field in the latter case.

For the first time, optical microscopic studies of the director field in pores with a spatially periodic diameter variation could be achieved by means of a nematic liquid crystal polymer that shows a glass-like nematic state at room temperature [52,53]. For fluorescence polarizing microscopy, the polymer was doped with N,N′-bis(2,5-di-tert-butylphenyl)-3,4,9,10-perylene-carboximide (BTBP). After filling the photonic crystal in vacuum, the sample was annealed in the nematic phase at 120 °C for 24 hours and subsequently cooled to room temperature, thereby freezing the director in the glassy state. The silicon wafer was dissolved in concentrated aqueous KOH solution and the remaining isolated polymer rods were washed and investigated by fluorescence confocal polarizing microscopy (FCPM). The transition dipole moment of the dichroic dye BTBP is oriented along the local director of the liquid crystal host.

The incident laser beam (488 nm, Ar^+) and the emitted light pass a polarizer, which implies that the intensity of the detected light scales as $I \propto \cos^4\alpha$ for an angle α between the local director and the electric field vector of the polarized light. Thus, the local fluorescence intensity indicates the local orientation of the liquid crystal director with very high sensitivity. For a template with homeotropic anchoring and a sine-like variation of the pore diameter between 2.2 μm and 3.3 μm at a modulation period of 11 μm, the FCPM images of the nematic glass needles (Figure 12.7) indicate an escaped radial director field with some characteristic features that differ from non-modulated pores. In the cylindrical cavities studied previously, point-like hedgehog and hyperbolic defects appear at random positions and tend to disappear after annealing, due to the attractive forces between defects of opposite topological charges. In contrast, the modulated pores stabilize a periodic array of disclinations. Moreover, disclination loops appear instead of point-like disclinations.

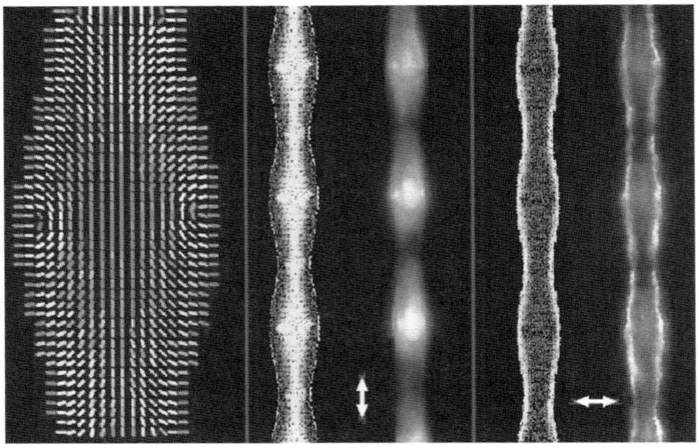

Figure 12.7 Left: Nematic escaped radial (ER) director field, calculated using the algorithm described in Refs. [55,56]. Centre: Theoretical and experimental fluorescent confocal polarizing microscopy (FCPM) images for polarized light with its electric field parallel to the tube axis. Right: Theoretical and experimental FCPM images for polarized light with its electric field perpendicular to the tube axis. For details, see Ref. [52].

A large variety of additional properties can be achieved by means of cholesteric liquid crystals [57]. Their ability to form helical structures offers the opportunity to combine an intrinsic periodicity with the periodicities of the solid substrate. In addition, the topology of their director fields is of fundamental interest. If the helix pitch is much smaller than the tube size, FCPM images show regular fingerprint lines indicating an undistorted helical structure (Figure 12.8, left). These lines are perpendicular to the pitch axis and their distance corresponds to one half of the pitch. However, if the pitch is comparable to the pore size or if the anchoring of the director

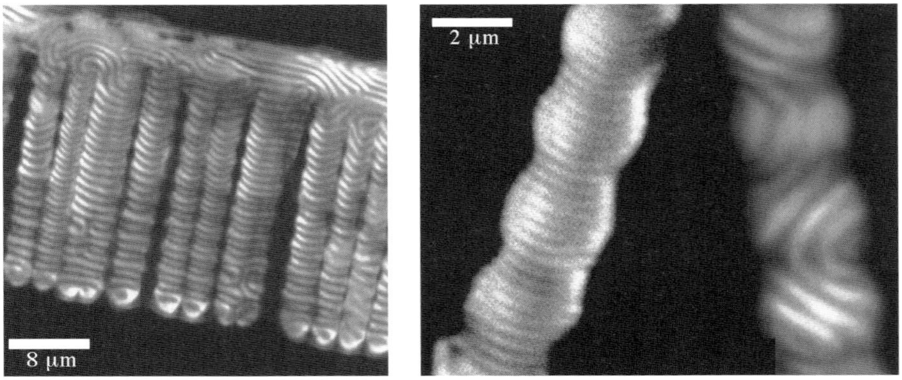

Figure 12.8 Left: FCPM image of cholesteric liquid crystal tubes. To the right: FCPM image of two cholesteric cylinders with different helix pitch and with perpendicular (left) and parallel anchoring (right), respectively. For further details, see Ref. [57].

is parallel to the interface, distorted director fields can appear that do not show a uniform pitch axis and may exhibit periodicities deviating from the intrinsic pitch (Figure 12.8, right).

12.2.3 Photonic Crystal Fibres

Liquid crystals are also useful in order to functionalise photonic crystal (PhC) fibres. In this study, we used PhC fibres [58–60] that were manufactured at the Institute for Physical High Technology (IPHT), Jena [61]. The fibres (Figure 12.9, left) exhibit a hollow core with diameter D_c surrounded by a hexagonal array of pores with a uniform diameter D_{cl}, where $D_{cl} < D_c$. The surfaces of the pores were coated with lecithin in order to achieve homeotropic anchoring and subsequently the pores were filled with the nematic liquid crystal mixture E9 (Merck), making use of capillary forces. The liquid crystal-filled PhC fibres were sandwiched between two glass plates coated with transparent indium tin oxide (ITO) electrodes and embedded in a refractive index-matched polymer. Investigation of equivalently treated single capillaries in a polarizing microscope indicates an escaped radial director field. Both the ordinary refractive index ($n_o = 1.5225$ at 633 nm) and the extraordinary refractive index ($n_e = 1.7765$ at 633 nm) of the liquid crystal are larger than the refractive index ($n_g = 1.4715$) of the glass composing the PhC fibre. Thus, the average refractive index of the liquid crystal core is larger than the average refractive index of the glass/liquid crystal cladding of the fibre and conventional index guiding can be expected. The spatial patterns of monochromatic (633 nm) radiation indicate multimode wave-guiding. By applying alternating electric fields, the transmission of the fibre can be switched off (Figure 12.9, centre). The transmission reappears again when the electric field is switched off. The time constants seem to show that the application of the field essentially destroys the uniform director field so that the low transmission in the field-on state can be attributed to scattering losses.

In order to extend the versatility of the use of PhC fibres, a new method of selective filling of single pores has been developed. For this purpose, a dielectric

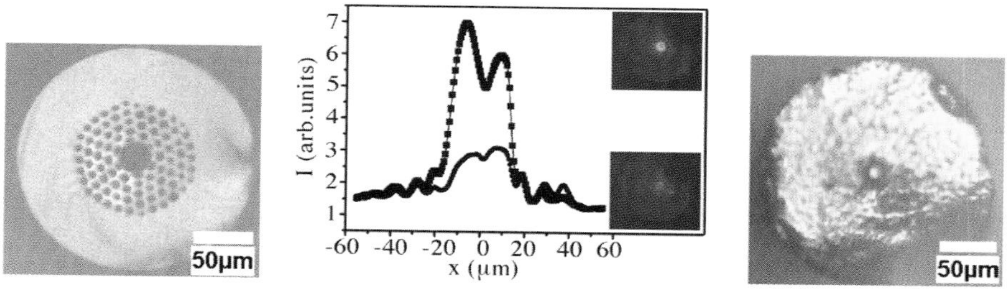

Figure 12.9 To the left: Photonic crystal fibre with a hollow core ($D_c = 26.4\,\mu m$) and a porous cladding ($D_{cl} = 6.7\,\mu m$). Centre: Transmission of this fibre as a function of position after infiltration with the nematic liquid crystal E9 (■) without external field and (•) at a field strength of $E = 200\,V/203\,\mu m \approx 1\,V/\mu m$ ($f = 1\,kHz$). To the right: Photonic crystal fibre of the same type as shown to the left after closing the small pores.

compound is deposited on the face of the fibre by thermal evaporation in high vacuum. If the dielectric film exhibits the appropriate thickness, the pores with small diameter are closed (Figure 12.9, to the right) and only the pore(s) with larger diameter can be filled by dipping the fibre into the liquid crystal. Observations in a polarizing microscope confirm that the liquid crystal infiltrates indeed the open pore, selectively. The closed pores, however, remain filled with air. This procedure opens a large variety of tailoring the photonic properties of PhC fibres. For example, polarization sensitive fibres or fibres with a true photonic band gap guiding and an active cladding could be fabricated. In addition, the advantage that glass is (at least partially) transparent in the ultraviolet spectral range can be used in order to generate sophisticated anchoring conditions by means of photo-induced alignment [62,63]. The investigation of various types of liquid crystal-filled fibres is the subject of forthcoming studies.

12.3 Discussion

Summarizing, the results described in Section 12.2 have shown that the dependence of the effective refractive index of liquid crystals on temperature and external fields can be used to alter the optical properties associated with photonic stop bands or defect modes in colloidal crystals, solid semiconductor photonic crystals and photonic crystal fibres, effectively. The investigation of these structures involves not only potential applications but also interesting fundamental physics. The confinement of liquid crystals to small cavities influences the threshold fields and switching times, but anchoring at a closed surface leads also to defects of the director field. These disclinations dominate the behaviour of the liquid crystal, they may interact according to their topological charges, and eventually lead to discontinuous topological transitions [4]. While studying micro-encapsulated or polymer-dispersed liquid crystals has motivated investigations on spherical or elliptical droplets long ago, the work on photonic crystals has additionally led to the consideration of liquid crystals confined to the tetrahedral or octahedral gaps of an opal [37] and to cylindrical pores with modulated diameter [52,53,57]. In the latter case, the director fields observed show similarities to configurations that are known to appear in spherical or perfect cylindrical cavities, but also some unique features such as an array of stabilized, ring-like defects [52].

Recent developments of novel materials, fabrication techniques and device concepts are likely to develop the use of liquid crystals in photonic crystals further and to lead to practical applications. With respect to possible substrates, the manufacturing of inverse colloidal crystals, has been greatly improved during the last few years, thereby making inverse opals with higher dielectric constants or higher filling fractions available that are, for example, made of silicon [64], germanium [65], gold [66] or titania [67]. Weak anchoring may help to increase the field-induced shift of the refractive index of liquid crystals filled into such structures [68]. Large changes of the photonic properties may be achieved in a ternary photonic crystal that is composed of two components with different dielectric constants ε_1, ε_2 and one additional component with tuneable dielectric constant ε_3. If ε_3 is adjusted to match either ε_1 or ε_2 or none of them, the lattice constant changes discontinuously, thereby opening and closing stop bands [69]. The electrochemical etching of solid substrates has also been considerably improved, recently. Extremely thorough control of the growth parameters makes it possible to fabricate silicon photonic structures where the pores are interconnected, thereby diminishing the silicon fraction and offering the possibility of an omni-directional photonic band gap [70].

However, the most important development steps towards functionalised integrated wave-guide structures are those fabrication techniques, which lead to well-controlled defect designs. Etching techniques for solid photonic crystals [47,48,70], direct writing of polymer structures by means of two-photon-induced polymerisation [71,72] and casting of such structures [73] can provide suitable substrates. In addition, selective filling [74–78] can be used to functionalise the structures by dyes, quantum dots, dielectric liquids, organic and inorganic glasses, or liquid crystals (Figure 12.9).

12.4 Conclusions

In conclusion, the infiltration of photonic crystals with liquid crystals proved to be a suitable method to achieve tuneable properties or active switching. In the year 2001 when the investigations described in Section 12.2 were started, there were doubts whether changes of the effective refractive index of a liquid crystal are sufficient to get useful alterations of photonic properties. The experiments that are performed until today, indicate that the spectral shifts can be quite reasonable but are more limited than one could optimistically guess: A simple reorientation of the optical axis of an always uniformly aligned liquid crystal is not possible if the liquid crystal is confined to sub-μm size cavities. Assuming a uniform orientation of the liquid crystal for different alignment directions is very useful to facilitate theoretical calculations and to provide visions of possible applications [79], but when put into practice more complicated director fields need to be considered. For example, the difference between the ordinary refractive index n_o and the extraordinary refractive index n_e – typically $\Delta n = n_e - n_o \approx 0.2$, and even in special materials $\Delta n \leq 0.3$ – can be effectively used if the light is linearly polarized, the liquid crystal is always uniformly aligned and its optical axis is reoriented by 90°. However, the difference between the effective indices of a uniform nematic state $[n_{eff}(T_1) = n_o]$ and the thermally induced isotropic state $[n_{eff}(T_2) \approx (1/3n_e^2 + 2/3n_o^2)^{1/2}]$, as observed in Ref. [49], corresponds to only one third of the birefringence, $n_{eff}(T_2) - n_{eff}(T_1) \approx 1/3\ (n_e - n_o) = 1/3\ \Delta n$. In addition, the relative spectral shift of a photonic stop band or the resonance frequency of a micro-cavity depends not only on the refractive index of the liquid crystal, but also on the refractive index of the passive substrate, so that averaging diminishes the spectral shift even further.

In spite of these limitations, the observed spectral shifts of some 10 nm or 20 nm induced by means of liquid crystals can nevertheless be large compared to the bandwidth of the resonance peak of a micro-cavity, provided that the quality factor Q is sufficiently high. Consequently, liquid crystals have been successfully applied in optically pumped tuneable III–V semiconductor lasers [80,81]. A thermally induced wavelength shift by 9 nm was achieved in a quantum dot/PhC waveguide laser with a line width of 11 nm ($Q \approx 90$) [80] and an electrically induced wavelength shift of the emission peak by 1.2 nm at 20 V/15 μm was obtained for point defects in a two-dimensional PhC laser with a line width of 0.6 nm ($Q \approx 2000$) [81]. In the latter case, the high quality factor of 2000 was found even when operated within an ambient refractive index of $n \approx 1.5$. For microcavities surrounded by air, quality factors Q of the order 1 million have been observed for photonic crystals [82,83], $Q > 10^8$ for toroid-shaped microresonators on a chip [84] and $Q > 10^9$ for silica microspheres [85], which seems to indicate that further improvements on the solid substrates may help to use the tuning capability that is possible with typical liquid crystal material parameters (i.e. shifts up to some 10 nm) very efficiently.

Acknowledgements

The authors would like to dedicate this work to Heinrich Marsmann (University of Paderborn) and Mary Neubert (Liquid Crystal Institute, Kent State University) in recognition of their work on ^{2}H-NMR spectroscopy and their remarkable help in this project. In addition, we would like to thank the co-workers of the liquid crystal research group at the University of Paderborn G. Jünnemann, G. Mertens and T. Röder for their valuable contributions and all our collaborators for their help and fruitful discussions. In particular, we would like to mention K. Huber, T. Kramer and R. Schweins (University of Paderborn) who worked with us on the preparation and characterization of colloidal particles, S. Matthias, C. Jamois, U. Gösele (Max-Planck-Institute for Microstructure Physics, Halle), S. Schweizer, R. B. Wehrspohn (University of Paderborn), S. Lölkes and H. Föll (University of Kiel) who fabricated and provided photonic crystals made of macroporous silicon, S. J. Picken (Delft University of Technology) who supported us with liquid crystalline polymers, as well as H. Bartelt, J. Kobelke, K. Mörl, and A. Schwuchow (Institute for Photonic Technology Jena) who collaborated with us on photonic crystal fibres. Support by the company E. Merck (Darmstadt) with liquid crystals as well as funding by the German Research Foundation (KI 411/4, SPP 1113) and the European Science Foundation (EUROCORES/05-SONS-FP-014) are gratefully acknowledged.

References

1 de Gennes, P.G. and Prost, J. (1993) *The Physics of Liquid Crystals*, 2nd edn, Clarendon Press, Oxford.

2 Chandrasekhar, S. (1992) *Liquid Crystals*, 2nd edn, Cambridge University Press, Cambridge.

3 Kumar, S. (ed.) (2001) *Liquid Crystals: Experimental Study of Physical Properties and Phase Transitions*, Cambridge University Press, Cambridge.

4 Crawford, G.P. and Žumer, S., (eds) (1996) *Liquid Crystals in Complex Geometries Formed by Polymer and Porous Networks*, Taylor & Francis, London.

5 Coles, H.J. and Pivnenko, M.N. (2005) *Nature*, **436**, 997–1000.

6 Kitzerow, H.-S. (2006) *Chem. Phys. Chem.*, **7**, 63–66.

7 Kitzerow, H.-S. (1991) *Mol. Cryst. Liq. Cryst.*, **202**, 51–83.

8 Heppke, G., Jérome, B., Kitzerow, H.-S. and Pieranski, P. (1989) *Liq. Cryst.*, **5**, 813–828.

9 Heppke, G., Kitzerow, H.-S. and Krumrey, M. (1985) *Mol. Cryst. Liq. Cryst. Lett.*, **2**, 59–65.

10 Reinitzer, F. (1888) *Monatshefte Chemie*, **9**, 421.

11 See, for example: Fergason, J.L. (1968) *Appl. Opt.*, **7**, 1729.

12 Goldberg, L.S. and Schnur, J.M. (1972) Tunable internal-feedback liquid crystal-dye laser, US Patent 3,771,065.

13 Finkelmann, H., Kim, S.T., Muñoz, A., Palffy-Muhoray, P. and Taheri, B. (2001) *Adv. Mater.*, **13**, 1069.

14 Cao, W., Muñoz, A., Palffy-Muhoray, P. and Taheri, B. (2002) *Nature Mater.*, **1**, 111.

15 Ozaki, M., Kasano, M., Ganzke, D., Haase, W. and Yoshino, K. (2002) *Adv. Mater.*, **14**, 306.

16 Joannopoulos, J.D., Meade, R.D. and Winn, J.N. (1995) *Photonic Crystals: Molding the Flow of Light*, Princeton University Press, Princeton, NJ.

17 Soukoulis, C.M. (ed.) (2001) *Photonic Crystals and Light Localization in the 21st Century, NATO Science Series, Ser. C: Mathematical and Physical Sciences*, **Vol. 563**, Kluwer Academic Publ., Dordrecht.

18 Johnson, S.G. and Joannopoulos, J.D. (2002) *Photonic Crystals: The Road from Theory to Practice*, Kluwer Academic Publishers, Boston.

19 Slusher, R.E. and Eggleton, B.J., (eds) (2003), *Nonlinear Photonic Crystals, Springer Series in Photonics*, Vol. **10**, Springer, Berlin.

20 Inoue, K. and Ohtaka, K. (eds) (2004) *Photonic Crystals: Physics, Fabrication and Applications, Springer Series in Optical Sciences*, Vol. **94**, Springer, Berlin.

21 Busch, K., Föll, H., Lölkes, S. and Wehrspohn, R.B. (eds) (2004) *Photonic Crystals: Advances in Design, Fabrication and Characterization*, Wiley-VCH, Weinheim.

22 Sakoda, K. (2005) *Optical Properties of Photonic Crystals, Springer Series in Optical Sciences*, 2nd edn, Vol. **80**, Springer, Berlin.

23 Lourtioz, J.-M., Benisty, H., Berger, V., Gerard, J.-M., Maystre, D. and Tchelnokov, A. (2005) *Photonic Crystals: Towards Nanoscale Photonic Devices*, Springer, Berlin.

24 Yasumoto, K. (ed.) (2006) *Electromagnetic Theory and Applications for Photonic Crystals, Series: Optical Science and Engineering*, Vol. **102**, CRC Press, Boca Raton.

25 Yablonovitch, E. (1987) *Phys. Rev. Lett.*, **58**, 2059.

26 John, S. (1987) *Phys. Rev. Lett.*, **58**, 2486.

27 Hornreich, R.M., Shtrikman, S. and Sommers, C. (1994) *Phys. Rev. B*, **49**, 10914–10917.

28 Kitzerow, H.-S. and Reithmaier, J.P. (2004) Tunable Photonic Crystals using Liquid Crystals, in *Photonic Crystals: Advances in Design, Fabrication and Characterization*, (eds K. Busch, H. Föll, S. Lölkes and R.B. Wehrspohn) Wiley-VCH, chap. 9, pp. 174–197.

29 Kitzerow, H.-S. (2002) *Liq. Cryst. Today*, **11** (4), 3–7.

30 Braun, Paul V. and Weiss, Sharon M. (eds) (2006) Tuning the Optic Response of Photonic Bandgap Structures III. *Proc. SPIE*, **6322**; Fauchet, P.M. and Braun, P.V. (eds) (2005) Tuning the Optic Response of Photonic Bandgap Structures II. *Proc. SPIE*, **5926**; Fauchet, P.M. and Braun, P.V. (eds) (2004) Tuning the Optic Response of

Photonic Bandgap Structures. *Proc. SPIE*, **5511**.

31 Busch, K. and John, S. (1998) *Phys. Rev. E*, **58**, 3896–3908.

32 Busch, K. and John, S. (1999) *Phys. Rev. Lett.*, **83**, 967.

33 John, S. and Busch, K. (1999) *J. Lightwave Technol.*, **17**, 1931 (2004). Electro-actively tunable photonic bandgap materials, United States patent 6813064.

34 Yoshino, K., Satoh, S., Shimoda, Y., Kawagishi, Y., Nakayama, K. and Ozaki, M. (1999) *Appl. Phys. Lett.*, **75**, 932.

35 Yoshino, K., Satoh, S., Shimoda, Y., Kawagishi, Y., Nakayama, K. and Ozaki, M. (1999) *Appl. Phys.*, **38**, 961.

36 Yoshino, K., Nakayama, K., Kawagiishi, Y., Tatsuhara, S., Ozaki, M. and Zakhidov, A.A. (1999) *Mol. Cryst. Liq. Cryst.*, **329**, 433.

37 Kang, D., Maclennan, J.E., Clark, N.E., Zakhidov, A.A. and Baughman, R.H. (2001) *Phys. Rev. Lett.*, **86**, 4052.

38 Leonard, S.W., Mondia, J.P., van Driel, H.M., Toader, O., John, S., Busch, K., Birner, A., Gösele, U. and Lehmann, V. (2000) *Phys. Rev. B*, **61**, R2389–R2392

39 Ho, K.M., Chan, C.T. and Soukoulis, C.M. (1990) *Phys. Rev. Lett.*, **65**, 3152–3155.

40 Sözüer, H.S., Haus, J.W. and Inguva, R. (1992) *Phys. Rev. B*, **45**, 13962–13972.

41 Garnett, J.C.M. (1904) *Philos. Trans. R. Soc. Lond.*, **203**, 385.

42 Müller, M., Zentel, R., Maka, T., Romanov, S.G. and Sotomayor-Torres, C.M. (2000) *Adv. Mat.*, **12**, 1499.

43 Mertens, G., Röder, Th., Schweins, R., Huber, K. and Kitzerow, H.-S. (2002) *Appl. Phys. Lett.*, **80**, 1885–1887.

44 Kitzerow, H.-S., Hoischen, A., Mertens, G., Paelke, L., Röder, T., Stich, N. and Strauß, J. (2002) *Polym. Preprints (Am. Chem. Soc.)*, **43**, 534–535.

45 Röder, T., Kramer, T., Huber, K. and Kitzerow, H.-S. (2003) *Macromol. Chem. Phys.*, **204**, 2204–2211.

46 Kramer, T., Röder, T., Huber, K. and Kitzerow, H.-S. (2005) *Polym. Adv. Technol.*, **16**, 38–41.

47 Müller, F., Birner, A., Gösele, U., Lehmann, V., Ottow, S. and Föll, H. (2000) *J. Porous Mater.*, **7**, 201.

48 Schilling, J., Müller, F., Matthias, S., Wehrspohn, R.B., Gösele, U. and Busch, K. (2001) *Appl. Phys. Lett.*, **78**, 1180.

49 Mertens, G., Röder, T., Matthias, H., Schweizer, S., Jamois, C., Wehrspohn, R., Neubert, M., Marsmann, H. and Kitzerow, H.-S. (2003) *Appl. Phys. Lett.*, **83**, 3036–3038

50 Mertens, G. (2004) *Anwendung von Flüssigkristallen für abstimmbare photonische Kristalle, Ph.D. dissertation*, University of Paderborn.

51 Mertens, G., Wehrspohn, R.B., Kitzerow, H.-S., Matthias, S., Jamois, C. and Gösele, U. (2005) *Appl. Phys. Lett.*, **87**, 241108.

52 Matthias, H., Röder, T., Wehrspohn, R.B., Kitzerow, H.-S., Matthias, S. and Picken, S.J. (2005) *Appl. Phys. Lett.*, **87**, 241105.

53 Kitzerow, H.-S., Mertens, G., Matthias, H., Marsmann, H., Wehrspohn, R.B., Matthias, S., Gösele, U., Frey, S. and Föll, H. (2005) *Proc. SPIE*, **5926**, 592605/1–10.

54 The MIT photonics band programme package was used to solve Maxwell's equations in a planewave basis. See, for example: Johnson, S.G. and Joannopoulos, J.D. (2001) *Opt. Express*, **8**, 173–190.

55 Dickmann, S. (1994) *Numerische Berechnung von Feld und Molekülausrichtung in Flüssigkristallanzeigen, Ph.D. thesis*, University of Karlsruhe.

56 Mori, H., Gartland, E.C.,Jr, Kelly, J.R. and Bos, P. (1999) *Jpn. J. Appl. Phys.*, **38**, 135–146.

57 Matthias, H., Schweizer, S.L., Wehrspohn, R.B. and Kitzerow, H.-S. (2007) Liquid Crystal Director Fields in Micropores of Photonic Crystals, J. Opt. A, Special Issue: Selected papers from NANO META-2007, 8–11 January 2007, Seefeld, Austria; *J. Opt. A: Pure Appl. Opt.*, 9 (9), S389–S395

58 Russell, P.St.J. (2003) *Science*, **299**, 358–362. (Review article).

59 Zolla, F., Renversez, G. and Nicolet, A. (2005) *Foundations of Photonic Crystal Fibres*, Imperial College Press, London.

60 Bjarklev, A. Broeng, J., and Bjarklev, A.S. (2003) *Photonic Crystal Fibres*, Kluwer Academic Publishers, Boston.

61 Bartelt, H., Kirchhof, J., Kobelke, J., Schuster, K., Schwuchow, A., Mörl, K., Röpke, U., Leppert, J., Lehmann, H., Smolka, S., Barth, M., Benson, O., Taccheo, S. and D'Andrea, C. (2007) *phys. stat. sol. (a)*, **204** (11), 3805 (this issue).

62 Vorflusev, V.P., Kitzerow, H.-S. and Chigrinov, V.G. (1997) *Appl. Phys. A*, **64**, 615–618.

63 Kitzerow, H.-S., Liu, B., Xu, F. and Crooker, P.P. (1996) *Phys. Rev. E*, **54**, 568–575.

64 Blanco, A., Chomski, E., Grabtchak, S., Ibisate, M., John, S., Leonard, S.W., Lopez, C., Meseguer, F., Miguez, H., Mondia, J.P., Ozin, G.A., Toader, O. and van Driel, H.M. (2000) *Nature*, **405**, 437.

65 van Vugt, L.K., van Driel, A.F., Tjerkstra, R.W., Bechger, L., Vos, W.L., Vanmaekelbergh, D. and Kelly, J.J. (2002) *Chem. Commun.*, **2002**, 2054–2055.

66 Wang, D., Salgueiriño-Maceira, V., Liz-Marzán, L.M. and Caruso, Frank (2002) *Adv. Mater.*, **2002**, 908–912.

67 Dong, W., Bongard, H.J., Tesche, B. and Marlow, F. (2002) *Adv. Mater.*, **2002**, 1457–1460.

68 Gottardo, S., Burresi, M., Geobaldo, F., Pallavidino, L., Giorgis, F. and Wiersma, D.S. (2006) *Phys. Rev. E*, **74**, 040702.

69 Takeda, H. and Yoshino, K. (2002) *Jpn. J. Appl. Phys., Part 2 (Lett.)*, **41**, L773–L776.

70 Matthias, S., Hillebrand, R., Müller, F. and Gösele, U. (2006) *J. Appl. Phys.*, **99**, 113102.

71 Serbin, J., Ovsianikov, A. and Chichkov Chichkov, B. (2004) *Opt. Express*, **12**, 5221–5228.

72 Deubel, M., Wegener, M., Linden, S., von Freymann, G. and John, S. (2006) *Opt. Lett.*, **31**, 805–807.

73 Tetreault, N., von Freymann, G., Deubel, M., Hermatschweiler, M., Pérez-Willard, F., John, S., Wegener, M., Ozin, G.A., *Adv. Mater.*, in press (DOI: 10.1002/adma.200501674).

74 Yang, J., Mingaleev, S.F., Schillinger, M., Miller, D.A.B., Shanhui, F. and Busch, K. (2005) *IEEE Photonics Technol. Lett.*, **17**, 1875–1877.

75 Richter, S., Steinhart, M., Hofmeister, H., Zacharias, M., Gösele, U., Gaponik, N., Eychmüller, A., Rogach, A.-L., Wendorff, J.-H., Schweizer, S.-L., von Rhein, A. and Wehrspohn, R. (2005) *Appl. Phys. Lett.*, **87**, 142107.

76 Röder, Th. and Kitzerow, H.-S. (2004) unpublished results.

77 Nielsen, K., Noordegraaf, D., Sørensen, T., Bjarklev, A. and Hansen, T.P. (2005) *J. Opt. A, Pure Appl. Opt.*, **7**, L13–L20.

78 Intonti, F., Vignolini, S., Turck, V., Colocci, M., Bettotti, P., Pavesi, L., Schweizer, S.-L., Wehrspohn, R. and Wiersma, D. (2006) *Appl. Phys. Lett.*, **89**, 211117.

79 Takeda, H. and Yoshino, K. (2003) *Phys. Rev. B*, **67**, 73106.

80 Schuller, C., Klopf, F., Reithmaier, J.P., Kamp, M. and Forchel, A. (2003) *Appl. Phys. Lett.*, **82**, 2767–2769.

81 Maune, B., Loncar, M., Witzens, J., Hochberg, M., Baer-Jones, T., Psaltis, D., Scherer, A. and Qiu, Y. (2004) *Appl. Phys. Lett.*, **85**, 360–362.

82 Asano, T., Song, B.-S. and Noda, S. (2006) *Opt. Express*, **14**, 1996–2002.

83 For a review on high-Q microcavities in photonic crystals see: Noda, S. (2006) *Science*, **314**, 260–261.

84 Armani, D.K., Kippenberg, T.J., Spillane, S.M. and Vahala, K.J. (2003) *Nature*, **421**, 925–928.

85 Vernooy, D.W., Ilchenko, V.S., Mabuchi, H., Streed, E.W. and Kimble, H.J. (1998) *Opt. Lett.*, **23**, 247–249.

13
Lasing in Dye-doped Chiral Liquid Crystals: Influence of Defect Modes

Wolfgang Haase, Fedor Podgornov, Yuko Matsuhisa, and Masanori Ozaki

13.1
Introduction

Photonic crystals (PCs) as a new class of optical materials exhibiting an ordered structure with a periodic dielectric constant attracted significant attention due to their fascinating properties and possible practical applications. For photonic crystals, the propagation of light in a certain frequency range is prohibited which, in turn, results in the appearance of photonic band gap (PBG) [1]. Two-dimensional (2D) and three-dimensional (3D) photonic crystals can show complete PBG which is not so for the one dimensional (1D) case. Nevertheless, the strong localization of electromagnetic modes demonstrated in 1D PCs within the photonic band gap or at the photonic band edge is especially useful for trapping photons and shaping the density of states. As result, they are perspective for potential applications as low threshold lasers or optical amplifiers [2].

The presence of a PBG affects the emission spectrum of dye molecules and as a result fluorescence is suppressed within the gap. However, near the band edges it is enhanced due to the high photonic density of states (DOS) [3]. In this case, the group velocity approaches zero, and the resulting long dwell times of emitted photons strongly promote stimulated emission. Hence, photonic crystals may be used as mirrorless resonators for laser emission.

Liquid Crystals (LCs) including chiral molecules have a self organized helical structure that is a 1D periodic structure and show characteristic optical properties [4]. In Cholesteric Liquid Crystals (CLCs) with helical structure, light propagating along the helical axis is selectively reflected depending on the polarization states if the wavelength of the light matches the optical pitch of the helical structure, which is a so-called selective reflection. The wavelength region in which the light cannot propagate is the stop band, which is considered as a 1D pseudo-bandgap. Lasing at the edge of the band has been reported e.g. in CLCs [5].

Chiral smectic LCs with a tilted structure show ferroelectricity and are called Ferroelectric Liquid Crystals (FLCs). Due to their fast response to the electric field, they have lot of potential for electro-optical applications [6]. FLCs too have a helical

Nanophotonic Materials: Photonic Crystals, Plasmonics, and Metamaterials. Edited by R.B. Wehrspohn, H.-S. Kitzerow, and K. Busch

ISBN: 978-3-527-40858-0

structure and show selective reflection due to their 1D periodic structure, similar to CLCs [7]. The studies on the enhancement of a nonlinear effect using the helical structure of FLCs have been performed [8] and the control of the stimulated emission and laser action in the helical structure of FLCs have been demonstrated [9].

Control of the photonic properties has been one of the advantages of fabricating a photonic device from chiral LCs and particularly from CLCs. This is realized by utilizing the external field sensitivity of the LC molecules or certain dopants to induce a change in the refractive indices or the helical pitch. Tuning the PBG and the lasing wavelength has been demonstrated upon thermally controlling the CLC pitch [10,11] or by fabricating a spatial modulation of the pitch [12,13]. On the other hand, stabilization of the helix was demonstrated in CLC standing films with doped photoreactive monomers which polymerize upon incidence of ultra-violet (UV) light [14]. An interesting tuning method is the utilization of external optical field. Photoinduced *trans-cis* isomerization in the azobenzene dye doped in CLCs was found to cause either an elongation or shortening of the CLC pitch.

Upon UV exposure, reversible modulation of the PBG was demonstrated in CLCs doped with various azobenzene derivatives [15,16]. A defect mode with high wavelength selectivity was realized in the usually non-transmitting PBG by introducing a structural defect in the perfect helical lattice.

While there are many possibilities for a medium to act as a defect such as an isotropic dielectric layer [17] or a phase shift in the LC director [18,19], an easily tunable defect mode is realized by introducing a different CLC material in the defect. In such a configuration, the defect-mode wavelength is determined by the contrast of the pitch lengths at the defect and bulk [20,21], and an optically tunable defect mode is realized if the CLC at the defect can selectively be pitch modulated by an external optical field. Such optical tuning of the photonic defect mode have already been demonstrated [22]. The proposed structure was fabricated by a laser-induced two-photon polymerization method, so that an unpolymerized CLC was left between two polymerized CLC layers, and then substituting the unpolymerized CLC by an azobenzene dye-doped CLC to act as the photoresponsive defect medium.

In this paper, another method to create structural defects, namely, by distributing the microbeads in the CLC has been shown. It has been demonstrated that the utilization of the multilayer structure results in great reduction of the threshold necessary for lasing of dye doped FLCs.

13.2
Experiment

At the first stage of our work, the influence of the structural defects on the lasing properties of chiral liquid crystals has been investigated. According to the analogy between the defects layers in band gap of semiconductors and similar effects in PBG structures, one can expect that the introduction of the structural defects will result in tremendous change in the emission spectrum. For the sake of simplicity experiments were carried out with cholesteric liquid crystals. Because light scattering of cholesterics is lower in comparison with those of FLCs, which results in significant reduction of pumping energy required to pass through the generation threshold. On the second stage of our research the confinement of FLC layer between multilayer dielectric coatings has been demonstrated, allowing a greater reduction in the pumping energy. Thus, making this geometry suitable for utilization of FLC as a lasing media.

13.2.1
Lasing in Cholesterics with Structural Defects

13.2.1.1 Preparation of Cholesterics

The cholesteric liquid crystal for the experiment was prepared using commercially available nematic MLC 2463 and the chiral dopant ZLI-811. Both of them were produced by Fa. Merck KGaA.

Because DCM dye was intended to be used as a lasing material, it was necessary to adjust the band gap in such a way that the stop band of the mixture should be between 532 nm and 610 nm, i.e., between the pumping beam wavelength and the maximum of the fluorescence spectrum of DCM. After series of experiments, it was found that the concentration of the chiral dopant ZLI-811 should be around 35 %wt. The prepared mixture was infiltrated in a cell with gap around 20 μm. The alignment layers provided homeotropical alignment of CLC. To determine the CLC pitch and the stop band, a fiber optic spectrometer OceanOptics HR2000 was utilized. The transmission spectrum at room temperature (25 °C) is demonstrated in Figure 13.1.

As one can clearly see from this figure, the prepared mixture satisfies the imposed requirements.

13.2.1.2 Cell Fabrication

For the investigations, experimental cells with 1.1 mm thick substrates as obtained from Merck were used. Onto the substrates the polyimide AL 2021 was spin coated

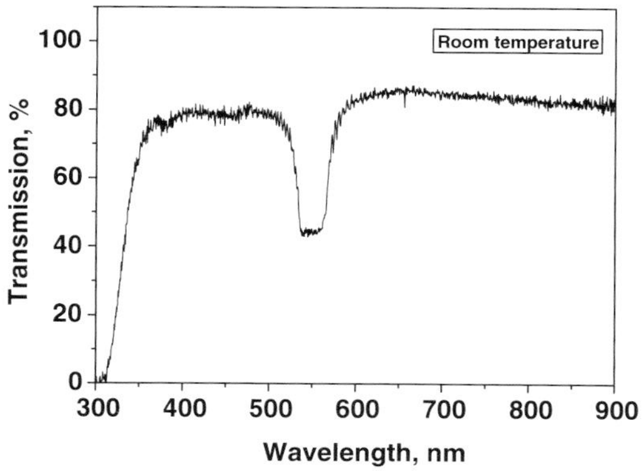

Figure 13.1 Transmission spectrum of the homeotropically aligned CLC cell. The cell gap is 20 μm. Measurement was done at room temperature (25 °C).

(3000 rpm) and baked at 185 °C for 2 hours. To get homeotropic aligned cells, the polymer layers were rubbed unidirectionally with a special tissue. The thickness of the cells was predetermined by glass microbeads with an average diameter of 20 μm dissolved in photocurable glue.

13.2.1.3 Preparation of CLC/TiO_2 Dispersion

To prepare the CLC dispersions, polydispersed TiO_2 particles were dispersed in acetone and sonificated for about an hour. The agglomeration of the particles was removed by filtering the solution through a filter having an average pore diameter of 1 μm. The amount of the TiO_2 particles was estimated as 0.1 wt%. Investigation under AFM showed that the diameter of the TiO_2 was distributed in a wide range with a maximum around 530 nm.

After preparation of dispersions, the texture of pure and doped mixture was investigated under a polarizing microscope. The texture of both the samples was practically identical. However, in case of CLC/TiO_2 dispersion, particles with a diameter of more than 800 nm could be observed but their amount was small.

At the final stage, the prepared dispersion was doped with DCM with concentration 1.5 wt%. In order to compare further results, the mixture was doped with DCM but without TiO_2 particles.

13.2.1.4 The Experimental Setup

For the experimental setup (Figure 13.2), a reflective configuration at 45° oblique incidence was used. The pump source was a frequency-doubled Q-switched Nd:YAG pulsed laser (purchased from Laser 2000) with wavelength $\lambda = 532$ nm, 6 ns pulse width, and 1 kHz repetition rate. The energy of the pumping beam was monitored by the energy meter (Thorlabs).

The diameter of the light beam was predetermined by a pinhole mounted just after the laser. The intensity of the pumping light was controlled by the attenuator which represents itself as a set of filters with different transmission coefficient.

To reduce the CLC Bragg reflection, the pumping light polarization was converted into a circular one with the help of a polarizer and a quarter wave plate in such a way

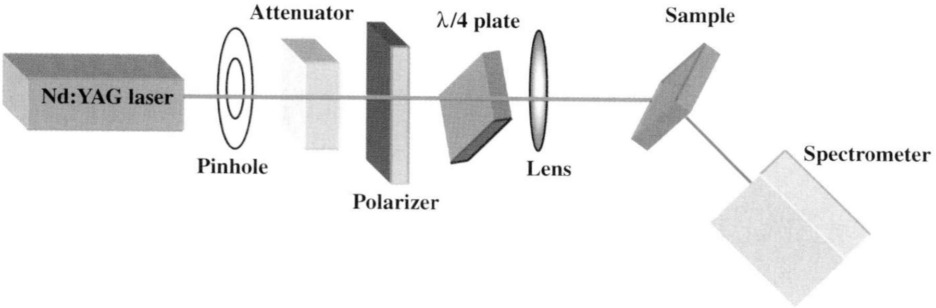

Figure 13.2 Experimental setup. The pumping beam is incident under an angle of 45° with respect to the cell normal.

that its handedness is counter to that of the CLC helix. The lens after this polarization controller focuses the light onto the spot with a diameter of around 200 μm. The emitted laser light was acquired with fiber-optic spectrometer (Ocean Optics, HR2000, resolution = 0.4 nm) along the direction perpendicular to the glass substrates.

13.2.1.5 **Experimental Results**

Using the above described setup, the emission spectra of the CLC/DCM and CLC/DCM/TiO_2 cells was investigated at room temperature. To avoid cell heating, the incident light beam was additionally modulated with a stroboscope operating at a frequency of 5 Hz. The energy of the pumping pulse was changed with neutral filters in range from 0.2 μJ/pulse to 10 μJ/pulse.

The threshold for lasing for the CLC/DCM cell was around 3 μJ/pulse. This relatively high value could be explained by the presence of texture defects inside the cell. The emission line, as expected, was located at long wavelength edge of the stop band.

The emission spectra for both type of cells is shown in Figure 13.3. As one can clearly see from this plot, the threshold for the CLC/DCM/TiO_2 is significantly higher than for CLC/DCM cell and equals around 4.5 μJ/pulse. Such behaviour could be explained by stronger light scattering in the CLC/DCM/TiO_2 cell which originated from the presence of the polydispersed TiO_2 particles which induced the texture defects.

The evolution of the emission spectra for the particles doped sample is clearly seen from plots (b), (d) and (f) (Figure 13.3). At low pumping energy (less than 4 μJ/pulse), the emitted light is the broad band fluorescence. After exceeding the threshold, the emission spectrum converts into two sharp lines with average width around 1.5 nm. This result is different from that of the pure CLC/DCM cell. This remarkable difference can be explained by the texture defects induced by the TiO_2 nanoparticles. In the polarizing microscope we clearly observed the multidomain structure of the CLC texture. The characteristic dimension of the domains was around 50–70 μm. So, we think that the second peaks appeared due to the illumination of, at least, two different domains by the pumping beams.

The other remarkable difference is that the emission spectra are shifted toward the higher wavelengths. This phenomenon is related to the increase of the average refraction index of the CLC/DCM/TiO_2 mixture in comparison with CLC/DCM one. According to the Maxwell–Garnet approach, the refractive index of the material should increase if another substance with significantly higher refractive index is added, which is exactly our case.

13.2.2
Lasing in Ferroelectric Liquid Crystals

The lasing in dye doped FLC cells has experimentally been demonstrated in works [9,23–25]. However, as it was shown, the threshold pumping energy was much higher in comparison with that of CLCs. This effect is related with inherent

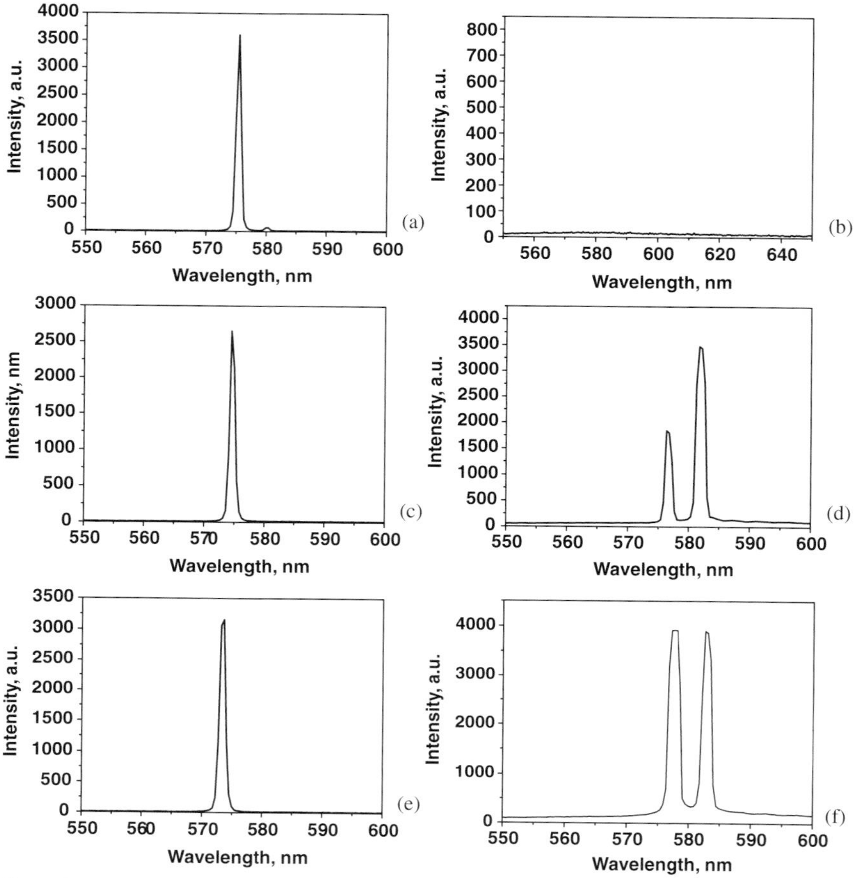

Figure 13.3 Lasing in CLC cells. Left hand column are samples doped with DCM only (concentration 1.5 wt%). Right hand column are samples doped with DCM (concentration 1.5 wt%) and TiO_2 microsphere (concentration 1 wt%). Pumping energy for (a) and (b) −3 μJ/pulse, for (c) and (d) −4.5 μJ/pulse, for (e) and (f) −5 μJ/pulse. The diameter of the illuminated area was around 200 μm.

properties of FLCs, i.e., high losses related to light scattering and lower depth of the stop band. In order to decrease the threshold energy, we have investigated laser action in FLC sandwiched between dielectric multilayers and achieved much lower lasing threshold compared with simple FLC without dielectric multilayers. The detailed results of this work are published in articles [26,27].

13.2.2.1 Sample Preparation

A dielectric multilayer, which consisted of five pairs, alternately stacked the SiO_2 and TiO_2 layers, deposited on a glass substrate (Figure 13.4). The refractive indices of SiO_2 and TiO_2 are 1.46 and 2.35, and their thicknesses are 103 nm and 64 nm, respectively.

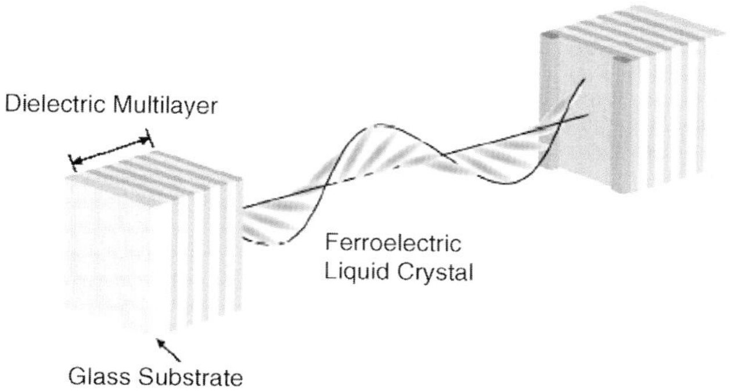

Figure 13.4 Ferroelectric liquid crystal cell operating in DHF mode. Both confining substrates contain dielectric multilayers.

The center wavelength of the stop band was adjusted to be 600 nm, and an excitation wavelength for lasing 532 nm was out of the stop band. The top surface of the dielectric multilayer was coated with polyimide (JSR, JALS-2021-R2) in order to obtain a homeotropic alignment.

The FLC compound used in this study was a multi component mixture having chiral smectic C (Sm C*) phase in a wide temperature range including the room temperature. The chemical content of the mixtures are summarized in Table 13.1.

As a laser dye dopant, DCM (Exciton) was doped in the FLC with concentration 0.76 wt%. The FLC was sandwiched between the multilayers using 16 μm spacers, as the helical axis was perpendicular to the multilayer surfaces. In order to make comparison, emission characteristics of simple FLC cell without the multilayer were also investigated. The transmission and emission spectra of simple FLC without dielectric multilayers has been investigated, which is shown with dashed lines in Figure 13.5. In the transmission spectrum, a drop of transmittance was observed at 595 nm, which was due to the stop band of the FLC. Decrease in the transmittance at shorter wavelength was attributed to absorption of the laser dye.

13.2.2.2 **The Experimental Setup**

The prepared samples were pumped with a Q-switched Nd:YAG laser. The wavelength, pulse duration and repetition rate were 532 nm, 8 ns and 10 Hz respectively. The pumping beam was incident perpendicular to the sample. The laser beam was focused onto the sample in such a way that the illumination area was around 0.2 mm^2. The emission spectrum of the samples was measured from the opposite side of the cells with a multichannel spectrometer Hamamatsu having the spectral resolution around 2 nm. The temperature of the sample was maintained by a temperature controller and was 28 °C.

13.2.2.3 **Experimental Results**

The emission spectrum of the simple cell at a pump energy of 54 mJ/cm^2 pulse is shown with the dashed line, which was dominated by broad spontaneous emission of

Table 13.1 Chemical content of the utilized FLC mixture.

Compound
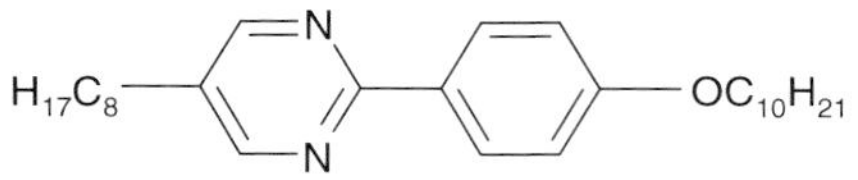
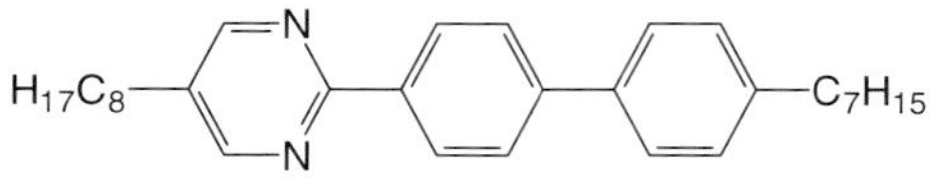
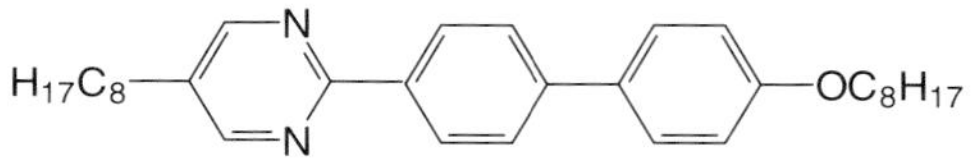
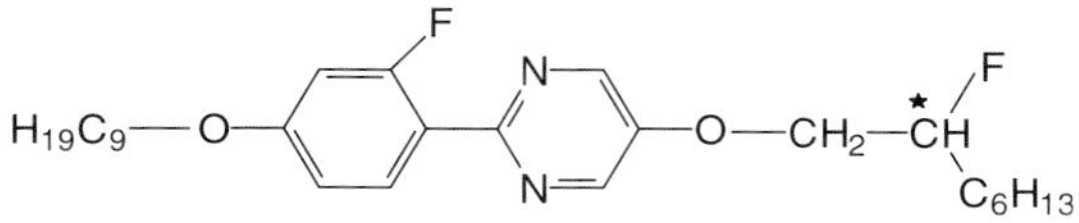
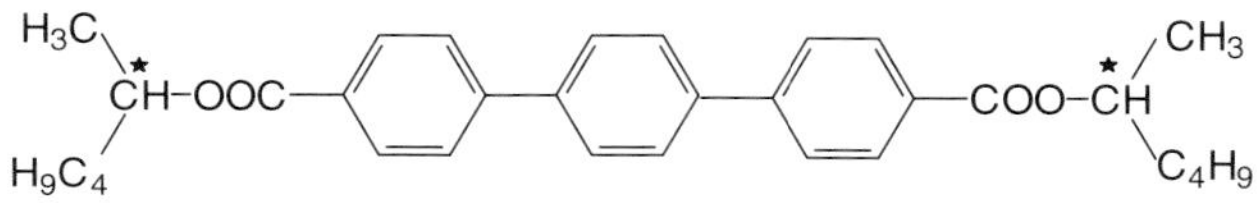
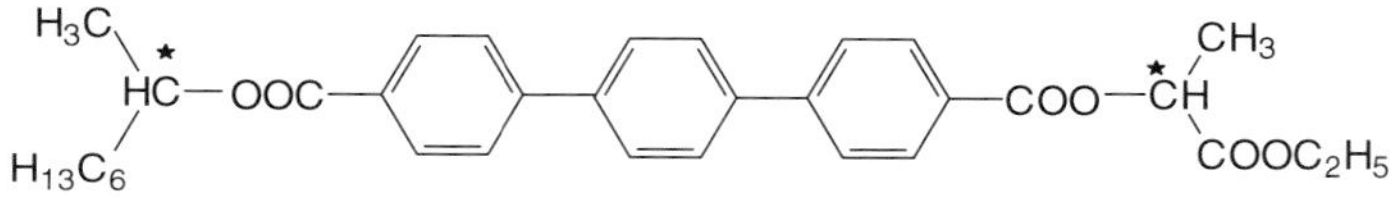

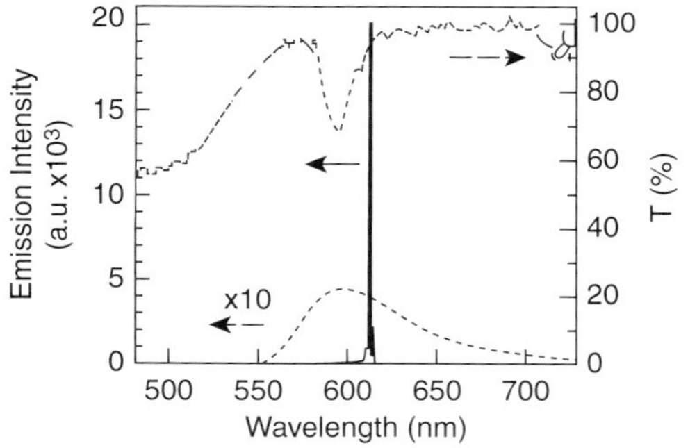

Figure 13.5 Emission and transmission spectra of simple FLC (dashed lines) and emission spectrum of FLC sandwiched between dielectric multilayers.

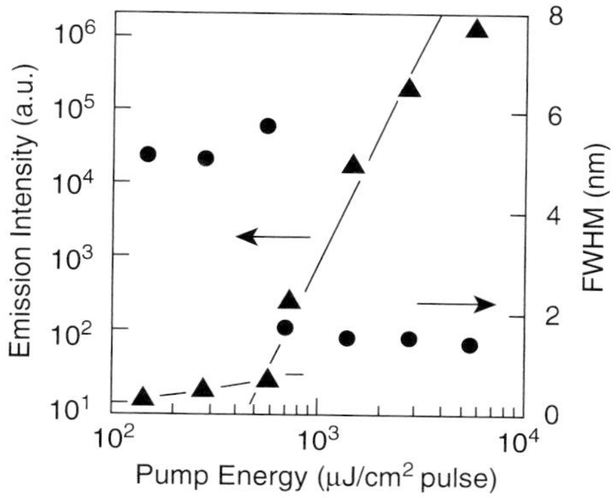

Figure 13.6 Intensity of the emitted light and emission linewidth (FWHM) vs. pump energy. The black circles stand for the emission line width (FWHM), the black triangles indicate the emission intensity.

the laser dye. The emission intensity increased as the pump energy increased; however, laser action was not observed at any pump energies up to the damage threshold of 580 mJ/cm^2 pulse. This result indicates that the lasing threshold is higher than the damage threshold in this material. The emission characteristics of FLC sandwiched between dielectric multilayers has been investigated, as shown in Figure 13.5. The solid line in Figure 13.5 shows the emission spectrum of the FLC with the dielectric multilayers at the pump energy of 1.4 mJ/cm^2 pulse.

Although many emission peaks appeared at low pump energy, above the threshold, a sharp emission line was observed, as shown in Figure 13.5, at 614 nm which corresponds to the long wavelength band edge of the FLC. Figure 13.6 shows the peak intensity and the full width at half maximum (FWHM) of the emission spectrum as a function of pump energy. Above the pump energy of 600 μJ/cm^2 pulse, the emission intensity increased drastically. Simultaneously, the FWHM of the emission peak decreased from 5 nm to less than the spectral resolution of 2 nm. This result indicated that single-mode laser action occurred above the threshold of 600 μJ/cm^2 pulse. It should be noted that the lasing threshold is 10^3 times lower than that of the simple FLC without the multilayer, which must be attributed to the sandwich structure with the dielectric multilayers. For estimation of this value, we suppose that, in principle, this threshold should exist, at least, theoretically and it is higher than the damage threshold. In other words, this is the rough estimation of the lower limit of this ratio.

This laser action can be explained by the double optical confinement of not only the band edge effect of the FLC but also the light localization effect of the photonic crystal composed of the dielectric multilayers. Addition of the optical confinement by the dielectric multilayer to that by the FLC must lead to lowering the lasing threshold of the FLC sandwiched between the dielectric multilayers compared with that of the simple FLC.

13.2.3
Conclusion

In this article we demonstrated the influence of the defect mode on the emission spectra of the dye doped chiral liquid crystal. At the first stage we investigated the lasing in cholesteric liquid crystal with defects introduced by the TiO_2 microbeads. It was revealed that due to scattering the threshold of the lasing increased at 30–40% and, in the same time, multimode laser action appeared.

At the second stage, the single-mode laser action in FLC sandwiched between dielectric multilayers at the band edge of the FLC was shown. The lasing threshold was hundred times lower than that of simple FLC because of the double optical confinement effect of the FLC and the dielectric multilayers.

Acknowledgement

This work was supported by DFG Priority Program "Photonic Crystals". The authors express their gratitude to Fa. Merck KGaA for the NLC mixture.

References

1 Yablonovitch, E. (1987) *Phys. Rev. Lett.*, **58**, 2059.
2 Dowling, J.P., Scalora, M., Bloemer, M.J. and Bowden, C.M. (1994) *J. Appl. Phys.*, **75**, 1896.
3 Painter, O., Lee, R.K., Scherer, A., Yariv, A., O'Brien, J.D., Dapkus, P.D. and Kim, I. (1999) *Science*, **284**, 1819.
4 de Gennes, P.G. and Prost, J. (1993) *The Physics of Liquid Crystals*, Oxford University Press, New York.
5 Morris, S.M., Ford, A.D., Pivnenko, M.N. and Coles, H.J. (2005) *J. Appl. Phys.*, **97**, 023103.
6 Beresnev, L.A., Chigrinov, V.G., Dergachev, D.I., Poshidaev, E.P., Fuenfchilling, J. and Schadt, M. (1989) *Liq. Cryst.*, **5**, 1171.
7 Hori, K. (1982) *Mol. Cryst. Liq. Cryst.*, **82**, L13.
8 Yoo, J., Choi, S., Hoshi, H., Ishikawa, K. and Takezoe, H. (1997) *Jpn. J. Appl. Phys.*, **36**, L1168.
9 Ozaki, M., Kasano, M., Ganzke, D., Haase, W. and Yoshino, K. (2002) *Adv. Mater.*, **14**, 306.
10 Funamoto, K. and Ozaki, M. and Yoshino, K. (2003) *Jpn. J. Appl. Phys.*, Part 2, **42**, L1523.
11 Taheri, B., Munoz, A.F., Palffy-Muhoray, P. and Twieg, R. (2001) *Mol. Cryst. Liq. Cryst.*, **358**, 73.
12 Huang, Y., Zhou, Y. and Wu, S.T. (2006) *Appl. Phys. Lett.*, **88**, 011107.
13 Huang, Y., Zhou, Y., Doyle, C. and Wu, S.T. (2006) *Opt. Express*, **14**, 1236.
14 Matsui, T., Ozaki, R., Funamoto, K., Ozaki, M. and Yoshino, K. (2002) *Appl. Phys. Lett.*, **81**, 3741.
15 Sackmann, E. (1971) *J. Am. Chem. Soc.*, **93**, 7088.
16 Kurihara, S., Matsumoto, K. and Nonaka, T. (1998) *Appl. Phys. Lett.*, **73**, 160.
17 Yang, Y.C., Kee, C.S., Kim, J.E., Park, H.Y., Lee, J.C. and Jeon, Y.J. (1999) *Phys. Rev. E*, **60**, 6852.
18 Ozaki, M., Ozaki, R., Matsui, T. and Yoshino, K. (2003) *Jpn. J. Appl. Phys.*, Part 1, **42**, 472.
19 Schmidke, J., Stille, W. and Finkelmann, H. (2003) *Phys. Rev. Lett.*, **90**, 083902.

20 Matsui, T., Ozaki, M. and Yoshino, K. (2004) *Phys. Rev. E*, **69**, 061715.

21 Yoshida, H., Lee, C.-H., Fujii, A. and Ozaki, M. (2006) *Appl. Phys. Lett.*, **89**, 231913.

22 Yoshida, H., Lee, C.-H., Miura, Y., Fujii, A. and Ozaki, M. (2007) *Appl. Phys. Lett*, **90**, 071107.

23 Kasano, M., Ozaki, M., Yoshino, K., Ganzke, D. and Haase, W. (2003) *Appl. Phys. Lett.*, **82**, 4026.

24 Ozaki, M., Kasano, M., Kitasho, T., Ganzke, D., Haase, W. and Yoshino, K. (2003) *Adv. Mater.*, **15**, 974.

25 Ozaki, M., Kasano, M., Haase, W. and Yoshino, K. (2005) *J. Soc. Electron. Mater. Eng.*, **14** (3), 103.

26 Matsuhisa, Y., Ozaki, R., Haase, W., Yoshino, K. and Ozaki, M. (2006) *Ferroelectrics*, **344**, 239.

27 Matsuhisa, Y., Haase, W., Fujii, A. and Ozaki, M. (2006) *Appl. Phys. Lett.*, **89**, 2011112.

14
Photonic Crystals based on Chiral Liquid Crystal

M. Ozaki, Y. Matsuhisa, H. Yoshida, R. Ozaki, and A. Fujii

14.1
Introduction

Photonic crystals (PCs) having a three-dimensional (3D) ordered structure with a periodicity of optical wavelength have attracted considerable attention from both fundamental and practical points of view, because in such materials novel physical concepts such as the photonic band gap (PBG) have been theoretically predicted and various applications of photonic crystals have been proposed [1,2]. Particularly, the study of stimulated emission in the PBG is one of the most attractive subjects, since, in the band gap, a spontaneous emission is inhibited and low-threshold lasers based on photonic crystals are expected [1,3–6]. So far intensive studies on one- and two-dimensional band-gap materials have been performed. In order to realize the photonic crystal, a large number of studies on a micro-fabrication based on a semiconductor processing technology [7–9] and a self assembly construction of nano-scale spheres [10,11] have been carried out.

Liquid crystals (LCs) including chiral molecule have a self-organized helical structure which can be regarded as a one-dimensional (1D) periodic structure (Figure 14.1). In such systems, there is a so-called stop band in which the light cannot propagate, which is considered as a 1D pseudo-bandgap. Lasing at the band edge has been reported in cholesteric liquid crystal (CLC) [12–19], chiral smectic liquid crystal [20–22], polymerized cholesteric liquid crystal (PCLC) [23–25] and cholesteric blue phase [26]. These laser actions in the 1D helical structure of the chiral liquid crystals are interpreted to be based on an edge effect of the 1D PBG in which the photon group velocity is suppressed [27].

On the other hand, the localization of the light based on the defect mode caused by the imperfection in the periodic structure has been expected as potential applications such as low-threshold lasers and microwaveguides [7–9,28,29]. The introduction of the defect layer into the periodic helical structure of the CLCs has been theoretically studied [30,31]. Especially, Kopp et al. have predicted the existence of a localized state for single circularly polarized light in the twist defect of the CLCs [31]. However, in

Nanophotonic Materials: Photonic Crystals, Plasmonics, and Metamaterials. Edited by R. B. Wehrspohn, H.-S. Kitzerow, and K. Busch

ISBN: 978-3-527-40858-0

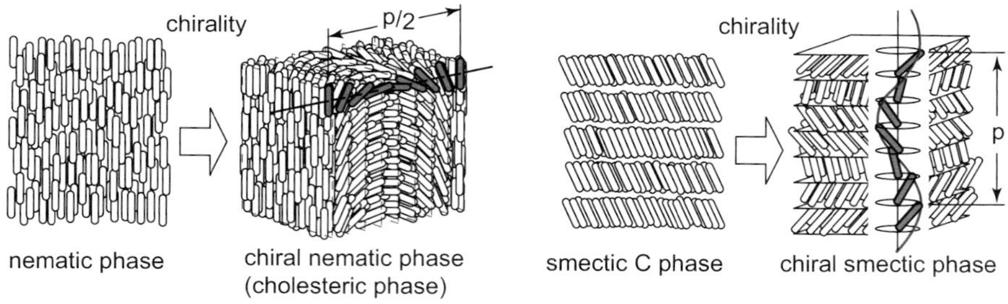

Figure 14.1 Molecular alignments of nematic and smectic liquid crystals. Upon adding chirality, helicoidal periodic molecular alignment is induced, whose periodicity can be tuned from hundreds nm to over several hundreds μm by changing chirality strength.

spite of the theoretical demonstration of the twist defect mode (TDM) in the CLC, evident experimental demonstration on such defect modes had not been carried out.

LCs have a large optical anisotropy and are sensitive to an external stress such as an electric field. Based on such optical anisotropy and field sensitivity, a tunable photonic crystal has been proposed in opal or inverse opal infiltrated with LC [32–36]. Although opals and inverse opals provide a simple and inexpensive approach to realize 3D PC using self-organization of colloidal particles [10,11], the introduction of defects into the 3D periodic structure is a problem that must be resolved. Not only 3D PCs but also 1D PCs are an attractive subject. Although, the 1D PC does not have a complete PBG, there are plenty of applications using extraordinary dispersion of the photon and localized photonic state in a defect layer. So far, intensive studies on 1D PC applications have been reported: air-bridge microcavities [7], the photonic band-edge laser [27], nonlinear optical diode [37] and the enhancement of optical nonlinearity [28,38]. Recently we have introduced a LC layer in a dielectric multilayer structure as a defect in the 1D PC [29,39,40], in which the wavelength of defect modes was controlled upon applying electric field in a basis of the change in optical length of the defect layer caused by the field-induced molecular reorientation of LC.

In this paper, we review our recent work on the photonic crystals based on the self-organized chiral liquid crystals and defect modes characteristic to the helical periodic structure.

14.2 Photonic Band Gap and Band Edge Lasing in Chiral Liquid Crystal

14.2.1 Laser Action in Cholesteric Liquid Crystal

In a periodically structured medium, when the Bragg condition is satisfied, reflected lights at each point interfere with each other. As a result, the light can not propagate and only the standing waves exist. This means the lights in a certain range of energy

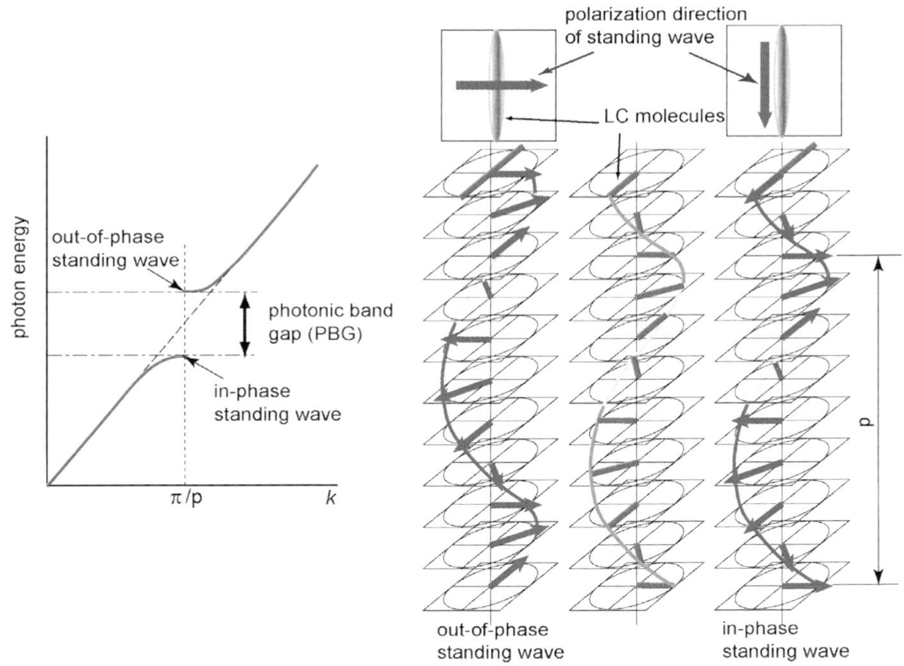

Figure 14.2 Schematic explanation of the appearance of photonic band gap in spiral periodic structure.

can not propagate in the medium, and the lights at both edges of this energy range have zero group velocity. This energy range is so called "photonic band gap".

In the helical periodic structure of the CLCs, the light propagating along the helical axis is selectively reflected depending on the polarization states if the wavelength of the light matches to the optical periodicity of the helical structure, which is a so-called selective reflection. In this case, there are two types of circularly-polarized standing waves with zero group velocity at the edges of the stop band as shown in Figure 14.2. Here, the rods indicate the molecular long axis of the CLC and the arrows show the polarization direction of the standing waves. For one of standing waves, the polarization direction of the light is always parallel to the molecular long axis (in-phase standing wave). This light feels extraordinary refractive index of the LC and has lower energy with respect to a travelling wave, which corresponds to the longer edge of the stop band. If we dope laser dye having a transition moment parallel to the long axis of LC, the polarization of the in-phase standing wave is parallel to the transition moment of the doped dye. This circularly polarized standing wave with lower energy effectively interacts with the laser medium and we can expect the low threshold laser action at the longer wavelength edge of the band gap.

Figure 14.3a shows reflection and emission spectra of the dye-doped CLC. The reflection band corresponds to the 1D PBG. The CLC compound with a left-handed

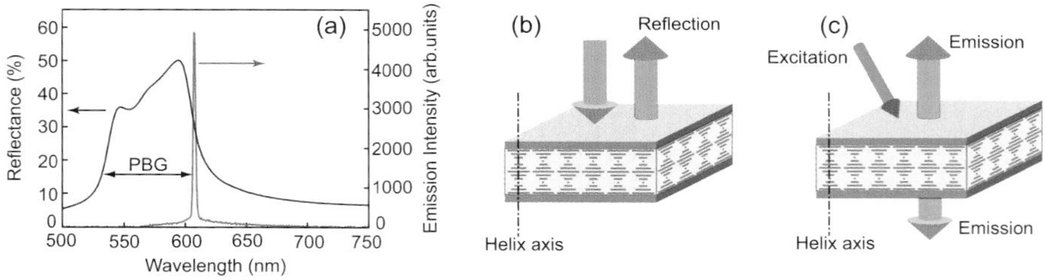

Figure 14.3 (a) Reflection and emission spectra of dye-doped CLC. A sharp lasing peak appears at the longer wavelength edge of the PBG at an excitation energy of 13 mJ/cm^2. (b) and (c) represent cell configuration for the reflection and emission spectra measurements, respectively.

helix was prepared by mixing a chiral dopant (S-811, Merck) and a nematic liquid crystal (E44, Merck). As a laser dye dopant in the CLC, [2-[2-4-(dimethylamino) pheny1]etheny1]-6-methy1-4H-pyran-4-ylidene propanedinitrile (DCM, Exciton) was used. The concentration of the dye was 1wt%. The sample was filled into a parallel sandwich cell, which consists of two glass plates. The CLC in the cell aligns their directors parallel to the glass plates, that is, the helical axis is perpendicular to the glass substrates as shown in Figure 14.3(b) and (c). The cell thickness of CLC is 9 μm. As an excitation source, a second harmonic light of a Q-switched Nd:YAG laser was used, whose wavelength and pulse width were 532 nm and 8 ns, respectively. The excitation laser beam irradiated the sample at an angle of about 45 degree with respect to the cell plate normal. At a low excitation energy, the emission spectrum is dominated by a broad spontaneous emission and the suppression of the emission due to the stop band of the CLC is observed. At a high excitation energy (>6 mJ/cm^2), the laser action is observed at the longer wavelength edge of the stop band, as shown in Figure 14.3(a). Emitted laser light is circularly polarized with the same handedness as the helix sense. With temperature increases, the lasing wavelength shifts toward a shorter wavelength, which corresponds to the shift of the stop band originating from the temperature dependence of the helical pitch of CLC.

14.2.2
Low-Threshold Lasing Based on Band-Edge Excitation in CLC

The laser action mentioned above is based on the suppression of photon group velocity and the enhancement of the photon density of states (DOS) at the lower energy edge of the PBG. On the other hand, at the higher energy edge of the PBG, an effective absorption should occur and a low-threshold laser action can be expected [41]. We have experimentally demonstrated the band-edge excitation effect on CLC lasing by matching the excitation laser wavelength to the higher energy edge of the PBG of a dye-doped CLC [42,43]. Figure 14.4(a) shows the lasing threshold of the dye-doped CLC for the right- and left-handed circularly polarized (RCP and LCP) pumping beams as a function of the wavelength of the high-energy band edge by changing the operating temperature. The CLC host was prepared by mixing nematic

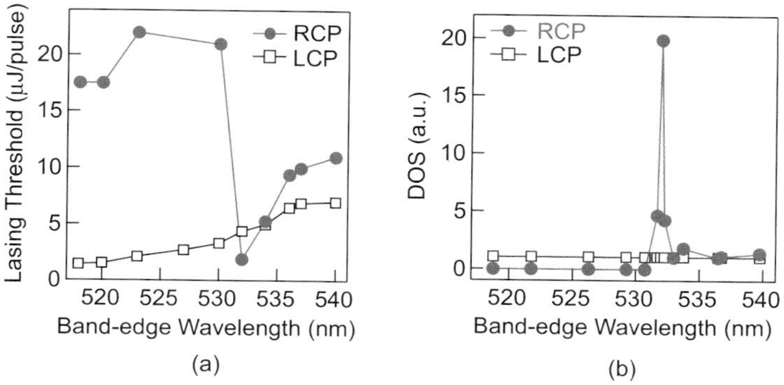

Figure 14.4 (a) Lasing threshold as a function of the wavelength of high-energy band edge for the LCP and RCP excitation beams. (b) The DOS as a function of wavelength of high-energy band edge for the LCP or RCP incident lights.

liquid crystal (BL006, Merck) with a right-handed chiral dopant (MLC6248, Merck). As a laser dye, PM597 (Exciton) was doped into the mixture. A second harmonic light (532 nm) of a Q-switched Nd:YAG laser was used as an excitation source. For the LCP excitation, the lasing threshold gradually increases with the redshift of the band edge. This originated from the position of the maximum spontaneous emission of the doped dye. Namely, PM597 dye exhibits its maximum spontaneous emission at 580 nm. Therefore, the emission efficiency should be highest when the high energy edge matches 510 nm because the band width of the CLC is about 70 nm. On the other hand, for the RCP excitation, the lasing threshold exhibits minimum when the shorter band edge of the PBG is located at the pumping wavelength of 532 nm. Moreover, the lasing threshold for the RCP excitation is about two times lower than that for the LCP one at 532 nm. This is attributed to the band-edge excitation effect for the CLC lasing.

The DOS of the CLC used in the above experiment for LCP or RCP incident lights (532 nm) was evaluated as a function of the wavelength of the high-energy band edge. The calculation was carried out using the 4×4 matrix method by changing the helix pitch. As is evident from Figure 14.4(b), the DOS for the RCP incident light is drastically enhanced when the high-energy edge coincides with the excitation wavelength (532 nm). This indicates that the high DOS caused by band-edge excitation enhances the lasing efficiency and decreases the lasing threshold as shown in Figure 14.4(a).

14.2.3
Laser Action in Polymerized Cholesteric Liquid Crystal Film

Liquid crystals are fluid materials and have to be filled in containers such as a sandwiched cell. This fluidity is unsuitable for a practical application. We have developed a solid film having CLC helicoidal molecular alignment for the use of

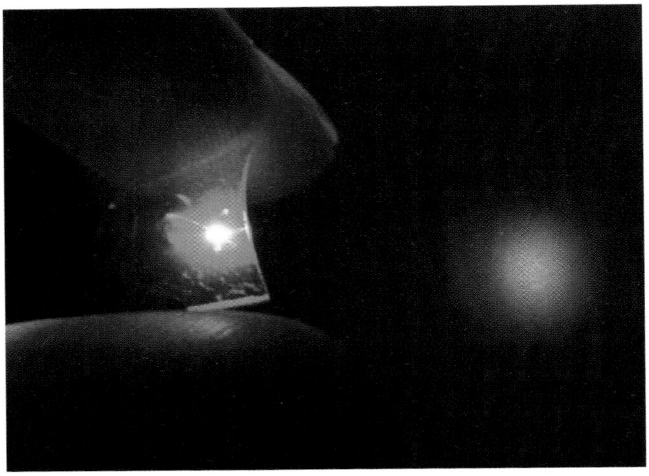

Figure 14.5 Lasing from bent film of polymerized CLC.

lasing [25]. For the fabrication of polymerized CLC (PCLC), photo-polymerizable CLC mixtures (Merck KGaA) were used. First, the monomer sample was inserted by a capillary action into the sandwiched cell that is composed of two glass plates. The CLCs in this cell align their director parallel to the glass plates, that is, the helical axis is perpendicular to the glass substrates. UV light irradiation was performed using a Xe lamp to induce the photo-polymerization of the UV-curable CLC monomer. After UV light irradiation, two glass substrates were removed and the free-standing PCLC film was obtained. Figure 14.5 shows the laser action from the PCLC film containing DCM of 0.4wt%. For the excitation light source, a second harmonic light of a Q-switched Nd:YAG laser was used. At high excitation energy ($>1.5\ mJ/cm^2$), laser action is confirmed at the edge of the stop band. This laser action is achieved without any substrates and is observed even when PCLC film is bent as shown in Figure 14.5. This suggests that 1D helical structure required for the laser action is maintained even in the deformed film. This flexibility may enable us to fabricate optical devices with new functionalities.

14.2.4 Electrically Tunable Laser Action in Chiral Smectic Liquid Crystal

Chiral smectic liquid crystals with a tilted structure show ferroelectricity, which are called ferroelectric liquid crystals (FLCs), and have an expected potential for electro-optic applications because of their fast response to an electric field. FLCs also have a helical structure and show the selective reflection due to their 1D periodic structure in an almost the same manner as the CLC [20]. The helix of the FLC can be easily deformed upon applying electric field and its response is fast because of the strong interaction between the spontaneous polarization and electric field. Therefore, a fast

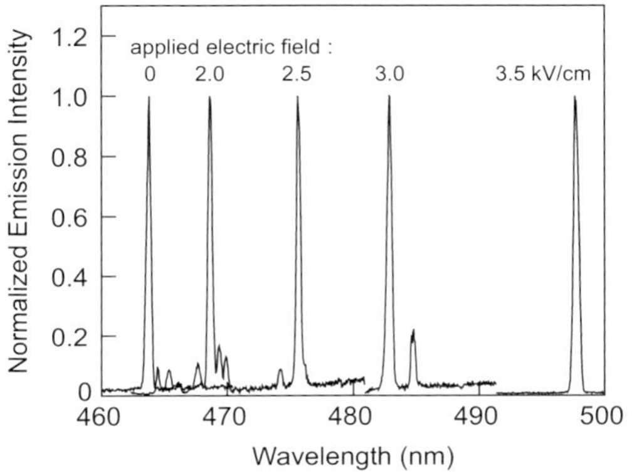

Figure 14.6 Lasing spectra of dye-doped FLC as a function of an applied electric field perpendicular to the helix axis.

modulation of the lasing wavelength upon applying electric field can be expected in a dye-doped FLC.

FLC has a spontaneous polarization *Ps* which points normal to the molecules and parallel to the smectic layers. When the electric field is applied parallel to the layer, for lower field *Ps* intends to point along the field direction and FLC molecules start to reorient to the direction normal to the field, resulting in the deformation of the helix. Above the threshold field, all FLC molecules orient to the same direction and the helix is unwound. At intermediate field strength, the deformation of the helix might cause the elongation of the periodicity of the helix.

Figure 14.6 shows the lasing spectra of the dye-doped FLC at high excitation energy (24 μJ/pulse) as a function of the applied electric field. It should be noted that lasing wavelength largely shifts toward longer wavelength with increasing the field, which corresponds to the shift of the selective reflection band. In spite of a low field (3.5 kV/cm), a wide tuning of the lasing wavelength was achieved [21].

The response of the electrooptical switching based on a slight deformation of the helix of a short pitch FLC is as fast as several μs, and the application to optical communications as well as to display devices has been proposed. The relaxation time τ of the helix deformation of FLC is represented by the following equation,

$$\tau \propto \frac{p^2 \gamma}{K},$$

where p is the helix pitch, γ is the rotational viscosity and K is the elastic constant. According to this relation, the response time is proportional to p^2, and high frequency modulation of the periodicity of the helix can be expected in a short pitch FLC. In deed, the electrooptic modulation device using a similar compound to the FLC material used in this study has a response of the order of several μs. Consequently, the fast modulation of the lasing is possible in the dye-doped FLC with a short pitch.

14.3
Twist Defect Mode in Cholesteric Liquid Crystal

Laser actions reported so far in chiral liquid crystals (Figure 14.7(a)) are observed at the edge wavelength of the stop band and are associated with the group velocity anomaly at the photonic band edge. On the other hand, low threshold laser action based on the photon localization at a defect in a periodic structure can also be expected. The introduction of the defect into the periodic helical structure of the CLCs has been theoretically studied. Especially, Kopp *et al.* have predicted the existence of a single circularly polarized localized mode in the twist defect of the CLCs (Figure 14.7 (b)) [31].

The PCLC film with the twist defect was prepared as follows [44]. The photo-polymerizable CLC monomer was spin-coated from a toluene solution on a glass substrate on which a polyimide (AL-1254) was coated and rubbed in one direction. In order to obtain a uniform planar alignment, the coated CLC was annealed at the temperature just below the clearing point. The CLC on the substrate aligns their director parallel to the glass plate, that is, the helical axis is perpendicular to the glass substrate. UV light irradiation was performed using a Xe lamp to induce the photo-polymerization of the UV-curable CLC monomer. Two PCLC films were put together

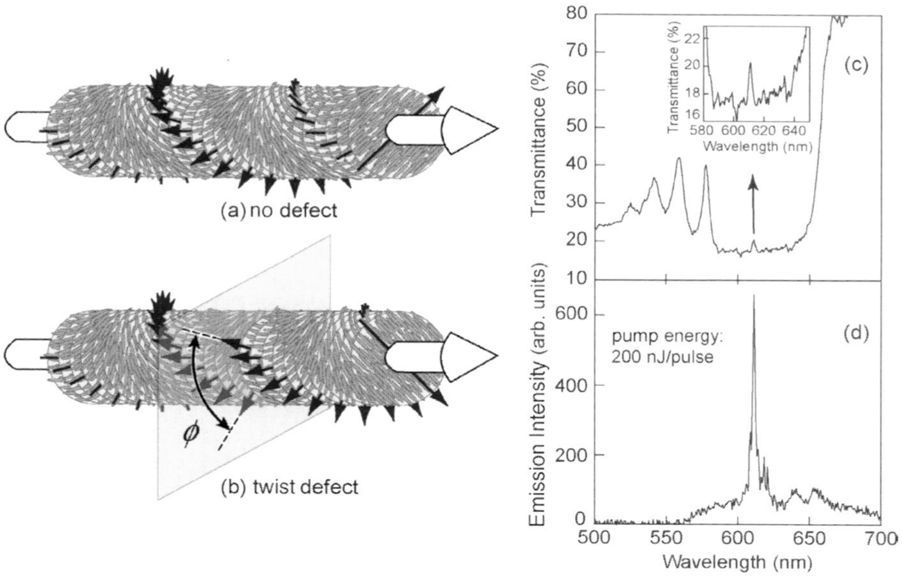

Figure 14.7 Helix structures without (a) and with (b) twist defect interface. (c) Transmission spectrum of the dye-doped double PCLC composite film with twist defect. (d) Emission spectrum of the double PCLC composite film with the defect at above the threshold pump pulse energy (200 nJ/pulse).

as the directors of liquid crystal molecules at an interface of these films make a certain angle ϕ. In other words, there is a discontinuous phase jump of the azimuthal angle of the helical structures between these PCLC films at the interface, which acts as a twist defect in the helicoidal periodic structure.

Figure 14.7(c) shows the transmission spectrum of the dye-doped double PCLC composite film containing a discontinuous defect interface. A stop band, 1D PBG, is confirmed in the spectral range from 580 nm to 640 nm. It should be noted that a sharp peak appears at 611 nm within the PBG, which might be related to the defect mode induced by the introduction of the twist defect interface.

We have performed theoretical calculation of the light propagation in PCLC films with and without twist defect interface, using a method of characteristic matrices. This method is a numerical analysis based on the Maxwell equation which can be used to quantitatively calculate the light propagation in the medium with refractive index varying along one direction. Assuming that the phase jump of the director angles ϕ at the interface of two PCLC layers is $4\pi/9$rad, the theoretically calculated results are in good agreement with the experimental ones. This result indicates that the sharp peak observed in the PBG corresponds to the TDM in the double PCLC composite film with the twist defect interface.

Figure 14.7(d) shows the emission spectrum of the dye-doped double PCLC composite film with the defect interface at pump energy of 200 nJ/pulse. For an excitation source, a second harmonic light of a mode-locked Nd:YAG laser (Ekspla, PL2201) was used. The pulse width, wavelength and pulse repetition frequency of the pump laser beam were 100 ps, 532 nm and 1 kHz, respectively. The illumination area on the sample was about 0.2 mm^2. At high excitation energy (200 nJ/pulse), laser action appears at 611 nm which is within the band gap and coincides with the TDM wavelength. The FWHM of the emission peak is about 2 nm, which is limited by the spectral resolution of our experimental setup. Above the threshold at a pump pulse energy of about 100 nJ/pulse, the emission intensity increases. The FWHM of the emission spectrum also drastically decreases above the threshold. These results confirm that laser action occurs above the threshold of the pump energy at the wavelength of the TDM in the PBG [44,45].

14.4
Chiral Defect Mode Induced by Partial Deformation of Helix

TDM based on the composite film of two PCLCs has been achieved. However, its wavelength can not be tuned by an external field such as the electric field or light. We have proposed a new type of defect mode in the helix which can be dynamically tuned by the external field [46]. Figure 14.8 shows schematic explanation of a photonic defect in CLC. If the periodicity (pitch) of the helix is partially changed as shown in Figure 14.8(b), that is, the pitch is partially squeezed or expanded, these irregularities in the periodic structure should act as a defect and cause the light localization. As a method to induce partial change in helix pitch, we suppose that the local modification of helical twisting power (HTP) is induced by a focused Gaussian laser light. Optical

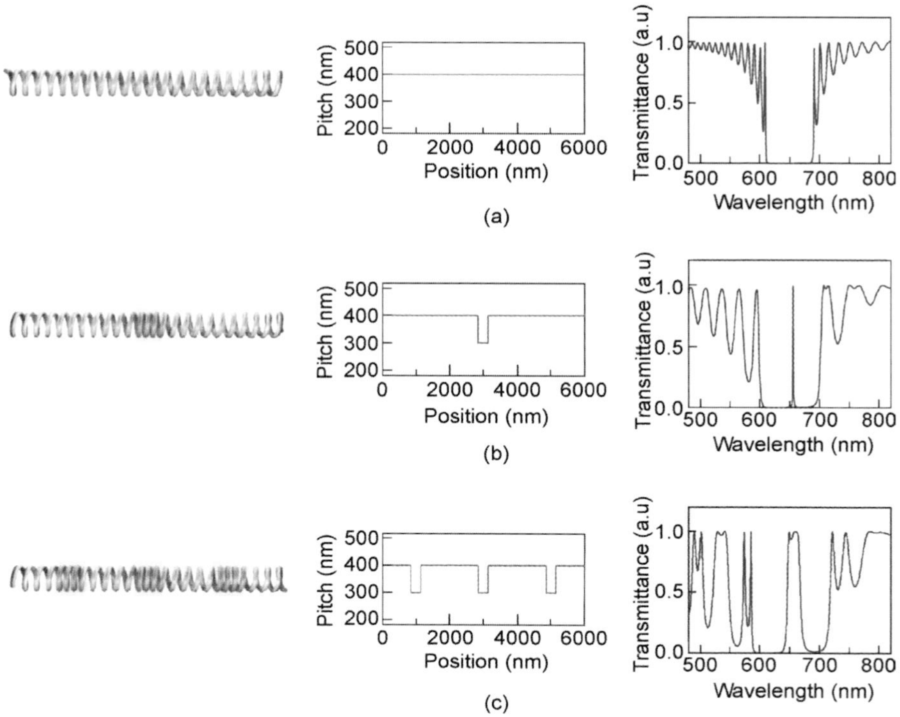

Figure 14.8 Pitch distributions and calculated transmission spectra for helical structures with (a) no defect, (b) single chiral defect and (c) multiple chiral defects.

control of HTP can be realized using photochemical effects of the doped azobenzene, nonlinear optical effects, or by simple heating. Photoinduced reversible control of HTP of CLC has been demonstrated in the CLC containing photochromic azobenzene, and applications to a reflection-type display devices, an optical shutter, an optical memory, and so on have been studied [47,48]. By the trans-cis photoisomerization of the doped azobenzene, HTP of the host CLC changes, so that photoinduced control of HTP can be realized.

On the other hand, when multiple chiral defects are introduced into the helix as shown in Figure 14.8(c), coupling between the modes confined in the defect layers leads to the formation of the defect band in the PBG [49].

We have also proposed a novel approach to introduce chiral defects (local modulation of the helix pitch) into the helix structure of CLC [50]. A schematic explanation of the fabrication procedure is shown in Figure 14.9(a). A 100 fs pulse of a Ti:Sapphire laser at wavelength of 800 nm and repetition rate of 80 MHz were focused on the sample cell by an objective lens with numerical number (N.A.) of 1.4.

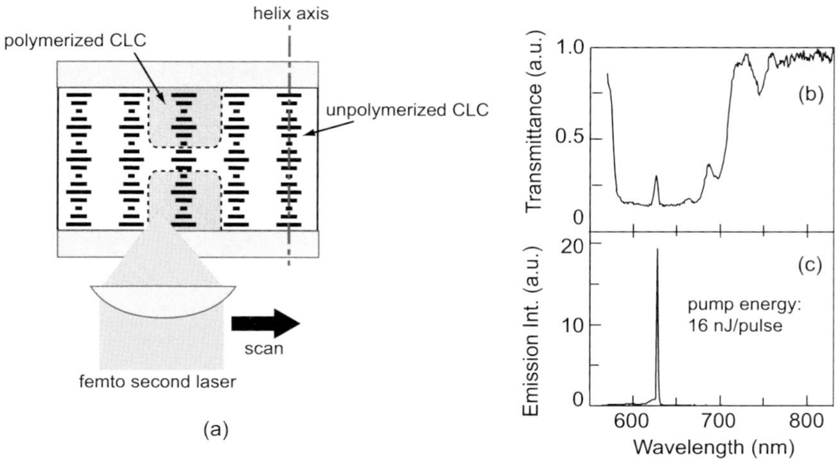

Figure 14.9 (a) Schematic explanation of the fabrication procedure of the PCLC with chiral defect based on local photo-polymerization using scanning confocal microscope. Transmission (b) and emission (c) spectra of the CLC with chiral defect.

The right-handed PCLC material doped with 1wt% of DCM dye was aligned homogeneously in a cell with a gap of 6–7 μm. The direct laser writing was performed with a confocal laser scanning microscope. The laser was scanned over an area of 146.2 μm × 146.2 μm, with a scan-line resolution of 2048 lines per scan-area. First, the laser light is tightly focused near the substrate surface in the CLC cell. Tow-photon polymerization occurs at the laser focusssing point and a locally polymerized PCLC thin film is obtained on the substrate surface. The sample is then flipped over and laser writing performed again near the opposite surface of the cell. As a result, a hybrid structure is fabricated in which an unpolymerized CLC region is left between two PCLC films on the cell surfaces. Moreover, in this system, the defect modes can be tuned by modulating the helix pitch in the defect layer upon changing temperature or irradiating with light [51,52].

Figure 14.9(b) shows the transmission spectra for right-handed circularly polarized light of the fabricated CLC defect structure. A single defect mode is observed within the selective reflection band of the CLC. The theoretical transmission spectrum was calculated using Berreman's 4×4 matrix [53], which showed a good agreement with the experimental result. Figure 14.9(c) shows the emission spectrum at high pumping energy of the CLC single-defect structure along with the corresponding transmission spectrum. Single mode laser action is observed at 628 nm, which corresponds to the defect mode wavelength. Lasing threshold for the defect mode structure is 16.7 mJ/cm^2, which is less than half of the threshold of CLC without defect structure. The reduction of the lasing threshold in the defect structure supports a high-Q cavity formed by the defect.

14.5
Tunable Defect Mode Lasing in a Periodic Structure Containing CLC Layer as a Defect

We have introduced a LC layer in a 1D PC as a defect, in which the wavelength of defect modes were controlled upon applying electric field in a basis of the change in optical length of the defect layer caused by the field-induced molecular reorientation of LC [29]. We also proposed a wavelength tunable laser based on an electrically controllable defect mode in a 1D dielectric periodic structure containing a dye-doped LC as a defect layer [39,40].

The introduction of periodic structure into a 1D PC as a defect is also interesting [54–57]. We have introduced a CLC layer in a 1D PC as a defect [58–61]. Figure 14.10(a) shows the theoretical transmission spectrum of a 10-pair multilayer without a CLC defect (solid line), and a simple CLC without a PC structure (dashed line). The PBG of the CLC was observed between 605 nm and 680 nm, which is inside that of the multilayer. Figure 14.10(b) shows the calculated transmission spectrum of a 1D hybrid photonic crystal (HPC) with a CLC defect. Many peaks appeared at regular intervals in the PBG of the HPC. These peaks are related to the defect modes

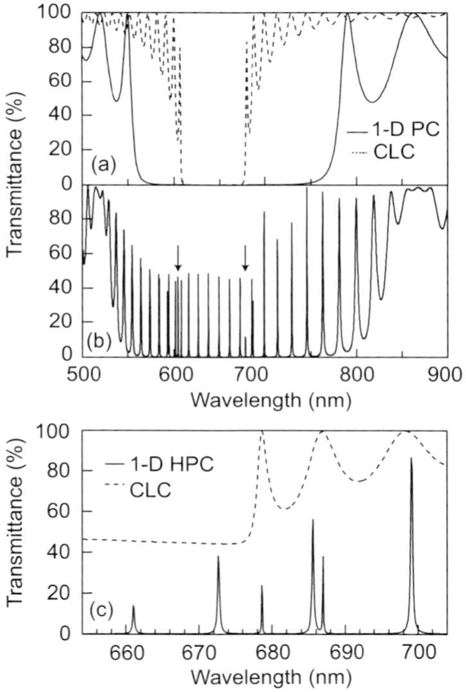

Figure 14.10 (a) Theoretical transmission spectra of 1-D PC without any defect (solid line) and CLC without PC structure (dashed line). (b) Theoretical transmission spectra of 1-D PC containing CLC as a defect. (c) Magnified transmission spectra corresponding to (b).

resulting from the introduction of the CLC defect. However, additional peaks were observed, as indicated by arrows, which disrupted the regular interval between the defect mode peaks at both band edges of the CLC.

The transmission spectra in Figure 14.10(b), around the longer edge of the PBG of the CLC, are shown magnified in Figure 14.10(c). Four main peaks due to the defect modes appeared at regular intervals (661 nm, 673 nm, 687 nm and 699 nm), although the peak at 687 nm splits. The splitting of the peak at 687 nm is attributed to the optical anisotropy of the CLC. Therefore, two kinds of defect mode corresponding to left- and right-handed circularly polarized lights could exist out of the PBG of the CLC. On the other hand, one additional peak was observed at 678.6 nm, which corresponds to the band-edge wavelength of the CLC. From detailed consideration of the polarization states of transmitted light, the additional peak was clearly distinguished from the other defect mode peaks. Such a peak was not observed in a 1D PC with a uniform defect such as an isotropic medium or nematic liquid crystals [29,43]. Namely, this peak is a defect mode peculiar to the helix defect in the 1D PC, and is associated with photon localization originating from the band-edge effect of the CLC helix. Note that this defect mode peak is very sharp and the full width at half-maximum (FWHM) of this peak was 0.05 nm, which is more than four times smaller than that of the other defect mode peaks (0.23 nm). From the peak width, the Q-factor of the additional mode at the band edge of the CLC was estimated to be 34000, which was much higher than that of the other defect modes.

In order to clarify the appearance of the high-Q defect mode in the double periodic structure, we have performed theoretical estimation of electric field distribution in three types of one-dimensional periodic structures described above using a finite difference time domain (FDTD) method. This method is numerical analysis based on the Maxwell differential form equations. In this calculation, light absorption is neglected in simulation space and the first Mur method [62] is applied as the absorbing boundary condition to absorb the outgoing light in simulation space edges. Figure 14.11 shows the calculated electric field distributions and refractive indices in two types of periodic structures. We assumed that the thickness, the helical pitch and the extraordinary and ordinary refractive indices of the CLC are 9.1 μm, 350 nm, 1.735 and 1.530, respectively. The wavelength of the incident light to the periodic structures corresponds to the high-Q defect mode wavelength. It should be noted that light is localized strongly in the double periodic structure and the maximum electric field intensity is more than 15 times as much as that of a simple CLC. Light is localized at the center of the CLC layer in the double periodic structure shown in Figure 14.11(a) and its field pattern is similar to that in the CLC shown in Figure 14.11(b), which indicates that light in the double periodic structure is confined by the band edge effect of CLC. Additionally light confinement is effectively enhanced by the outer periodic structure because the wavelength of light is within the PBG of the outer periodic structure. Namely, from the contribution of both band edge effect of CLC and defect mode effect of the outer periodic structure, light is localized strongly in the double periodic structure.

We have investigated the laser action in a 1D HPC with a CLC defect. As a laser dye dopant, DCM was compounded in a CLC, whose concentration was 1wt%. The PBG

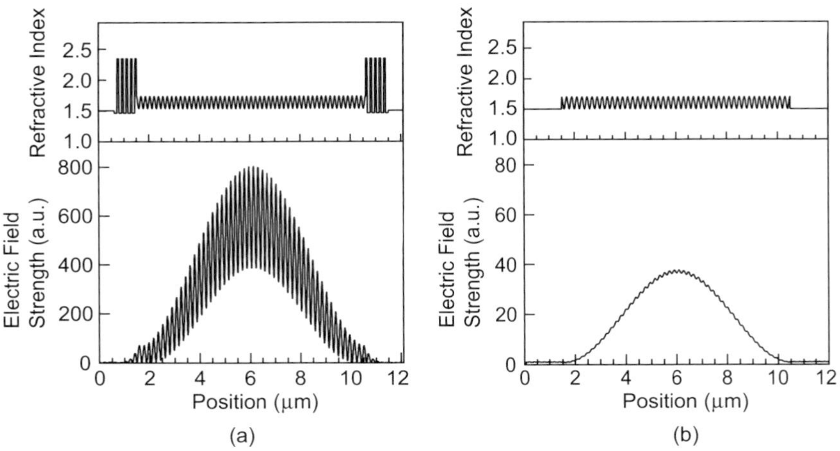

Figure 14.11 Calculated electric field strength in the cell of 1D PC containing CLC defect (a) and a simple CLC (b).

of the CLC shifts with a temperature change, which is attributed to the temperature dependence of the helical pitch of the CLC. The band edge of the CLC was adjusted to 644 nm by temperature regulation in this experiment because of the emission wavelength window of the laser dye. The pumping wavelength was 532 nm. The excitation laser beam irradiated the sample perpendicularly to the glass surface, whose illumination area on the sample was about 0.2 mm^2. The emission spectra from the 1D HPC were measured using a CCD multichannel spectrometer. At a low

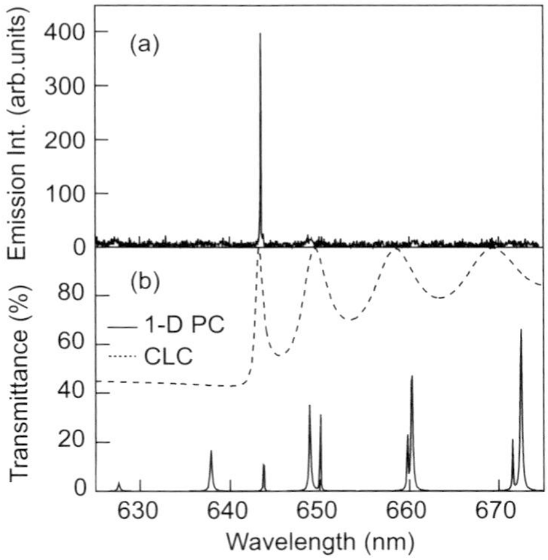

Figure 14.12 (a) Emission spectrum of 1D PC containing dye-doped CLC defect layer. (b) Theoretical transmission spectra corresponding to the emission spectrum.

pumping energy, the emission peaks appear, which were attributed to spontaneously emitted light passing out through narrow spectral windows owing to the defect modes. At a high pumping energy of 18 nJ/pulse, as shown in Figure 14.12(a), only one sharp lasing peak appeared at 643.5 nm. The calculated transmission spectrum of this system is also shown in Figure 14.12(b). Comparing Figure 14.12(a) and (b), the lasing peak coincides with the wavelength of the peculiar defect mode. Note that the laser action was single-mode based on one additional mode, although many modes exist because of the high Q-factor. The threshold of laser action in the 1D HPC with a CLC defect was lower than that in simple CLC without a 1D PC [58]. This result is attributed to strong optical confinement due to the high Q-factor of the additional mode. Similar results have been confirmed in the 1D PC containing ferroelectric liquid crystal as a defect [63].

14.6 Summary

Two types of laser actions were demonstrated using the CLC. They originated from the band edge effect of the one-dimensional photonic band gap and the defect mode within the band gap. We experimentally demonstrated the TDM in the 1D PBG of the CLC film having a twist defect which was a discontinuity of the director rotation around the helix axis at an interface of two PCLC layers. The laser action based on the TDM was also observed in the dye-doped PCLC composite film with the twist interface. We also proposed a new type of defect mode based on the chiral defect in which the partial modulation of the helix pitch was induced. In the new chiral defect mode, the tuning of the mode frequency could be expected by partial modulation of the helical twisting power. In order to realize chiral defect in the CLC, we also proposed a novel approach using a direct laser writing technique with femtosecond pulse laser. We fabricated a local modification of the CLC helix pitch based on a two-photon polymerization. Finally, we proposed double periodic structure in which the defect has also periodicity, and we demon-strated a hybrid photonic crystal (HPC) structure based on a combination of a 1D PC and CLC defect. In the HPC, a single-mode laser action with low pumping threshold is observed, which is based on the defect mode with a high Q-factor peculiar to the CLC defect having periodic structure.

Acknowledgements

This work is partially supported by a Grant-in-Aid for Scientific Research from the Japan Ministry of Education, Culture, Sports, Science and Technology. We would like to show our acknowledgements to Prof. W. Haase and Prof. S. T. Wu for a fruitful discussion and Merck KGaA for providing the photo-polymerizable CLC materials.

References

1 Yablonovitch, E. (1987) *Phys. Rev. Lett.*, **58**, 2059.
2 John, S. (1987) *Phys. Rev. Lett.*, **58**, 2486.
3 Joannopoulos, J.D., Meade, R.D. and Winn, J.N. (1995) *Photonic Crystals, Molding the Flow of Light*, Princeton Univ. Press, Princeton, New York.
4 Martorell, J. and Lawandy, N.M. (1990) *Phys. Rev. Lett.*, **6**, 1877.
5 Yoshino, K., Lee, S.B., Tatsuhara, S., Kawagishi, Y., Ozaki, M. and Zakhidov, A.A. (1998) *Appl. Phys. Lett.*, **73**, 3506.
6 Yoshino, K., Tatsuhara, S., Kawagishi, Y., Ozaki, M., Zakhidov, A.A. and Vardeny, Z.V. (1999) *Appl. Phys. Lett.*, **74**, 2590.
7 Foresi, J.S., Villeneuve, P.R., Ferrera, J., Thoen, E.R., Steinmeyer, G., Fan, S., Joannopoulos, J.D., Kimerling, L.C., Smith, H.I. and Ippen, E.P. (1997) *Nature*, **390**, 143.
8 Painter, O., Lee, R.K., Scherer, A., Yariv, A., O'Brien, J.D., Dapkus, P.D. and Kim, I. (1999) *Science*, **284**, 1819.
9 Noda, S., Tomoda, K., Yamamoto, N. and Chutinan, A. (2000) *Science*, **289**, 604.
10 Yoshino, K., Tada, K., Ozaki, M., Zakhidov, A.A. and Baughman, R.H. (1997) *Jpn. J. Appl. Phys.*, **36**, L714.
11 Vlasov, Y.A., Luterova, K., Pelant, I., Honerlage, B. and Astratov, V.N. (1997) *Appl. Phys. Lett.*, **71**, 1616.
12 Kopp, V.I., Fan, B., Vithana, H.K. and Genack, A.Z. (1998) *Opt. Lett.*, **23**, 1707.
13 Taheri, B., Munoz, A.F., Palffy-Muhoray, P. and Twieg, R. (2001) *Mol. Cryst. Liq. Cryst.*, **358**, 73.
14 Alvarez, E., He, M., Munoz, A.F., Palffy-Muhoray, P., Serak, S.V., Taheri, B. and Twieg, R. (2001) *Mol. Cryst. Liq. Cryst.*, **369**, 75.
15 Araoka, F., Shin, K.C., Takanishi, Y., Ishikawa, K., Takezoe, H., Zhu, Z. and Swager, T.M. (2003) *J. Appl. Phys.*, **94**, 279.
16 Funamoto, K., Ozaki, M. and Yoshino, K. (2003) *Jpn. J. Appl. Phys.*, **42**, L1523.
17 Ozaki, M., Kasano, M., Funamoto, K., Ozaki, R., Matsui, T. and Yoshino, K. (2003) *Proc. SPIE*, **5213**, 111.
18 Furumi, S., Yokoyama, S., Otomo, A. and Mashiko, S. (2003) *Appl. Phys. Lett.*, **82**, 16.
19 Shirota, K., Sun, H.B. and Kawata, S. (2004) *Appl. Phys. Lett.*, **84**, 1632.
20 Ozaki, M., Kasano, M., Ganzke, D., Haase, W. and Yoshino, K. (2002) *Adv. Mater.*, **14**, 306.
21 Ozaki, M., Kasano, M., Ganzke, D., Haase, W. and Yoshino, K. (2003) *Adv. Mater.*, **15**, 974.
22 Kasano, M., Ozaki, M., Ganzke, D., Haase, W. and Yoshino, K. (2003) *Appl. Phys. Lett.*, **82**, 4026.
23 Finkelmann, H., Kim, S.T., Munoz, A., Palffy-Muhoray, P. and Taheri, B. (2001) *Adv. Mater.*, **13**, 1069.
24 Schmidtke, J., Stille, W., Finkelmann, H. and Kim, S.T. (2002) *Adv. Mater.*, **14**, 746.
25 Matsui, T., Ozaki, R., Funamoto, K., Ozaki, M. and Yoshino, K. (2002) *Appl. Phys. Lett.*, **81**, 3741.
26 Cao, W., Munoz, A., Palffy-Muhoray, P. and Taheri, B. (2002) *Nat. Mater.*, **1**, 111.
27 Dowling, J.P., Scalora, M., Bloemer, M.J. and Bowden, C.M. (1994) *J. Appl. Phys.*, **75**, 1896.
28 Hattori, T., Tsurumachi, N. and Nakatsuka, H. (1997) *J. Opt. Soc. Am.*, **14**, 348.
29 Ozaki, R., Matsui, T., Ozaki, M. and Yoshino, K. (2002) *Jpn. J. Appl. Phys.*, **41**, L1482.
30 Yang, Y.C., Kee, C.S., Kim, J.E., Park, H.Y., Lee, J.C. and Jeon, Y.J. (1999) *Phys. Rev. E*, **60**, 6852.
31 Kopp, V.I. and Genack, A.Z. (2002) *Phys. Rev. Lett.*, **89**, 033901.
32 Yoshino, K., Satoh, S., Shimoda, Y., Kawagishi, Y., Nakayama, K. and Ozaki, M. (1999) *Jpn. J. Appl. Phys.*, **38**, L961.
33 Yoshino, K., Shimoda, Y., Kawagishi, Y., Nakayama, K. and Ozaki, M. (1999) *Appl. Phys. Lett.*, **75**, 932.

34 Busch, K. and John, S. (1999) *Phys. Rev. Lett.*, **83**, 967.

35 Shimoda, Y., Ozaki, M. and Yoshino, K. (2001) *Appl. Phys. Lett.*, **79**, 3627.

36 Ozaki, M., Shimoda, Y., Kasano, M. and Yoshino, K. (2002) *Adv. Mater.*, **14**, 514.

37 Tocci, M.D., Bloemer, M.J., Scalora, M., Dowling, J.P. and Bowden, C.M. (1995) *Appl. Phys. Lett.*, **66**, 2324.

38 Dumeige, Y., Vidakovic, P., Sauvage, S., Sgnes, I. and Levenson, J.A. (2001) *Appl. Phys. Lett.*, **78**, 3021.

39 Ozaki, R., Matsui, T., Ozaki, M. and Yoshino, K. (2003) *Appl. Phys. Lett.*, **82**, 3593.

40 Ozaki, R., Matsuhisa, Y., Ozaki, M. and Yoshino, K. (2004) *Appl. Phys. Lett.*, **84**, 1844.

41 Belyakov, V.A. (2006) *Mol. Cryst. Liq. Cryst.*, **453**, 43.

42 Matsuhisa, Y., Huang, Y., Zhou, Y., Wu, S.T., Takao, Y., Fujii, A. and Ozaki, M. (2007) *Opt. Express*, **15**, 616.

43 Matsuhisa, Y., Huang, Y., Zhou, Y., Wu, S.T., Ozaki, R., Takao, Y., Fujii, A. and Ozaki, M. (2007) *Appl. Phys. Lett.*, **90**, 091114.

44 Ozaki, M., Ozaki, R., Matsui, T. and Yoshino, K. (2003) *Jpn. J. Appl. Phys.*, **42**, L472.

45 Schmidtke, J., Stille, W. and Finkelmann, H. (2003) *Phys. Rev. Lett.*, **90**, 083902.

46 Matsui, T., Ozaki, M. and Yoshino, K. (2004) *Phys. Rev. E*, **69**, 061715.

47 Sackmann, E. (1971) *J. Am. Chem. Soc.*, **93**, 7088.

48 Lee, H.K., Doi, K., Harada, H., Tsutsumi, O., Kanazawa, A., Shiono, T. and Ikeda, T. (2000) *J. Phys. Chem. B*, **104**, 7023.

49 Yoshida, H., Ozaki, R., Yoshino, K. and Ozaki, M. (2006) *Thin Solid Films*, **509**, 197.

50 Yoshida, H., Lee, C.H., Matsuhisa, Y., Fujii, A. and Ozaki, M. (2007) *Adv. Mater*, **19**, 1187.

51 Yoshida, H., Lee, C.H., Fujii, A. and Ozaki, M. (2006) *Appl. Phys. Lett.*, **89**, 231913.

52 Yoshida, H., Lee, C.H., Miura, Y., Fujii, A. and Ozaki, M. (2007) *Appl. Phys. Lett.*, **90**, 071107.

53 Berreman, D.W. (1973) *J. Opt. Soc. Am.*, **63**, 1374.

54 Ozaki, R., Sanda, T., Yoshida, H., Matsuhisa, Y., Ozaki, M. and Yoshino, K. (2006) *Jpn. J. Appl. Phys.*, **45**, L493.

55 Ozaki, R., Matsuhisa, Y., Yoshida, H., Yoshino, K. and Ozaki, M. (2006) *J. Appl. Phys.*, **100**, 023102.

56 Song, M.H., Park, B., Shin, K.-C., Ohta, T., Tsunoda, Y., Hoshi, H., Takanishi, Y., Ishikawa, K., Watanabe, J., Nishimura, S., Toyooka, T., Zhu, Z., Swager, T.M. and Takezoe, H. (2004) *Adv. Mater.*, **16**, 779.

57 Hwang, J., Song, M.H., Park, B., Nishimura, S., Toyooka, T., Wu, J.W., Takanishi, Y., Ishikawa, K. and Takezoe, H. (2005) *Nat. Mater.*, **4**, 383.

58 Matsuhisa, Y., Ozaki, R., Ozaki, M. and Yoshino, K. (2005) *Jpn. J. Appl. Phys.*, **44**, L629.

59 Matsuhisa, Y., Ozaki, R., Yoshino, K. and Ozaki, M. (2006) *Appl. Phys. Lett.*, **89**, 101109.

60 Matsuhisa, Y., Ozaki, R., Yoshino, K. and Ozaki, M. (2006) *Thin Solid Films*, **509**, 189.

61 Matsuhisa, Y., Ozaki, R., Takao, Y. and Ozaki, M. (2007) *J. Appl. Phys.*, **101**, 033120.

62 Mur, G. (1981) *IEEE Trans. Electromagn. Compat.*, **23**, 377.

63 Matsuhisa, Y., Haase, W., Fujii, A. and Ozaki, M. (2006) *Appl. Phys. Lett.*, **89**, 201112.

15
Tunable Superprism Effect in Photonic Crystals

F. Glöckler, S. Peters, U. Lemmer, and M. Gerken

15.1
Introduction

Since the introduction of the photonic crystal (PhC) concept in 1987 most of the attention was drawn to exploit the photonic bandgap (PBG) of such structures to confine light, to guide light, or to inhibit or amplify spontaneous emission properties. Another remarkable effect that occurs in periodic media is the superprism effect referring to a largely enhanced angular dispersion in comparison to bulk materials. This effect does not occur inside the bandgap, but in a region where light propagation is allowed. Due to the effect a small change in the wavelength of an incoming beam results in a large change of its group propagation angle.

When operating a superprism structure in the region of strong angular dispersion, small changes in the structure's optical properties also result in large changes of a beam's group propagation angle. By actively controlling the optical properties, a miniaturized beam steering device can be realized. Following this concept, a number of different designs has been investigated, both by simulation and experimentally, during the last 15 years. The ideal device would have a size on the order of the wavelength of the operating light and would be power efficient. For many applications, especially in optical communications, additionally fast switching speeds in the picosecond range are required. These goals have been pursued using two different approaches. First, larger effects can be obtained by optimizing the design of the PhC. Second, the effect utilized for changing the optical properties can be improved, which often corresponds to improving the material.

In the next four sections, we will give an overview over the research on tunable PhC superprism structures of the last 15 years. Section 15.2 starts with a short review of the superprism effect itself mentioning important publications and giving some explanations concerning the origin and the limitations of the superprism effect. While relatively little research has been performed on tunable superprism structures, more results have been achieved in the two closely related fields of on-off-switching of optical signals and of optical modulators. These two fields draw much attention due to their

Nanophotonic Materials: Photonic Crystals, Plasmonics, and Metamaterials. Edited by R.B. Wehrspohn, H.-S. Kitzerow, and K. Busch

ISBN: 978-3-527-40858-0

promises for all-optical circuitry. As many of the concepts proposed and realized for such devices are also usable for tunable superprism structures they will be reviewed in Section 15.3. In Section 15.4, we outline important results achieved in the field of tunable superprism structures. Finally, in Section 15.5 we present a detailed theoretical analysis of our own approach for all-optical superprism switching combining one-dimensional thin film structures with optically nonlinear organic materials. We first present the current state in nonlinear organic materials and then evaluate the performance of a thin film superprism device incorporating nonlinear organic layers.

15.2 The Superprism Effect

As early as 1987 John mentioned the possibility of changed refraction properties in periodic dielectric structures [1]. In the same year, Zengerle published research he had performed almost a decade earlier discussing many aspects of the peculiar properties of light propagating in one-dimensional (1D) and two-dimensional (2D) periodic dielectric structures including the large angular dispersion [2,3]. Dowling theoretically examined these properties and proposed to utilize them for "ultra-refractive" optics [4]. In 1996, Lin realized a 2D structure that exhibited a greatly enhanced dispersion in the millimeter wavelength range and called it "highly dispersive photonic bandgap prism" [5]. He used alumina-ceramics rods arranged in a triangular 2D lattice. Later on, Kosaka showed in a semiconductor-dielectric three-dimensional (3D) photonic crystal structure an angular beam steering of 50 degrees when detuning the wavelength of an incoming beam by only 1% between 990 nm and 1000 nm, a 500-fold larger angular dispersion than in a regular glass prism [6–8]. These numbers motivate the labeling of the strong angular dispersion in PhC structures as superprism-effect. Kosaka's results ignited the hope to realize miniaturized high dispersion devices. These are promising for dispersion compensation or light deflection devices, above all for compact devices to replace waveguide gratings in dense wavelength-division multiplexing applications. In 2002, Wu *et al.* confirmed the superprism effect observing a 10 degree angular steering over a wavelength range from 1290 nm to 1310 nm, which corresponds to an angular dispersion of 0.5 degrees/nm in a triangular 2D PhC [9]. Baumberg *et al.* later demonstrated the superprism effect in silicon-based 2D structures in the optical wavelength region [10]. In 2003 Prasad *et al.* predicted a superprism effect in 3D macroporous polymer PhCs. They showed that an angular dispersion as high as 14 degrees/nm should be feasible, even for the small refractive indices of polymers [11].

15.2.1 Origin of the Superprism Effect

The reason why superprism phenomena occur in periodic media can most conveniently be seen when looking at the first Brillouin zone of such a structure. For explanation purposes a 2D triangular lattice is chosen (Figure 15.1a). In Figure 15.1b,

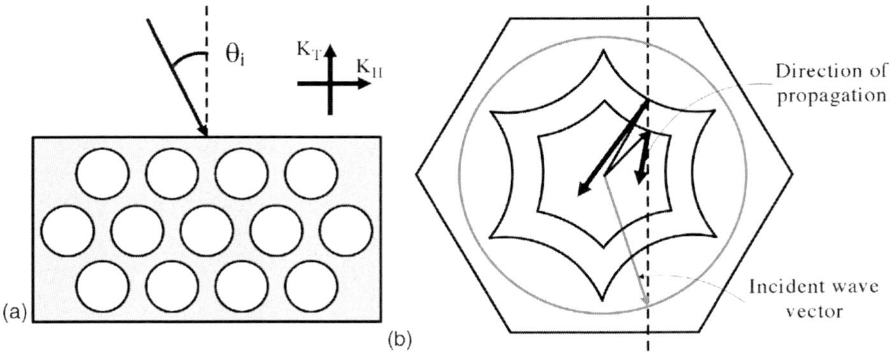

Figure 15.1 a) Schematic drawing of a triangular 2D PhC. Light is incident on the surface of the crystal under an angle θi. b) The first Brillouin zone with two equi-frequency surfaces is drawn. The propagation direction inside the PhC is determined graphically.

the first Brillouin zone of such a photonic crystal is sketched (this is a schematic drawing for the sake of explanation, not a calculated curve). The solid black lines mark two equi-frequency curves. In contrast to a bulk material, where these lines have ellipsoidal shape, one recognizes immediately the strong distortion to starlike shapes. The propagation direction of an incoming beam of light can be deduced from this diagram in the following way. First, the tangential component of the wavevector of the incident beam has to be conserved to fulfill momentum conservation. Therefore, we can find for a given frequency of light and incident angle of the beam the corresponding wavevector inside the PhC at the intersection point of a parallel to the $k_\perp$-axis and the equi-frequency curves. As the propagation direction is parallel to the group velocity v_g that is defined by

$$v_g = \nabla_k \omega(k), \qquad (15.1)$$

we find the propagation direction graphically by drawing the normal vector at the intersection point on the dispersion curve. The strong deviation from a spherical shape leads to large changes in the normal vector of the dispersion curve with small changes of the wavevector of the incoming beam. This can be illustrated by introducing an equi-incident-angle curve, that is a curve containing all points that belong to beams of equal incidence angle, but different wavelength. When looking at the intersection point of the first and second dispersion curve and comparing the respective propagation directions, the large effect for the group propagation angle becomes apparent.

15.2.2
Performance Considerations for Superprsim Devices

In the first publications that theoretically or experimentally investigated superprism structures, high expectations for future devices containing these structures were stated. These were supported by the large angular dispersion in the region of

1 degree/nm compared to 0.006 degrees/nm for a conventional prism and 0.1 degrees/nm for a diffraction grating. Later work showed that the performance of superprism structures has to be analyzed not only in terms of angular dispersion, but also with respect to other parameters such as the bandwidth, the width of the operating beam, the position of the focus, and the device size. Baba *et al.* investigated in a series of papers the performance of superprism structures when using an incident beam that is focused on the input face of the photonic crystal. He calculated that for this mode of operation, the size of the beam waist of the incoming light needs to be chosen large resulting in a large device size. For a sample structure with a resolution of 0.4 nm over 56 resolved wavelength points, Baba determined for a beam waist of 115 μm a device size of more than 10 cm [12]. A further drawback of superprism structures is that only the lower order photonic bands can be used. In contrast to diffraction gratings, for higher order bands a high diffraction efficiency towards a unique direction is not given. The multiplicity of bands would lead to heavy distortions of the beam. Baba *et al.* proposed optimized structures later [13]. Gerken *et al.* demonstrated experimentally that more compact superprism devices are possible for wavelength multiplexing and demultiplexing applications by considering an incident beam focused on the detector or output waveguide array not on the photonic crystal input surface [14,15]. The number of resolvable wavelength channels is proportional to the product of angular dispersion and device bandwidth demonstrating that high angular dispersion alone does not guarantee a high performance device.

15.2.3
Bragg-Stacks and Other 1D Superprisms

Though the term "superprism" effect was coined in connection with 2D and 3D PhC structures, it should be emphasized that 1D photonic crystals also show superprism effects. Periodic structures with alternating areas of low and high index of refraction, i.e. Bragg-stacks, have been used for more than 30 years for numerous purposes from anti-reflection coatings to filters for optical communications. Gerken *et al.* investigated the superprism effect in 1D thin film stacks in several publications [16,17]. Additionally, it was pointed out, that nonperiodic structures also show the superprism effect. This fact opens many new degrees of freedom in designing such structures tailored for certain applications. An effective angular dispersion of 30 degrees over a 20 nm wavelength range around 835 nm was demonstrated in a 13.4 μm thick nonperiodic 1D stack allowing for a four channel multiplexing device [14]. By tailoring the dispersion properties in nonlinear structures to achieve constant dispersion values beam distortions are reduced significantly. Also, step-like dispersion characteristics can be achieved which are useful for communication devices [15].

15.2.4
Current State in Superprism Structures

Summarizing the above mentioned developments the superprism effect has been demonstrated experimentally in 1D, 2D, and 3D structures. While early proof-of-principle papers focus on obtaining large angular dispersion [2–11], the relevance of the bandwidth over which this dispersion is achieved as well as problems due to beam divergence were addressed in more recent work [20]. For 1D thin film structures, detailed studies of the bandwidth-dispersion product and beam divergence issues exist [15,18]. In the case of 2D structures for planar integration similar considerations on bandwidth and coupling can be found in [12,20], though not in a quantified manner as for 1D thin film stacks. This work demonstrates that large super-prism structures are necessary to realize a dense wavelength division multiplexer (DWDM). This result is in agreement with the theoretical work of Miller on the fundamental limits for optical components [21] stating that a minimum device size is necessary for the separation of a specific number of optical pulses. The current focus lies on realizing superprism structures for coarse WDM [19] and beam steering in sensors and communication [18]. For both 1D and 2D superprism systems first steps towards devices have been made and work is ongoing. In the case of 3D systems, the field remains in the state of proof-of-principle experiments as fabrication issues and design problems pose much harder problems than for thin films stacks and integrated planar PhCs. Thus, predictions of device performances are purely speculative.

15.3 Tunable Photonic Crystals

The work on tunable PhC structures mainly involves two tasks. First, the geometry of the device has to be chosen to maximize the effect, and second, an appropriate tuning mechanism has to be designed. The first point is mainly a question of the PhC structure's properties and normally involves the development of appropriate fabrication techniques, which is especially demanding on the nanometer scale required for optical wavelengths. The quest for the best switching mechanism is rather situated in the field of material sciences and chemistry, as it requires the modification of certain properties of a material. In the following sections, different approaches for tuning PhC structures will be presented sorted by the switching effect they utilize. At the end of each subsection, a short summary of the development in the respective field is given. As considerable parts of the work consists of proof-of-principle experiments, an extensive analysis of the effects is often missing. Additionally, the field is not aiming or converging towards a single application. Therefore, comparison of different concepts with respect to a single figure of merit is difficult. Nevertheless, there are several important parameters for evaluation of different concepts. First, required energy consumption for achieving a certain wavelength shift of bandgap or defect modes should be considered. Second, the switching speed of the structure is crucial for the scope of possible applications. Third, complexity of the fabrication and integration is an issue. The following approaches will be evaluated with respect to these parameters if possible.

15.3.1
Liquid Crystals

In a considerable part of the research on tunable PBG structures in the last years liquid crystals are deployed as active media. Due to their strong birefringence, refractive index changes of $\Delta n = 0.2$ and more are feasible, a value hardly reached by any other mechanism in the optical domain with reasonably low switching energies. Additionally, liquid crystals are comparably easy to integrate into PBG structures. A more detailed survey of the developments in the field of liquid crystal based tunable PhCs up to 2004 is given in a review paper by Kitzerow [22].

The liquid-crystal approach was proposed by Busch and John in 1999, when they theoretically predicted a tunable shift of the bandgap of an inverse opal structure coated with liquid crystals. They suggested to utilize the large temperature dependence of the liquid-crystal's refractive index when changing from isotropic to nematic phase [23]. The preferred order of the liquid crystals alters and the large birefringence causes the refractive index change. This prediction was soon afterwards confirmed by experiment. In the following years, numerous modifications of this original concept were presented. Prominently amongst them is the infiltration of the pores of a PBG structure with liquid crystals. Following this path, Yoshino *et al.* [25] and Leonard *et al.* [24] tuned the bandgap of an artificial opal structure and a macro-porous silicon PhC, respectively. Yoshino achieved a 70 nm shift of the stop band edge at 4.3 μm, Leonard a 60 nm shift at 5.7 μm. Work on thermal tuning is ongoing, Mertens *et al.* incorporated a 2D defect layer forming a cavity in a 3D PhC to obtain a tunable defect mode by 20 nm at 7.375 μm wavelength [26].

Thermal tuning of liquid crystals is experimentally inexpensive, but is not a method of choice when fast and accurate tuning is required. Fortunately, liquid crystals offer other nonlinear effects that can be used to overcome these problems. An operating principle often employed is tuning by an applied voltage, here E. Kosmidou proposed a 2D PhC structure containing liquid crystal filled defects [27]. A theoretical tuning of the defect modes covering the whole C- and L-band with 4 V operation voltage and narrow linewidth was calculated. For 1D PBG structures, liquid-crystal filled defects and even defect waveguides in dielectric layers have been shown [28]. In the field of three dimensional PBGs, most of the published work deals with structures that were fabricated using "classical" methods (self assembly, direct laser beam writing and holography) and where the voids have afterwards been filled with liquid crystals.

Though quite straightforward to realize, the infiltration method has the drawback of decreasing the refractive index contrast of the PhC by replacing an air filled void by liquid crystals (refractive index between 1.6 and 2.1). The decrease of the refractive index contrast reduces the transmission and reflection efficiency of the PhC. Therefore, a number of approaches has been proposed to overcome this drawback. One way to avoid the refractive index contrast decrease is to use metallic structures [29]. Another method to overcome this problem was presented by McPhail *et al.* who doped liquid crystals into a polymer host and then fabricated a woodpile PhC structure out of this guest-host material [30]. They showed an optically induced 65 nm stop band shift achieved with low laser powers, though at a timescale of 60 minutes. Escuti *et al.* introduced another single step method for fabrication of PBG structures activated by liquid crystals [31]. They used a mixture of photosensitive monomers, a photo-initiator, and liquid crystals for holographic lithography. In addition to the fabrication of a three-dimensional PhC from polymer, the liquid crystals formed regular droplets due to a mass transport phenomenon driven by the intensity distribution inside the material during the lithography process. Resulting structures contain composite areas of polymer and of liquid crystals. Gorkahli *et al.* used this approach to create quasi-crystals with different odd-fold symmetries containing liquid crystals and analyzed their properties when tuned electrically [32,33].

To summarize, liquid crystals have been used successfully to tune the properties of photonic crystals. Especially the strong tuning by applied voltage in liquid crystals makes the combination of PBG structures and liquid crystals promising for switching applications. The inherent drawback of liquid crystals is their speed limitation. The large mass that has to be moved when using orientation effects sets the limit of these processes to a maximum of several tens of megahertz. Therefore, liquid crystal PhCs are good candidates for reconfigurable devices or sensors but will not play a role for optical communication. A great deal of the published work on tuning by liquid crystals especially in the case of 3D PhCs deals with the task of achieving high bandgap or defect mode shifts by modification of geometry and active material. The exploitation of liquid crystals in PhCs for devices and their combination with other elements for more complex functions is still at the beginning.

15.3.2
Tuning by Pockels Effect

Besides the electrical tuning of liquid crystals, some effort has been undertaken to change a PBG structure's properties electrically. Especially the Pockels effect, i.e. the manipulation of the electron distribution by an applied voltage, is of interest, as for this effect the inherent switching frequencies are in the gigahertz range. As lithium niobate ($LiNbO_3$) is the standard electro-optic material nowadays and possesses a relatively high electro-optic coefficient, PhCs made from $LiNbO_3$ have been proposed. Unfortunately, $LiNbO_3$ is difficult to structure, such that an elaborated focused ion beam process had to be involved [34]. Roussey *et al.* found that for triangular 2D PhCs in $LiNbO_3$ promising results can be obtained. They shifted the stop band gap of this structure by 160 nm at 1550 nm wavelength, though at quite high voltages of 80 V [35]. This value of 2.5 nm/V is by a factor 312 bigger than the bulk value of $LiNbO_3$ (0.008 nm/V).

Another interesting group of electro-optic materials is constituted by the polymers (see below). Schmidt *et al.* presented a 2D PhC slab waveguide with dye-functionalized polymethyl methacrylate as the active material for realizing a modulator [36]. In the 1D domain, Gan *et al.* realized a large resonance shift of 0.75 nm/V over a broad wavelength range of 50 nm in the C-band [37]. Their structure comprises an electro-optically active polymer processed by a sol–gel technique into a Fabry–Pérot-type cavity. Though they achieved large shifts at low driving voltages, fast switching with this type of structure is doubtful as slow piezo-electric effects contribute largely to the overall effect.

Further interesting electro-optic material systems are ferro-electric ceramics, for example lead lanthanum zirconate titanate [(Pb,La) (Zr,Ti)O_3, PLZT]. B. Li *et al.* demonstrated the possibility to create an infiltrated inverse opal structure with ferro-electrics and showed an electro-optic bandgap shift of several nanometers at high voltages of 700 V [38].

Long before the advent of tunable photonic crystals, tunable Fabry–Pérot-type structures have drawn much attention. Most of these concepts aim at tunable filters for the C-band range for optical communication, thus supplying a tuning range of 50 nm [39]. Here we find electrically tunable structures in different material systems such as III–V-semiconductors [40] and liquid crystals [41]. Another class of devices are micro-mechanical structures. Here, interesting approaches involving air-gap Fabry–Pérot-filters or membranes from periodic semiconductor layers have been demonstrated [42]. A good amount of work has also been done on thermally or thermo-electrically tunable structures. B. Yu *et al.* showed for example a device with a thermal resonance shift of 1.63 nm/K [43].

Another promising material system was presented by Zhang *et al.* They created PhCs out of CdSe quantum dots by self-assembly in spin-coated films. For an electric field of 80 kV/cm they obtained a refractive index change of 1.8% [44]. In comparison with liquid crystal infiltrated structures, tunable PhCs using the Pockels effect exhibit smaller bandgap shifts with voltage. Therefore, comparatively high voltages of several tens of volts to hundreds of volts are required for tuning in the nanometer to tens of nanometers range. Additionally, traditional electro-optic materials are difficult to structure, making the fabrication of 2D or 3D structures a hard task. Here, organic electro-optical materials are an interesting alternative showing large electro-optic effects and being simple to process (see Section 15.5.1.2).

15.3.3
All-Optical Tuning

The ultimate goal of nonlinear switching in photonic devices is the optical tuning of the device's properties. Here, the material's properties are changed by incoming light thus manipulating light by light and aiming at all-optical data-processing. Some of the optically induced effects have extremely low time constants and allowing for switching rates between 10^{15} and 10^{16} Hz in case of the optical Kerr-effect. This refers to the intensity dependent change of the refractive index of a material. Other effects utilized are the tuning via carrier injection or thermal tuning via absorption. The latter methods lead to lower switching rates. Unfortunately, this effect is extraordinarily weak in most present-day materials. Therefore, all published devices need high pump powers, large interaction distances, or both. X. Hu *et al.* demonstrated all-optical switching in 2D defect-layer PhC [45]. For a 5 nm shift of the resonance peak, 18.7 GW/cm^2 of pumping power were required at 1064 μm. Switching times of less than 10 ps were observed. Although Kerr-active materials are improving constantly, only a combination of structure and material enhancements will result in a useful all-optical switching device exploiting the Kerr effect. One possible way towards tunable elements using Kerr-nonlinearities are the so called slow-light waveguides [46]. Here, a low group propagation velocity multiplies the effect of the changes of optical properties inside the structure by enhancing the interaction time between the pump and the signal beam shortening the operation length of such devices. Slow light waveguides could, for example, comprise coupled defects in a 2D PhC [47]. Using carrier injection impressive tuning rates have been achieved.

Bristow *et al.* performed pump-probe experiments on 2D PhCs from AlGaAs and found large changes in reflectivity due to carrier induced effects [48]. They calculated a refractive index change of $\Delta n = -0.019$, with a lifetime of less than 10 ps. The high absolute value of Δn stems from the generation of free carriers inside the semiconductor material when pumped at wavelengths located at the tail of the absorption band. To achieve the essential short carrier lifetimes for fast switching, the large surface area of the PhC plays a crucial role, as it opens effective non-radiative recombination channels for free carriers. Following this approach, Fushman *et al.* recently showed a 20 GHz modulation of a GaAs PhC cavity containing InAs quantum dots [49]. They additionally achieved this at low pumping powers, using a 3 ps pulse at 750 nm with only 60 fJ. This is remarkable, as fast all-optical tuning often suffers from impracticably high pump power consumption due to small effects. The key ingredients of Fushman's approach are a quite high quality cavity ($Q = 2000$) and a short lifetime of the generated carriers. This combines to a shift of the structure's resonance of up to one and a half linewidths. The essential short carrier lifetime is achieved by the small modal volume and the large surface area of the PhCs used. The idea of overcoming small effects by enhancing the local field inside a structure plays an important role for PhC switching devices. Asano *et al.* fabricated cavities with quality factors of up to 10^6, and they predicted a further increase [50]. A completely different geometry, though also a structure utilizing a high quality factor, was presented by Noll *et al.* [51]. They demonstrated a waveguide switch based on a 1D resonator with a rod defect whose resonance frequency can be shifted by a control beam. An alternative way for reducing the carrier lifetime is sweeping them out by an applied voltage. This approach is similarly adopted for the first silicon Raman lasers [52].

The last above mentioned method, the thermal tuning, suffers from speed limitations. The refractive index change with temperature works no faster than on the order of microseconds [49]. In summary, fast all-optical tunable elements containing PBG structures will need considerable improvement to find their way to commercial devices. Optical computers are still far away as the necessary high integration and low switching energies may not be reached due to material limitations. Optical communication devices ranging down to optical interconnects are in closer reach, as here higher switching rates may be achieved at the expense of higher power consumption. Additionally, the requirements for integration are not as high as for optical computing. In terms of the employed tuning effect, the carrier injection approach is currently favorable to Kerr-nonlinear materials as larger effects can be obtained. However, when proceeding the several hundreds of gigahertz or terahertz range, carrier injection approaches will reach their limit. The most advanced structures base on 2D PhCs, employing both silicon and III–V-semiconductors. As material parameters cannot be changed here, geometrical optimization of the structures is performed. Here, 2D planar PhCs containing high-Q cavities are momentarily the most widespread approach. First steps towards photonic circuitry have been achieved with these structures.

15.3.4
Other Tuning Mechanisms

A number of more or less "exotic" tuning mechanisms have been presented and will be mentioned here for completeness. First, different tuning schemes based on thermal effects have been proposed. Though not under the label of photonic crystals, Weissman *et al.* achieved tunable resonances over a broad spectral range using polymer particles in an ordered matrix [53]. Park and Lee presented in 2004 a mechanically tunable PhC [54]. This PhC consists of a periodic high index material embedded in a low-index polymer matrix. The matrix polymer (poly-dimethylsiloxane) is flexible and can therefore be stretched and released by MEMS-actuators up to 10% in the Γ–M-direction. By stretching the structure, a large change in both the optical properties and the periodicity of the PhC follows, thus yielding a large change in the transmission and reflection characteristics. Kim *et al.* proposed a similar approach involving a strain-tunable 2D PhC on a piezoelectric substrate [55], for a 1D structure Wong *et al.* showed similar results [56].

Other groups experiment with systems employing the magneto-optical effect. Khartsev *et al.* presented a tunable PBG structure made from alternating layers of $Bi_3Fe_5O_{12}$ and $Gd_3Ga_5O_{12}$ [57]. A completely different approach is followed by D. Erickson *et al.*, who combine nanophotonics and nanofluidics by integrating nanofluidic channels in 2D PhCs. By infiltrating different fluids into the PhC, they achieve refractive index changes of up to 0.1 [58].

15.4
Tunable Superprism Structures

In the previous sections, numerous approaches for tuning the properties of PBG structures have been presented. Most of them are also applicable for tunable superprism structures. Despite this fact, to our knowledge, only one actual measurement of the tunable superprism effect in PBG structures has been demonstrated to date.

Theoretical investigations have been performed by Summers for 2D triangular PhCs forming a slab waveguide. Here, the concept of infiltrated liquid crystals was considered. Summers stated that for guided modes the beam steering was only on the order of 10 degrees when tuning the refractive index between 1.6 and 2.1. To enhance the structure's performance, he proposed a 1D super-lattice on top of the slab waveguide. Calculations showed that this measure increased the beam steering of the device to 50 degrees by creating new allowed guided modes in the slab. Li *et al.* theoretically investigated the superprism effect in a 2D structure formed by cylindrical rods from strontium barium niobate (SBN) [59]. They also obtained large beam steering of 30 degrees by simulating the electro-optic effect for an applied field of up to 1.4×10^7 V/m. For a PhC formed by periodic inclusions in a stretchable polymer, Park *et al.* calculated by FDTD-simulation an achievable change in the propagation direction of light of 75 degrees. For a ferro–electric system forming a 2D PhC, Okamura *et al.* predicted a high electro-optic tunability that results in a beam steering of 40 degrees [60]. The all-optical switching properties of different 2D PhCs were investigated in several publications by Panoiu *et al.* Assuming a Kerr medium with a nonlinear refractive index of $n_2 = 3 \times 10^{-16}$ m^2/W, they predicted beam steering ranges of several tens of degrees by applying a laser beam with a pump intensity of 100 W/μm^2 [61]. Scrymgeour *et al.* simulated the properties of PLZT-photonic crystals and obtained similar predictions for the beam steering for an applied field of 6 V/μ m [62].

This year, we showed thermal tuning of the superprism effect in a 1D thin film stack [64]. To our knowledge, this is the first experimental demonstration of a tunable superprism structure.

15.5 1D Hybrid Organic–Anorganic Structures

We propose a tunable superprism structure realized by the combination of non-periodic thin film stacks and nonlinear organic materials as active media. Though higher dimensional PhCs may exhibit stronger dispersive effects, there are nevertheless a number of reasons for using 1D stacks. First, superprism phenomena have been measured experimentally in dielectric thin film stacks in good agreement with theory. Gerken showed in several publications that dielectric thin film stacks are versatile structures for realizing highly dispersive devices with tailored properties. The key step is to go from strictly periodic 1D designs (Bragg-Stacks) to nonperiodic ones. Here, high angular dispersion still occurs, but the number of degrees of freedom in design are greatly enhanced.

An important advantage of 1D structures over 2D and 3D ones is the mature thin film deposition technology providing high material quality and production accuracy in the nanometer range. Here, especially the production of 3D structures in the optical range is more challenging. Directly related to this fact are the low cost of thin film deposition compared for example to electron beam writing or direct laser beam writing.

15.5.1 Survey of Optically Nonlinear Organic Materials

The use of organic materials as active media has several reasons. First, in many cases organic materials show stronger nonlinear effects than inorganic systems, which is especially true for the electro-optic effect and thermal effects as will be shown in detail in the next two sections. In the case of all-optical nonlinearities organic materials also show promising developments, though they are not as far developed as the electro-optic and thermo-optic materials.

Here we consider organic materials in the form of polymers and small molecules. These classes of organic materials offer relatively easy processing by spin-casting or evaporation at low temperatures and relatively low vacuum levels, respectively. Although there exist a number of interesting organic crystals, the latter lack these advantages in fabrication and are therefore not discussed here.

Another advantage of organic materials is that they are modifiable by chemical synthesis. This offers more possibilities in changing the material properties than for example in the case of semiconductor materials.

15.5.1.1 Thermo-Optic Organic Materials

Organic materials, especially polymers, possess large thermal expansion coefficients. This fact poses a difficulty in fabrication, especially when combining organic materials with inorganic ones. In terms of tunability, this fact comes useful, as the large expansion coefficients are also accompanied by large alterations in the refractive index of the polymer. A decrease rate in the refractive index, i.e., a thermo-optic coefficient dn/dT, in the range of 10^{-4}/K is observed for most polymers. In PPV-derivates even values of 10^{-3}/K are obtained [65]. These values are at least one order of magnitude larger than the coefficients of other optical materials as, for example, optical glasses. Therefore, polymers are well suited for tuning with temperature. As the thermal conductance of polymers is low, devices with low switching energies can be realized. However, the excitation and relaxation times of thermal effects are not faster than microseconds and are often in the range of milliseconds to seconds. Therefore, no fast switching devices based on thermal effects are possible.

15.5.1.2 Electro-optic Organic Materials

The most rapid development for optically nonlinear organics has taken place in the field of electro-optic materials. The linear electro-optic effect denotes the change in index of refraction in dependence on an applied electric field. For small changes, which covers all commercially relevant effects, it can be approximated linearly by

$$\Delta n_i = -\frac{n_i^3 r_{iik} E_k}{2}, \qquad (15.2)$$

where n is the unperturbed refractive index, E is the electric field strength, and r is the electro-optic coefficient. Being proportional to the refractive index change, the electro-optic coefficient is the characteristic parameter for the electro-optic activity of a material. The subscripts in 15.2 refer to the Cartesian directions. r_{iik} is a tensor element as the electro-optic coefficient is derived from the electro-optic susceptibility that connects the first order term of the polarization vector inside the material with the applied field. For an electro-optic material, a high electro-optic coefficient at the operating wavelength is required.

Low absorption and scattering losses, and a high glass transition temperature are further important parameters. To assess the electro-optic activity, the standard material of today's electro-optic industry, $LiNbO_3$, is the benchmark. At 1300 nm, for $LiNbO_3$ a r_{113} of 30.8 pm/V is found. Since the 1970s research on organic electro-optic systems is ongoing, but only in the early 1990s sufficient theoretical understanding of the origin of molecular hyper-polarizabilities and their connection to the macroscopic nonlinearity in polymers evolved to allow for the synthesis of more powerful materials [66]. The cornerstones of this theory were the explanation of the relationship between bond length alternation and bond order alternation in the carbon chains of the molecules and the nonlinear effects. Now, different groups pushed the electro-optic coefficients of organic systems to ten times the values of disperse red, which was until recently the standard organic electro-optic material [67]. In 1996, Marder *et al.* synthesized a conjugate molecule with r_{333} of 55 pm/V [68]. Some years later, the incorporation of so called CLD-chromo-phores brought the polymers to values exceeding 130 pm/V leaving $LiNbO_3$ far behind [69]. Recently, electro-optic coefficients of 500 pm/V were announced and leading scientists in the field consider 1000 pm/V possible in the near future [70].

A second, important improvement concerns the temperature stability of polymers. Low glass transition temperatures and related fast performance degradation rendered early electro-optic polymers unsuitable for device fabrication. This problem was addressed parallel to the improvement of the electro-optic coefficients. Today stable, high performance electro-optic polymers are available and already used in commercial devices.

15.5.1.3 **All-optical Organic Materials**

Kerr-active organic materials for all-optical switching have also improved, though not as rapidly as electrooptic ones. For all-optical devices there has not been a convergence towards a single group of materials yet. Besides the speed of the effect, a number of other parameters is important for evaluating the performance of a Kerr-material. Of course, the nonlinear refractive index change due to the Kerr effect should be large, this is expressed by

$$n = n_0 + \Delta n = n_0 + n_2 I. \qquad (15.3)$$

Here, n_0 denotes the unperturbed refractive index and I the intensity of the incident light. The nonlinear optical Kerr coefficient n_2, normally expressed in cm^2/W, is the macroscopic parameter for Kerr activity. Stegeman introduced two figures of merit that classify the absorption properties of Kerr media and are widely used in publications. At the one hand, the linear absorption must be weak in comparison to the nonlinearity, which is expressed by

$$X = \frac{\Delta n}{\alpha\lambda} > 1. \qquad (15.4)$$

Here α is the linear absorption, λ is the wavelength, and Δn is given in Eq. (15.3). Typically, an intensity at which the dependence of the refractive index change drops significantly below a linear behavior is considered. Since two photon absorption processes can also appear Kerr-like, concurrent two photon absorption processes have to be regarded as well. This is quantified by the second figure of merit, which should be smaller than unity

$$T = \frac{2\beta_2\lambda}{n_2} < 1, \tag{15.5}$$

with β_2 being the two photon absorption. If conditions 15.4 and 15.5 are met, a nonlinear phase shift of 2π can be achieved before the light intensity drops below $1/e$ in the material. A phase shift of this size is required for many on-off-switching devices as well as for tunable superprisms.

Two promising candidates for organic Kerr media are azobenzene molecules [72] and β-carotenoids [71]. The nonlinear refractive indices range, depending on the exact shape of the molecule and the spectral region, between $10^{-14}\,cm^2/W$ and $10^{-12}\,cm^2/W$ in the regions where the Stegeman-conditions are met. Though some III–V-semiconductors offer the same order of magnitude in the nonlinear effect, these materials normally have a much higher absorption than organic materials. Optical glasses range at least one order of magnitude below these values.

Especially the azo-dyes, which can be covalently bonded to host polymers such as polymethyl methacrylate or polysterene, have drawn much attention. Besides high nonlinear effects, their absorption peak between 400 nm and 500 nm makes them interesting, as this opens two spectral regions where fast, non-absorbing nonlinear action can take place. The first is between 600 nm and 900 nm and the second is between 1200 nm and 1800 nm, covering optical communication wavelengths. For the β-carotenoids Marder *et al.* showed molecular nonlinear coefficients exceeding the values for fused silica by a factor of 35.

Last year, Hochberg *et al.* showed terahertz all-optical amplitude modulation in a hybrid silicon-polymer system [73]. It consists of a silicon ridge waveguide in a Mach-Zehnder-geometry embedded in a cladding made of the nonlinear active material. Such examples demonstrate both, the high performance of Kerr-active polymers and their compatibility in fabrication with optical glasses or semiconductors. Though first results have been shown, progress in the performance of polymer materials is still needed to decrease the necessary pump powers for optical switching. Power levels of standard laser diodes in optical communications are currently too low, whereas Erbium-doped fiber amplifiers (EDFAs) already provide power levels nearly high enough for inducing a sufficient Kerr-effect.

15.5.2
Numerical Simulation of a Doubly Resonant Structures for All-Optical Spatial Beam Switching

We numerically investigate the properties of devices for actively shifting the exit position of a beam of light inside a 1D multilayer superprism stack. The device schematic is depicted in Figure 15.2. The beam shifting is achieved by changing the optical properties of the stack and exploiting the large group propagation angle dispersion due to the superprism effect. In the following, we presume the induction

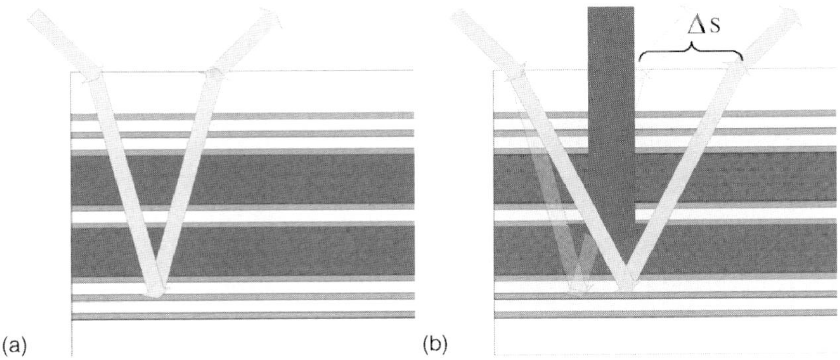

Figure 15.2 Device schematic of an optically pumped thin film stack for signal beam displacement. The light gray arrows represent the signal beams, the dark gray arrow the pump beam, respectively. Dark gray layers symbolize high index material and light gray layers are low index materials. The active material is placed either in the defects or replaces one of the low-index layers between the defects.

of a change in the optical properties by a pump beam to tune the output position of a beam focused into the device.

The design consists of a stack of $\lambda/4$-layers with two $\lambda/2$-defects. Here, the layers are designed for pump light of 870 nm wavelength that enters the stack from the normal direction with respect to the surface of the thin film stack. The signal beam is incident under an angle of 45 degrees. The stack is operated in reflection. The defects are placed in such way that a symmetrical stack is formed with the layer sequence $(HL)^5H2L(HL)^5H2L(HL)^5H$. Here H denotes high index $\lambda/4$-layers at the wavelength of 870 nm at normal incidence. Using this value a thickness of $H = 106$ nm and $L = 149$ nm is obtained. As low index and high index dielectrics, silica and tantala, are assumed respectively. In the wavelength range of interest, their respective refractive indices are 1.45 and 2.06. Furthermore, nonlinear organic layers were placed at different positions inside the stack to evaluate the performance of the device in terms of the achievable shift with respect to the pump power. For the simulation an organic material with an unperturbed refractive index of 1.55 and a nonlinear coefficient of $10^{-12}\,\text{cm}^2/\text{W}$ were assumed and Eq. (15.3) was used to determine the refractive index as a function of intensity.

The calculations were performed by a two step transfer-matrix calculation [17]. First, the propagation of the pump light in the device was simulated and the resulting changes in the optical properties of the stack with intensity were determined iteratively until a self-consistent solution was found. All calculations were performed for the plane-wave case. For the studied designs this is justified for a beam radius larger than 15 μm corresponding to an angular range of 1.9° [74]. For obtaining the highest possible intensity, the pump beam radius was set to 15 μm. Then, the propagation of the signal beam was simulated in a second simulation run using the structure with modified optical properties. Here, a sufficiently small signal power level was assumed such that changes in the optical properties due to the signal beam can be neglected. The pump beam is assumed to be focussed in the center of the active layer. Due to the small layer thickness pump beam divergence within the layer may be neglected. The probe beam is assumed to overlap ideally with the pump beam. For the experimental realization of the device a refocussing of the probe beam after exiting the stack is necessary.

15.5.2.1 **Beam Shifting for Two Active Cavities**

The first case under consideration is that both λ/2-defects comprise the active organic material. For this case, the dependence of the beam shift with respect to the peak power of the pump pulse is depicted in Figure 15.3. A significant beam shift is seen for peak powers greater than 300 W and for a pump power of 10000 W a beam displacement of 19 μm is observed at the signal wavelength of 787 nm. Also depicted is the beam displacement for a detuning from the designed signal wavelength. A detuning of the signal wavelength by 0.5 nm reduces the beam displacement to half the value.

15.5.2.2 **Beam Shifting for One Active Cavity**

Figure 15.4 shows the results for only one active λ/2-defect. In Figure 15.4, the left hand side and right hand side pictures depict the case where the lower and respectively the upper λ/2-defect is the active one. Here, an earlier onset of beam shift at 90 W peak power and a higher achievable beam shift of almost 30 μm for a pump power of 10 000 W can be seen for the lower active cavity.

15.5.2.3 **Beam Shifting for Active Coupling Layers**

Finally, the influence of the layers between the λ/2-defects is considered(Figure 15.5). For this case, calculations for a structure where one low-index dielectric layer is

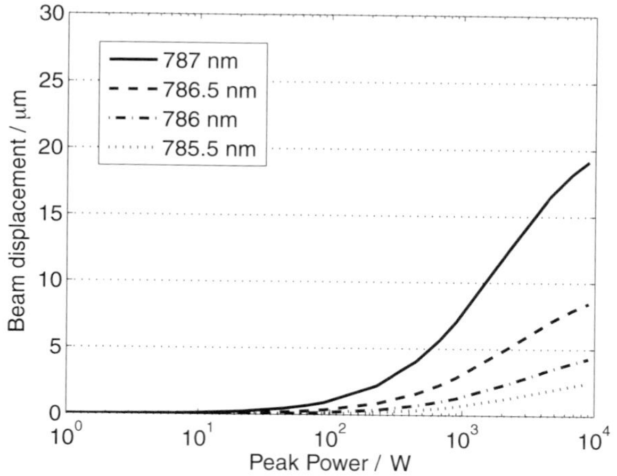

Figure 15.3 Beam displacement over peak pump power for a design with two active λ/2-defects for four different signal wavelengths.

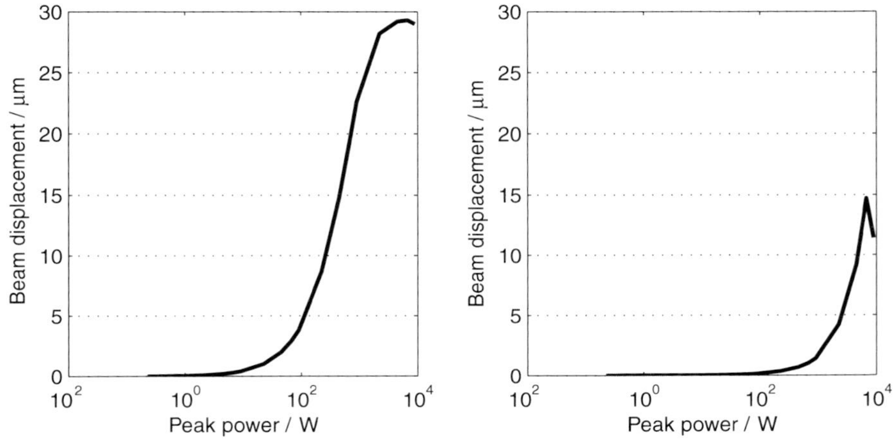

Figure 15.4 Beam displacement over peak pump power for a design with one active λ/2-defect (Right diagram: upper defect active; left diagram: lower defect active).

replaced by an active organic layer are carried out. In the following L1 denotes the case where the uppermost layer is active, L2 the case where the next lower one is active and so on. Here, a performance advantage of designs where the lower layers are active is observed. In terms of total beam shift and pump powers, these designs stay behind the active defect designs as the best performance of the coupled layer designs only yields a 15 µm beam displacement.

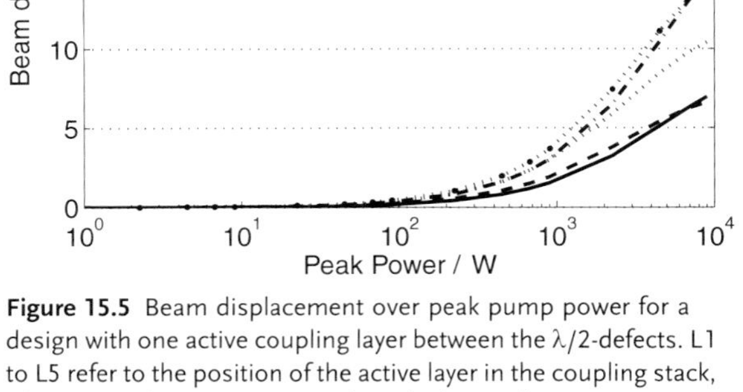

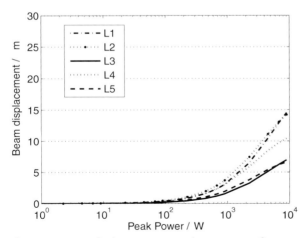

Figure 15.5 Beam displacement over peak pump power for a design with one active coupling layer between the λ/2-defects. L1 to L5 refer to the position of the active layer in the coupling stack, where L1 is the uppermost and L5 the lowest layer.

15.6
Conclusions and Outlook

In conclusion, tunable superprism PhC structures combine high angular dispersion with optically active media to achieve spatial beam switching. While active tuning of the band gap position has been demonstrated theoretically and experimentally by many authors, only little research has been performed on tunable superprism devices. Theoretical studies of tunable superprism structures demonstrated that such devices are in reach for proof-of-principle experiments using different geometries and switching mechanisms.

Due to the size of the effect and the comparatively straightforward fabrication methods the liquid crystal approach is a promising candidate for experimental access. For demonstration of higher switching rates electro-optic materials are favorable because of large nonlinear effects and high switching speeds. Here, especially polymer materials will play an important role. In this class of materials high nonlinearities are combined with easy processing. Polymers can be evaporated or spin-cast and afterwards patterned by lithography, hot embossing, or other nanoimprint techniques. Additionally, they are processable on most substrates. These properties are due to the fact that polymers do not require a high amount of order (though for electro-optic materials a breaking of the centro-symmetry is necessary) or even crystal structure to exhibit large effects. This gives them an advantage over other electro-optic materials such as $LiNbO_3$ or PLZT-ceramics.

We concluded with a theoretical investigation of a one-dimensional hybrid organic-inorganic all-optical tunable superprism stack. This structure combines the accurate thin-film deposition technology with large and fast all-optical nonlinear effects achieved in organic materials. Simulation results demonstrate that a significant spatial beam shifting of 30 μm at pump power of 1.5 W is possible using a single active layer. The necessary switching energies can be achieved using a short-pulse laser system for a proof-of-principle experiment. More research is necessary to further decrease the switching energies and to obtain commercially viable tunable superprism devices.

References

1 John, S. (1987) *Phys. Rev. Lett.*, **58**, 2486.

2 Zengerle, R. (1979) Lichtausbreitung in ebenen periodischen Wellenleitern, Dissertation 1979/23 12, Universität Stuttgart.

3 Zengerle, R. (1987) *J. Mod. Opt.*, **34** (12), 1589.

4 Dowling, J.P. and Bowden, C.M. (1994) *J. Mod. Opt.*, **41** (2), 345.

5 Lin, S.Y., Hietela, V.M., Wang, L. and Jones, E.D. (1996) *Opt. Lett.*, **21** (21), 1771.

6 Kosaka, H., Kawashima, T., Tomita, A., Notomi, M., Tamamura, T., Stato, T. and Kawakami, S. (1998) *Phys. Rev. B*, **58** (16), R10096.

7 Kosaka, H., Kawashima, T., Tomita, A., Notomi, M., Tamamura, T., Stato, T. and Kawakami, S. (1999) *Appl. Phys. Lett.*, **74** (10), 1370.

8 Kosaka, H., Kawashima, T., Tomita, A., Notomi, M., Tamamura, T., Stato, T. and Kawakami, S. (1999) *J. Lightwave Technol.*, **17** (11), 2032.

9 Wu, L., Mazilu, M., Karle, T. and Krauss, T.F. (2002) *IEEE J. Quantum Electron.*, **38** (7), 915.

10 Baumberg, J.J., Perney, N., Netti, M., Charlton, M., Zoorob, M. and Parker, G. (2004) *Appl. Phys. Lett.*, **85** (3), 354.

11 Prasad, T. Colvin, V. and Mittleman, D. (2003) *Phys. Rev. B*, **67** (16), 165103.

12 Baba, T. and Matsumoto, T. (2002) *Appl. Phys. Lett.*, **81** (13), 2325.

13 Matsumoto, T. and Baba, T. (2004) *J. Lightwave Technol.*, **22** (3), 917.

14 Gerken, M. and Miller, D.A.B. (2003) *IEEE Photonics Technol. Lett.*, **15** (8), 1097.

15 Gerken, M. and Miller, D.A.B. (2004) Optics East 2004, Philadelphia, PA, SPIE Conference, p. 5597.

16 Nelson, B.E., Gerken, M., Miller, D.A.B., Piestun, R., Lin, C.-C. and Harris, J.S. Jr. (2000) *Opt. Lett.*, **25** (20), 1502.

17 Gerken, M. and Miller, D.A.B. (2003) *Appl. Opt.*, **42** (7), 1330.

18 Gerken, M. and Miller, D.A.B. (2005) *Appl. Opt.*, **44** (16), 3349.

19 Jugessur, A., Wu, L., Bakhtazad, A., Kirk, A., Krauss, T. and De La Rue, R. (2006) *Opt. Express*, **14** (4), 1632.

20 Wu, L., Mazilu, M., Gallet, J.-F. and Krauss, T.F. (2003) *Photonics Nanostruct.*, **1**, 31.

21 Miller, D.A.B. (2007) *J. Opt. Soc. Am. B*, 76778.

22 Kitzerow, H.-S. and Reithmaier, J.P. (2004) in: *Tunable Photonic Crystals using Liquid Crystals in Photonic Crystals: Advances in Design, Fabrication and Chracterization*, (eds K. Busch, H. Föll, S. Lölkes, and R.B. Wehr-spohn), Wiley-VCH.

23 Busch, K. and John, S. (1999) *Phys. Rev. Lett.*, **83**, 967.

24 Leonard, S.W., Mondia, J.P., van Driel, H.M., Toader, O., John, S., Busch, K., Birner, A., Gösele, U. and Lehmann, V. (2000) *Phys. Rev. B*, **61** (4), R2389.

25 Yoshino, K., Shimoda, Y., Kawagishi, Y., Nakayama, K. and Ozaki, M. (1999) *Appl. Phys. Lett.*, **75** (5), 932.

26 Mertens, G., Wehrspohn, R.B., Kitzerow, H.-S., Matthias, S., Jamois, C. and Gösele, U. (2005) *Appl. Phys. Lett.*, **87**, 241108.

27 Kosmidou, E.P. and Kriezis, E.E. and Tsiboukis, T.D. (2005) *IEEE J. Quantum Electron.*, **41** (5), 657.

28 Ozaki, R., Ozaki, M.M. and Yoshino, K. (2003) *Jpn. J. Appl. Phys. B*, **42** (6), L669.

29 Kee, C.-S., Lim, H., Ha, Y.-K., Kim, J.-E. and Park, H.-Y. (2001) *Phys. Rev. B*, **64**, 085114.

30 McPhail, D., Straub, M. and Gu, M. (2005) *Appl. Phys. Lett.*, **86** (5), 051103.

31 Escuti, M.J., Qui, J. and Crawford, G.P. (2003) *Opt. Lett.*, **28** (7), 522.

32 Gorkhali, S.P., Qui, J. and Crawford, G.P. (2005) *Appl. Phys. Lett.*, **86**, 011110.

33 Gorkhali, S.P., Qui, J. and Crawford, G.P. (2006) *J. Opt. Soc. Am. B*, **23** (1), 149.

34 Roussey, M., Bernal, M., Courjal, N. and Baida, F. (2005) *Appl. Phys. Lett.*, **87**, 241101.

35 Roussey, M., Bernal, M., Courjal, N., van Labeke, D. and Baida, F. (2006) *Appl. Phys. Lett.*, **89**, 241110.

36 Schmidt, M., Eich, M., Huebner, U. and Boucher, R. (2005) *Appl. Phys. Lett.*, **87**, 121110.

37 Gan, H., Zhang, H., deRose, C.T., Norwood, R.A., Peyghambarian, N., Fallahi, M., Lou, J., Chen, B. and Yen, A. (2006) *Appl. Phys. Lett.*, **89**, 041127.

38 Li, B., Zhou, J., Li, L., Wang, X.J., Liu, X.H. and Zi, J. (2005) *Appl. Phys. Lett.*, **83** (23), 4704.

39 Sadot, D. and Boimovich, E. (1998) *IEEE Commun. Mag.*, **36** (12), 50.

40 Patel, J.S., Saifi, M.A., Berreman, D.W., Lin, C., Andreakis, N. and Lee, S.D. (1990) *Electron. Lett.*, **57** (17), 1718.

41 Patel, J.S., Saifi, M.A., Berreman, D.W., Lin, C., Andreakis, N. and Lee, S.D. (1990) *Appl. Phys. Lett.*, **57** (17), 1718.

42 Tayebati, P., Wang, P.D., Vakhshoori, D. and Sacks, R.N. (1998) IEEE Photonics Technol. *Lett.*, **10** (3), 394.

43 Yu, B. Pickrell, G. and Wang, A. (2005) *Photonics Technol. Lett.*, **16** (10), 2296.

44 Zhang, F., Zhang, L., Wang, Y. and Claus, R. (2005) *Appl. Opt.*, **44** (19), 3969.

45 Hu, X., Gong, Q., Liu, Y., Cheng, B. and Zhang, D. (2005) *Appl. Phys. Lett.*, **87**, 231111.

46 Soljačič M. and Joannopoulos, J.D. (2004) *Nature Mater.*, **3**, 211.

47 Yariv, A., Xu, Y., Lee, R.K. and Scherer, A. (1999) *Opt. Lett.*, **24**, 711.

48 Bristow, A.D., Wells, J., Fan, W.H., Fox, A.M., Skolnick, M.S., Whittaker, D.M., Tahraoui, A., Krauss, T.F. and Roberts, J.S. (2003) *Appl. Phys. Lett.*, **83** (5), 851.

49 Fushman, I., Waks, E., Englund, D., Stoltz, N., Petroff, P. and Vučkovió J. (2007) *Appl. Phys. Lett.*, **90**, 091118.

50 Asano, T., Song, B.S., Akahane, Y. and Noda, S. (2006) *IEEE J. Sel. Top. Quantum Electron., Part 1*, **12** (6), 1123.

51 Moll, N., Harbers, R., Mahrt, R.F. and Bona, G.L. (2006) *Appl. Phys. Lett.*, **88** (17), 171104.

52 Rong, H., Jones, R., Liu, A., Cohen, O., Hak, D., Fang, A. and Paniccia, M. (2005) *Nature*, **433**, 725.

53 Weissman, J.M., Sunkara, H.B., Tse, A.S. and Asher, S.A. (1999) *Science*, **274**, 959.

54 Park, W. and Lee, J. (2004) *Appl. Phys. Lett.*, **85** (21), 4845.

55 Kim, S. and Gopalan, V. (1999) *Appl. Phys. Lett.*, **78** (20), 3015.

56 Wong, C.W., Rakich, O.T., Johnson, S.G., Qi, M.H., Smith, H.I., Jeon, Y., Barbastathis, G., Kim, S.G., Ippen, E.P. and Kimerling, L.C. (2004) *Appl. Phys. Lett.*, **84** (8), 1242.

57 Khartsev, S.I. and Grishin, A.M. (1999) *Appl. Phys. Lett.*, **87**, 122504.

58 Erickson, D., Rockwood, T., Emery, T., Scherer, A. and Psaltis, D. (2006) *Opt. Lett.*, **31** (1), 59.

59 Li, J., Lui, N.-H., Feng, L., Liu, X.-P. and Chen, Y.-F. (2007) *J. Appl. Phys.*, **102**, 013516.

60 Okamura, S., Mochiduki, Y., Motohara, H. and Shiosaki, D. (2004) *Integr. Ferroelect.*, **69**, 303.

61 Panoiu, N.C., Bahl, M. and Osgood, R.M. Jr. (2004) *J. Opt. Soc. Am. B*, **21** (8), 1500.

62 Scrymgeour, D., Malkova, N., Kim, S. and Gopalan, V. (2003) *Appl. Phys. Lett.*, **82** (19), 3176.

63 Alagappan, G., Sun, X.W., Yu, M.B., Shum, P. and de Engelsen, D. (2006) *IEEE J. Quantum Electron.*, **42** (4), 404.

64 Glöckler, F., Peters, S., Lemmer, U. and Gerken, M. (2007) Spatial beam switching by thermal tuning of a hybrid organic–inorganic thin-film stack. OSA Topical Meeting on Optical Interference Coatings 2007,WD3.

65 Town, G.E., Vasdekis, A., Turnbull, G.A. and Samuel, I.D.W. (2005) Temperature tuning of a semiconducting-polymer DFB laser. Proc. LEOS 2005, 763.

66 Marder, S.R., Gorman, C.B., Meyers, F., Perry, J.W., Bourhill, G., Bredas, J.L. and Pierce, B.M. (1994) *Science*, **265**, 632.

67 Marder, S.R., Cheng, L.T., Tiemann, B.G., Friedeli, A.C., Blanchard-Desce, M., Perry, J.W. and Skindhoj, J. (1994) *Science*, **263**, 511.

68 Ahlheim, M., Barzoukas, M., Bedworth, P.V., Blanchard-Desce, M., Fort, A., Hu, Z.-Y., Marder, S.R., Perry, J.W., Runser, C., Staehlin, M. and Zysset, B. (1996) *Science*, **271**, 335.

69 Dalton, L. (2002) *Adv. Polym. Sci.*, **158**, 1.

70 Dalton, L., Robinson, B., Jen, A., Ried, P., Eichinger, B., Sullivan, P., Akelaitis, A., Bale, D., Haller, M., Luo, J., Liu, S., Liao, Y., Firestone, K., Bhatambrekar, N., Bhattacharjee, S., Sinness, J., Hammond, S., Buker, N., Snoeberger, R., Lingwood, M., Rommel, H., Amend, J., Jang, S.-H., Chen, A. and Steier, W. (2005) *Proc. SPIE*, **5935**, 593502.

71 Marder, S.R., Torruellas, W.E., Blanchard-Desce, M., Ricci, V., Stegeman, G.I., Gilmour, S., Brédas, J.-L., Li, J., Bublitz, G.U. and Boxer, S.G. (1997) *Science*, **276**, 1233.

72 Brzozowski, L. and Sargent, E.H. (2001) *J. Mater. Sci., Mater. Electron.*, **12**, 483.

73 Hochberg, M., Baehr-Jones, T., Wang, G., Shearn, M., Harvard, K., Luo, J., Chen, B., Shi, Z., Lawson, R., Sullivan, P., Jen, A.K.Y., Dalton, L. and Scherer, A. (2006) *Nature Mater.*, **5**, 703.

74 Gerken, M. and Miller, D.A.B. (2004) *J. Lightwave Technol.*, **22** (2), 612.

III
Photonic Crystal Fibres

Nanophotonic Materials: Photonic Crystals, Plasmonics, and Metamaterials. Edited by R.B. Wehrspohn,
H.-S. Kitzerow, and K. Busch

ISBN: 978-3-527-40858-0

16
Preparation and Application of Functionalized Photonic Crystal Fibres

H. Bartelt, J. Kirchhof, J. Kobelke, K. Schuster, A. Schwuchow, K. Mörl, U. Röpke, J. Leppert, H. Lehmann, S. Smolka, M. Barth, O. Benson, S. Taccheo, and C. D'Andrea

16.1
Introduction

The development of microstructured and photonic crystal fibres (PCFs) during recent years has considerably extended the potential of functional properties of optical fibres [1,2]. The introduction of holey structures in and around the core region of an optical fibre enabled new and even extreme optical transmission properties compared with conventional optical fibres for applications in signal processing, fibre lasers and amplifiers, for new broad-band light sources or for remote fibre sensing techniques. Besides the variations in hole size and hole distance or in the number of hole layers in pure silica glass systems, additional design and functional flexibility is achieved by the introduction of doped materials or by using non-silica glasses. Figure 16.1 shows the general optical guiding principles in photonic crystal fibres compared with a standard single mode fibre. The light propagation in PCFs can be achieved by index guiding with a higher effective index core or band gap guiding with a lower effective index core in PCFs. Figure 16.1(c) shows an air core design. Another possibility to prepare a band gap guiding fibre is to surround a pure silica core with a higher index profiled doped silica glass rod package. We describe the investigations of this PCF type in Section 16.4.

In the following sections, we will first describe shortly the typical preparation technique for such silica-based microstructured and photonic crystal fibres. Specific propagation properties (index guiding) will be discussed concerning, e.g., attenuation, mode field and dispersion. As examples for specialized PCFs we will then investigate fibres using the combination of holey structures and highly doped regions, discuss properties of solid band gap fibres and consider some aspects of non-silica PCFs. As examples of specific applications for such PCFs in the linear as well as in the nonlinear regime, results from spectral sensing and for super continuum generation are presented.

Nanophotonic Materials: Photonic Crystals, Plasmonics, and Metamaterials. Edited by R.B. Wehrspohn, H.-S. Kitzerow, and K. Busch

ISBN: 978-3-527-40858-0

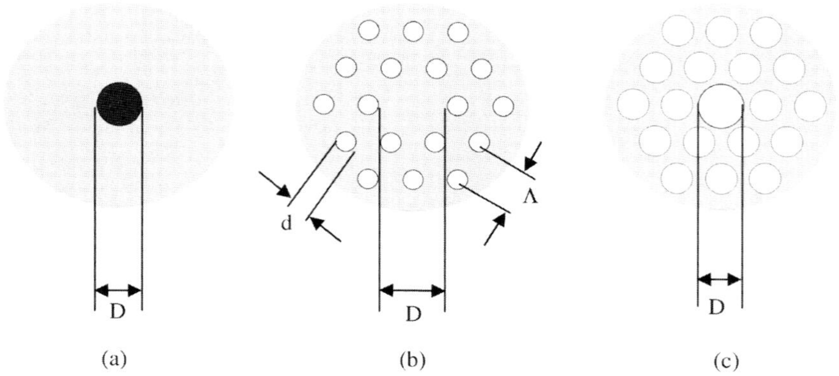

Figure 16.1 Principle of light guiding by optical fibres: (a) SMF – standard single mode fibre (germanium doped core), (b) index guided PCF, (c) photonic band gap fibre; geometrical parameters. d – hole diameter, Λ – pitch, D – core diameter.

16.2 General Preparation Techniques for PCFs

The preparation of microstructured or photonic crystal fibres requires a multitude of processing steps. The direct fibre extrusion technology (as used e.g. for polymer PCFs) left out of consideration here, typically a preform has to be prepared which is then drawn to the final fibre. For the preform manufacturing process we used the stack-and-draw method [3].

Besides the holey structure, also the material properties strongly influence the final fibre properties. Whereas the preparation of pure silica PCFs allows commercially available high silica tubes to be accessed (e.g. Heraeus Suprasil F 300), the fabrication of the doped PCF preforms requires additional technological steps for material modification. We used the MCVD technology for the preparation of special PCF preform components, capillaries and thin rods. The outer diameter of rods and capillaries was typically 1 mm. The inner diameter of capillaries was adapted between 0.5 mm and 0.9 mm. The stack arrangement for index guiding PCF was made in hexagonal symmetry with mostly uniform sizes of central rod and capillaries. The modification of optical properties by material dopance, e.g. for variation of refractive index profile, for material-induced dispersion behaviour or for nonlinear properties was accomplished by the utilization of germanium-doped silica layers in silica substrate tubes.

16.3 Silica-Based PCFs with Index Guiding

Silica is a well developed and investigated basis material for the preparation of optical fibres. It is commercially available in high quality and permits to obtain fibres with extremely low attenuation. The material impurities of the silica type F300 used is at the ppb-level for transition elements and in the sub-ppm range for OH contamination. This material is also the starting point for the preparation of capillaries with germanium-doped layers by the MCVD technology. We concentrate in this section on the properties of index guiding PCFs, where the core region has an effective refractive index higher than the cladding region due either to lower air content or doping. The spectral loss and the mode propagation behaviour are influenced by the holey design and can be described for index-guided PCFs by models such as in [3].

16.3.1
Specific Properties of Pure Silica PCFs

Typical index guiding PCF structures with five air hole rings are shown in Figure 16.2. The different fibres were prepared from the same preform by variation of the fibre drawing conditions. These typical examples of index guiding PCFs consist of a silica core surrounded by a hexagonally arranged holey cladding. The variation of the number of cladding hole rings can influence the confinement loss behaviour and the effective index of the cladding region. Generally, an increasing ring number depresses the mode leaking and the fibre loss. However, most important are the geometrical parameters of the holey package: hole diameter d and pitch Λ (hole-to-hole centre distance). In such an arrangement the cross section of the core is usually given by the pitch parameters, i.e. the core can be built up by one, seven, nineteen etc. rod elements, substituting capillaries. In addition, the pitch relations between core and cladding can be changed by modifying the drawing conditions.

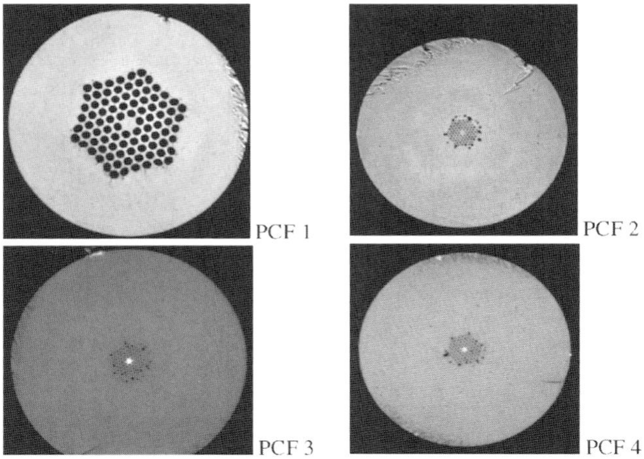

Figure 16.2 Cross section micrograph of five-ring air hole PCFs, prepared from an identical preform: PCF 1: 12 µm core diameter and $d/\Lambda = 0.8$ (top left), PCF 2: 5.0 µm core diameter and $d/\Lambda = 0.4$ (top right), PCF 3: 3.6 µm core diameter and $d/\Lambda = 0.3$ (bottom left), PCF 4:2.0 µm core diameter and $d/\Lambda = 0.5$ (bottom right).

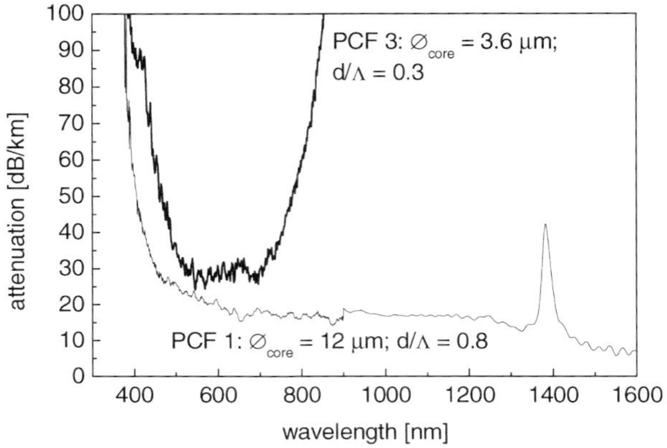

Figure 16.3 Attenuation spectra of material-identical PCFs, influenced by the structural design parameters.

This PCF type with a five-ring air hole design is a useful model system to investigate basic propagation properties. For our model fibres the core size varies from 12 μm to 2 μm, the fibre diameter lies between 200 μm and 90 μm and the d/Λ ratio varies between 0.3 and 0.8.

The attenuation of the final fibre is strongly defined by the material properties, but will also depend on the hole structure and the field confinement. Figure 16.3 shows the different transmission properties of two different PCFs with equal material base. PCF 1 has a pitch $\Lambda = 10$ μm and $d/\Lambda = 0.8$. This corresponds to a numerical aperture (NA) of 0.16 and to a cut-off wavelength of about 2.5 μm. In the investigated spectral range the fibre operates in multimodal regime. The OH absorption peak corresponds to 0.75 ppm caused by material impurity limits of Suprasil F300 and technology-based OH contaminations. The small core fibre PCF 3 with a pitch of $\Lambda = 2.1$ μm and $d/\Lambda = 0.5$ shows an infrared edge in the transmission window at a wavelength of about 800 nm induced by bending loss effects. The short wavelength edge corresponds to the expected cut-off-wavelength of about 500 nm. PCF 2, 3, 4 give also a considerable variation in the values of the numerical apertures (0.1, 0.14, 0.25) caused by the design parameters Λ and d/Λ. The corresponding cut-off wavelengths are 1.05 μm for PCF 2 and close to 650 nm for PCF 4. In general, a higher air content in the cladding region increases the numerical aperture and shifts the cut-off wavelength to smaller values for a constant core cross section.

The parameters d, Λ play also an important role for the dispersion behaviour. This is also determined by material effects and geometrical characteristics.

The adjustment of PCF design parameters allows modification of the zero dispersion point and the dispersion slope on a much larger scale than possible with conventional fibres. One aspect important for application is a shift of the zero dispersion wavelength to extremely short wavelengths, e.g. to the VIS or UV region, which is not possible with conventional solid fibres.

Figure 16.4 shows the change of dispersion behaviour in the case of the different five-ring fibre designs in comparison to bulk silica material dispersion. The PCF 3

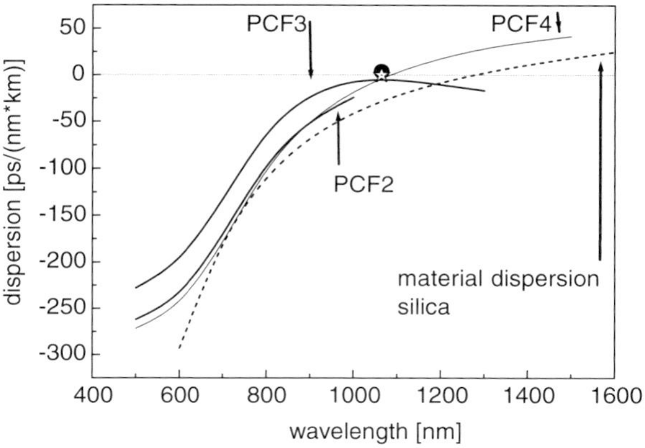

Figure 16.4 Calculated dispersion behaviour of the fibre designs PCF 2, PCF 3, PCF 4, and measured dispersion values of PCF 3 (star) and PCF 4 (filled circle) at 1064 nm compared with the material dispersion of silica (dotted line).

enables an undercut of the zero dispersion line at all investigated wavelengths. PCF 4 gives a zero dispersion shift to shorter wavelengths compared to bulk silica. With appropriate PCF design the zero dispersion point can be shifted further into the visible wavelength range, which is of special interest for non-linear applications (e.g. for supercontinuum sources, short wavelength Raman lasers).

16.3.2
PCF with Very Large Mode Field Parameter (VLMA-PCF)

Achieving a very large effective mode area is a key goal in current research on microstructured as well as on solid fibres for high power fibre amplifiers and lasers. The interest is driven by the need for scaling up the optical power or pulse energy and at the same time to limit the light intensity in the fibre core to prevent nonlinear effects such as Raman and Brillouin impairments and to avoid thermal damages. The enlargement of core size in conventional fibres degrades the beam quality and reduces the guiding performance, so that it is difficult to maintain a stable fundamental mode [4]. The necessary low index contrast between core and cladding is difficult to obtain in a conventional fibre. In PCFs it is possible to control the necessary low numerical aperture very precisely by the hole sizes and the hole structure.

Figure 16.5 shows a relatively simple hole-assisted fibre design, which nevertheless gives a stable and large single mode field.

The design consists of a core cross section area of about 2800 μm^2, the size of (relatively large) holes is about 47 μm. The overall diameter of the fibre is 594 μm.

Figure 16.6 shows the measured fundamental mode in grey scales (left) in comparison to the modelled fundamental mode (right). The intensity profiles of the mode scanned through the holes and through the bridges, respectively, are shown in Figure 16.7

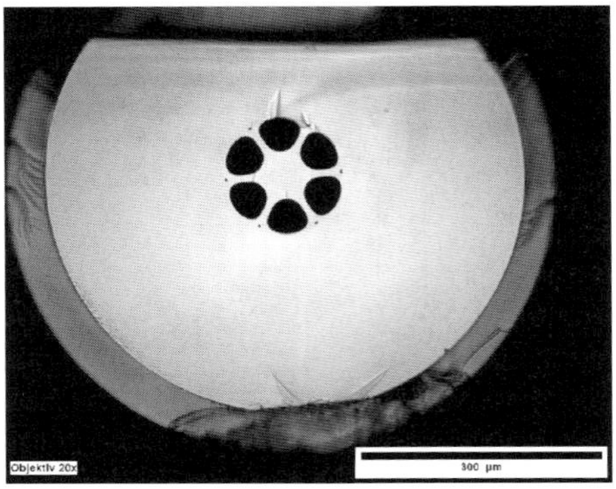

Figure 16.5 Cross section micrograph of the VLMA-PCF (core diameter: 60 µm, width of bridges: 9 µm, total fibre diameter: 594 µm).

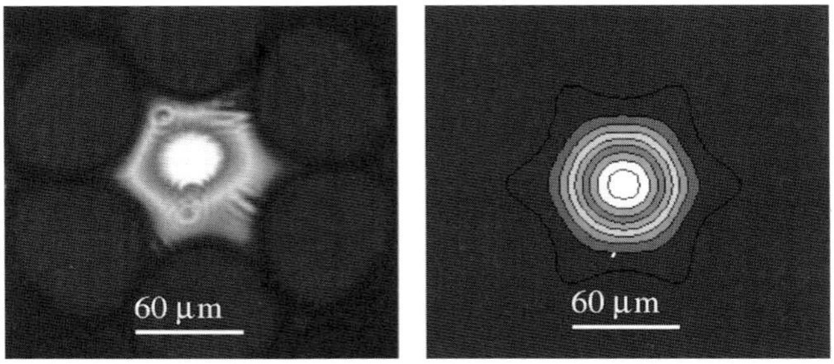

Figure 16.6 Mode field of the fundamental mode demonstrated in grey scales; left: experimental, right: simulation using a finite element method.

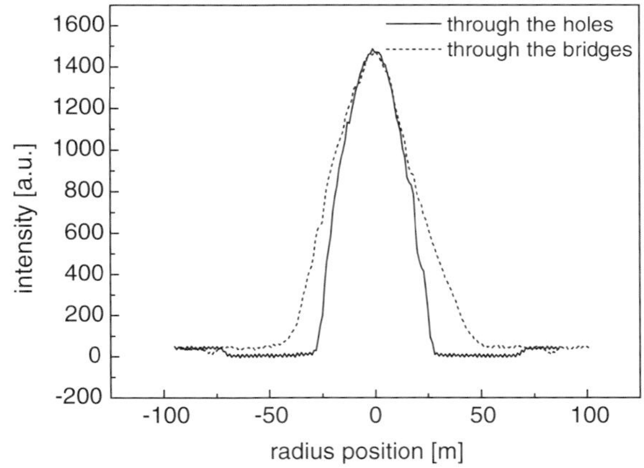

Figure 16.7 Measured profiles of the fundamental mode from the fibre shown in Figure 16.5.

From these results, one can derive a mode field diameter of about 50 μm and an effective mode area of close to 2000 μm^2. This is an improvement in mode field area by a factor of about 4 compared to conventional solid large mode area fibres. Simulations by means of a finite element method show good agreement with the experiments concerning the wave field diameter (Figure 16.6).

16.3.3
Doped Silica PCF with Germanium-Doped Holey Core

Until now, only pure silica material was considered in the design and in the preparation of PCFs. In this section we want to include the option of doped layers in a PCF. As an example, we consider a special large mode area PCF as shown in Figure 16.8. The core of this fibre consists of seven holes, where the refractive index was changed by doping the core capillaries with germanium. The geometrical parameters are: total core diameter 8 μm, diameter of the cladding and core holes 3.7 μm and 1.6 μm, respectively. The holey cladding shows a diameter of 58 μm, the total fibre diameter is 129 μm. The germanium oxide concentration in the MCVD-deposited core layers of about 0.18 μm thickness corresponds to about 10 mol% and the phosphorus oxide concentration is about 1 mol%.

The cladding is arranged in a 3-air-ring design with phosphorus fluorine-doped, index-matched profiled capillaries (preparation details of this fibre are described in [5]).

Microstructured optical fibres of such design are advantageous for sensing applications due to their holey structure in the light guiding core region, where a sample material can be infiltrated. The overlap between the guided mode wave field and the holey area of the analyte offers the possibility of an increased sensitivity in comparison to conventional fibres. In difference to the mostly used evanescent field absorption sensors (EFAS), based on attenuated total reflection (ATR), with a small overlapping of the evanescent wave only at the core-cladding interface (corresponding to the glass-analyte interface), the concept with microstructured fibres should make it possible to overcome analyte concentration limits, poor sensitivity per length or restrictions in the spectral range.

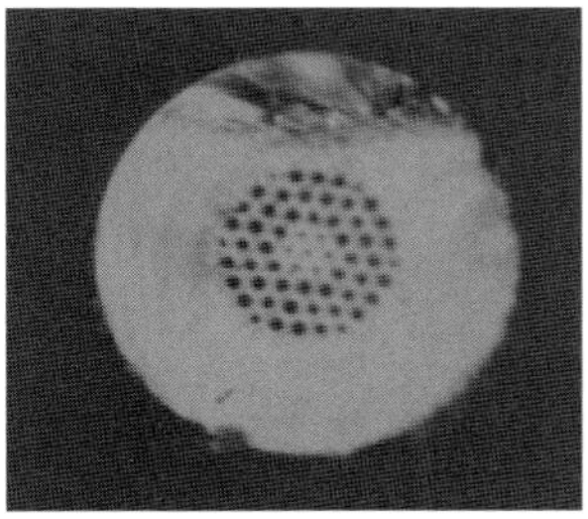

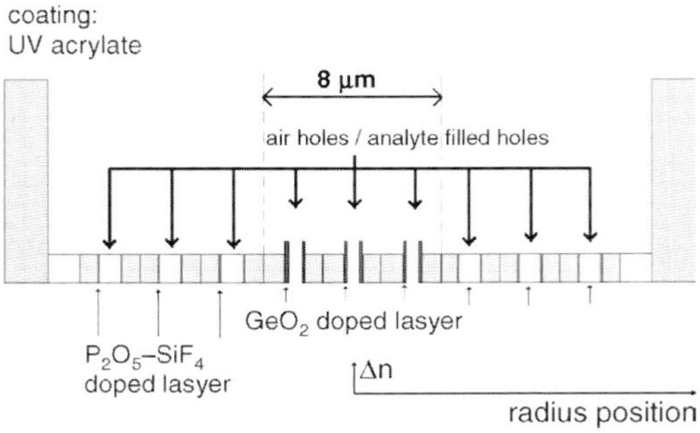

Figure 16.8 Micrograph and refractive index profile scheme of the germanium-doped holey core PCF.

In order to quantify the interaction between the light and the possible sample material, we have modelled the guided mode field and the fraction of the electric field intensity in the holes. For this purpose we numerically solve Maxwell's equations using either a finite element method (FEM) or a plane wave expansion method (PWEM). For the FEM calculation it is assumed that the electric field of the guided modes, which propagate in the z direction, is of the form of Eq. (16.1):

$$E(r) = E(x, y)\, e^{i\beta z}, \tag{16.1}$$

$$\beta = n_{\mathrm{eff}} 2\pi/\lambda. \tag{16.2}$$

In Eq. (16.1), β is the propagation constant, which depends on the effective refractive index n_{eff} of the guided mode and the vacuum wavelength λ.

The simulated structure is shown in Figure 16.9(a). As an approximate value for the refractive index of silica in this wavelength region we assume $n_{\mathrm{SiO2}} = 1.45$, while the germanium-doped core material has a refractive index of $n_{\mathrm{Ge}} = 1.46$. The diameter of the cladding and core holes is 3.7 μm and 1.6 μm, respectively. For the sample material we have assumed a refractive index of $n_{\mathrm{nonane}} = 1.405$ (this corresponds to the sample material nonane as discussed in Section 16.6). With the FEM, the electric field

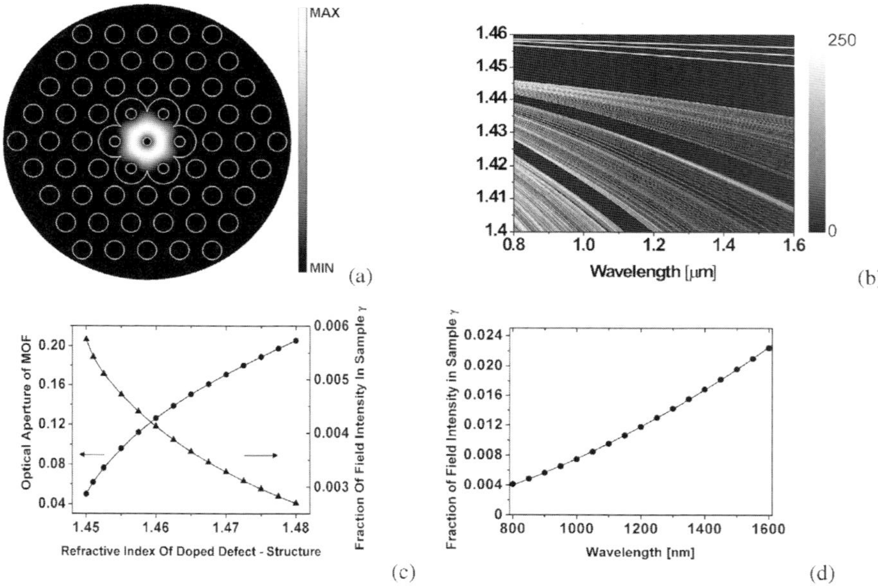

Figure 16.9 (a) Simulated electric field intensity distribution of the fundamental mode in a germanium-doped PCF filled with nonane at 800 nm. (b) Mode structure: the top white lines represent the first guided modes, while the density of modes for the cladding structure is shown as a grey gradient. (c) The numerical aperture (circles) and the fraction of the electric field intensity in the sample volume (triangles) for the fundamental guided mode of the PCF at 800 nm as a function of the refractive index of the core. (d) Fraction of the electric field intensity in the sample volume of the germanium-doped PCF as a function of the wavelength of the light. The core has a refractive index of $n = 1.46$.

intensity distribution for the fundamental guided mode is calculated at 800 nm (see Figure 16.9(a)), and a confinement of the light in the core of 99.6% is estimated.

For a more detailed characterization of the light confinement in the core, the energy bands of this structure are calculated by the PWEM [6]. The upper white lines in Figure 16.9(b) represent the first guided modes of the Ge-doped core. Furthermore, the density of modes for the surrounding cladding structure is plotted. As can be seen, the effective refractive index of the core (n_{core}^{eff}) is significantly higher than that of the cladding ($n_{cladding}^{eff}$). This can be attributed to the doping of the silica in the core with germanium and in addition to a higher fraction of air in the cladding region in comparison to the core (due to the reduced diameter of the inner holes).

Equation (16.3) gives a correlation between effective indices of core and cladding and the corresponding numerical aperture (NA):

$$NA = sin\left[\frac{\pi}{2} - \text{arc } sin\left(\frac{n_{cladding}^{eff}}{n_{core}^{eff}}\right)\right]. \tag{16.3}$$

Results for the numerical aperture are shown in Figure 16.9(c). With growing NA the tolerance of the light confinement in the core against disturbances (e.g. fibre bending, inhomogeneous fibre profile) increases, and thus the losses in the microstructured optical fibre decreases. This can be attributed to a weaker coupling of the core modes to the cladding modes. By raising the core refractive index from 1.45 to 1.46 through doping with germanium, the corresponding NA is increased by a factor of 2.5. On the other hand, the interaction of the light with the sample material decreases as the field is stronger confined in the higher dielectric.

This can be seen in Figure 16.9(c), where the fraction γ of the electric field intensity in the sample volume (i.e. in the holes) is shown as a function of the refractive index of the Ge-doped material. When n_{Ge} is raised to 1.46 by doping with germanium, γ and consequently the sensing efficiency is reduced by 40% compared to an undoped core. Thus, extremely high germanium concentrations are not reasonable for sensing applications, and a trade-off between robust light guiding and suitable sensitivity has to be found.

Furthermore, we have investigated the dependence of the sensing efficiency on the wavelength of the light. As can be seen in Figure 16.9(d), the evanescent field intensity in the sample volume increases with longer wavelengths.

This has to be taken into account for quantitative absorption measurements, since the shape of the acquired absorption spectra is distorted due to the wavelength-dependent detection efficiency of the fibre.

Experimental results of spectral attenuation behaviour of n-nonane filled PCF are described in Section 16.1.

16.3.4
Highly Germanium-Doped Index Guiding PCF

Highly germanium-doped PCFs (up to 36 mol%) with small cores and high numerical aperture are especially interesting for optical nonlinear applications: supercontinuum

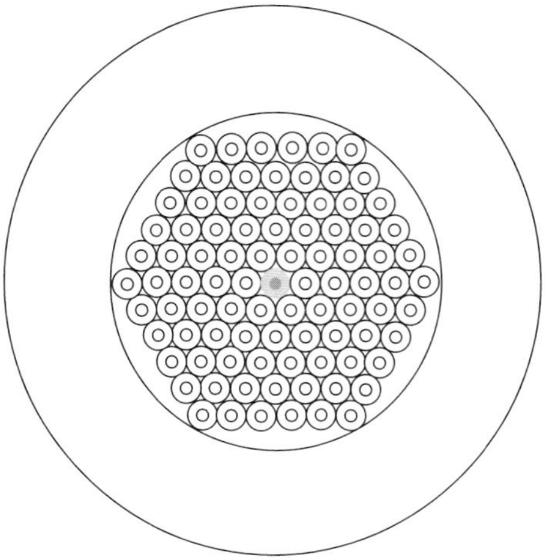

Figure 16.10 Design scheme of the highly germanium doped index guiding PCF: dark gray: highly germanium doped area, light gray: total cross section of the doped core, rings surrounding the core: silica capillary package, overcladded by a silica tube.

generation, Raman lasers and amplifiers and Bragg grating writing (for fibre Bragg sensors). High germanium concentration strongly increases the nonlinearity and the photosensitivity compared with undoped silica. A five-air-ring PCF with small, highly germanium-doped core was drawn for such applications. The fibre, shown in Figures 16.10 and 16.11, was prepared with different diameters: 200 μm, 125 μm, 85 μm to vary the core diameter in a wide range: 10.2 μm, 6.6 μm, 3.3 μm. The germanium-doped cross sections are 7.4 μm, 3.3 μm and 1.8 μm, respectively. The hole-pitch ratios in the cladding cover a range of about 0.87 to 0.9. The maximum GeO_2 concentration of the core was 36 mol%, corresponding to a refractive index difference to silica of about 5×10^{-2}. The high germanium concentration increases

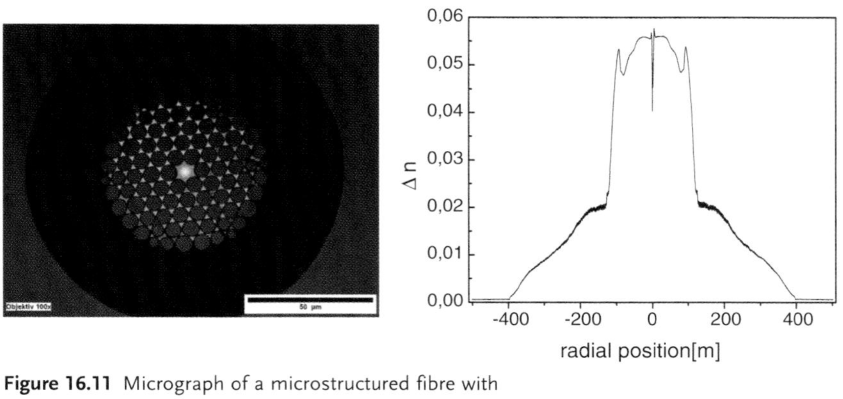

Figure 16.11 Micrograph of a microstructured fibre with highly germanium doped core (left) and core refractive index profile (right).

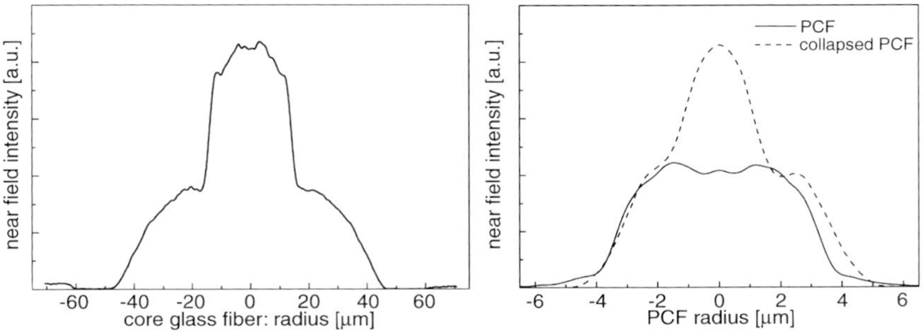

Figure 16.12 Near-field distributions of the compact core glass fibre (left) and the collapsed and holey PCF (right).

the thermal expansion coefficient by a factor up to 8 compared with the surrounding silica. The viscosity of the highly germanium-doped region during the stretching and drawing procedures is about one order of magnitude lower than the isothermal viscosity of the surrounding silica. This effect has to be considered for the drawing process.

Figure 16.12 shows that the slope of intensity distributions of highly air-fractioned PCF ($d/\Lambda = 0.88$) and completely collapsed fibre ($d/\Lambda = 0$) show differences in the central core region. However, the shape of the near-field pattern of the collapsed PCF is very similar to a solid multimodal core glass fibre with a diameter of the germanium-doped region of 97 μm and a central highly germanium-doped area of 27 μm in their near-field pattern. One main reason for the differences between the holey PCF and the collapsed PCF is obviously the effect of effective index depression of the air clad with a high air fraction.

The drawn highly germanium-doped PCF with a high air fraction ($d/\Lambda = 0.88$) shows a relatively high loss (minimum ca. 50 dB/m at 1 μm wavelength, see Figure 16.13). For comparison, the solid multimode fibre drawn separately from the

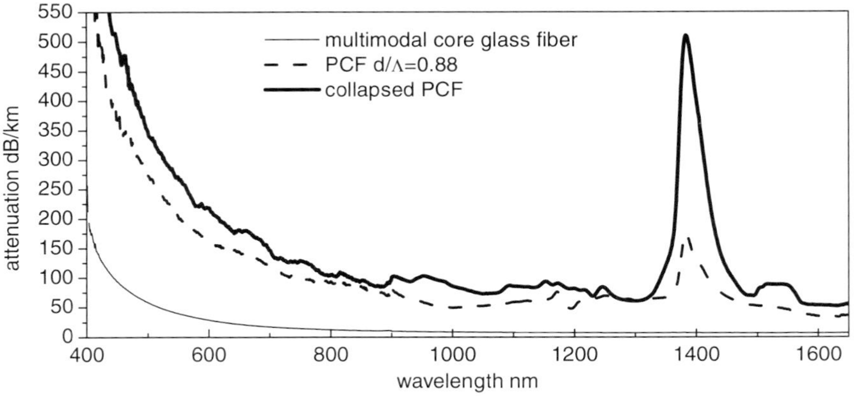

Figure 16.13 Attenuation spectra of the highly Ge-doped PCF and the compact fibres of the core glass and the collapsed PCF.

germanium-doped core rod shows a low spectral attenuation (6.8 dB/km at 1.55 μm). The highly air-fractioned PCF and the collapsed PCF are similar in their background loss behaviour. Therefore we conclude that the capillary package arrangement has a significant influence on the attenuation increase, e.g. by generation of scattering defects.

Both PCF based fibres show an increased hydroxide contamination, compared with the core glass fibre. The collapsed PCF corresponds to an OH concentration of 11 wt.ppm, whereas the air-fractioned PCF corresponds to 2.7 wt.ppm. Sources for OH contamination are silica surface-located hydroxide groups and humidity in capillary holes and interstitial volumes. The approximated diffusion lengths of hydroxides during the drawing process are 3 μm [7], i.e. the OH distribution over the holey cladding bridges of 1 μm or smaller is equalized. The local OH contamination distribution of the collapsed PCF is highly concentrated, and the PCF shows a more expanded distribution of solid OH-containing silica material over the cladding cross section. In consequence, the evanescent OH absorption interaction of the collapsed PCF is larger by about a factor of 4 than that of the holey PCF.

16.4 Photonic Band Gap Fibres

In recent years, solid-core photonic band gap fibres (SC-PBG) have attracted great interest. In comparison to the hollow-core photonic band gap fibre, these fibres should be easier to fabricate, implement additional functionality (e.g. laser gain, Bragg reflection) and integrate into all fibre systems by splicing [8]. Here we consider the specific possibilities to design the spectral and dispersion characteristics of such SC-PBG fibres.

The cladding of a SC-PBG fibre comprises a periodic array of uniformly high-index elements in a low-index background material (e.g. silica). A low index core is formed by substitution of one or more of the high-index elements in the centre of the array by the background material (i.e. defect of the photonic crystal structure). It was theoretically and experimentally proved that the optical properties of such structures are mainly due to the properties of the high-index elements [9,10]. Especially the positions of the band gaps in the spectrum are related to the cut-off wavelengths of each single element. It can be concluded therefore that the spectral and dispersion properties can be designed by a proper choice of the refractive index profile of the rods used as high index elements. So far, most investigations of SC-PBG fibres refer to step index and parabolic index profiles of the single elements [8,11]. It is expected that a SC-PBG fibre consisting of high-index elements with the index profile shown in Figure 16.11 (right) will both significantly modify the dispersion properties of these structure elements and change their mode structure und consequently the band structure of the photonic crystal fibre.

To understand the influence of the high-refractive-index packaging elements on the band gap spectrum, we will discuss at first the cut-off wavelengths of a single high-index element with the special index profile as shown in Figure 16.11 (right) and

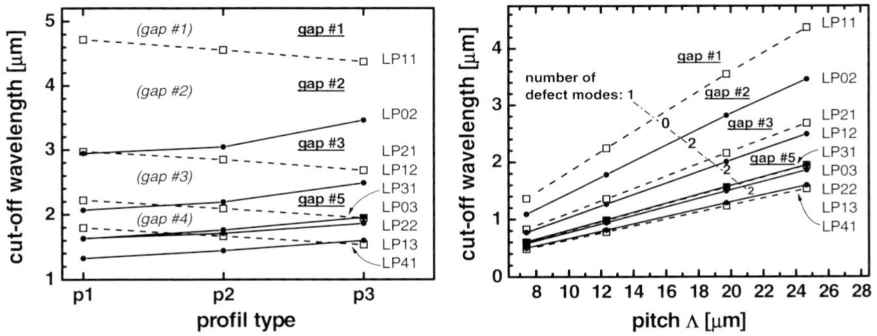

Figure 16.14 LP-mode cut-off of high-index elements: comparison of the cut-off wavelengths of two conventional elements (p1: step index with $r = 4.5\,\mu m$, p2: parabolic index profil with $r = 6.5\,\mu m$) and the special high-index element (p3: $r = 13\,\mu m$), (left); spectral position of the gaps (cut-off-free regions) as a function of the pitch, which is proportional to the radius r of the high-index elements (right).

compare it with the results for a more conventional step-index and parabolic profile. For the sake of comparison, the effective index difference of the different profiles was chosen to obtain cut-off wavelengths in the same wavelength region (Figure 16.14 left). The parabolic profile p2 has the benefit of wider gaps in higher gap numbers because of a (nearly) degenerated mode index for defined groups of modes. From the cut-off wavelengths of the step index profile p1 we obtain approximately the same gap spectrum. The cut-off spectrum of the new profile p3 differs qualitatively. The high-index value in the centre of the index profile increases the effective index of the modes with strong light intensity in the central region and decreases the effective index of the other modes.

These considerations were confirmed by computer simulations (software MIT-Photonic-Bands). They allow a deeper insight into the mode fields within the gaps and at the boundary of the gaps. Figure 16.15(a), (b) show the results concerning the band gaps of the crystal structure and the defect modes within gap #3 when

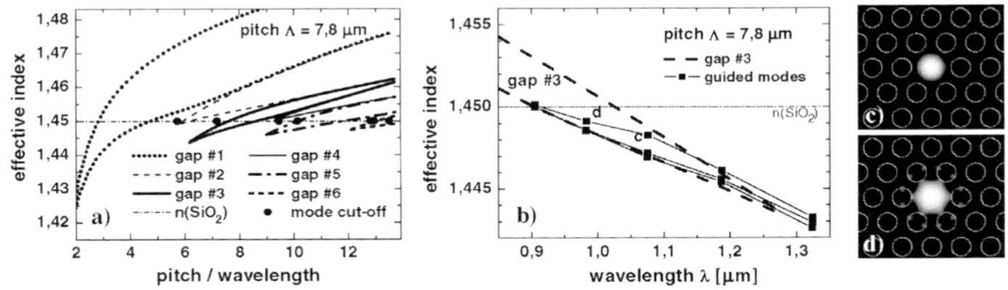

Figure 16.15 Simulation results for SC-PBG structure using special profile type p3: (a) gap map of the crystal fibre structure; (b) defect modes within the gap #3; (c) and (d) field distributions of the fundamental defect mode within gap #3 at points indicated in (b) as c and d, respectively.

high-index elements with profile type p3 are used. Two guided modes, the LP_{01}- and LP_{11}-defect modes, are found in the gap. The intensity distribution of the fundamental mode is shown in Figure 16.15(c), (d) for two locations c and d in the gap map Figure 16.15(b). The field distributions within the high-index elements correlate with the LP_{02} or LP_{21} modes, in agreement with the cut-off considerations. The computer simulation shows that mode indices smaller than 1.45 close the gap #4, as expected from Figure 16.14 (left). It can be noted, that this is also true for gap #2, which forms a wide gap region above 1.45. This result supplements the cut-off considerations, which states necessary but not sufficient conditions of gap formation.

To fabricate fibre samples, the spectrum of the gaps must be adopted to the spectral region used for measurement and applications. Figure 16.14 (right) gives an overview of the expected position of the gaps in the spectrum as a function of pitch, which can be adjusted in the fibre drawing process.

Based on the special index-profiled rods of type p3, we prepared a SC-PBG fibre with a cladding design of five highly germanium-doped rod rings and a single pure silica defect core (Figure 16.16(a)). The fibre was drawn to different diameters in order to shift the estimated band gap edges. For the investigation of the spectral distribution of the gaps, the pitch was varied between 4.5 μm and 11 μm.

We investigated experimentally the band gap properties of the fabricated fibres. We used a sample with a length of 11 cm and launched a spot of light with a diameter of approx. 4.5 μm in the defect at the input facet. A halogen lamp and a monochromator were used as light source. Figure 16.16(b) shows the transmission spectrum of the excited defect mode. The output field distributions at the wavelengths of the two power peaks are shown in Figure 16.16(c), (d). By analyzing the characteristics of the field distributions within the high-index elements on both sides of the transmission windows we were able to identify the number of gaps as indicated in Figure 16.16. Even though the fibre structure looks uniform, the loss of the defect mode is relatively high, and the gap spectrum is disturbed. From an experimental analysis of the coupling between the high-index elements we found out that especially the cut-off wavelengths of the single high index elements in the prepared fibre differ because of a

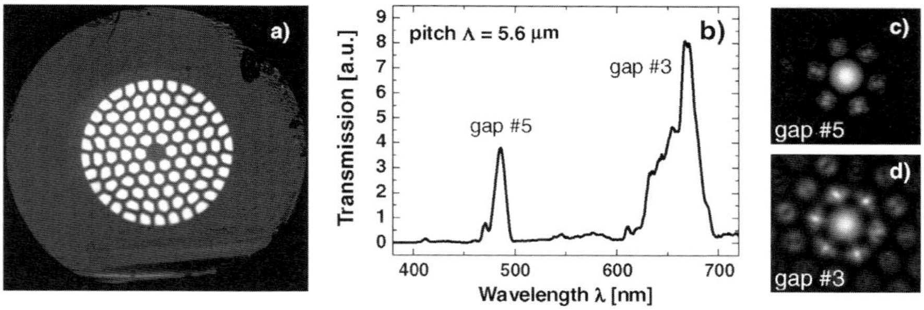

Figure 16.16 Experimental results of the SC-PBG-fibre with special index profile p3: (a) micrograph of a fibre cross section with pitch 5.6 μm; (b) spectral distribution of the gaps by defect transmission; (c) and (d) measured field distribution of the fundamental defect mode in the gaps #5 and #3, respectively.

stochastic elliptic deformation of the originally circular index profiles of the rods utilised. A further improvement of the technology should enable a better low-loss band gap guiding and would make possible the investigation of the dispersion properties.

16.5 Non-Silica PCF

The use of non-silica glasses for index guiding microstructured fibres is also of special interest for nonlinear applications such as supercontinuum generation, Raman amplifiers and lasers or optical switching. Various glass compositions have been discussed and tested regarding their applicability for photonic crystal fibres. Most investigations were concentrated on oxide glasses with highly polarizable glass components (e.g. lead silicate glasses, tellurite, bismuth borate, antimony oxide glasses) [12–14]. Also in the focus of research are non-oxide glasses like sulphides and selenides, since they offer chances for efficient stimulation of nonlinear processes. However, many aspects of microstructured fibres made of these mostly multicomponent materials are insufficiently investigated until now. Open questions refer to optical damage thresholds, pump power stability, photodarkening or optically induced chemical glass change.

A very essential requirement is the necessity to be able to prepare low loss fibres. Most non-silica-based microstructured fibres presented until now are characterized by a background loss level in a range of a few up to tens of dB/m [15,16].

Therefore the material optimization demands both a glass composition with low optical loss (i.e. high purity and crystallization stability) and the possibility to adopt the developed highly nonlinear glass in a microstructured fibre design with high air fraction. In a first step we tested different high refractive index glasses with respect to their applicability to the fibre drawing process. Different silicate-based glasses with high lanthanum oxide content were melted. Its concentration was increased up to 26 mol% to receive a high nonlinearity. Different additive components (alumina and boron oxide) were tested to improve the glass stability. The refractive index was about 1.66, n_2 can be approximated to be about half an order of magnitude higher than fused silica [17]. Highly alumina-codoped glasses with low boron content show a non-sufficient glass stability. However, by an increase in boron concentration above 20 mol%, the crystallization tendency can be broken. In consequence, the background loss can be decreased by orders of magnitudes at least in unstructured fibres. A minimum attenuation was achieved with 1.2 dB/m at a wavelength of 1.2 μm (see Figure 16.17), which represents a well applicable value and belongs to the best values reported until now.

The preparation of highly non-silica PCFs causes multiple technological challenges. Two approaches for preform manufacturing were tested by different groups: extrusion [13] and stack and draw methods [18]. Drawbacks of the extrusion method are impurity effects from the tools and extrusion tool-caused limits in fabrication designs of the PCFs. The stack and draw method is not limited by these disadvantages, but requires the manufacturing technologies for rods and tubes (capillaries) of

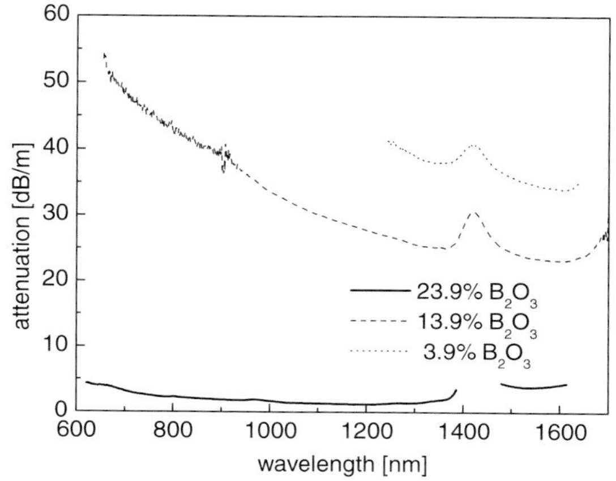

Figure 16.17 Loss spectra of unstructured fibres drawn from high lanthanum oxide containing silica glasses [(0.836 − x) SiO_2 − 0.164 La_2O_3 − xB_2O_3] with different boron contents.

compatible glasses concerning thermo-chemical properties. The commercial availability of suitable tubes and rods is limited. Optical glass rods are typically produced in non-long-distance fibre-optical quality. Besides advantages in preparation (lower processing temperature compared with silica, better sintering caused by low surface tension of multicomponent glasses) there are two crucial problems for low-loss fibre manufacturing: the insufficient purity of the raw materials used in large scale, and impurity effects caused by tool-based contaminations of the extruded rods.

A first lead silicate-based five-air-ring PCF, prepared by drawing of a pre-sintered preform, is shown in Figure 16.18, right. As expected it has a relative high background loss level in the 10 dB/m range. Absorption bands of transition metal and hydroxide impurities with dB/m intensities can be identified (see Figure 16.18,

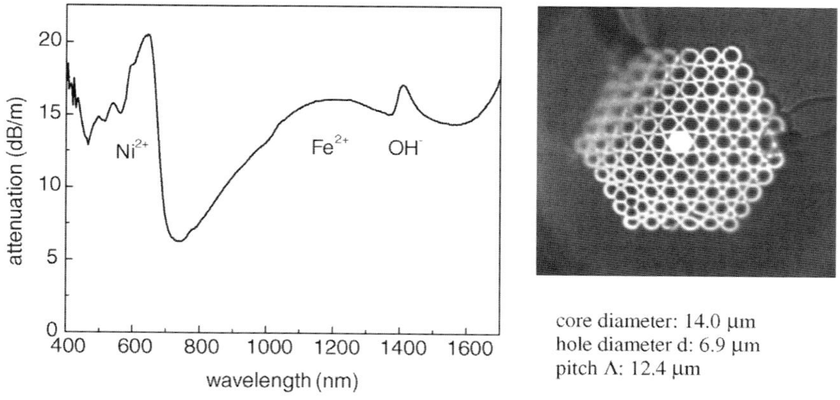

Figure 16.18 Loss spectrum and micrograph/geometrical parameters of the acrylate-coated microstructured lead silicate fibre.

left). Long-distance applications require a background loss reduction by at least two orders of magnitude. This can be expected from the use of highly purified glass material components and by an improved technology for rod and tube preparation.

16.6 Selected Linear and Nonlinear Applications

The specific properties of PCFs have opened up many new fields of applications. We will now briefly discuss as examples two of such applications making use of either the specific linear or the specific nonlinear propagation properties in such fibres. The first application example uses the holey structure of PCFs for an evanescent wave spectral sensing system. Similar concepts also offer the chance to build tunable fibre modules, e.g. by using liquid crystals for filling microstructured fibres [19]. The second example is related to supercontinuum light emission in structured optical fibres.

16.6.1 Spectral Sensing

The holey structure of PCFs offers the opportunity to introduce specific gases or liquids near the fibre core structure. In the extreme case of band gap fibres even the core itself may be filled. For index guiding fibres, the overlapping field in the holey structures can help to detect the material properties in the filled holey regions. Due to a close distance to the guiding fibre core, the effect of modifying the spectral transmission property can be increased in comparison to conventional evanescent wave fibre sensors. We have tested different PCFs concerning their sensitivity to material absorption for liquid hydrocarbons. As a suitable analyte we specifically used n-nonane, which may act as a model substance for petrol or gasoline fractions.

The analyte was filled into the capillaries of the microstructured fibres for fibre lengths of about 1.5 m by a pressurizing method. The pressure difference between the end faces of the fibres was 1 bar, and a filling time of seven hours was estimated from Hagen-Poiseuille's law considering a capillary diameter of about 3 μm. White light from a halogen lamp was launched in both, air and nonane filled PCFs. The cleaved end faces were butt-coupled with a germanium doped fibre of $NA = 0.17$ and core diameter 9 μm. The bending radius of the PCF was about 0.3 m. Minor bending variations of about ±10 cm have practically no influence on the attenuation spectrum. The transmitted light is measured with an Optical Spectrum Analyzer. The fluid n-nonane with $n = 1.405$ enables a stable index guiding in the filled PCFs.

The analysis of absorption variations in the 1120 nm to 1270 nm wavelength region allows the systematic sensing of hydrocarbons. The CH_3 second overtone band absorbance at 1194 nm and the CH_2 second overtone band absorbance at 1210 nm correspond to the concentrations of the hydrocarbon groups [20]. An information about the mean chain length of normal linear hydrocarbons (e.g. in lube base oils) is available by calibration.

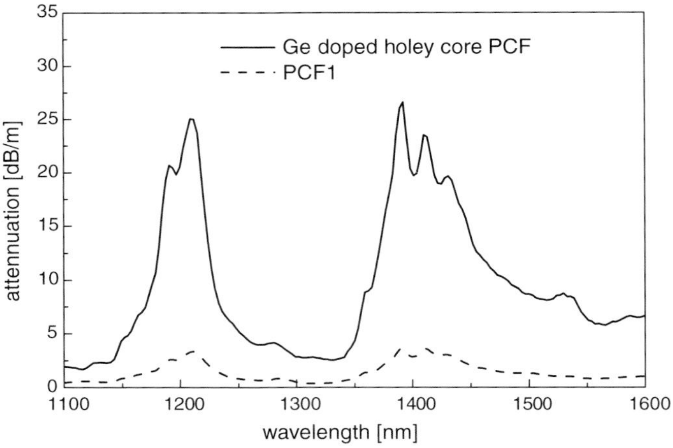

Figure 16.19 Spectra of additional loss of the n-nonane filled Ge-doped PCF (shown in Figure 16.8) compared with the filled silica PCF 1 (see Figure 16.2).

In Figure 16.19 the attenuation spectra are shown for a pure silica PCF in comparison with a PCF using germanium-doped structures. The spectra exhibit the additional attenuations related to the losses of the air filled PCFs. For a comparison of loss behaviour of the air filled silica PCF see Section 16.3.1, Figure 16.3. The air filled Ge-doped PCF has a similar spectral loss behaviour (42 dB/km at wavelength 1.2 μm). The doped PCF achieves a sensitivity which is higher by one order of magnitude because of stronger overlapping field distributions with the analyte volume. This demonstrates how the design flexibility in PCFs allows the improvement of the specific sensing functionality of optical fibres.

16.6.2
Supercontinuum Generation

A specific property of PCFs is the ability to generate broad spectra using combined effects of a high nonlinear coefficient and the possibility for tailoring the dispersion characteristics. One of the greatest challenges using these properties is to generate light in the visible wavelength region. This spectral range is very useful in particular for biomedical applications such as fluorescence spectroscopy, optical microscopy, and optical coherence tomography. In particular, we address here the challenge to extend the output wavelength of a Ti:Sapphire laser, a common tool for researchers because of its large tuneability (700–1050 nm) and the short temporal pulses, to the visible wavelength region. A recent method is to use high-order mode propagation in a microstructured fibre exploiting the dispersion curve shift that occurs as the mode order increases [21,22]. Here we consider a new approach where the zero dispersion wavelength is applied for higher-order modes instead for the fundamental mode. The fibre for this application was drawn from the same preform as the fibres in Figure 16.2, but under modified conditions. The diameters of the core and the holes of the PCF

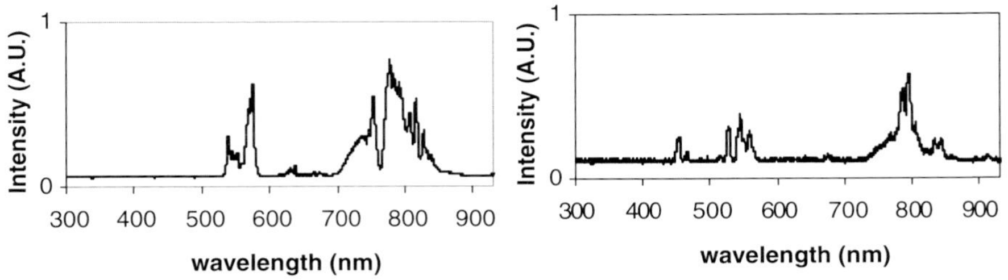

Figure 16.20 Spectral power distribution of generated supercontinua: $L = 35$ cm, $P = 266$ mW (left) and $L = 15$ m, $P = 285$ mW (right). L and P are the fibre length and the incident power, respectively. The total output power is about 180 mW (left) and 20 mW (right), respectively.

were 3.2 μm and 2.4 μm with a bridge width of 0.5 μm ($d/\Lambda = 0.83$). The zero dispersion wavelength for this fibre was estimated to be 960 nm at the fundamental mode and should be lower for higher-order modes. For short wavelength light generation the output of a standard 80 MHz repetition rate Ti:Sapphire delivering sub 100-fs pulses was focussed into the PCF with a maximum incident power of 285 mW.

We tested several fibre lengths from 35 cm to 30 m and several pump configurations, where the light was focussed into different order modes. Spectra were recorded with two optical spectrometers covering the 200 nm to 2400 nm spectral range. We observed that the generated visible spectrum was sensitive to the launching condition and to the fibre length. The best results were obtained at a pump power of 285 mW and with a 15 m long fibre. Figure 16.20 shows examples of the obtained output spectra. It was possible to tune the wavelengths from 600 nm down to 420 nm by varying the input power. We measured 670 μW at 570 nm, 400 μW at 480 nm and a few tens of μW at 420 nm by filtering the output signal with a 12 nm FWHM gaussian bandpass filter.

Figure 16.21 presents the observed field pattern for different visible wavelengths. The output was again tuned covering the entire visible region from 600 nm down to 410 nm with an output power sufficient to perform fluorescence lifetime spectroscopy measurements [22]. This demonstrates that high-order mode propagation in a suitably designed PCF may extend the output spectrum of a Ti:Sapphire laser to the visible region and may replace more standard sliced continuum spectrum techniques.

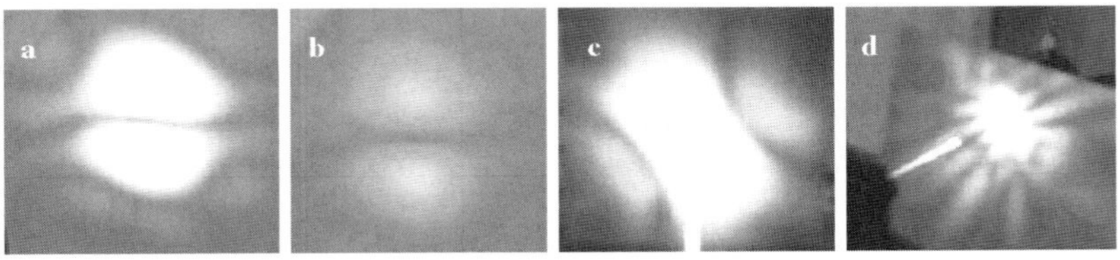

Figure 16.21 Output field pattern for different launching conditions: (a) $L = 35$ cm, $P = 205$ mW; (b) $L = 35$ cm, $P = 266$ mW; (c) $L = 80$ cm, $P = 296$ mW; (d) $L = 15$ m, $P = 285$ mW. Cases (b) and (d) refer to Figure 16.20.

16.7 Conclusions

The combined use of the design parameters of a periodically arranged holey structure and of dopant material modifications offers a high flexibility for tailoring the optical properties in photonic crystal fibres and to adapt their properties to different applications. Basic requirements for suitable applications are a sufficiently low spectral loss and stable mode guiding in the prepared fibres. We have shown that such properties are possible not only for pure silica PCF but also for doped and non-silica PCFs. The achieved minimum loss is in the range of tens of dB/km and allows applications with hundreds of meters of interaction length. We prepared different PCFs with specific mode field patterns (PCFs with extremely small and very large cores). Both types of fibres have been implemented for single mode propagation with proper holey cladding design. Further modifications of the linear (and nonlinear) material index allow additional variations in the mode field design and enable the application of optical nonlinear properties. The practical use of non-silica PCFs is, up to now, limited by their attenuation effects. The usable lengths in the range of a few meters are smaller compared to silica-based PCFs. All-solid band-gap fibres, prepared with extremely high germanium-doped claddings, represent a recently developed new PCF type. Simulations show that, besides the typical parameters d/Λ and Λ/λ, also the mean refractive index difference in the stacking elements plays an important role for spectral band gap structuring. As examples of the wide range of possible applications of PCFs, a spectral method for fibre sensing of hydrocarbons and the generation of supercontinuum light down to the blue wavelength region has been presented.

Acknowledgements

Funding of the work on photonic crystal fibres, on microstructured fibres and on active fibres by the German Science Foundation within the priority program "photonic crystals" (SPP 1113) is gratefully acknowledged.

References

1 Russell, P.St.J. (2003) *Science*, **299**, 358–362.

2 Lægsgaard, J. and Bjarklev, A. (2006) *J. Am. Ceram. Soc.*, **89** (1), 2–12.

3 Kirchhof, J., Kobelke, J., Schuster, K., Bartelt, H., Iliew, R., Etrich, C. and Lederer, F. (2004) in *Photonic Crystal Fibres, Photonic Crystals: Advances in Design, Fabrication, and Characterization*, (eds K. Busch, St. Lölkes, R.B. Wehrspohn and H. Föll, Wiley-VCH, pp. 266–288.

4 Wong, W.S., Peng, X., McLaughlin, J.M. and Dong, L. (2005) *Opt. Lett.*, **30** (21), 2855–2857.

5 Kobelke, J., Bartelt, H., Kirchhof, J., Unger, S., Schuster, K., Mörl, K., Aichele, C. and Oh, K. (2005) DGaO-Proceedings ISSN 1614–8436, paper Nr. B40.

6 Johnson, S.G. and Joannopoulos, J.D. (2001) *Opt. Express*, **8**, 173–190.

7 Kirchhof, J., Kleinert, P., Radloff, W. and Below, E. (1987) *phys. stat. sol. (a)*, **101** (2), 391–401.

8 Bouwmans, G., Bigot, L., Quiquempois, Y., Lopez, F., Provino, L. and Douay, M. (2005) *Opt. Express*, **13**, 8452–8459.

9 Litchinitser, N., Dunn, S., Usner, B., Eggleton, B., White, T., McPhedran, R. and de Sterke, C. (2003) *Opt. Express*, **11**, 1243–1251.

10 Argyros, A., Birks, T., Leon-Saval, S., Cordeiro, C.M.B. and Russell, P. St. J. (2005) *Opt. Express*, **13**, 2503–2511.

11 Luan, F., George, A.K., Hedley, T.D., Pearce, G.J., Bird, D.M., Knight, J.C. and Russell, P.S.J. (2004) *Opt. Lett.*, **29**, 2369–2371.

12 Ebendorff-Heidepriem, H., Petropoulos, P., Asimakis, S., Finazzi, V., Moore, R., Frampton, K., Koizumi, F., Richardson, D. and Monro, T. (2004) *Opt. Express*, **12**, 5082–5087.

13 Kiang, K.M., Frampton, K., Monro, T.M., Moore, R., Tucknott, J., Hewak, D.W., Richardson, D. and Rutt, H.N. (2002) *Electron. Lett.*, **38** (12), 546–547.

14 Feng, X., Mairaj, A.K., Hewak, D.W. and Monro, T.M. (2004) *Electron. Lett.*, **40** (3), 167–169.

15 Feng, X., Monro, T., Petropoulos, P., Finazzi, V. and Hewak, D.W. (2003) *Opt. Express*, **11**, 2225–2230.

16 Petropoulos, P., Ebendorff-Heidepriem, H., Finazzi, V., Moore, R., Frampton, K., Richardson, D. and Monro, T.M. (2003) *Opt. Express*, **11**, 3568–3573.

17 Fournier, J.T. and Snitzer, E. (1974) *IEEE J. Quantum Electron.*, **QE-10** (5), 473–475.

18 Sato, F., Sakamoto, A., Yamamoto, S., Kintaka, K. and Nishii, J. (2005) *J. Ceram. Soc. Jpn.*, **113** (5), 373–375.

19 Kitzerow, H.S., Lorenz, A. and Matthias, H., (2007) *phys. stat. sol.(a)*. **204** (11), 3754–3767.

20 Chung, H. and Ku, M.-S. (2003) *Appl. Spectrosc.*, **57** (5), 545–550.

21 Efimov, A., Taylor, A., Omenetto, F., Knight, J., Wadsworth, W. and Russell, P. (2003) *Opt. Express*, **11**, 910–918.

22 D'Andrea, C., Ferrari, R., Bassi, A., Taccheo, S., Cubeddu, R., Schuster, K. and Kobelke, J. CLEO Europe 2007, Munich, Germany 2007, paper CD-1613.

17
Finite Element Simulation of Radiation Losses in Photonic Crystal Fibers

Jan Pomplun, Lin Zschiedrich, Roland Klose, Frank Schmidt, and Sven Burger

17.1
Introduction

Photonic crystals consist of a periodic arrangement of materials with different refractive indices, usually a dielectric material and air with a period on the scale of the wavelength [1]. This structure allows to control light propagation on veryshort distances. Due to their small size photonic crystal structures can be used to fabricate integrated optical components like resonators and waveguides. Photonic crystal fibers [2,3] are an important application example which utilizes the light guiding principles of photonic crystals. They are used e.g. in nonlinear applications [4] for high power transmission of light in optical astronomy.

Radiation losses from PhC fibers are one of the loss mechanism which potentially restrict their technological application [5]. They have therefore to be controlled by proper component design. Due to the complex geometry and large number of glass air interfaces the accurate computation of light propagation in photonic crystal structures is a challenging task [6]. To achieve a realistic model of a real device its finite size and the exterior domain have to be taken into account. The coupling of a photonic crystal fiber to the exterior will always lead to unwanted losses which with a proper fiber design are very small. For numerical computation transparent boundary conditions are used to take into account the surrounding material. Even for simple geometries the accurate computation of radiation losses can become difficult and many numerical methods can fail especially for systems where these losses are very small [7].

In this paper we review an approach to this problem using adaptive finite element algorithms and we apply this approach to accurate computation of radiation losses in hexgonal and kagome-structured PhC fibers. Computational results agree well with experimental transmission spectra measured in [5]. In the context of topics related to the DFG priority programme photonic crystals, we have applied adaptive finite element algorithms to a variety of nano-optical problems [6,8–15].

Nanophotonic Materials: Photonic Crystals, Plasmonics, and Metamaterials. Edited by R.B. Wehrspohn, H.-S. Kitzerow, and K. Busch

ISBN: 978-3-527-40858-0

17.2 Formulation of Propagation Mode Problem

For the analysis of photonic crystal fibers we start with the derivation of the mathematical formulation of the propagation mode problem of the electric and magnetic field. The geometry of a waveguide system like a photonic crystal fiber is invariant in one spatial dimension along the fiber. Here we choose the z-direction. Then a propagating mode is a solution to the time harmonic Maxwell's equations with frequency ω, which exhibits a harmonic dependency in z-direction:

$$\boldsymbol{E} = \boldsymbol{E}_{\text{pm}}(x, y) \exp(ik_z z), \quad \boldsymbol{H} = \boldsymbol{H}_{\text{pm}}(x, y) \exp(ik_z z). \tag{17.1}$$

$\boldsymbol{E}_{\text{pm}}(x, y)$ and $\boldsymbol{H}_{\text{pm}}(x, y)$ are the electric and magnetic propagation modes and the parameter k_z is called propagation constant. If the permittivity ε and permeability μ can be written as:

$$\varepsilon = \begin{bmatrix} \varepsilon_{\perp\perp} & 0 \\ 0 & \varepsilon_{zz} \end{bmatrix} \quad \text{and} \quad \mu = \begin{bmatrix} \mu_{\perp\perp} & 0 \\ 0 & \mu_{zz} \end{bmatrix}, \tag{17.2}$$

we can split the propagation mode into a transversal and a longitudinal component:

$$\boldsymbol{E}_{\text{pm}}(x, y) = \begin{bmatrix} \boldsymbol{E}_{\perp}(x, y) \\ E_z(x, y) \end{bmatrix}. \tag{17.3}$$

Inserting (17.1) with (17.2) and (17.3) into Maxwell's equations yields:

$$\begin{bmatrix} P\nabla_{\perp}\mu_{zz}^{-1}\nabla_{\perp}\cdot P - k_z^2 P\mu_{\perp\perp}^{-1}P & -ik_z P\mu_{\perp\perp}^{-1}P\nabla_{\perp} \\ -ik_z\nabla_{\perp}\cdot P\mu_{\perp\perp}^{-1}P & \nabla_{\perp}\cdot P\mu_{\perp\perp}^{-1}P\nabla_{\perp} \end{bmatrix} \begin{bmatrix} \boldsymbol{E}_{\perp} \\ E_z \end{bmatrix} = \begin{bmatrix} \omega^2\varepsilon_{\perp\perp} & 0 \\ 0 & \omega^2\varepsilon_{zz} \end{bmatrix} \begin{bmatrix} \boldsymbol{E}_{\perp} \\ E_z \end{bmatrix}, \tag{17.4}$$

with

$$P = \begin{bmatrix} 0 & -1 \\ 1 & 0 \end{bmatrix}, \quad \nabla_{\perp} = \begin{bmatrix} \partial_x \\ \partial_y \end{bmatrix}. \tag{17.5}$$

Now we define $\tilde{E}_z = k_z E_z$ and get:

$$A \begin{bmatrix} \boldsymbol{E}_{\perp} \\ \tilde{E}_z \end{bmatrix} = k_z^2 B \begin{bmatrix} \boldsymbol{E}_{\perp} \\ \tilde{E}_z \end{bmatrix} x \in R^2, \tag{17.6}$$

with

$$A = \begin{bmatrix} P\nabla_{\perp}\mu_{zz}^{-1}\nabla_{\perp}\cdot P - \omega^2\varepsilon_{\perp\perp} & -iP\mu_{\perp\perp}^{-1}P\nabla_{\perp} \\ 0 & \nabla_{\perp}\cdot P\mu_{\perp\perp}^{-1}P\nabla_{\perp} - \omega^2\varepsilon_{zz} \end{bmatrix}, \tag{17.7}$$

$$B = \begin{bmatrix} P\mu_{\perp\perp}^{-1}P & 0 \\ i\nabla_{\perp}\cdot P\mu_{\perp\perp}^{-1}P & 0 \end{bmatrix}. \tag{17.8}$$

Equation (17.6) is a generalized eigenvalue problem for the propagation constant k_z and propagation mode $\boldsymbol{E}_{\text{pm}}(x, y)$. We get a similar equation for the magnetic field

$\boldsymbol{H}_{\mathrm{pm}}(x, y)$ exchanging ε and μ. For our numerical analysis we define the effective refractive index n_{eff} which we will also refer to as eigenvalue:

$$n_{\mathrm{eff}} = \frac{k_z}{k_0} \quad \text{with} \quad k_0 = \frac{2\pi}{\lambda_0}, \tag{17.9}$$

where λ_0 is the vacuum wavelength of light.

17.3
Discretization of Maxwell's Equations with the Finite Element Method

For the numerical solution of the propagation mode problem Eq. (17.6) derived in the previous section we use the finite element method [16] which we sketch briefly in the following. We start with the curl curl equation for the electric field. Since we want to solve an eigenvalue problem we are looking for pairs $\boldsymbol{E}$ and k_z such that:

$$\nabla_{k_z} \times \frac{1}{\mu} \nabla_{k_z} \times \boldsymbol{E} - \frac{\omega^2 \varepsilon}{c^2} \boldsymbol{E} = 0 \quad \text{in} \quad \Omega, \tag{17.10}$$

$$\left(\frac{1}{\mu} \nabla_{k_z} \times \boldsymbol{E}\right) \times \boldsymbol{n} = \boldsymbol{F} \quad \text{given on } \Gamma \text{ (Neumann boundary condition)} \tag{17.11}$$

holds, with $\nabla_{k_z} = [\partial_x, \partial_y, ik_z]^T$. For application of the finite element method we have to derive a weak formulation of this equation. Therefore we multiply (17.10) with a vectorial test function $\boldsymbol{\Phi} \in V = H$ (curl) [16] and integrate over the domain Ω:

$$\int_\Omega \left\{ \overline{\boldsymbol{\Phi}} \cdot \left[\nabla_{k_z} \times \frac{1}{\mu} \nabla_{k_z} \times \boldsymbol{E} \right] - \frac{\omega^2 \varepsilon}{c^2} \overline{\boldsymbol{\Phi}} \cdot \boldsymbol{E} \right\} \mathrm{d}^3 r = 0, \quad \forall \boldsymbol{\Phi} \in V, \tag{17.12}$$

where bar denotes complex conjugation. After a partial integration we arrive at the weak formulation of Maxwell's equations:

Find $\boldsymbol{E} \in V = H$ (curl) such that

$$\int_\Omega \left\{ (\overline{\nabla_{k_z} \times \boldsymbol{\Phi}}) \cdot \left(\frac{1}{\mu} \nabla_{k_z} \times \boldsymbol{E} \right) - \frac{\omega^2 \varepsilon}{c^2} \overline{\boldsymbol{\Phi}} \cdot \boldsymbol{E} \right\} \mathrm{d}^3 r = \int_\Gamma \overline{\boldsymbol{\Phi}} \cdot \boldsymbol{F} \, \mathrm{d}^2 r, \quad \forall \boldsymbol{\Phi} \in V. \tag{17.13}$$

We define the following bilinear functionals:

$$a(\boldsymbol{w}, \boldsymbol{v}) = \int_\Omega (\overline{\nabla_{k_z} \times \boldsymbol{w}}) \cdot \left(\frac{1}{\mu} \nabla_{k_z} \times \boldsymbol{v} \right) - \frac{\omega^2 \varepsilon}{c^2} \overline{\boldsymbol{w}} \cdot \boldsymbol{v} \, \mathrm{d}^3 r, \tag{17.14}$$

$$f(\boldsymbol{w}) = \int_\Gamma \overline{\boldsymbol{w}} \cdot \boldsymbol{F} \, \mathrm{d}^2 r. \tag{17.15}$$

Now we discretize this equation by restricting the space V to a finite dimensional subspace $V_h \subset V$, $\dim V_h = N$. This subspace and therewith the approximate solution

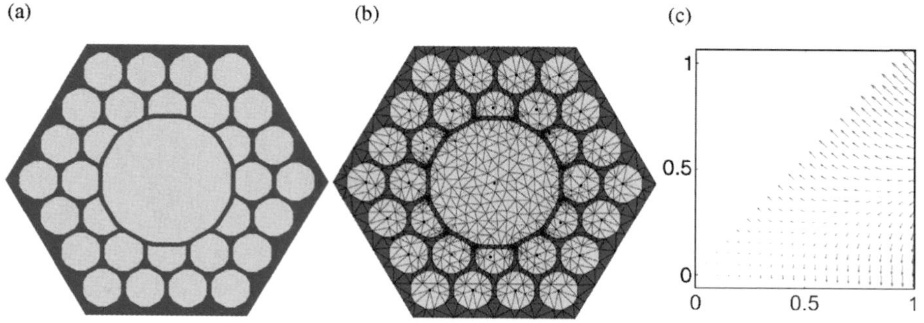

Figure 17.1 (a) Computational domain and (b) triangulation of hollow-core photonic crystal fiber; (c) example of vectorial ansatz function on a reference patch.

are constructed as follows. One starts with a computational domain Ω see Figure 17.1(a). This domain is subdivided into small patches, e.g. triangles in 2D and tetrahedrons in 3D, Figure 17.1(b). On these patches vectorial ansatz functions v_i are defined. Usually on each patch the ansatz funtions v_i form a basis of a polynomial function space of a certain degree p [16]. The approximate solution $\boldsymbol{E}_h$ for the electric field is a superposition of these ansatz functions of all patches:

$$\boldsymbol{E}_h = \sum_{i=1}^{N} a_i v_i. \tag{17.16}$$

Together with (17.16), (17.17) the discrete version of Maxwell's Eq. (17.15) reads:

$$\sum_{i=1}^{N} a_i a(v_i, v_j) = f(v_j), \quad \forall j = 1, \ldots, N, \tag{17.17}$$

which is a linear system of equations for the unknown coefficients a_i:

$$A \cdot \boldsymbol{a} = \boldsymbol{f},$$

with

$$A_{ij} = a(v_i, v_j), \quad f_j = f(v_j), \quad \boldsymbol{a} = \begin{pmatrix} a_1 \\ \cdots \\ a_N \end{pmatrix}. \tag{17.18}$$

The matrix entries $a(v_{i,} v_j)$ arise from computing integrals (17.14). In practice these integrals are evaluated on a reference (unit) patch. Such a reference patch together with a vectorial ansatz function is shown in Figure 17.1(c).

In the above sketch we assumed for simplicity that boundary conditions (17.11) are known for the electric field $\boldsymbol{E}$. However here we want to take the infinite exterior into account. Therefore the eigenvalue problem (17.6) has to be solved on an unbounded domain $\mathbb{R}^2$. This leads to the computation of leaky modes which enable us to estimate radiation losses. Since our computational domain still has to be of finite size, we apply so-called transparent boundary conditions to $\partial\Omega$. We realize these boundary conditions with the perfectly matched layer (PML) method [17]. Details about our numerical implementation are described in [18]. The propagation constant k_z (and the effective refractive index) becomes complex and the corresponding mode is damped according to $\exp(-\Im(k_z)\,z)$ while propagating along the fiber, see Eq. (17.1).

Applying the finite element method to propagating mode computation has several advantages [19,20]. The flexibility of triangulations allows the computation of virtually arbitrary structures without simplifications or approximations, as illustrated in Figures 17.1(b) and 17.7(b). By choosing appropriate ansatz functions $v_i(x, y)$ for the solution of Maxwell's equations, physical properties of the electric field like discontinuities or singularities can be modeled very accurately and do not give rise to numerical problems. Such discontinuities often appear at glass/air interfaces of photonic crystal fibers, see Figure 17.2. Adaptive mesh-refinement strategies lead to very accurate results and small computational times. Furthermore the FEM approach converges with a fixed convergence rate towards the exact solution of Maxwell-type problems for decreasing mesh width (i.e. increasing number N of sub-patches) of the triangulation. Therefore, it is easy to check if numerical results can be trusted [16].

Especially for complicated geometrical structures the finite element method is better suited for mode computation than other methods. In contrast to the plane wave expansion (PWE) method, whose ansatz functions are spread over the whole computational domain (plane waves) the FEM method uses localized ansatz functions, see Figure 17.1(c). In order to expand a solution with discontinuities as shown in Figure 17.2(a) large number of plane waves would be necessary using the PWE method. This leads to slow convergence and large computational times [12].

For accurate and fast computation of leaky eigenmodes we have implemented several features into the FEM package JCMsuite. As we will see later e.g. high-order edge elements, a-posteriori error control and adaptive and goal-oriented mesh refinement increase the convergence of numerical results dramatically.

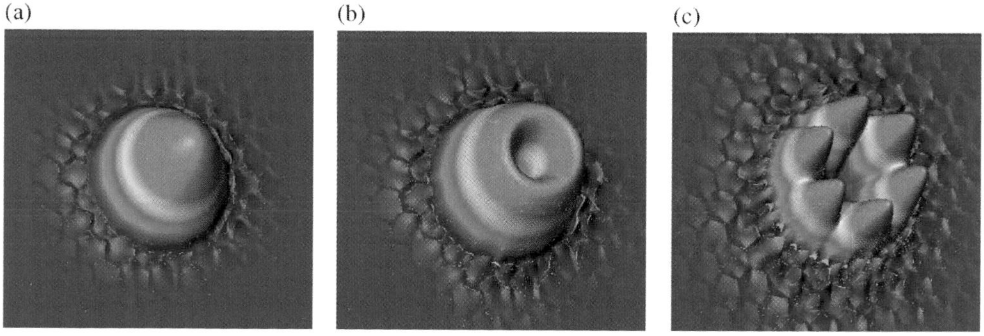

Figure 17.2 First, second and fourth fundamental core modes of HCPCF illustrated in Figure 17.3(a) – Parameters: $\Lambda = 1550$ nm, $r = 300$ nm, $w = 50$ nm, $t = 170$ nm, $\lambda = 589$ nm, see Figure 17.7(a) for definition of parameters.

17.4
Computation of Leaky Modes in Hollow Core Photonic Crystal Fibers

The only simplification we make for the computation of leaky modes is to extend the fiber cladding to infinity and thereby neglect its finite size. This is justified if the cladding of the fiber is much larger than the microstructured core, and no light entering the cladding is reflected back into the core, which is usually the case. We investigate two different types of hollow-core photonic crystal fibers (HCPCF). The first type has a hollow core which corresponds to 19 omitted hexagonal cladding cells Figure 17.3(a) the second type is kagome-structured [5], Figure 17.3(b). The cross sections of the HCPCFs depicted in Figure 17.3 have a C_{6V} invariance. Therefore it is possible to take only a fraction of the layout as computational domain. At the artificial inner boundaries appropriate boundary conditions have to be stated, setting the tangential magnetic and electric field to 0 respectively. Table 17.1 shows the first and second eigenvalue for the full half and quarter fiber used as computational domain. The geometrical parameters are given in Figure 17.2. The computed values are identical up to the chosen accuracy. The number of unknowns and computational time is of course much smaller for half and quarter fiber cross section.

The imaginary and real parts of the eigenvalues given in Table 17.1 differ by up to 11 orders of magnitude. The complex eigenvalue leads to a dampening of the mode according to

$$|\boldsymbol{E}|^2 \propto \mathrm{e}^{-2\Im(k_z)z} \tag{17.19}$$

see Eq. (17.1). Computing the fiber without transparent boundary conditions but setting the tangential component of the electric field to 0 at the outer boundary we get $n_{\mathrm{eff}}^1 = 0.99826580015$ for the first eigenvalue, i.e. a real eigenvalue. Comparing to Table 17.1 we see that taking into account the finite size of the fiber and coupling to the exterior does not change the real part. However when analyzing radiation losses of a fiber, the imaginary part of the effective refractive index n_{eff} is the quantity of interest. Computation of leaky modes with only small losses is therefore a multi-scale

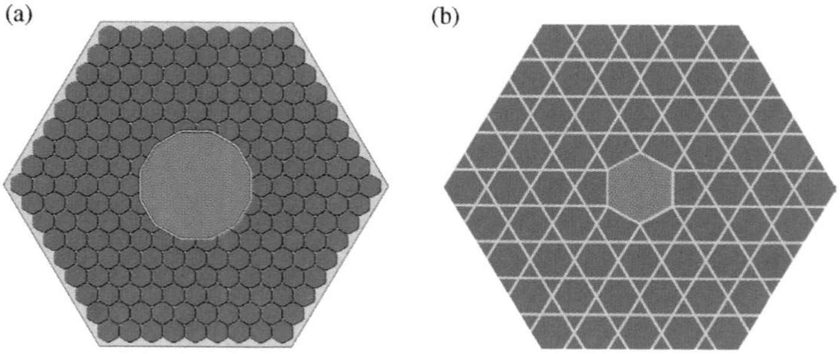

Figure 17.3 (a) Geometry of 19-cell HCPCF and (b) kagome-structured fiber used for mode computation.

Table 17.1 First, second and third eigenvalue computed with full, half and quarter fiber as computational domain.

	Unknowns	1st eigenvalue	2nd eigenvalue
full fiber	3070641	$0.99826580337 + 8.9297 \times 10^{-12}i$	$0.992141777 + 1.3312 \times 10^{-10}i$
half fiber	1567377	$0.99826580252 + 8.9311 \times 10^{-12}i$	$0.992141758 + 1.3314 \times 10^{-10}i$
quarter fiber	781936	$0.99826580254 + 8.9311 \times 10^{-12}i$	$0.992141754 + 1.3315 \times 10^{-10}i$

problem which is numerically very difficult to handle. In the next section we will introduce a goal-oriented error estimator which controls the grid refinement in order to obtain an accurate imaginary part as fast as possible.

17.5 Goal Oriented Error Estimator

The finite element method enables us to refine patches of the unstructured grid only locally. We use error estimators to control the refinement process of the grid. Usually such an error estimator is defined by minimizing a target functional $j(\boldsymbol{E})$, i.e. it is *goal-oriented*. The target functional depends on the solution of the electric field $\boldsymbol{E}$. Since we are interested in radiation losses this will be the imaginary part of the propagation constant [15]. From Maxwell's equations applied to a waveguide structure one can derive the following expression for the imaginary part of the propagation constant:

$$\Im(k_z) = \frac{P_{\partial\Omega}(\boldsymbol{E})}{2P_{\Omega}(\boldsymbol{E})}, \tag{17.20}$$

where

$$P_{\Omega}(\boldsymbol{E}) = \frac{1}{2\Re(\omega)}\Im\left(\int_{\Omega}\left[\overline{\boldsymbol{E}} \times \frac{1}{\mu}\nabla_{k_z}\boldsymbol{E}\right]\cdot\hat{n}_z\right), \tag{17.21}$$

$$P_{\partial\Omega}(\boldsymbol{E}) = \frac{1}{2\Re(\omega)}\Im\left(\int_{\partial\Omega}\left[\overline{\boldsymbol{E}} \times \frac{1}{\mu}\nabla_{k_z}\boldsymbol{E}\right]\cdot\hat{n}_{xy}\right) \tag{17.22}$$

is the power flux of the electric field through the cross section Ω of the computational domain and the in plane power flux across the boundary $\partial\Omega$ of the computational domain respectively. Equation (17.20) also shows us that the imaginary part of k_z reflects radiation leakage from the photonic crystal fiber to the exterior. The right hand side of (17.20) is the target nonlinear functional $j(\boldsymbol{E})$ which is used for the control of the error estimator. The finite element mesh should now be adapted such that

$j(E_h) - j(E)$ is minimized, where E_h is the finite element solution and E the exact solution. This task can be embedded into the framework of optimal control theory [21]. We define the trivial optimization problem:

$$j(\boldsymbol{E})-j(\boldsymbol{E}_h) = \min_{\Psi \in H(\mathrm{curl})} \{j(\Psi)-j(\boldsymbol{E}_h) : a(\Phi, \Psi) = f(\Phi)\, \forall \Phi \in H(\mathrm{curl})\}. \tag{17.23}$$

This formulation is trivial because the restriction simply states that Ψ is a solution of Maxwell's equation. The minima of (17.23) correspond to stationary points of the Lagrangian density

$$\mathcal{L}(\boldsymbol{E}, \boldsymbol{E}^*) = j(\boldsymbol{E})-j(\boldsymbol{E}_h)+f(\boldsymbol{E}^*)-a(\boldsymbol{E}^*, \boldsymbol{E}),$$

where $\boldsymbol{E}^*$ denotes the 'dual' variable (Lagrangian multiplier). Hence, we seek the solution $(\boldsymbol{E}, \boldsymbol{E}^*)$ to the Euler–Lagrange system

$$a(\Phi, \boldsymbol{E}) = f(\Phi) \quad \forall \Phi \in H(\mathrm{curl}), \tag{17.24}$$

$$a(\boldsymbol{E}^*, \Phi) = j'(\boldsymbol{E}; \Phi) \quad \forall \Phi \in H(\mathrm{curl}). \tag{17.25}$$

Equation (17.24) is the variational form of the original Maxwell's equations. In the dual Eq. (17.25) the target functional $j(\boldsymbol{E}) = k_z$ appears in form of its linearization $j'(\boldsymbol{E})$. A finite element discretization of the Euler–Lagrange system yields a supplemental discrete problem

$$a(\boldsymbol{E}_h^*, \Phi_h) = j'(\boldsymbol{E}_h; \Phi_h) \quad \forall \Phi_h \in V_h.$$

To quantify the error of the finite element solution $\boldsymbol{E}_h$ we introduce the primal and dual residuals:

$$\rho(\boldsymbol{E}_h; \cdot) = f(\cdot) - a(\cdot, \boldsymbol{E}_h), \tag{17.26}$$

$$\rho(\boldsymbol{E}_h^*; \cdot) = j'(\boldsymbol{E}_h; \cdot) - a(\boldsymbol{E}_h^*, \cdot). \tag{17.27}$$

The residuals quantify the error inserting the approximate solution into the exact non-discretized Maxwell's equations. For the exact solution one finds $\rho(\boldsymbol{E}_h; \boldsymbol{E}) = 0$ and $\rho(\boldsymbol{E}_h^*; \boldsymbol{E}) = 0$. These residuals are computed for each patch and only patches with the largest residuals are refined. More details about the mathematical formulation and implementation can be found in [15]. A closer look to $\rho(\boldsymbol{E}_h;\cdot)$ which is also used for field-energy based adaptive refinement will be given in the next section.

17.6
Convergence of Eigenvalues Using Different Error Estimators

In this section we will look at the convergence of the computed eigenvalues. As computational domain we use the HCPCF depicted in Figure 17.3(a). We compare three different refinement strategies namely uniform refinement and two different adaptive goal-oriented refinement strategies. In an uniform refinement step each triangle is subdivided into 4 smaller ones. In an adaptive refinement step a predefined resdiuum ρ of the numerical solution is minimized by refining only triangles with largest values for this residuum. The two different adaptive strategies we are using differ by the chosen residua. The first one only uses (17.26). We will refer to it as *strategy A*. The second *strategy B* which was introduced in the previous section uses (17.27) in addition to (17.26).

Let us have a closer look at the common residuum (17.26) of both strategies. We start with the weak formulation of Maxwell's Eq. (17.13) on a sub patch Ω_i of our triangulation and reverse the partial integration:

$$0=\int_{\Omega_i}\overline{\Phi}\cdot\left[\nabla_{k_z}\times\frac{1}{\mu}\nabla_{k_z}\times\boldsymbol{E}-\frac{\omega^2\varepsilon}{c^2}\boldsymbol{E}\right]d^3r-\int_{\Gamma_i}\overline{\Phi}\cdot\left[\left(\left[\frac{1}{\mu}\nabla_{k_z}\times\boldsymbol{E}\right]\right)\times\boldsymbol{n}\right]d^2r, \tag{17.28}$$

where Γ_i is the boundary of the sub patch Ω_i and $[(1/\mu)\nabla_{k_z}\times\boldsymbol{E}]$ is the difference of the electric field $\boldsymbol{E}$ and the permeability μ on both sides of this boundary. Since Eq. (17.28) holds for arbitrary Φ the terms in brackets in both integrals have to vanish. For an exact solution to Maxwell's equations the first term vanishes because this is the Maxwell equation itself and the second term because the tangential component of $(1/\mu)\nabla_{k_z}\times\boldsymbol{E}$ is continuous across a boundary. Since we approximate the exact solution by $\boldsymbol{E}_h$ both terms will generally not vanish. Therefore we define the residuum ρ_i of sub patch Ω_i:

$$\begin{aligned} r_i(\boldsymbol{E}_h;\boldsymbol{E}_h)=h_i^2\int_{\Omega_i}\left|\nabla_{k_z}\times\frac{1}{\mu}\nabla_{k_z}\times\boldsymbol{E}_h-\frac{\omega^2\varepsilon}{c^2}\boldsymbol{E}_h\right|^2 d^3r\\ +h_i\int_{\Gamma_i}\left|\left(\left[\frac{1}{\mu}\nabla_{k_z}\times\boldsymbol{E}_h\right]\right)\times\boldsymbol{n}\right|^2 d^2r,\end{aligned} \tag{17.29}$$

where h_i is the size of the sub patch. For the solution $\boldsymbol{E}_h=\boldsymbol{E}$ of Maxwell's equations we find $\rho_i(\boldsymbol{E}_h;\boldsymbol{E})=0$. The residuum is therefore a measure how well the approximation $\boldsymbol{E}_h$ fulfills the exact Maxwell's equations.

Figure 17.4 shows the relative error of the fundamental eigenvalue n_{eff}^1 in dependence on the number of unknowns of the FEM computation for different refinement strategies and finite element degrees p. The real part of the effective refractive index converges for all finite element degrees and all refinement strategies, see Figure 17.4(a), (c), (e). For higher finite element degrees we find faster convergence. For the lowest finite element order and coarsest grid (with $\approx$16000 unknowns) we have a relative error of $\approx 10^{-5}$. This error decreases down to 10^{-11} for fourth order elements with 4×10^6 unknowns. Figure 17.4(b), (d), (f) show that the imaginary part

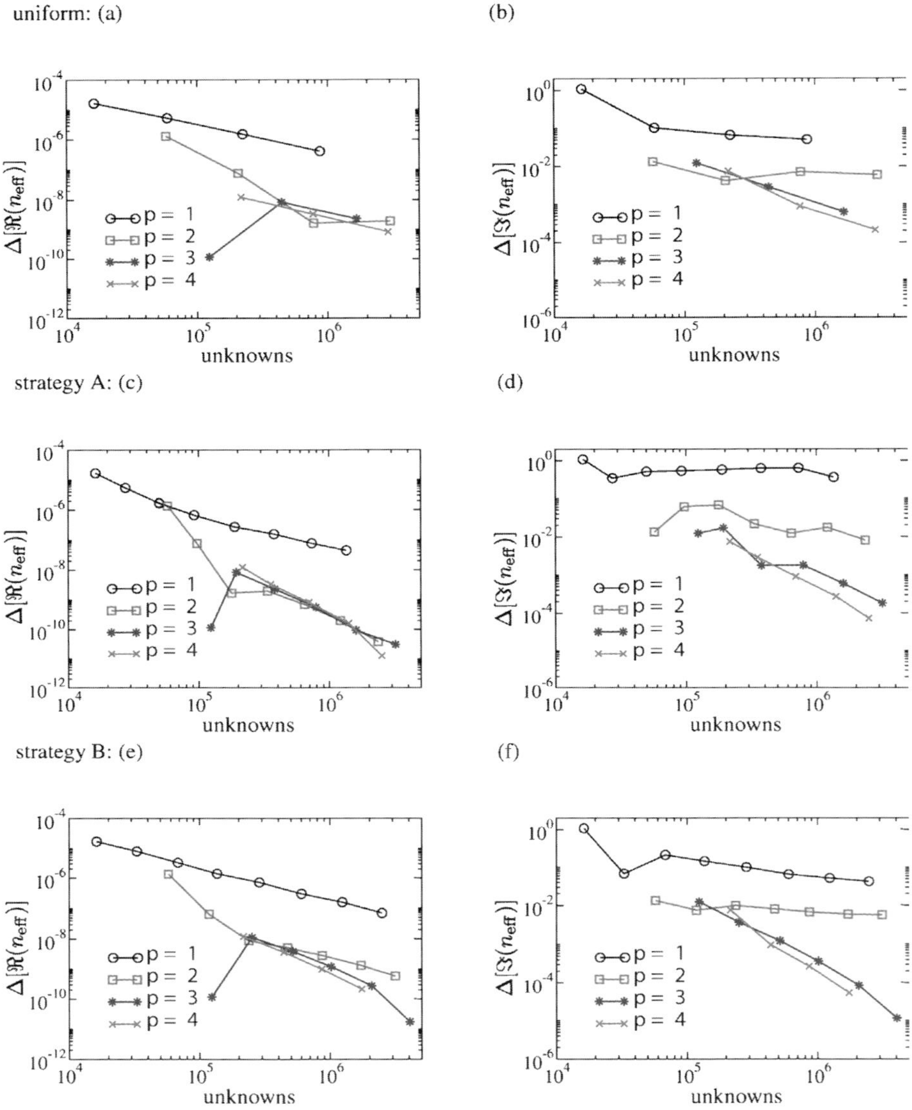

Figure 17.4 Relative error of fundamental eigenvalue in dependence on number of unknowns of FEM computation for different refinement strategies and finite element degrees p. Parameters: $\Lambda = 1550$ nm, $r = 300$ nm, $w = 50$ nm, $t = 170$ nm, 6 cladding rings, wavelength $\lambda = 589$ nm. Adaptive refinement strategy A minimizes residuum (17.26), strategy B minimizes residua (17.26) and (17.27).

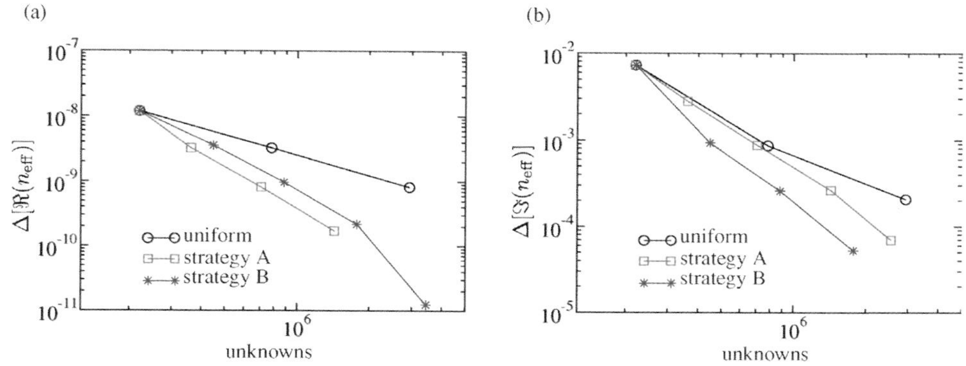

Figure 17.5 Relative error of fundamental eigenvalue in dependence on number of unknowns of FEM computation for different refinement strategies and finite element degree $p = 3$. Parameters: $\Lambda = 1550$ nm, $r = 300$ nm, $w = 50$ nm, $t = 170$ nm, 6 cladding rings, wavelength $\lambda = 589$ nm. Adaptive refinement strategy A minimizes residuum (17.26), strategy B minimizes residua (17.26) and (17.27).

converges much slower. For uniform refinement and finite element degrees of $p = 1, 2$ we do not find convergence at all. The adaptive strategies A and B refinement also lead to poor convergence for low finite element degrees. For an accurate computation of small losses therefore high finite element degrees are necessary. For a relative error of $10^{-3} = 0.1\%$ we already need $\approx 10^6$ unknowns for $p = 3, 4$. In Figure 17.5(a) and (b) the different refinement strategies are compared for fixed finite element order $p = 3$. For the real part we find fastest convergence for adaptive strategy A. As explained before the corresponding error estimator refines those triangles where the electric field has a large deviation from the exact solution. The error estimator of strategy B also uses this residuum but furthermore the residuum of the dual problem, Eqs. (17.26) and (17.27). Therefore it also converges faster than uniform refinement but slower than strategy A. For the imaginary part the adaptive refinement strategy B shows fastest convergence. Strategy A shows almost no benefit for the first two refinement steps in comparison to uniform refinement. For the exact result for the convergence plots we used the most accurate result ($p = 4$) for the real and imaginary part obtained from strategy A and B respectively.

Figure 17.6 shows the benefit of high order finite elements. Here the adaptive refinement strategy A was used corresponding to Figure 17.4(c), (d) but with finite element degree up to order $p = 7$. While the real part of the eigenvalue converges very fast even with order $p = 2$ the imaginary part can be computed much more accurate with higher order finite elements. Compared to $p = 2$ using orders greater than $p = 4$ the relative error is 2 orders of magnitude smaller for the same number of unknowns. Computational times for an 2.6 GHz AMD Opteron processor system are also shown.

A comparison of the convergence and computation efficiency between our finite element package and the MIT Photonic-Bands (MPB) package (plane wave expansion method) is presented in [22] where bloch modes of photonic crystal structures were

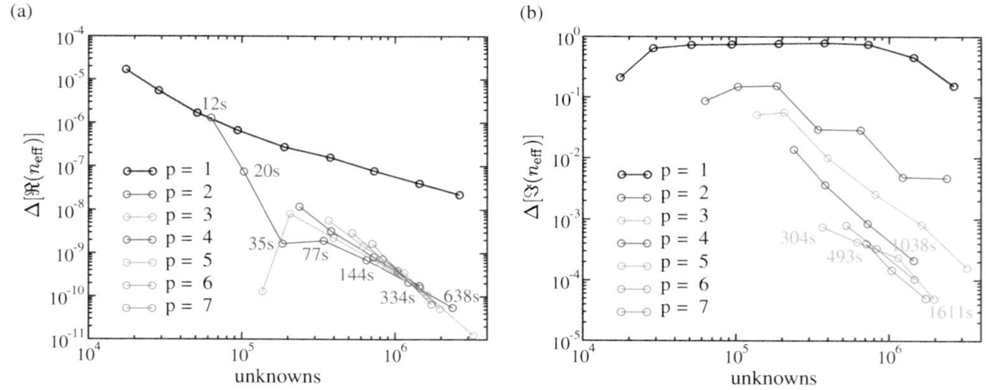

Figure 17.6 Relative error of fundamental eigenvalue in dependence on number of unknowns of FEM for adaptive refinement and different finite element degrees p. Parameters: $\Lambda = 1550$ nm, $r = 300$ nm, $w = 50$ nm, $t = 170$ nm, 6 cladding rings, wavelength $\lambda = 600$ nm. Computational times are shown.

computed. The FEM computations showed a much higher convergence rate. Results of the same accuracy could be computed over 100 times faster then with the MPB package. In [12] FEM computations of guided modes in HCPCFs are also compared to plane wave expansion (PWE) computations. Eigenmodes which were computed in about a minute with our finite element package took several hours with the PWE method.

17.7 Optimization of HCPCF Design

Since we are enabled to compute radiation losses very accurately we can use our method to optimize the design of photonic crystal fibers to reduce radiation losses. The basic fiber layout is a 19-cell core with rings of hexagonal cladding cells, Figure 17.3(a). Since these cladding rings prevent leakage of radiation to the exterior for core guided modes we expect that an increasing number of cladding rings reduces radiation leakage and therefore reduces $\Im(n_{\text{eff}})$. This is confirmed by our numerical simulations shown in Figure 17.8. The radiation leakage decreases exponentially with the number of cladding rings and thereby the thickness of the photonic crystal structure. This behavior agrees with the exponential dampening of light propagating through a photonic crystal structure with frequency in the photonic band gap. For our further analysis we fix the number of cladding rings to 6. The free geometrical parameters are the pitch Λ, hole edge radius r, strut thickness w, and core surround thickness t depicted in Figure 17.7(a) together with the triangulation (b). Figure 17.8 shows the imaginary part of the effective refractive index in dependence on these parameters. For each scan all but one parameter were fixed. For the strut thickness w and the hole edge radius r we find well-defined optimal values which minimize $\Im(n_{\text{eff}})$. For pitch Λ and core surround thickness t a large number of local minima and maxima can be seen. Now we want to optimize the fiber

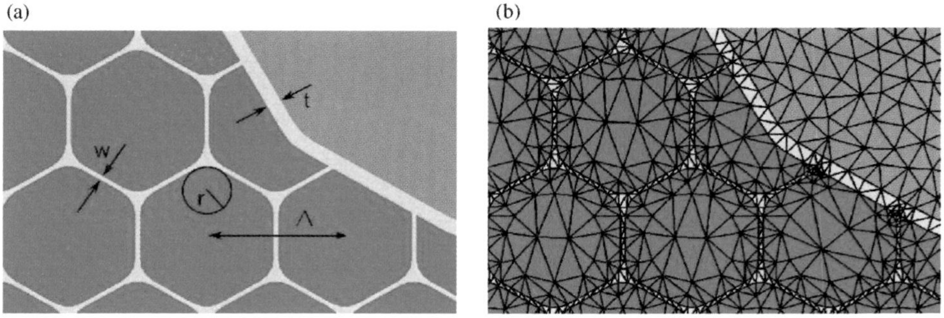

Figure 17.7 (a) Geometrical parameters describing HCPCF: pitch Λ, hole edge radius r, strut thickness w, core surround thickness t; (b) detail from a triangulation of HCPCF. Due to the flexibility of triangulations all geometrical features of the HCPCF are resolved.

design using multidimensional optimization with the Nelder–Mead simplex method [12]. To reduce the number of optimization parameters we fix the hole edge radius to the determined minimum at $r = 354$ nm since its variation has the weakest effect on $\Im(n_{\text{eff}})$. For optimization we have to choose starting values for Λ, t and w. Since the simplex method searches for local minima we have to decide in which local minimum of Λ and t we want to search. We choose $t = 152$ nm since here $\Im(n_{\text{eff}})$ has a global minimum and $\Lambda = 1550$ nm since the bandwidth of this minimum is much larger than for the global minimum at $\Lambda = 1700$ nm. Optimization yields a minimum value of $\Im(n_{\text{eff}}) = 5 \times 10^{-15}$ 1/m for the imaginary part of the effective refractive index. The corresponding geometrical parameters are $\Lambda = 1597$ nm, $w = 38$ nm, $t = 151$ nm.

17.8
Kagome-Structured Fibers

In [5] attenuation spectra of large-pitch kagome-structured fibers have been measured experimentally. Figure 17.9(a), (b) show the first and fourth fundamental core mode of such a 19-cell kagome-structured fiber. The corresponding layout is given in Figure 17.9(c). In this section we compute attenuation spectra numerically. Since we only take into account radiation losses we do not expect quantitative agreement with experimental measurements. Discussions about other loss mechanism can be found in [12,5]. We fix the geometrical parameters of the fiber and search for leaky eigenmodes in a wavelength interval. In the experiment core modes could be excited selectively. Here we also use the imaginary part of the fundamental core mode for the attenuation spectra. A small imaginary part then corresponds to low losses and therefore high transmission. Furthermore we look at the confinement of the computed modes. Therefore we compute the energy flux of the mode within E_{core} and outside E_{strut} the hollow core. A well confined mode then has a confinement $E_{\text{core}}/E_{\text{strut}}$ close to 1. We expect that well confined modes have small losses.

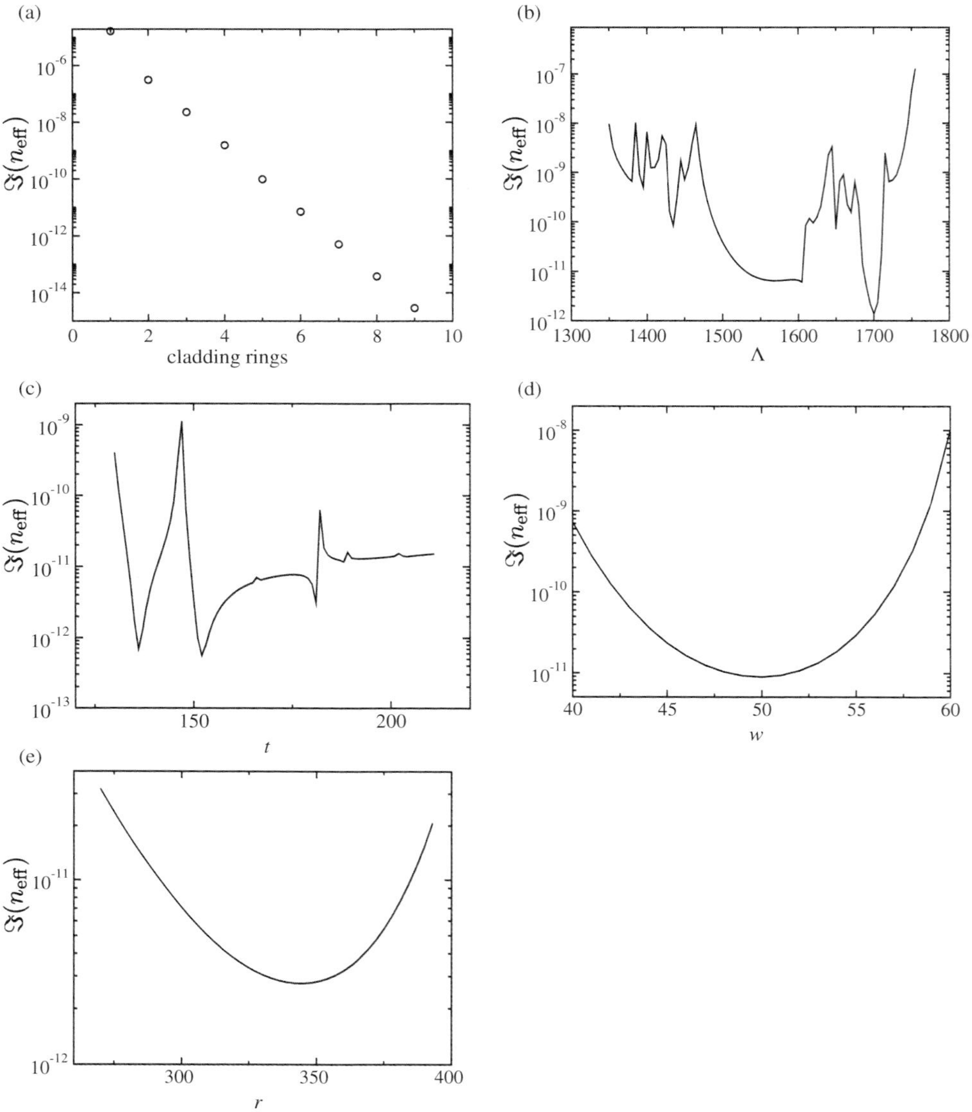

Figure 17.8 Imaginary part of effective refractive index $\Im(n_{\text{eff}})$ in dependence on: (a) number of cladding rings, (b) pitch Λ, (c) core surround thickness t, (d) strut thickness w, (e) hole edge radius r. Parameters: $\Lambda = 1550$ nm, $r = 300$ nm, $w = 50$ nm, $t = 170$ nm, 6 cladding rings, wavelength $\lambda = 589$ nm.

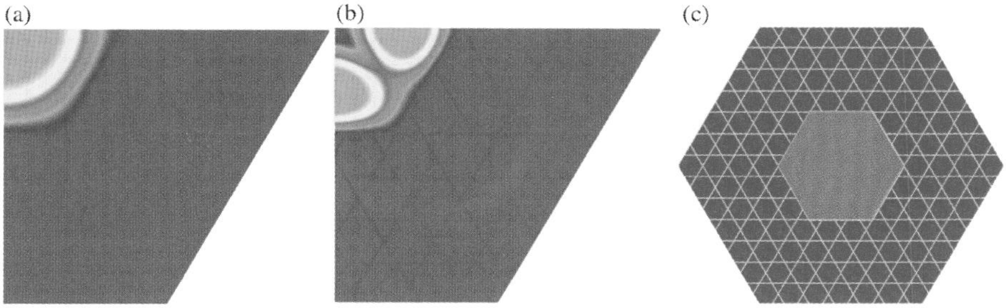

Figure 17.9 (a) Fundamental and (b) fourth core mode of 19-cell kagome-structured HCPCF (c). Parameters according to Table 17.2.

For simulation we use layouts corresponding to [5] shown in Figures 17.3(b) and 17.9(c) with parameters given in Table 17.2.

Figure 17.10(a) shows the real part of the effective refractive index of the fundamental mode in dependence on the wavelength. Since the core is filled with air it is close to 1 and hardly changes. Figure 17.11 shows the imaginary part of the eigenvalues which is proportional to the losses according to Eq. (17.19). For the 19-cell fiber we find a region of low transmission in the wavelength interval 950–1050 nm. In [5] a region of low transmission spans from 850–1050 nm. For the 1-cell fiber we find a peak of high attenuation at $\lambda = 700$ nm and high attenuation from 1200–1400 nm. A dip in the transmission spectrum can also be found in the corresponding experimental measurement at 630 nm. The range of the low transmission band however differs from the simulated. Experimentally it reaches from 800–1250 nm. An important loss mechanism is coupling of the fundamental mode to interface modes at the glass air interfaces of the fiber [5,12]. This is not taken into account in our simulations and could explain the mentioned disagreement. The regions of very high attenuation correspond to poorly confinement modes shown in Figure 17.13.

We notice that both attenuation spectra are very noisy. Since our computed results have converged we assume that this is no numerical artifact. To further investigate the large number of local extrema we zoom into the 19-cell spectrum from 0.715–0.735 nm where a very large local peak can be seen, see Figure 17.10(b). To explain the resonance in the attenuation spectrum at 725 nm we look at the field distribution within the kagome structure.

Table 17.2 Layouts of kagome-structured fibers used for mode computation.

Layout	Hollow-core	Pitch Λ (μm)	Strut width (μm)
A	19-cell	10.9	0.51
B	1-cell	11.8	0.67

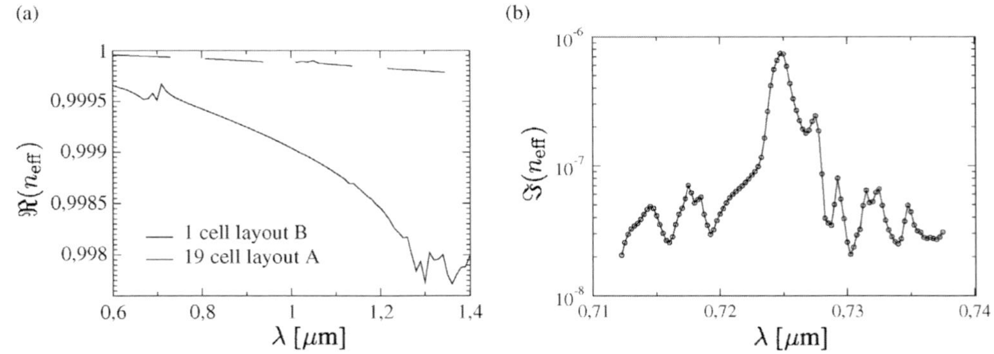

Figure 17.10 (a) Real part of effective refractive index of fundamental leaky mode in dependence on wavelength λ for 19-cell and 1-cell kagome-structured fiber. Parameters: see Table 17.2 – (b) zoom into Figure 17.11(a).

Figure 17.12 shows the fundamental eigenmode for $\lambda = 719.25$ nm where we find very low attenuation and for $\lambda = 725$ nm at the local peak attenuation. The intensity of the eigenmodes in Figure 17.12(a) and (b) is shown for the same pseudo color range. The field distribution for $\lambda = 725$ nm is more intense in the glass struts which connect the core and the cladding of the fiber. Light from the fundamental core mode is coupled much stronger into the first neighbouring triangles of the kagome structure for $\lambda = 725$ nm. They could be seen as resonators coupled to the hollow core and being excited by the core mode. Better coupling then leads to more light leaving the core and therefore higher attenuation. The coupling into the complicated kagome structure depends very sensitively on the wavelength and the shape of the mode.

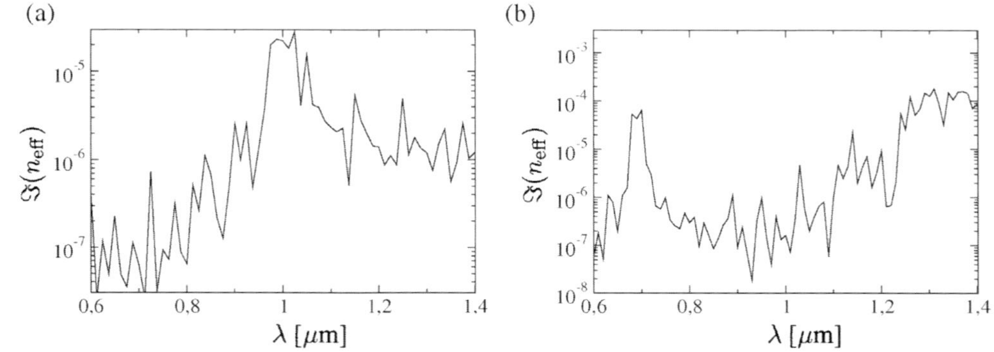

Figure 17.11 Imaginary part of effective refractive index of fundamental leaky mode in dependence on wavelength λ for (a) 19-cell and (b) 1-cell kagome-structured fiber. Parameters: see Table 17.2.

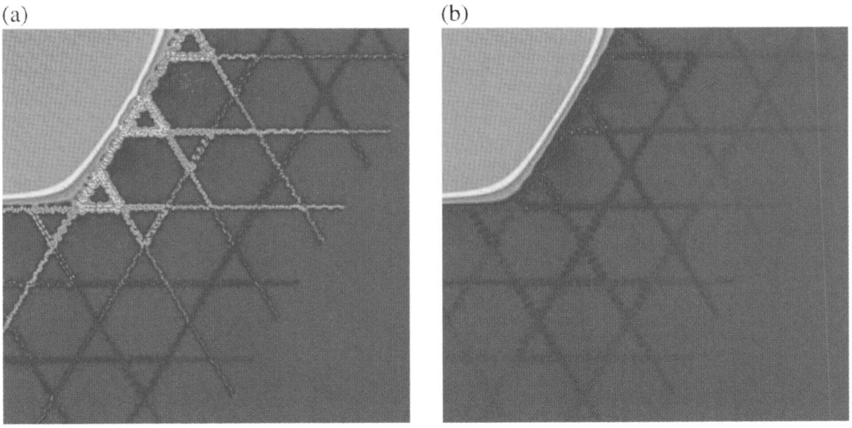

Figure 17.12 Pseudo color image of intensity of fundamental core mode for (a) $\lambda = 725$ nm and (b) $\lambda = 719.25$ nm. Parameters: Layout A, Table 17.2.

Finally Figure 17.13(a) shows the confinement of the fundamental mode in dependence on the wavelength. Regions of low loss correspond to regions of high $\Im(n_{\text{eff}})$, compare Figure 17.11.

17.9 Conclusion

We have investigated radiation losses in photonic crystal fibers. We have shown that with our FEM analysis we can efficiently determine the complex eigenvalues to a very

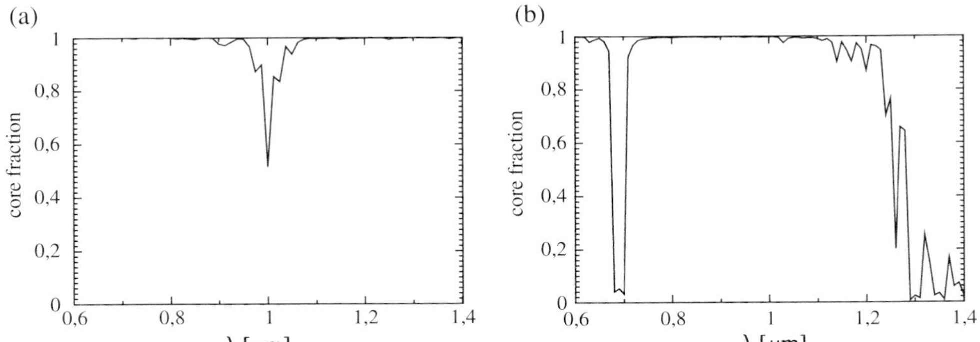

Figure 17.13 (a) Fraction of electric field intensity of fundamental leaky mode located inside hollow core in dependence on wavelength λ for (a) 19-cell and (b) 1-cell kagome-structured fiber. Parameters: see Table 17.2.

high accuracy. The finite size of the photonic crystal fibers was taken into account by transparent boundary conditions realized with the PML method.

Computing leaky propagation modes in HCPCFs eigenvalues with real and imaginary part differing by over 10 orders of magnitude were found. The relative error of the imaginary part was thereby about 7 orders of magnitude larger than the relative error of the real part. In order to precisely compute the much smaller imaginary part of the eigenvalues special techniques were implemented and applied to 2 different types of HCPCFs, namely HCPCFs with hexagonal cladding cells and kagome-structured fibers. A goal-oriented error estimator was introduced which focused on the accurate computation of a target functional, in our case the imaginary part of the eigenvalue. After each computation on a refinement level this error estimator was used to adaptively refine thegrid. In a convergence analysis it was shown that the goal-oriented error estimator led to faster convergence of the quantity of interest compared to uniform refinement or standard field energy based refinement strategies. Also the usage of high order finite elements significantly increased the accuracy of the imaginary part.

Since the imaginary part of a fiber could be computed in about 10 minutes with a relative error smaller than 10^{-3} it was possible to use the finite element method to automatically optimize a fiber design with respect to pitch, strut thickness, cladding meniscus radius and core surround thickness in order to minimize radiation losses. Furthermore attenuation spectra of a 1-cell and a 19-cell kagome-structured fiber were computed and compared to experimental results. It was shown that high losses were connected to poorly confined modes within the hollow-core. Furthermore the appearance of a large number of local minima and maxima in the attenuation spectra was explained by analyzing the intensity distribution of the core modes.

Acknowledgements

We acknowledge support by the DFG within SPP 1113 under contract number BU 1859/1-1. We thank Ronald Holzlöhner and John Roberts for fruitful discussions.

References

1 Joannopoulos, J.D., Meade, R.D. and Winn, J.N. (1995) *Photonic Crystals: Molding the Flow of Light*, 1st ed., Princeton University Press.

2 Cregan, R.F., Mangan, B.J., Knight, J.C., Russel, P.St.J., Roberts, P.J. and Allan, D.C. (1999) *Science*, **285** (5433), 1537–1539.

3 Russell, P.St.J. (2003) *Science*, **299** (5605), 358–362.

4 Benabid, F., Knight, J.C., Antonopoulos, G. and Russel, P.St.J. (2002) *Science*, **298** (5592), 399–402.

5 Couny, F., Benabid, F. and Light, P.S. (2006) *Opt. Lett.*, **31** (24), 3574–3576.

6 Pomplun, J., Holzlöhner, R., Burger, S., Zschiedrich, L. and Schmidt, F. (2007) *Proc. SPIE*, **6480**, 64800M.

7 Bienstman, P. Selleri, S. and Rosa, L. (2006) Modelling lossy photonics wires: a

mode solver comparison, in *Proc. OWTNM 05*, p. 5.

8 Burger, S., Klose, R., Schädle, A., Schmidt, F. and Zschiedrich, L. (2005) FEM modelling of 3D photonic crystals and photonic crystal waveguides, in *Integrated Optics: Devices, Materials, and Technologies IX, Proc. SPIE*, (eds Y. Sidorin and C.A. Wächter), **5728**, 164–173.

9 Burger, S., Klose, R., Schädle, A., Schmidt, F. and Zschiedrich, L. (2006) Adaptive FEM solver for the computation of electromagnetic eigenmodes in 3d photonic crystal structures, in *Scientific Computing in Electrical Engineering*, (eds A.M. Anile, G. Ali G. and Mascali), Springer Verlag, Berlin, Germany, pp. 169–175.

10 Burger, S., Zschiedrich, L., Klose, R., Schädle, A., Schmidt, F., Enkrich, C., Linden, S., Wegener, M. and Soukoulis, C.M. (2005) Numerical investigation of light scattering off split-ring resonators, Metamaterials, (eds T. Szoplik, E., Özbay, C.M. Soukoulis and N.I. Zheludev), *Proc. SPIE*, **5955**, 18–26.

11 Enkrich, C., Wegener, M., Linden, S., Burger, S., Zschiedrich, L., Schmidt, F., Zhou, C., Koschny, T. and Soukoulis, C.M. (2005) *Phys. Rev. Lett.*, **95**, 203901.

12 Holzlöhner, R., Burger, S., Roberts, P.J. and Pomplun, J. (2006) *J. Eur. Opt. Soc.*, **1** (06011), 1–7.

13 Kalkbrenner, T., Håkanson, U., Schädle, A., Burger, S., Henkel, C. and Sandoghdar, V. (2005) *Phys. Rev. Lett.*, **95**, 200801.

14 Linden, S., Enkrich, C., Dolling, G., Klein, M.W., Zhou, J., Koschny, T., Soukoulis, C.M., Burger, S., Schmidt, F. and Wegener, M. (2006) *IEEE J. Sel. Top. Quantum Electron.*, **12**, 1097–1105.

15 Zschiedrich, L., Burger, S., Pomplun, J. and Schmidt, F. (2007) *Proc. SPIE*, **6475**, 64750H.

16 Peter, Monk (2003) *Finite Element Methods for Maxwell's Equations*, Oxford University Press, Oxford, UK.

17 Bérenger, J. (1994) *J. Comput. Phys.*, **114** (2), 185–200.

18 Zschiedrich, L., Burger, S., Klose, R., Schädle, A. and Schmidt, F. (2005) Jcmmode: an adaptive finite element solver for the computation of leaky modes, *Integrated Optics: Devices, Materials, and Technologies IX, Proc. SPIE*, (eds Y. Sidorin and C.A. Wächter), **5728**, 192–202.

19 Cucinotta A., Selleri, S. and Vincetti, L. (2002) *IEEE Photonics Technol. Lett.*, **14**, 1530–1532.

20 Brechet, F., Marcou, J., Pagnoux, D. and Roy, P. (2000) *Opt. Fiber Technol.*, **6**, 181–191.

21 Becker, R. and Rannacher, R. (2000) An optimal control approach to a posteriori error estimation in finite element methods, in *Acta Numerica*, (ed. A. Iserles), Cambridge University Press, Cambridge, UK, pp. 1–102.

22 März, R., Burger, S., Golka, S., Forchel, A., Herrmann, C., Jamois, C., Michaelis, D. and Wandel, K. (2004) Planar high index-contrast photonic crystals for telecom applications, *Photonic Crystals – Advances in Design, Fabrication and Characterization* (ed. Kurt Busch, Stefan Lölkes, Ralf B. Wehrspohn, Helmut Föll), Wiley-VCH, Weinheim, Germany, pp. 308–329.

IV
Plasmonic and Metamaterials

Nanophotonic Materials: Photonic Crystals, Plasmonics, and Metamaterials. Edited by R.B. Wehrspohn, H.-S. Kitzerow, and K. Busch

ISBN: 978-3-527-40858-0

18
Optical Properties of Photonic/Plasmonic Structures in Nanocomposite Glass

H. Graener, A. Abdolvand, S. Wackerow, O. Kiriyenko, and W. Hergert

18.1
Introduction

Photonic device fabrication in glass was investigated intensively during the last years. Three-dimensional structures like waveguides [1], couplers [2], photonic crystals [3] and other structures can be created by use of nonlinear materials processing with near-IR femtosecond laser pulses. A refractive index change inside of such structures of $\Delta n \sim 10^{-3}$–10^{-2} can be achieved [1].

Glass containing metallic nanoparticles is a promising material for various photonic applications due to the unique optical properties mainly resulting from the strong surface plasmon resonance (SPR) of the metallic nanoparticles. The details of this resonance can be modified to a large extend by varying the size, shape and concentration of the inclusions. Furthermore the metal itself and the dielectric constant of the glass influences the resonance [4]. First estimates show, that in glass containing metallic nanoparticles refractive index changes $\Delta n \sim 1$ are possible.

Recent progress in structuring this type of material by ultrashort laser pulses [5,6] and electric fields [6] opens promising possibilities due to the large index contrast to tailor new nanodevices and optical elements with interesting linear and nonlinear properties [7].

18.2
Experimental Investigations

Under the influence of moderate temperatures and high electric fields, silver nanoparticles embedded in a glass matrix can be dissolved [8,9]. Under certain conditions a percolated silver layer in the depth of the glass can be observed [10,11]. By means of a structured electrode (e.g. Si photonic crystals with a metal cover) a spatial modulated particle dissolution can be achieved resulting in a metal

Nanophotonic Materials: Photonic Crystals, Plasmonics, and Metamaterials. Edited by R.B. Wehrspohn, H.-S. Kitzerow, and K. Busch

ISBN: 978-3-527-40858-0

nanoparticle distribution with a two-dimensional periodicity which is inverse to the electrode structure [12].

Besides of the 2D structure of such systems the photonic characteristics strongly depend on the properties of an individual nanoparticle containing cell. Here the cell shape especially in the third dimension and the particle shape and distribution within the cell have to be considered. Experiments are performed to consider how the formation of the structures can be controlled by external parameters like applied voltage, temperature and duration of the experiment. In a first series of experiments it is shown that the cell shape strongly depends on the parameters of the dissolution process. For all experiments described in the following glass samples of a thickness of 2 mm with a particle containing layer of an overall thickness of approximately 6 μm were used. The filling factor had a gradient starting with 0.7 near the surface [12]. Figure 18.1 shows electron microscope pictures of three samples processed under different conditions. The pictures show cross-sections nearly perpendicular to the sample surface. For all experiments a 2D structured Si electrode with a period of 2 μm

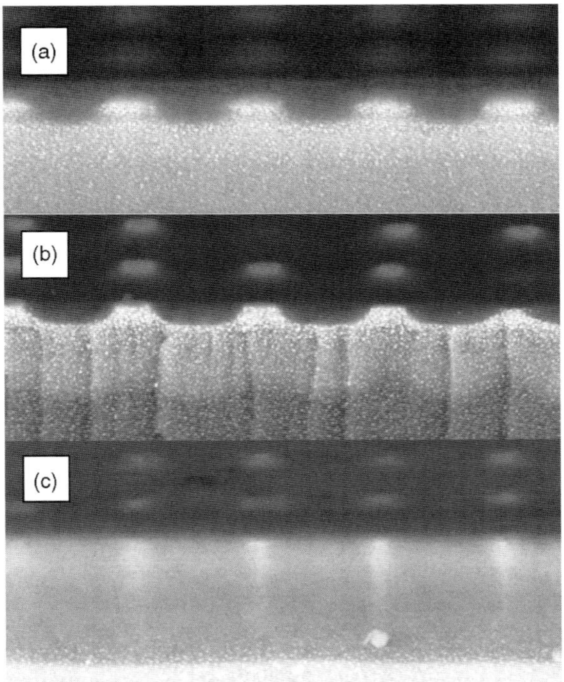

Figure 18.1 Electron microscope pictures of three structured samples. The anode was a 2D-square lattice Si photonic crystal with 2 μm periodicity: (a) voltage of 400 V applied for 30 min; sample temperature 250 °C, (b) voltage of 600 V applied for 30 min; sample temperature 190 °C, (c) voltage 600 V applied for 30 min; sample temperature 250 °C.

on top of the glass samples was used. The bright dots represent silver nanoparticles. The grey areas are due to (ionic) silver. In all pictures the 2 μm period of the electrode can be seen. For the experiment shown in Figure 18.1(a) a voltage of 400 V at a temperature of 250°C was applied for 30 min. The particle dissolution is restricted to a layer of typically 1 μm thickness beneath the Si part of the electrode, which is in direct contact to the glass surface. The size of the remaining particle cell corresponds to the hole size of the electrode.

Enhancement of the applied voltage and reduction of the temperature (Figure 18.1(b)) leads to a decrease of the depth of the dissolution volume and simultaneously the surface area of the particle containing cell shrinks. This leads to nearly plan-convex shaped particle free volumes. In Figure 18.1c also a voltage of 600 V is applied for 30 min, but the temperature is increased to 250°C. The thickness of the dissolution volume increases and the size of the particle containing cell shrinks simultaneously. As a result roughly square shaped cells with ~500 nm diameter and 3 μm length are generated. We can conclude from the experimental results, that various shapes of structural elements can be produced. It should be noted as well that the details of the shape do not only depend on the external parameters, but also on the particle concentration and its gradient in the sample.

The details of the optical properties of such structures will depend on size, shape and the local concentration of the nanoparticles. Of special interest are systems with dicroitic optical properties. If the base material used for the experiments would contain ellipsoidal nanoparticles homogeneously oriented, new properties could be expected. Such material can be processed by a thermomechanical treatment of samples with originally spherical particles [13,14]. By this treatment the thickness of the particle containing layer is reduced to roughly 1 μm. Polarization resolved absorption spectra are shown in Figure 18.2. Here p-pol means that the polarization direction of the probe light is parallel to the semi-major axis and s-pol is in the direction of the semi-minor axis. The color of the material is typically green for unpolarized and p-polarized illumination. The color changes to yellow for s-polarized illumination. For sample b) in Figure 18.2 the color of p-polarized illumination is blue changing to orange for s-polarized light.

The dicroitic samples can be structured by the dissolution method [8,9]. Figure 18.3 shows two microscope pictures (100 × objective magnification) as an example. For the structuring process a 8 μm line spacing (4 μm line thickness) grating was used as electrode. The left picture is obtained with p-pol the right one with s-pol illumination. The structure can clearly be seen in both pictures, but the color is different indicating that the optical parameters strongly depend on the polarization of the illumination.

There is no reason why this line like structure cannot be scaled down. We will use a 3600 mm^{-1} grating to try to generate a 300 nm line structure.

Such dicroitic samples can also be structured using a Si photonic crystal as an electrode by applying a voltage of 400 V at 190°C for 30 min with a structured electrode as described above.

Figure 18.4 shows microscope images of the sample after the treatment. If the sample is illuminated with light having a polarization parallel to the semi-major axis of the particles a blue shining 2 × 2 μm structure is visible with a rather high contrast.

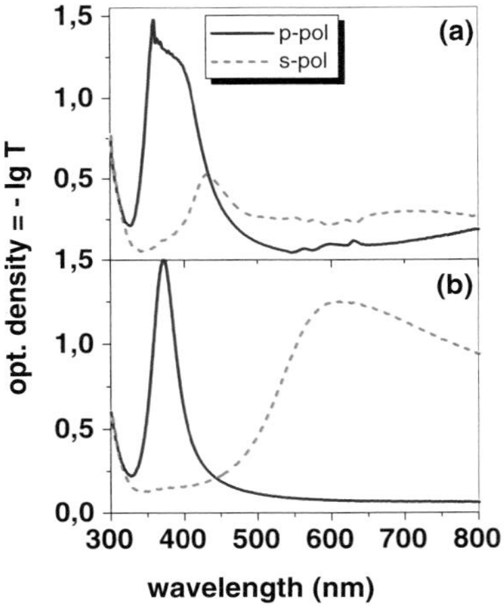

Figure 18.2 Polarization resolved absorption spectra of two different dicroitic glass samples containing ellipsoidal silver nanoparticles. The different spectra are due to a different distribution of sizes (volumes) and aspect ratios. These examples show that the base material can be produced with a large variety of optical properties.

If the polarization is rotated by 90° the color changes to yellow and the structure nearly disappears. Using femto-second laser pulses the structures can further be modified [5,6]. Addressing a well-defined nanoparticle containing cell the particles can either be destroyed or changed in shape.

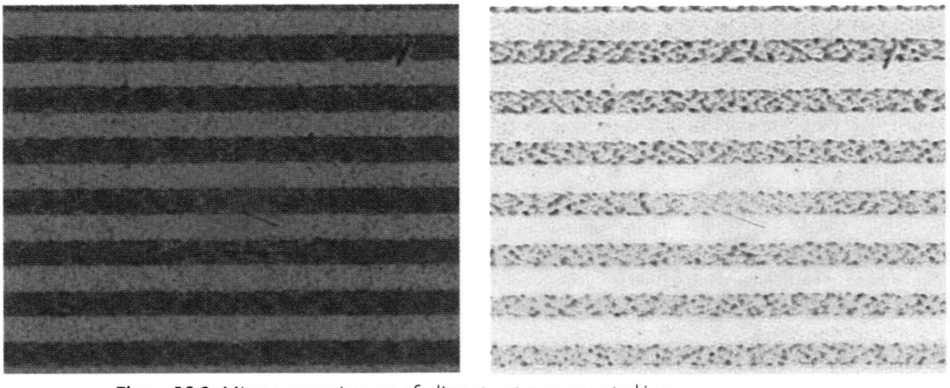

Figure 18.3 Microscope pictures of a line structure generated in a dicroitic glass. The two pictures only differ in the polarization direction of the incident light. The base material (dark lines) show spectra like those shown in Figure 18.2a.

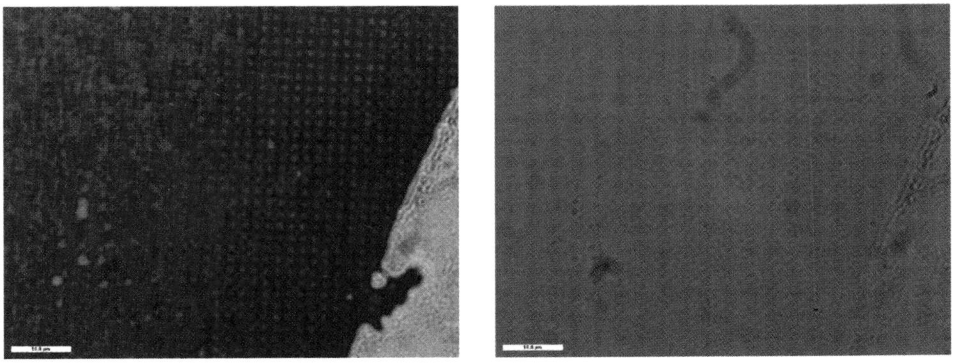

Figure 18.4 Polarized microscope pictures of a structured dicroitic sample; for the left part the polarization of the light was parallel to the semi-major axis of the nanoparticle; for the right part the polarization was rotated by 90°.

18.3 Calculation of Effective Permittivity

The photonic/plasmonic structures under investigation are characterized by a layer of glass containing Ag-nanoparticles on a glass substrate. The system is patterned by the dissolution technique. The size of the remaining building blocks and the irregular distribution of the particles prevent the treatment of the particles as single scatterers. Therefore, we want to describe the building blocks by an effective frequency dependent permittivity $\varepsilon_{eff}(\omega)$. The dielectric properties of such nano-composite materials are usually described by means of effective medium theories (EMT). A series of such theories has been developed. The most successful approaches are the Maxwell–Garnett theory (MGT) [15] and the Bruggeman effective medium approximation (BEMA) [16]. Whereas the MGT theory is able to describe qualitatively the surface-plasmon resonances for metal-dielectric composites, BEMA is not. In contrast to MGT, BEMA includes the percolation among the inclusions for filling factors of $f \geq 1/3$. The effective permittivity in MGT is given by the permittivity of the host ε_h, the permittivity of the inclusions ε_i and the filling factor f.

$$\varepsilon_{eff}^{MGT} = \varepsilon_h \frac{(\varepsilon_i + 2\varepsilon_h) + 2f(\varepsilon_i - \varepsilon_h)}{(\varepsilon_i + 2\varepsilon_h) - f(\varepsilon_i - \varepsilon_h)}. \qquad (18.1)$$

We assume that the metallic particles are plasma spheres, i.e. the optical properties of the single sphere are described by a dielectric function

$$\varepsilon_i(\omega) = \varepsilon_b + 1 - \frac{\omega_p^2}{\omega^2 + i\gamma\omega}, \qquad (18.2)$$

where ω_p is the free electron bulk plasma frequency and γ describes the damping of electron oscillations ($\gamma = 1/\tau$, τ – relaxation time) and ε_b is in principle the contribution of interband transitions and all other non-conduction electron contributions to the dielectric constant. Values of $\varepsilon_h = 2.3$, $\omega_p = 9.2$ eV, $\gamma = 0.5$ eV and $\varepsilon_b = 4.2$ are used [18]. The nanocomposite is controlled by the parameters filling factor and radius of the nanoparticles. Those parameters are known from experiment. The mean diameter of the particles is $\cong$20 nm [12].

Even if one has complete information about the permittivity of the constituents and the geometry, the calculation of $\varepsilon_{eff}(\omega)$ remains difficult. Therefore the estimation of upper and lower bounds of the permittivity is of interest. The derivation of such bounds is given by Milton [19]. A bound implies a restriction of $\varepsilon_{eff}(\omega)$ to a certain region of the complex plane. The region depends on what is known about the composite. A construction of bounds, illustrated here for values of $\varepsilon_1 = 1 - 7i$, $\varepsilon_2 = 10 - i$ is shown in Figure 18.5. Without any information on the microgeometry $\varepsilon_{eff}(\omega)$ is confined to the moon-like shaped pink region. Knowing the volume fraction, we get a restriction to the green area. If we know also that our ensemble is statistically isotropic and the dimension $d = 2$ or $d = 3$, respectively, the values of $\varepsilon_{eff}(\omega)$ are restricted to the yellow region. MGT contains all this kind of information. The result from MGT for the given ε-values of the constituents is indicated by a black dot in Figure 18.5. We will calculate the effective permittivity by means of a finite element method (FEM) directly. The results will be compared with Maxwell–Garnett theory. In both calculations statistical isotropic ensembles are used. Therefore the results of our FEM calculations have to be also inside or on the bounding arcs of the yellow region in Figure 18.5. The discussion of the bounds is valid in the quasi-static limit, which is

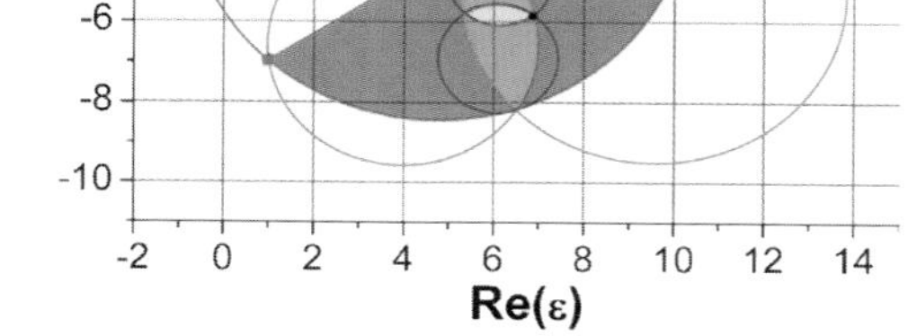

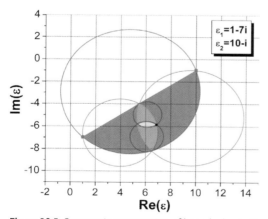

Figure 18.5 Systematic construction of bounds due to the information about the constituents. (pink – no information on the microgeometry, green – information on filling fractions, yellow – statistically isotropic structure, dimension) The black dot at the crossing of the two blue circles indicates the MGT result.

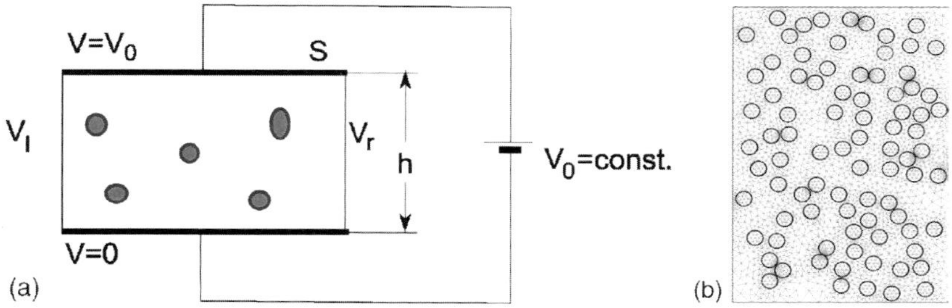

Figure 18.6 (a) Calculation of the effective permittivity with a plate capacitor arrangement. (b) FEM mesh for 85 particles in two dimensions which leads to 8157 elements and 16478 knots.

approximately fulfilled in our case. A quasi-static situation will be also assumed for the following considerations.

The finite element method can be used to calculate the permittivity for a given structure of the nanocomposite [20]. The FEM software package FEMLAB was used [21]. We consider particles in a matrix in two and three dimensions. For the estimation of $\varepsilon_{eff}(\omega)$ a plate capacitor with area S and the distance h of the plates is considered (cf. Figure 18.6(a)).

The capacitor is filled with the composite and a voltage V_0 is applied. To avoid edge effects, periodic boundary conditions are applied at the left and righthand side. The energy stored in a capacitor for a dielectric with real and constant permittivity ε is given by

$$W = \frac{1}{2}\varepsilon_0\varepsilon\frac{S}{h}V_0^2 . \tag{18.3}$$

The energy stored in the capacitor would be the same, if the composite of constituents having real and constant permittivity is replaced by a homogeneous effective medium with a suitable ε_{eff}. The calculation of the energy W stored in the capacitor containing the composite allows us therefore to calculate the effective permittivity. In such a case the energy W follows from the minimization of the functional F with respect to the potential $\phi(\mathbf{r})$ calculated for the composite system (here for two dimensions) from

$$W = F[\phi(\mathbf{r})] = \tfrac{1}{2}\varepsilon_0 \int_{\Omega} \varepsilon(\mathbf{r})[\nabla\phi(\mathbf{r})]^2 d^2\mathbf{r}, \tag{18.4}$$

where $\varepsilon(\mathbf{r})$ is the local dielectric constant, the integral extends over the surface of the capacitor Ω and $\phi(\mathbf{r})$ is the local electrostatic potential. The local electrostatic potential is calculated from the following boundary-value problem:

$$\nabla \cdot [\varepsilon(\mathbf{r})\, \nabla\phi(\mathbf{r})] = 0. \tag{18.5}$$

In the system under consideration one of the constituents is a lossy, dispersive material. In this case W is complex and the imaginary part of this quantity describes the losses. This part determines at the end the imaginary part of the effective permittivity. The Eqs. (18.4) and (18.5) retain there form for this special case if a complex phasor formulation is used (cf. [17]).

A typical mesh of the FEM method is shown in Figure 18.6(b). A direct contact of the particles is avoided by construction. Because of the finite size of the systems used in the calculation of $\varepsilon_{eff}(\omega)$, we have to perform an ensemble-average over many realizations of the composite. Typically several hundred realizations of the systems are performed.

A method was constructed to generate random ensembles of particles for arbitrary filling fractions. If the filling fraction of the particles is low, a random sequential addition process is used. The particles are randomly placed in the simulation cell one after the other. If the position of a particle is in conflict with already existing particles the particle will be discarded. This process has a saturation behavior. In two dimensions saturation is reached for $f^{\text{sat}} \cong 0.55$. Another process, mainly used for higher filling factors is based on the dynamics in a system of hard discs or spheres. We start with particles on a regular lattice. To each particle a randomly chosen velocity is assigned. The dynamics of the system is considered by a direct calculation of the scattering events of the particles. Several hundreds of scattering events are taken into account to generate a random ensemble. Periodic boundary conditions are applied. To be sure, that our systems have the appropriate statistical properties, the radial distribution functions (RDF) are calculated and compared with radial distribution functions calculated from the Percus–Yevick approximation to the Ornstein–Zernike equation describing hard disc or hard sphere gases [22]. The radial distribution function is the most basic statistical descriptor which provides structural information about the equilibrium system. The RDF is the probability of finding one particle center at a given distance r from the center of a reference particle. Figure 18.7 shows the radial distribution function for a hard sphere system with $f = 0.25$. For a low filling factor only the peak of the nearest neighbor shell appears. For higher filling factors an oscillatory behavior develops. The periodicity of the peaks is governed by the diameter of the particles.

The calculated $\varepsilon_{eff}(\omega)$ for a filling factor of $f = 0.25$ for two and three dimensions are given in Figure 18.8. Interesting for applications is the long wavelength side of the plasmon resonance. The calculation shows, that the real part of $\varepsilon_{eff}(\omega)$ is larger than the result from MGT. Also the imaginary part is increased compared to MGT leading to larger absorbtion in this wavelength region.

The difference between MGT and the FEM calculation can be seen clearly in the representation given in Figure 18.9. The values of $\varepsilon_{eff}(\omega)$ are plotted in the complex plane for wavelengths between $\lambda = 300$ nm and $\lambda = 1600$ nm. The differences in this representation are caused by the shift in the peak position of the FEM result for $\varepsilon_{eff}(\omega)$ with respect to the MGT and the increase of the imaginary part.

Finally it can be checked, if the results can be found in the bounds predicted by Miltons theory and described in Figure 18.5. The Figure 18.10(a) and (b) show for two different wavelengths how the results of the FEM calculations are related to the bounds. The blue lines correspond to the bounds of the yellow region in Figure 18.5. The form of the bounds are of course strongly influenced by the ε-values of the

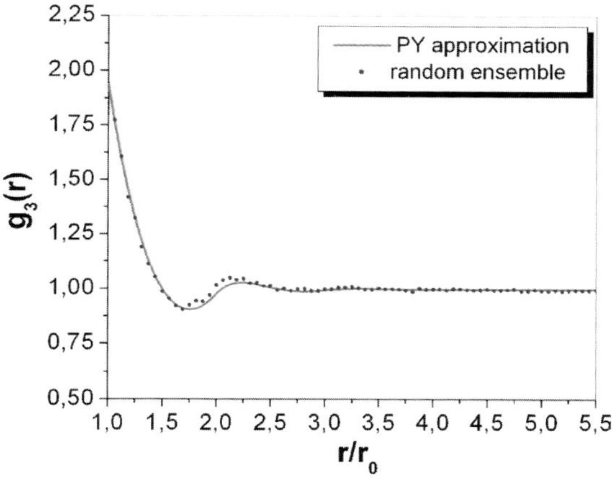

Figure 18.7 Radial distribution function for a 3D ensemble of spheres with a filling fraction $f = 0.25$.

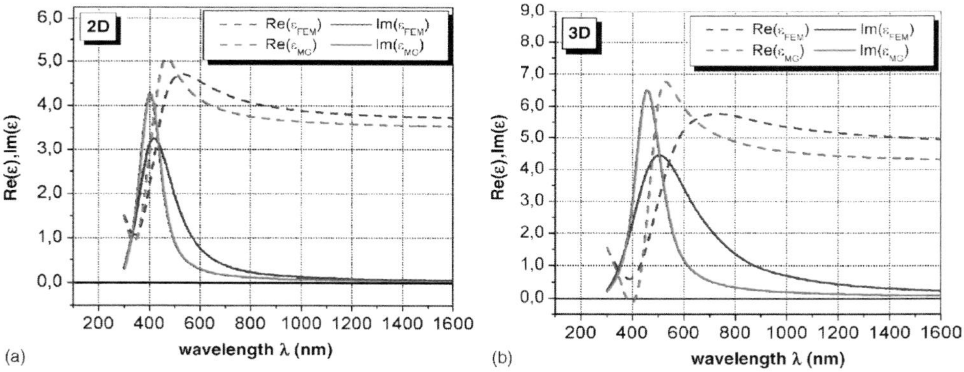

Figure 18.8 $\varepsilon_{eff}(\omega)$ calculated by FEM method. (a) Real part and imaginary part of $\varepsilon_{eff}(\omega)$ for a 2D system of discs for $f = 0.25$, compared with results of MGT. (b) Real part and imaginary part of $\varepsilon_{eff}(\omega)$ for a 3D system of spheres for $f = 0.25$, compared with results of MGT.

constituents. (cf. [19]) Even if the calculated values are near the bounds, all calculated values are inside the allowed region.

The method of calculation allows also other considerations. It can be shown, that the results of the FEM calculations are very close to the MGT results for low filling factors. It is also possible to study the influence of order/disorder on ε_{eff}. Figure 18.11 shows the difference in ε_{eff} between metallic nanoparticles arranged on a cubic lattice and a disordered structure with the same filling factor. It can be concluded from Figure 18.11 that disorder leads to higher $\Re\varepsilon_{eff}$ and $\Im\varepsilon_{eff}$ in the interesting wavelength region.

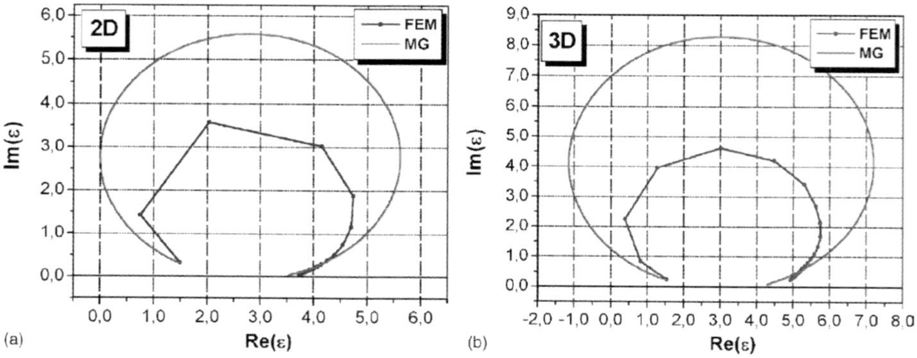

Figure 18.9 $\varepsilon_{eff}(\omega)$ from $\lambda = 300$ nm to 1600 nm in the complex plane. (a) 2D system; (b) 3D system.

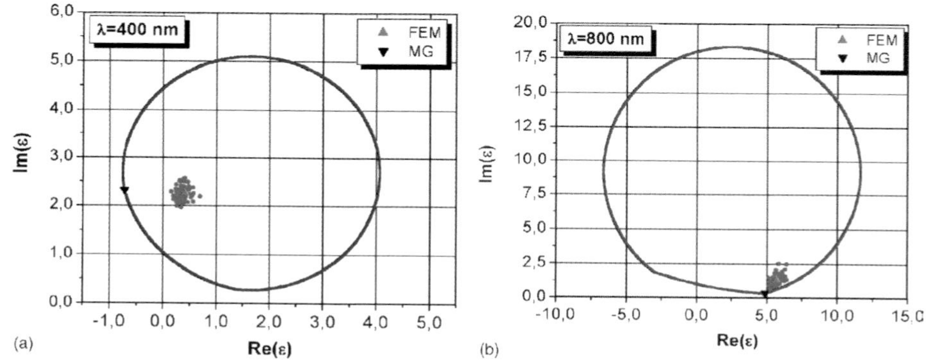

Figure 18.10 $\varepsilon_{eff}(\omega)$ from MGT and FEM calculations for the 3D system and bounds (blue lines). The results for 100 FEM calculations of $\varepsilon_{eff}(\omega)$ are indicated by green dots. Also the mean value is given. (a) $\lambda = 400$ nm; (b) $\lambda = 800$ nm.

18.3.1
Extensions of the Method

The method to calculate the effective permittivity can be extended in several ways. From experiment it is known, that under intense illumination with a laser beam the metal particles are deformed to rotational ellipsoids. The semi-major axis of the ellipsoids is oriented parallel to the polarization of the laser beam. The method of calculation of ε_{eff} can be extended to random ensembles of rotational ellipsoids with a preferred direction of the semi-major axis.

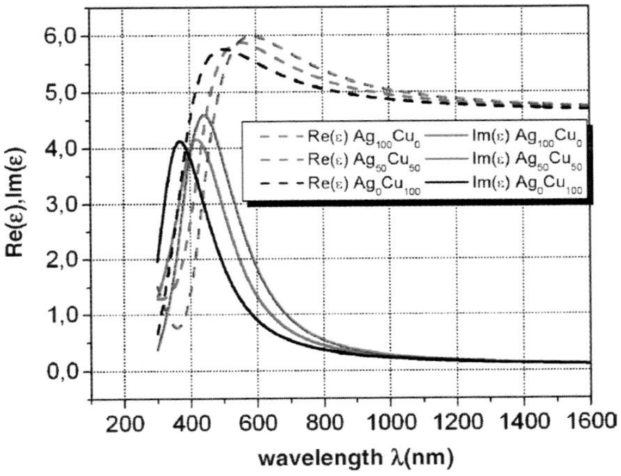

Figure 18.11 Difference in $\varepsilon_{eff}(\omega)$ between an ordered structure of metallic nanoparticles and a disordered arrangement of the same filling factor of $f = 0.25$.

It was found experimentally that the concentration of the silver particles decreases from the surface into the bulk exponentially. If random sequential addition can be used to create the ensembles, the probability distribution of particle positions can be transformed accordingly to take this effect into account.

Garcia *et al.* [23] considered mixing rules in a self-consistent fashion in the framework of EMT to predict the plasmonic behavior including peak positions, shifts and and shapes in ternary nanocomposites. The extension to multi-component systems in our approach is straightforward. The random ensembles are generated as described. Instead of assuming that all the particles have the same properties, we assign to the particles the properties of the constituents we want to consider. The method is not restricted to a special form of $\varepsilon(\omega)$ for the metallic nanoparticles. Also experimental information could be used here.

18.4 Summary

The above reported results show that glass doped with metal nanoparticles is a very promising material for photonic applications. The basic optical material parameters can be varied to a large extend by using optimized size, shape and concentration of the nanoparticles. The material can easily be structured down to a length scale of visible light and the shape of the individual structure cell can be adopted to special purposes.

Finite element methods (FEM) allow to model the dielectric properties of the material. This is the basis to model photonic devices efficiently. Optimal designs for waveguide structures in such a nanocomposite material can be calculated.

Of special interest is the possibility of the generation of uniformly oriented ellipsoidal metal nanoparticles in the structured material. First estimates show that the index contrast can be increased using birefringence the in glass containing ellipsoidal particles.

Acknowledgement

The authors thank the Schwerpunktprogramm 1113 "Photonische Kristalle" of the DFG (Deutsche Forschungemeinschaft) for financial support.

References

1 Minoshima, K., Kowalevicz, A.M., Hartl, I., Ippen, E.P. and Fujimoto, J.G. (2001) *Opt. Lett.*, **26**, 1516.

2 Minoshima, K., Kowalevicz, A.M., Ippen, E.P. and Fujimoto, J.G. (2002) *Opt. Express*, **10**, 645.

3 Sun, H., Xu, Y., Juodkazis, S., Sun, K., Watanabe, M., Matsuo, S., Misawa, H. and Nishii, J. (2001) *Opt. Lett.*, **26**, 325.

4 Kreibig, U. and Vollmer, M. (1995) *Optical Properties of Metal Clusters*, Springer, Berlin.

5 Kaempfe, M., Rainer, T., Berg, K.-J., Seifert, G. and Graener, H. (1999) *Appl. Phys. Lett.*, **74**, 1200; (2000) *Erratum: Appl. Phys. Lett.* **77**, U1 .

6 Seifert, G., Kaempfe, M., Berg, K.-J. and Graener, H. (2001) *Appl. Phys. B*, **73**, 355.

7 Podlipensky, A.V., Gebenev, S., Seifert, G. and Graener, H. (2004) *J. Lumin.*, **109**, 135.

8 Deparis, O., Kazansky, P.G., Abdolvand, A., Podlipensky, A., Seifert, G. and Graener, H. (2004) *Appl. Phys. Lett.*, **85**, 872.

9 Podlipensky, A., Abdolvand, A., Seifert, G., Graener, H., Deparis, O. and Kazansky, P.G. (2004) *J. Phys. Chem. B*, **108**, 17699.

10 Abdolvand, A., Podlipensky, A., Seifert, G., Deparis, O., Kazansky, P.G. and Graener, H. (2005) *Opt. Express*, **13**, 1266.

11 Sancho-Parramon, J., Abdolvand, A., Podlipensky, A., Seifert, G., Graener, H. and Syrowatka, F. (2006) *Appl. Opt.*, **45**, 8874.

12 Abdolvand, A., Podlipensky, A., Matthias, S., Syrowatka, F., Gösele, U., Seifert, G. and Graener, H. (2005) *Adv. Mater.*, **17**, 2983.

13 Porstendorfer, J., Berg, K.-J. and Berg, G. (1999) *J. Quant. Spectrosc. Radiat. Transf.*, **63**, 479.

14 These glasses are sold as porlarizers by CODIXX AG Barleben, Germany.

15 Garnett, J.C.M. (1904) *Philos. Trans. R. Soc. Lond.*, **2043**, 385.

16 Bruggeman, D.A.G. (1935) *Ann. Phys.*, **416**, 665.

17 Krakovský, I. and Myroshnychenko, V. (2002) *J. Appl. Phys.*, **92**, 6743; Sareni, B., Krähenbühl, L. and Beroual, A. (1996) *J. Appl. Phys.*, **80**, 4560.

18 Podlipensky, A., Abdolvand, A., Seifert, G. and Graener, H. (2005) *Appl. Phys. A, Mater. Sci. Process*, **80**, 1647.

19 Milton, G.W. (1981) *J. Appl. Phys.*, **52**, 5286; (1980) *Appl. Phys. Lett.*, **37**, 300.

20 Myroshnychenko, V. and Brosseau, C. (2005) *Phys. Rev. E*, **71**, 016701; (2005) *J. Appl. Phys.*, **97**, 044101.

21 FEMLAB 3.2, Comsol, Stockholm Sweden. (2006).

22 Uehara, Y., Ree, T. and Ree, F.H. (1979) *J. Chem. Phys.*, **70**, 1876; Wertheim, W.S. (1964) *J. Math. Phys.*, **5**, 643; Wertheim, M.S. (1963) *Phys. Rev. Lett.*, **10**, 321; Yuste, S.B. and Santos, A. (1991) *Phys. Rev. A*, **43**, 5418.

23 Garcia, H., Trice, J., Kalynaraman, R. and Sureshkumar, R. (2007) *Phys. Rev. B*, **75**, 045439.

19
Optical Properties of Disordered Metallic Photonic Crystal Slabs

D. Nau, A. Schönhardt, A. Christ, T. Zentgraf, Ch. Bauer, J. Kuhl, and H. Giessen

19.1
Introduction

In recent times, a lot of effort has been devoted to examine the optical properties of dielectric photonic crystals. The idea for these crystals arose several years ago, when they were discussed as materials to control radiative properties [1] or to localize light [2]. Especially the proposal to use photonic crystal structures for novel applications brought about fascinating concepts [3]. The idea behind these crystals is a perfect periodic variation of the dielectric constant, where the periodicity is on the order of the wavelength of light [4,5]. Such an arrangement can cause Bragg scattering of electromagnetic waves, resulting in stop bands in their electromagnetic transmission characteristics.

One specific problem arises when working with photonic crystals. Theory and device concepts always deal with perfect periodic structures, where the different constituents are arranged on perfect lattices. However, such crystals are artificially fabricated materials. Especially when working in the visible spectral range, the fabrication requirements often reach the limits of the utilized lithographic methods. Consequently, real photonic crystals can show strong deviations from the perfect structure [6]. Of course, such disorder directly influences the optical properties of real crystals [7]. Not only do possible applications require a detailed knowledge about the influence of disorder in these artificial structures. A fundamental point of view, the already interesting optical properties of photonic crystals show further intriguing effects in the presence of disorder. Typical examples are disorder-induced modifications of photon states and of the transmission [2,8–10].

The subclass of *metallic* photonic crystals has gained a lot of interest recently [11]. In metal-based structures, one of the dielectric constituents is replaced by a metal. One possibility to fabricate such structures is the periodic arrangement of metallic nanostructures on top of a dielectric waveguide slab [12]. This metallic photonic crystal slab (MPCS) belongs to the class of crystals that provide simultaneously photonic and electronic resonances in the same spectral range. A strong coupling

Nanophotonic Materials: Photonic Crystals, Plasmonics, and Metamaterials. Edited by R.B. Wehrspohn, H.-S. Kitzerow, and K. Busch

ISBN: 978-3-527-40858-0

between the resonances (i.e., particle plasmon and quasiguided slab mode) due to the formation of a polariton-type quasiparticle comes along with a pronounced polaritonic bandstructure [13,14].

This paper summarizes our work about the optical properties of disordered metallic photonic crystal slabs [15–18]. The paper starts with the description of the samples and the different disorder models, which are characterized in detail (Section 19.2). Section 19.3 presents experimental and numerical results of the linear optical properties of disordered MPCS, their bandstructure is described in Section 19.4.

19.2
Sample Description and Disorder Models

We fabricated the samples by using electron-beam lithography. As this method allows to precisely control shape and position of the fabricated nanostructures within an area of about $100 \times 100\ \mu m^2$, we were able to artificially introduce disorder with very high accuracy. This process provides a powerful tool to implement artificial disorder with a well defined type and strength into the samples [15–18]. When measuring the optical properties of such structures, a direct relation between disorder type and amount on the one hand and optical properties on the other hand can be found.

The MPCS consisted of metallic nanowires that were arranged on top of an indium-tin oxide (ITO) layer that was deposited on a glass substrate [see Figure 19.1 (a)]. The ITO-layer had a thickness of 140 nm, the nanowires had a height of 20 nm and a width of 100 nm. The optical properties were determined with a linear white-light transmission setup. It consisted of a halogen lamp that produced linearly polarized white-light at an aperture angle of below 0.2°, and the transmitted light was

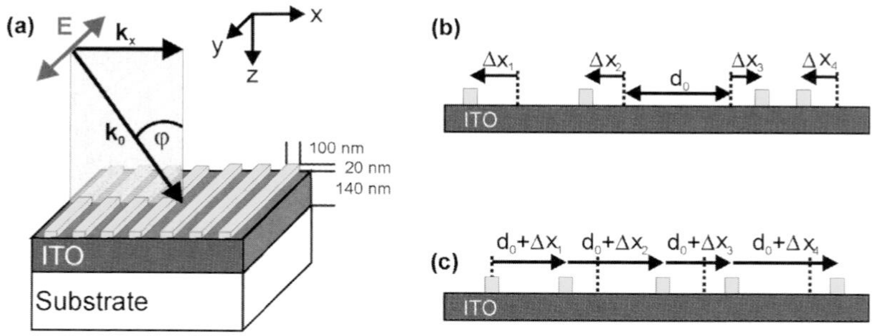

Figure 19.1 (a) Metallic photonic crystal slab consisting of a gold grating on top of a dielectric waveguide layer. E indicates the direction of the electrical field polarization, φ denotes the angle of light incidence. (b) and (c) Schemes of uncorrelated and correlated disorder, respectively. Dotted lines indicate the center positions of the perfect grating with period d_0.

analyzed with a spectrometer [13]. With this setup we were able to vary the polar angle φ of light incidence. For the transmission measurement (see Section 19.3) we set $\varphi = 0$, the bandstructure was determined at an oblique angle $\varphi \neq 0$ (see Section 19.4).

In this context, disorder means a variation of the positions of the nanowires on top of the waveguide layer (positional disorder), the width and the height of the nanowires are kept fixed. Two different disorder models with different next-neighbor correlations are considered [15]. In the first, the positions of neighboring nanowires are uncorrelated. Their positions are shifted with respect to their positions in the perfectly ordered system [see Figure 19.1(b)]. Starting at position x_0, the position of nanowire i is given by

$$x_i = x_0 + i d_0 + \Delta x_i, \tag{19.1}$$

where d_0 is the period of the perfect grating and Δx_i is the variation of the i-th position. This model resembles a "frozen phonon".

In the second model, the position of a nanowire is related to the position of the preceding one. Hence, the positions x_i and x_{i-1} of neighbors are correlated [Figure 19.1(c)], which resembles long-range disorder [19]. Therefore, x_i includes all variations of the preceding nanowires,

$$x_i = x_{i-1} + d_0 + \Delta x_i = x_0 + i d_0 + \sum_{n=1}^{i} \Delta x_n. \tag{19.2}$$

The variations Δx follow a uniform or a normal distribution with a full-width at half-maximum w. Giving w as a fraction of d_0 quantifies the disorder amount $a[\%] = w/d_0 \cdot 100$.

To precisely study the influence of different disorder types on the optical properties of MPCS, we fabricated a series of samples with uniform uncorrelated, uniform correlated, normal uncorrelated and normal correlated disorder, where the amount of disorder was increased in steps of 10%. In a previous analysis we found that only the disorder type (correlated or uncorrelated) affects the optical properties of the MPCS [16]. The distribution of the Δx does not alter the principal observations. However, considered disorder type and distribution are specified throughout the text. For further details about disorder type and distribution see [16].

Statistical methods allow to point out the differences of the disorder models [16]. According to the definition of the positional variations, the mean value $\overline{\Delta x}$ of all variations of the positions is

$$\overline{\Delta x} = \frac{1}{N} \sum_{i=1}^{N} \Delta x_i = 0. \tag{19.3}$$

Together with Eqs. (19.1) and (19.2), the averaged position x_N of nanowire N is found to be identical for n different realizations of correlated (C) and uncorrelated (U) disorder samples:

$$\overline{x_N^{\mathrm{U}}} = \overline{x_N^{\mathrm{C}}} = \frac{1}{n} \sum_{j=1}^{n} x_{N,j} = x_0 + N d_0. \tag{19.4}$$

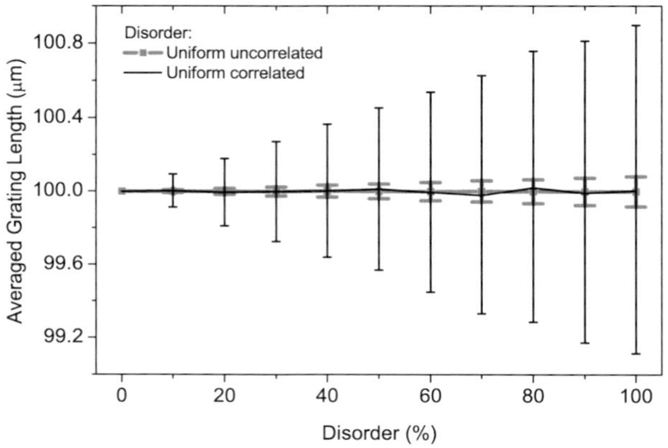

Figure 19.2 Averaged grating length $\bar{x}_N$ of samples for different amounts of uniform uncorrelated and uniform correlated disorder. The simulations were performed for 251 nanowires and a period of 400 nm, the statistics was performed for 2000 different arrays.

However, their standard deviations are different,

$$s_N^{\mathrm{U}} = \sqrt{\overline{(\Delta x_N)^2}}, \qquad s_N^{\mathrm{C}} = \sqrt{\frac{1}{n}\sum_{j=1}^{n}\left(\sum_{i=1}^{N}\Delta x_{i,j}\right)^2}.$$

Therefore $\sigma_N^{\mathrm{U}}\ ?\ \sigma_N^{\mathrm{C}}$, saying that correlated and uncorrelated disorder cause the same averaged grating length but with different deviations from this length. Hence, the long-range order is reduced stronger in a system with correlated disorder than in a system with uncorrelated disorder. The numerical results in Figure 19.2 underline these considerations. Here, the averaged grating length $\overline{x_N}$ is plotted together with the standard deviation σ_N for different disorder types and amounts.

Another possibility to characterize different models of disorder is to calculate the *two-point correlation function* (TPCF) [16]

$$D(d) = \sum_{i=0}^{N}\sum_{j=0}^{N}\delta(d-(x_j-x_i)). \tag{19.5}$$

$D(d)$ gives the statistics of the distance d between any two nanowires in the arrangement [20] and is the autocorrelation of the positions x of the nanowires.

The calculation is done as follows: Starting at a certain nanowire, the distances to all other wires are calculated and plotted in a histogram. The same is then done for all other structures, resulting in a modified histogram, giving the aforementioned statistics. For a perfect periodic arrangement, the distance between any two wires is a multiple integer of the lattice period. Hence, the TPCF consists of δ-peaks at multiple integers of the lattice period. With $N+1$ nanowires and d_0 being the lattice constant,

$$D(d) = \sum_{i=0}^{N}\sum_{j=0}^{N}\delta(d-(j-i)d_0). \tag{19.6}$$

That means that the TPCF of a perfect grating is simply the sum of perfectly ordered combs of δ-peaks, centered at different points and resulting in a perfectly ordered comb. Since the size of a real array with nanowires is finite, the δ-peaks do not have the same amplitudes: they rather decrease their height for larger distances. Therefore, the resulting comb appears as a triangular shape. Figure 19.3 shows the calculated TPCF for a perfect periodic assembly of nanowires. The TPCF is plotted for different scales of the x-axis. In the top figure that shows all occurring distances, the peaks form a broad band due to their large density. Clearly, a reduction of the amplitudes towards the sides can be seen. Zooming into the TPCF reveals their character of single δ-peaks as mentioned above.

If we perform the TPCF-calculation for disordered structures, the characteristics of uncorrelated and correlated disorder become clearer [16]. In the uncorrelated model,

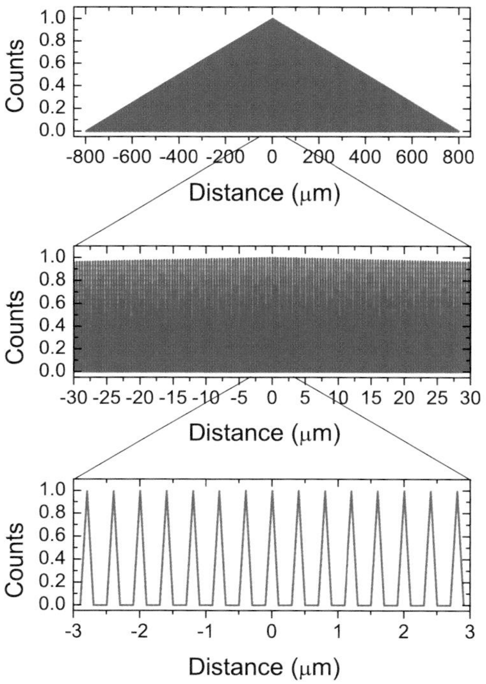

Figure 19.3 Two-point correlation function of a perfect periodic structure. From top to bottom the scale of the x-axis is decreased to reveal details. Calculation for 2000 grating points; the period was set to $d_0 = 400$ nm.

the positions of the nanowires are varied around their original grid positions. Hence, the distance between two certain nanowires i and j has increased by $\Delta x_i + \Delta x_j$ when compared with their distance $(j-i)d_0$ in the perfect grid. As a result, the TPCF-peak at $(j-i)d_0$ shows some broadening that is determined by the positional variation. This holds for all nanowires, which means that the TPCF-peaks at multiple integers of the lattice period are simply broadened homogeneously by the sums of the variations of all nanowires. At the same time, the amplitudes of the peaks reduce. The TPCF is given by

$$D(d) = \sum_{i=0}^{N}\sum_{j=0}^{N} \delta(d - (j-i)d_0 + \Delta x_i - \Delta x_j). \tag{19.7}$$

The result describes the summation of combs whose δ-peaks at $(j-i)d_0$ are shifted by $\Delta x_i - \Delta x_j$ with respect to each other. The TPCF of a structure with uncorrelated disorder is therefore a comb of broadened δ-peaks. Since the FWHM of the distribution of Δx_i and Δx_j increases for increasing disorder, the broadening of the δ-peaks increases and the amplitudes decrease. Figure 19.4 shows the calculated TPCF for a structure with increasing uniform uncorrelated disorder (left panel). Again we observe broad bands containing peaks, but their amplitudes decrease and their widths increase for increasing disorder. Visually spoken, the band of δ-peaks is squeezed together when compared to the case of perfect order. Due to the small

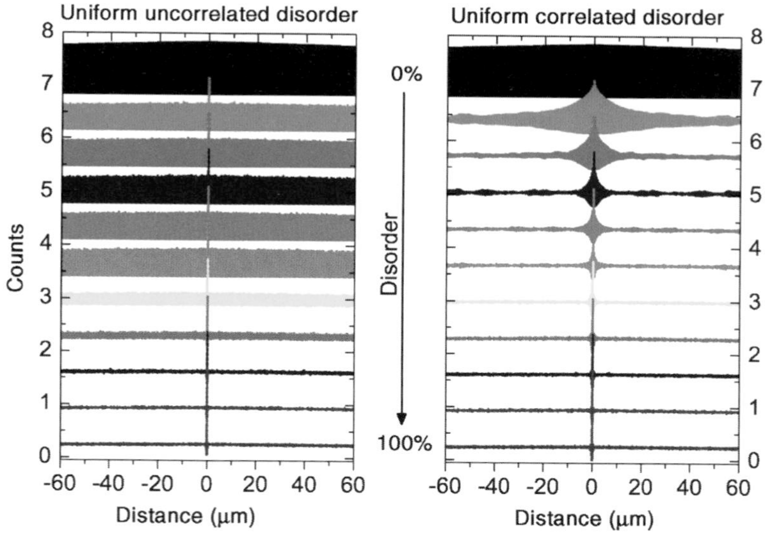

Figure 19.4 Two-point correlation function for structures with different amounts of uniform uncorrelated (left panel) and uniform correlated disorder (right panel). The calculation was done for structures with 2000 nanowires, the period d_0 of the perfect crystal was 400 nm.

period of $d_0 = 400$ nm, the single δ-shaped peaks can not be observed in the broad distance-range presented in this graph.

The TPCF of a crystal with correlated disorder looks quite different. Due to the next-neighbor correlation of the nanowires' positions, varying the position of a certain wire influences the positions of all following wires. Hence, starting at a certain nanowire, the distance to following nanowires includes the variations of all nanowires in-between. Therefore, the distance between two nanowires i and j deviates stronger from their distance in the perfect crystal as more nanowires lie in-between them (i.e., the larger the difference $j - i$ is). Calculating the TPCF for such a system shows that the TPCF-peaks broaden more strongly if the distances are larger. At the same time, the amplitudes for larger distances are decreased more strongly than those for smaller distances. The TPCF is given by

$$D(d) = \sum_{i=0}^{N}\sum_{j=0}^{N} \delta\left(d - (j - i)d_0 + \sum_{n=1}^{i} \Delta x_n - \sum_{m=1}^{j} \Delta x_m\right). \qquad (19.8)$$

Here, the positions of the δ-peaks at $x_0 + (j - i)d_0$ are shifted by the sums of the variations Δx_n and Δx_m, which are not constant but rather changing. Therefore, the δ-peaks are broadened and additionally shifted, which leads to a strongly modified TPCF of systems with long-range disorder.

The TPCF for a structure with correlated disorder is plotted in Figure 19.4 (right panel). It shows a strong decline with large distances, and increasing the disorder causes this decline to become stronger.

Comparing the TPCF for various disorder models clarifies the differences of these models. Peaks at small distances characterize short-range ordering, whereas peaks at long distances give information about the long-range arrangement of the grating structure. In the uncorrelated model the peaks reduce their amplitudes homogeneously for small and large distances. Therefore, short-range and long-range characteristics are modified equally. Hence, uncorrelated disorder only reduces the quality of the grating arrangement. The grating itself is not "destroyed", which means that the long-range order of the arrangement is preserved. This does not hold for the correlated disorder model, where the peaks at larger distances reduce their amplitudes stronger than those at smaller distances. Here, the long-range order is destroyed faster than the short-range order. The sharp peak at the origin of the TPCF arises from the distance of each point to itself. This peak therefore appears in both disorder models for all disorder strengths.

Further information about the physical nature of the considered disorder models can be deduced from a Fourier analysis of the spatial arrangement of the nanowires [16,18], as this analysis is similar to the results obtained from diffraction experiments [21]. The spatial arrangement of all nanowires in a fabricated sample is considered and the Fourier transform of this arrangement is calculated. Figure 19.5 shows results for exemplary arrangements with normal uncorrelated and normal correlated disorder, for simplicity we concentrate around the first reciprocal lattice vector $g_0 = 2\pi/d_0$ ($d_0 = 400$ nm). It is found that correlated and uncorrelated disorder have a different influence on the amplitudes $|A|$ and the momenta k in the Fourier

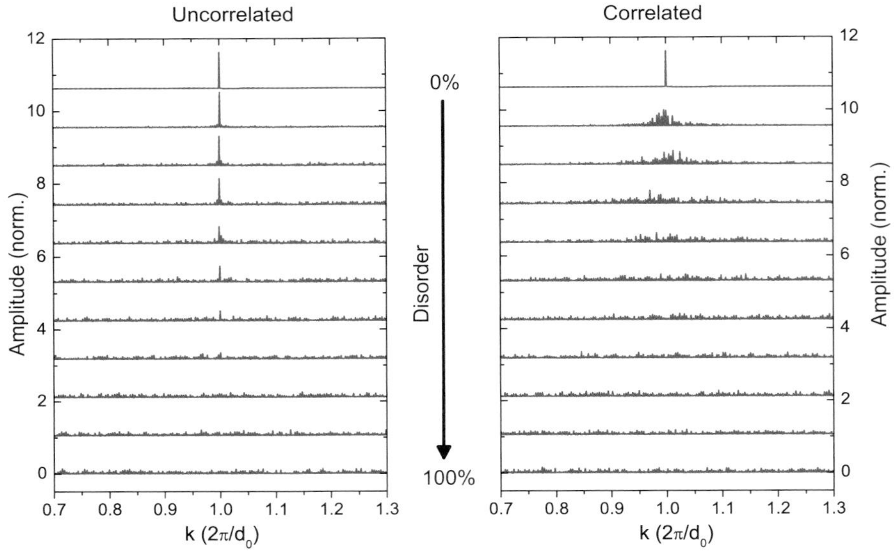

Figure 19.5 Fourier analysis of exemplary structures with normal uncorrelated and normal correlated disorder. The disorder was increased in steps of 10%. Each structure consisted of 1000 nanowires with $d_0 = 400$ nm. The curves are shifted for clarity.

spectrum. In the case of no disorder, a sharp δ-shaped peak appears at $k = g_0$. This peak is retained for increasing uncorrelated disorder, although its amplitude is continuously reduced. The peak completely vanishes at an disorder amount of 80%. Correlated disorder also diminishes the amplitude of this peak, but for smaller amounts of disorder: It vanishes for an amount of about 50%. Furthermore, additional peaks arise at $k \neq g_0$, causing an inhomogeneous broadening in the Fourier spectrum. These differences are the key to understand the spectral characteristics of the various disorder models.

To better understand these results, we make use of the similarity between a Fourier analysis of a spatial arrangement of nanostructures and the results from diffraction experiments on these nanoscatterers [21]. We can use these results to explain the Fourier analysis of the disorder types. The diffraction pattern of perfect crystals with period d_0 shows sharp peaks at multiple integers of the reciprocal lattice vector $g_0 = 2\pi/d_0$. This corresponds to the Fourier peak at $k = g_0$. In a thermal hot solid, the positions of the atoms are varied around their original grid positions resulting in reduced amplitudes of the diffraction peaks. This behaviour can be modelled with the Debye–Waller factor in solid state physics [21] and was already transferred to metallic photonic crystals [22]. For a variation Δx^2 of the nanowires' positions, the scattered intensity I_{Sca} for the reciprocal lattice vector k is given by

$$I_{Sca} = I_0 \exp(-\tfrac{1}{3}\langle \Delta x^2 \rangle k^2). \tag{19.9}$$

I_0 is the scattered intensity without disorder. The phenomena observed in the Fourier analysis of correlated disordered samples resemble diffraction experiments with liquids and amorphous materials. These are systems in which some ordering takes place on a short-range scale; however, they are disordered on a long-range scale [23]. Their diffraction patterns show broad maxima that are not clearly separated from each other [24,25]. Similar observations can be made in the Fourier analysis of systems with correlated disorder (see Figure 19.5). The scattered intensity is given by [21]

$$I_{\text{Sca}} = Nf^2\left[1 + \sum_{m \neq n} (\sin(kx_{mn}))/kx_{mn}\right]. \tag{19.10}$$

I_{Sca} contains the number N of atoms and the atomic form factor f.

These results indicate that our models describe in fact systems with different disorder types. Uncorrelated disorder is similar to thermally excited phonons in a solid lattice, and correlated disorder resembles liquid or amorphous systems.

19.3 Transmission Properties

The linear optical properties of MPCS were determined at normal light incidence with $\varphi = 0$ (see Section 19.2). The extinction ($-\ln(T)$, T: transmission) of *ordered* MPCS is characterized by complex lineshapes in TE and TM polarization. In TE polarization, the extinction shows a highly asymmetric Fano-form, caused by the excitation of the TE quasiguided mode inside the dielectric layer [13,14]. In TM polarization, the additional particle plasmon couples to the TM quasiguided mode. The extinction of the resulting plasmon-waveguide-polariton is characterized by two pronounced peaks (lower and upper polariton branch) [12–14,26].

The linear optical properties of *disordered* MPCS are found to show a distinctively different behavior [15,16,18]. Extinction spectra of samples with different types and amounts of disorder, together with results obtained by theory (see later), are plotted in Figure 19.6. The typical extinction characteristics of ordered MPCS are found for the case of no disorder. In TE polarization, a highly asymmetric Fano-shaped resonance is observed. However, increasing uncorrelated disorder reduces the amplitude of this resonance and causes a vanishing at about 70% disorder. The width of the resonance is not affected. This amplitude reduction goes along with a slight shift of the resonance to lower energies. Correlated disorder reduces the resonance amplitude more drastically, resulting in a vanishing at about 50% disorder. Additionally, a strong broadening of the resonance is observable which completely destroys the asymmetric Fano-lineshape of the resonance. In TM polarization, the two peaks of the coupled system of quasiguided mode and particle plasmon appear in the spectrum for no disorder.

The differences in energy and the form of the resonances are caused by the fact that the MPCS were fabricated on different samples in different fabrication processes.

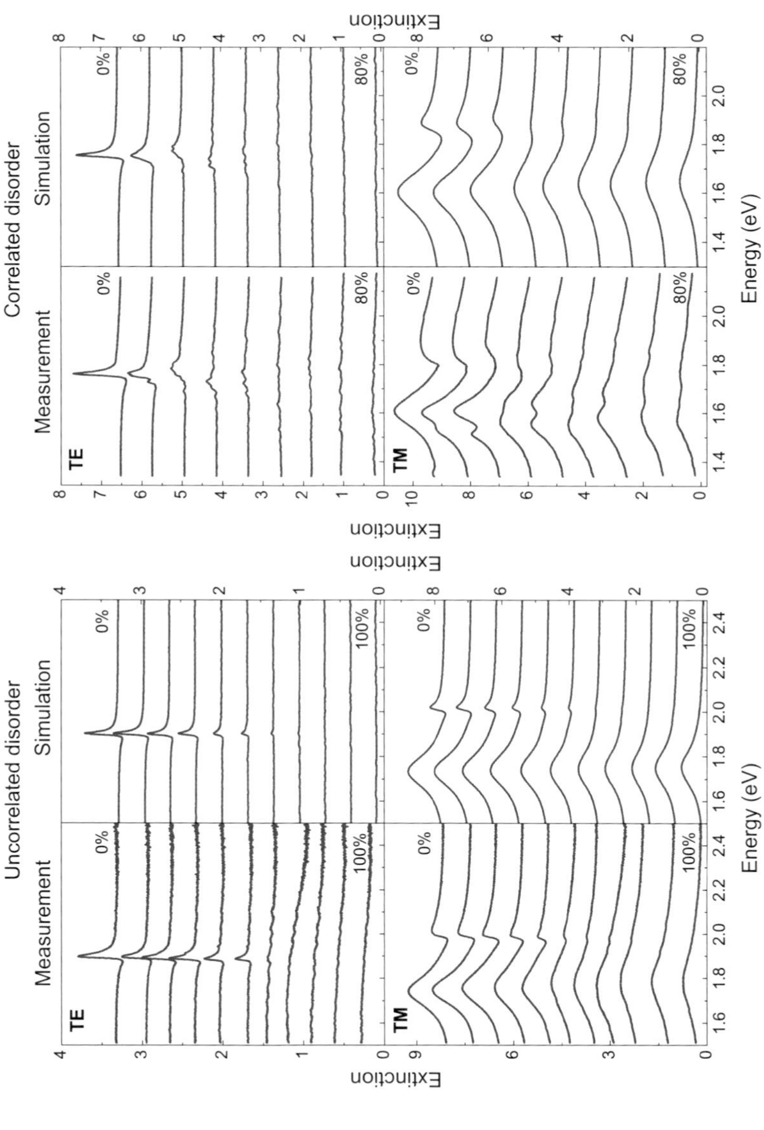

Figure 19.6 Measured and simulated extinction spectra for different disordered samples in TE and TM polarization. Left panel: results for normal uncorrelated disorder, $d_0 = 430$ nm. Right panel: results for normal correlated disorder, $d_0 = 475$ nm. The disorder was increased in steps of 10%. The individual spectra are shifted for clarity.

Deviations in the gold evaporation process are responsible for varying heights of the nanowires on the order of a few nanometers and hence for varying energies of the particle plasmon. Differences of the periods of the MPCS lead to different energies of the quasiguided mode.

Uncorrelated disorder reduces the amplitude of the upper polariton branch, causing a broadened particle plasmon peak for maximum disorder. We also note a decreased energy separation between the branches for increasing disorder. In the case of correlated disorder the shape of both polariton branches are influenced, they both show a broadening. For maximum disorder, again a single broadened plasmon peak arises.

These results can be modelled and understood by using a theoretical approach presented in [18]. The simulation of the optical properties of the samples is done in several steps. First, the spatial arrangement of all nanowires in a fabricated sample is considered and the Fourier transform of this arrangement is calculated (see Section 19.2 and Figure 19.5). In the second step, we consider the dispersion relation of the MPCS. The TE and TM quasiguided modes depend on the lateral wavevector, and they are given by the solutions of transcendental equations [27]. In TM polarization, the quasiguided mode at energy $E_{\mathrm{wg}}(k_j)$ couples to the plasmon at constant energy E_{pl}. This results in a typical polariton dispersion with two dispersion branches in ordered MPCS [13,14]. For normal light incidence, the period, or, more generally, the Fourier spectrum determines the energies of the excited resonances [22]. Their amplitudes are given by the amplitudes of the Fourier components and hence the lattice structure [28]. Combining Fourier analysis and dispersion of the sample yields a transformation of the Fourier peaks from k-space into energy space: each Fourier component at k_j excites resonances with energies $E_j = E(k_j)$ and amplitudes $I_j \propto |A_j|^2$.

To finally calculate the optical spectra of the samples, we plug the energies E_j and amplitudes $|A_j|$ of the excited resonances into suitably chosen lineshape functions. In TE polarization, the extinction $\alpha_j^{\mathrm{TE}}(E)$ is characterized by a Fano-type lineshape [13,29]. In TM polarization, the absorption of the system can be modelled with two coupled Lorentzian oscillators [26]. For simplicity, we identify the extinction with this absorption [26]:

$$\alpha^{\mathrm{TM}}(E) = \alpha_N \frac{4\gamma_{\mathrm{pl}}^2 E^2 [E^2 - E_{\mathrm{wg}}^2 - (q_{\mathrm{wg}}/q_{\mathrm{pl}}) E_{\mathrm{c}}^2]^2}{1/\hbar^2 [(E^2 - E_{\mathrm{pl}}^2)(E^2 - E_{\mathrm{wg}}^2) - E_{\mathrm{c}}^4]^2 + 4\gamma_{\mathrm{pl}}^2 E^2 (E^2 - E_{\mathrm{wg}}^2)^2}, \tag{19.11}$$

with α_N as a scaling factor depending on the density of the excited pairs of plasmon and quasiguided mode, and E_l, γ_l, and q_l ($l = \mathrm{pl}$, wg) as resonance energies, homogeneous half-widths, and oscillator strengths of the uncoupled system. The waveguide-plasmon coupling strength is denoted as E_{c}^2. Thus, each Fourier component at k_j yields an extinction $\alpha_j^{\mathrm{pol}}(E)$, where 'pol' denotes the polarization (TE or TM). The total extinction $\alpha^{\mathrm{pol}}(E)$ of the disordered samples is the sum over all $\alpha_j^{\mathrm{pol}}(E)$:

$$\alpha^{\mathrm{pol}}(E) = \sum_j \alpha_j^{\mathrm{pol}}(E). \tag{19.12}$$

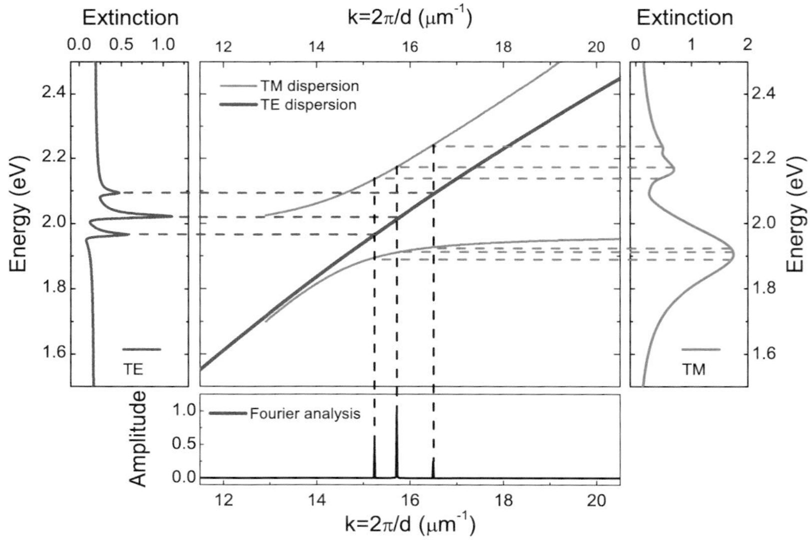

Figure 19.7 Transformation of exemplary Fourier peaks from *k*-space into energy space via the dispersion relations. Schematic lineshape functions are added in energy space to yield $\alpha^{TE}(E)$ and $\alpha^{TM}(E)$.

The necessary TE and TM parameters are determined by fitting the individual lineshape functions to the measured extinction spectra of the perfect samples. They are then used as disorder-independent parameters in the calculation of $\alpha_j^{pol}(E)$ and $\alpha^{pol}(E)$. Adapting the polariton picture to Eq. (19.11) means that each Fourier component excites simultaneously a plasmon and a quasiguided mode as a pair of coupled resonances in TM polarization. Hence, I_j affects directly the density-dependent factor α_N of these pairs in Eq. (19.11). We set $\alpha_N/\alpha_0 = I_j = |A_j|^2$, where α_0 is the maximum extinction as determined from the fits of the perfect samples. An illustration of the model is presented in Figure 19.7 with spatial Fourier analysis, TE and TM dispersion, and the resulting extinction spectra in energy space.

The comparison in Figure 19.6 shows a good agreement of experimental and simulated results. Both the peak reduction and the peak broadening are reproduced well in the simulations. With the model described above we are able to completely understand and explain the observations. In TE polarization, the extinction is simply a convolution of the spatial Fourier spectrum (see Figure 19.5 as an example) with the Fano lineshape. For correlated disorder, the arrangement of the nanowires acts as a superposition of gratings with different periods. Therefore, multiple quasiguided modes are excited in a wide energy range [30], causing a broadening of the extinction peak.

The spatial Fourier analysis also helps to understand the observations in TM polarization. In the case of uncorrelated disorder, the reducing Fourier peak amplitude gives evidence for a decrease of the polariton contribution to the spectrum. The increasing background in the Fourier spectrum increases the contribution of plasmons coupled to waveguides of arbitrary energy. As a result, a simple plasmon appears at large disorder amounts. For correlated disorder, the broadening of both polariton branches originates from the excitation of multiple resonances at energies varying in a certain, finite spectral range.

Comparing measurements and simul ations in TM polarization shows that some smaller details are not well reproduced. We account the omission of near-field coupling of the individual nanowires to be responsible for this effect. This coupling presumably causes the strong broadening of the pure plasmon resonance at large disorder and the pronounced maxima and minima at moderate disorder.

19.4 Bandstructure

The results in Section 19.3 were obtained at a normal light incidence. Varying the polar angle of light incidence φ allows to determine the bandstructure of the MPCS [14]: The energies of the polariton resonances are plotted as a function of φ or of the momentum k_x with $k_x = k_0 \sin \varphi$.

In contrast to normal light incidence, two quasiguided modes are excited for $\varphi \neq 0$ in both polarisations: symmetric and antisymmetric quasiguided mode. They are both solutions of the transcendental wave equation in dielectric waveguide systems [27]. Due to symmetry reasons, the antisymmetric mode can not be excited at normal light incidence; it only appears in systems with broken symmetry at $\varphi \neq 0$ [31]. As a result, the extinction spectra in TE and TM show an additional quasiguided mode (not shown here).

In TM polarization, the bandstructure of *ordered* MPCS is characterized by the interaction of particle plasmon, symmetric quasiguided mode, and antisymmetric quasiguided mode [13]. The bandstructure consists of three pronounced bands that correspond to upper, middle, and lower polariton branch. The bands are separated by stop gaps being caused by the strong coupling in the MPCS. In TE polarization, the excitation of only symmetric and antisymmetric quasiguided modes leads to the dispersion of the quasiguided slab modes with two separated bands [14].

Note that at $k_x = 0$ (i.e., normal light incidence) the antisymmetric quasiguided modes are not excited (see above). The corresponding TE and TM bands therefore display a missing data point.

In this section we discuss the bandstructure of *disordered* MPCS. Different disorder models show pronounced differences in their effects on the bandstructure [17]. Results for a sample with uniform uncorrelated disorder are shown in Figure 19.8 (left panel) for increasing disorder. For no disorder, the bandstructure of the polariton can be observed. Increasing uniform uncorrelated disorder reduces the gaps between the bands. The splitting between the middle and upper polariton branches vanishes for a disorder amount of about 60%, and the splitting between upper and lower branch reduces continuously. However, the bandstructure itself is retained and not destroyed by this type of disorder. Even for large disorder amounts of up to 70% the bands of the different polariton branches are distinguishable. For still larger amounts

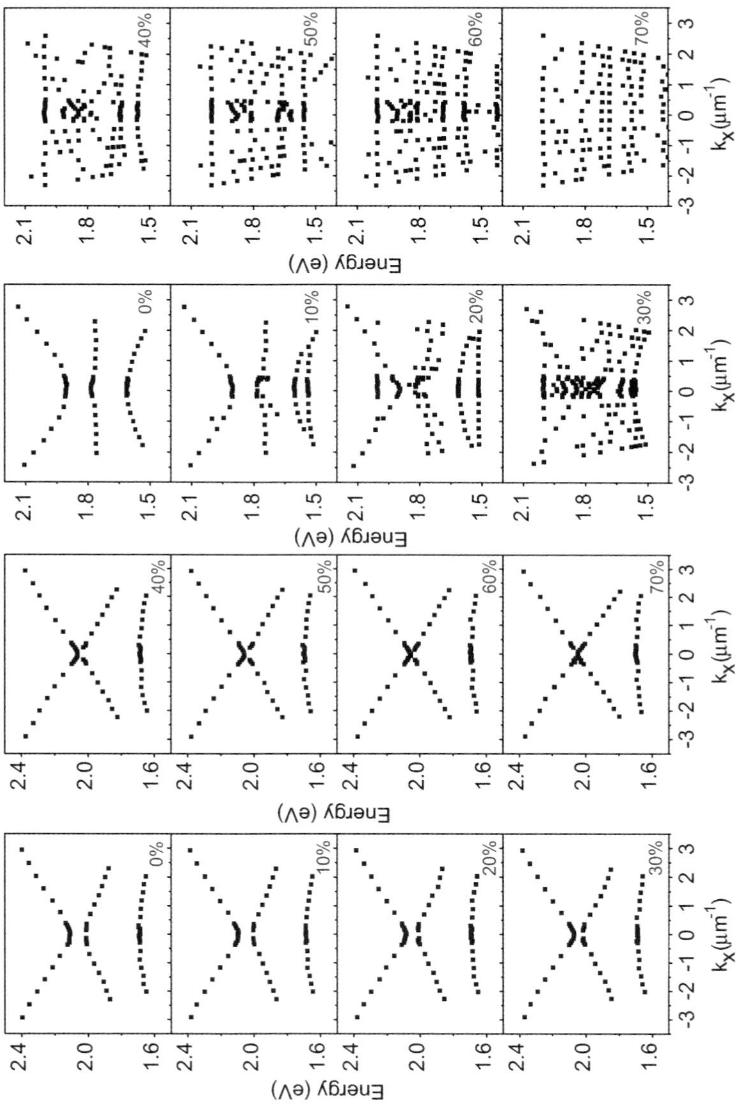

Figure 19.8 Polaritonic bandstructure of a sample with increasing uniform uncorrelated disorder (left panel) and with increasing normal correlated disorder (right panel) in TM polarization. The dots denote peaks in the extinction spectra.

(80% and more) the quasiguided modes are not excited any more (see Section 19.3). Thus, only the plasmonic polariton branch appears (not shown here).

The dispersion of a sample with Gaussian correlated disorder is shown in the right panel of Figure 19.5. For no disorder, the bandstructure is again that of the polariton in ordered MPCS.

As stated in Section 19.3, bands appear at slightly different energies when compared to the sample with uniform uncorrelated disorder due to different detunings of plasmon and quasiguided mode in the samples.

Increasing disorder starts to wash out the bandstructure, and an amount of 30% and higher completely destroys the polariton branches. No bandstructure is retained, and further structures appear, causing the original polariton branches to be inhomogeneously broadened.

The polariton dispersion of the *ordered* MPCS can be tailored by determining the eigenvalues of an effective Hamiltonian H_{eff} (see [13] for details). It includes the energy E_0 of the TM_0 quasiguided modes at $k_x = 0$, the energy E_{pl} of the individual wire plasmons, the stop-band half-width V_1 in the 1-dim photonic crystal slab, and the coupling energy V_2 of quasiguided mode and wire plasmon. Absorption and losses are taken into account by introducing finite half-widths to the resonances of plasmon and quasiguided mode. With Γ as half-width of the plasmon, its energy E_{pl} is replaced by $E_{\text{pl}} - i\Gamma$. The same holds for the quasiguided modes, whose half-width γ modifies their energies to $E_0 \pm c\, k_x - i\gamma$. The radiative losses of the quasiguided modes are modelled with a complex photonic band gap. With γ_1 being the radiative damping, V_1 is replaced by $V_1 - i\gamma_1$. We obtain the following matrix to calculate H_{eff} [13]

$$\begin{pmatrix} E_0 + V_1 - i(\gamma+\gamma_1) & \tilde{c}\, k_x & \sqrt{2}V_2 \\ \tilde{c}\, k_x & E_0 - V_1 - i(\gamma-\gamma_1) & 0 \\ \sqrt{2}V_2 & 0 & E_{\text{Pl}} - i\Gamma \end{pmatrix}. \tag{19.13}$$

It was found that the coupling strength of plasmon and quasiguided mode is modified by disorder and defects [15,22]. This can be understood intuitively as V_2 can be determined from the spatial overlap of the electrical fields of plasmon $\boldsymbol{E}_{\text{pl}}(\boldsymbol{r})$ and quasiguided mode $\boldsymbol{E}_{\text{WG}}(\boldsymbol{r})$. In a system with positional disorder, this overlap can be drastically reduced. To quantitatively determine the decrease of V_2, we calculate V_2 in a simple model. Here, the modified coupling strength obeys

$$V_2^{\text{dis}} = \frac{1}{V_2} \int_{-\infty}^{\infty} \boldsymbol{E}_{\text{WG}} \cdot \boldsymbol{E}_{\text{Pl}}\, \mathrm{d}\boldsymbol{r}. \tag{19.14}$$

We normalize it to the coupling strength V_2 of the ordered MPCS. The modification of V_2 is then taken into account by replacing V_2 in Eq. (19.13) by $V_2^{\text{mod}} = V_2 V_2^{\text{dis}}$. The electrical field of the plasmon $\boldsymbol{E}_{\text{pl}}$ is mainly localized at the positions of the nanowires. Introducing disorder varies these positions. Hence, the coupling of plasmon and quasiguided mode is reduced in disordered systems due to a changing spatial overlap and a therefore modified V_2 [see Figure 19.9 (a)]. Approximating $\boldsymbol{E}_{\text{WG}}$ by a cosine-type oscillation with period d_0

$$E_{\text{WG}}(x, \varphi) = \cos\left((2\pi x - \varphi)/d_0\right) \tag{19.15}$$

and assuming $\boldsymbol{E}_{\mathrm{pl}}$ to be a non-zero constant inside the nanowires and vanishing outside

$$E_{\mathrm{Pl}}(x) = \begin{cases} 1 : x \in \text{nanowire}, \\ 0 : x \notin \text{nanowire} \end{cases} \tag{19.16}$$

we estimate the change of V_2 by using the spatial arrangement of the nanowires. The results are shown in Figure 19.9(b). V_2 decreases with increasing disorder, however, the models differ in the dependence of V_2 on the degree of disorder. Decreasing V_2 reduces the coupling of plasmon and quasiguided mode, causing a reduction of the polariton splitting in the dispersion.

With this Hamiltonian we are able to simulate the polariton dispersion of a MPCS with uncorrelated disorder. All parameters in Eq. (19.13) are adapted to yield the correct results for no disorder (see also the parameters in [13]). For larger disorder amount, only V_2^{mod} was changed according to Eq. (19.14). It should be noted that also the stop-band half-width V_1 of the 1-dim photonic crystal slab is influenced by disorder (see later). However, we neglect this effect here.

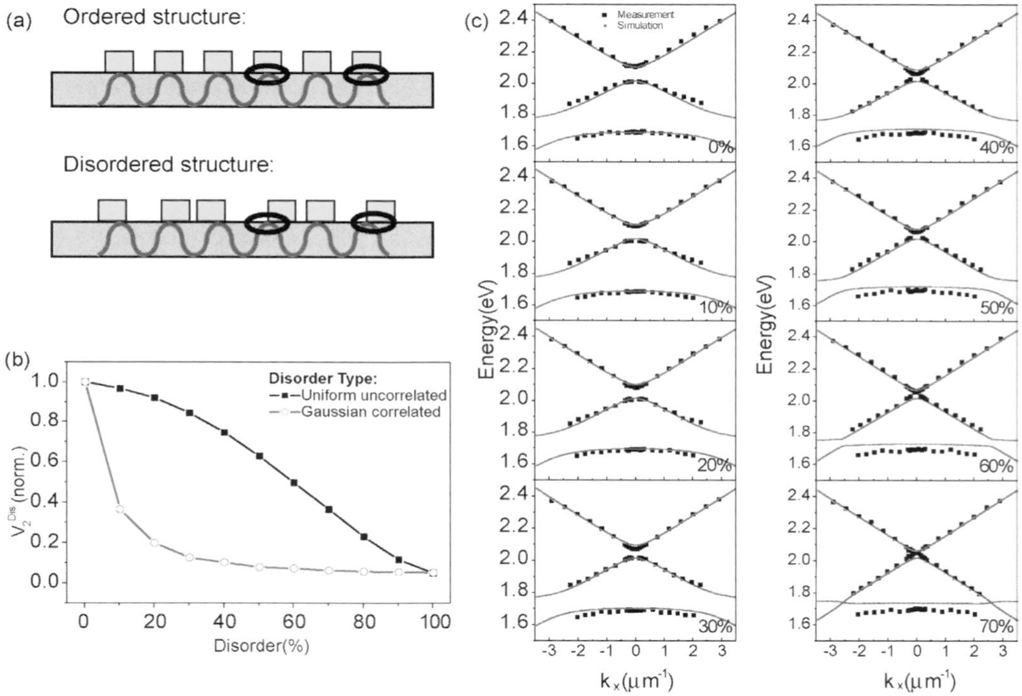

Figure 19.9 (a) Spatial overlap of the electrical fields of plasmon and quasiguided mode in ordered and disordered MPCS. (b) Normalized modified coupling strength V_2^{mod} as a function of disorder and for different disorder types. (c) Dispersion of the sample with uniform uncorrelated disorder: experiment and simulation in TM polarization. The bandstructure is plotted for increasing amounts of disorder.

The polariton bandstructure of a sample with uncorrelated disorder agrees well in experiment and simulation, as can be seen in Figure 19.9(c). Increasing uncorrelated disorder retains the dispersion branches of the polariton, and the splittings between the bands are reproduced nicely by the simulations. Deviations in the simulated results from the measurements are presumably caused by not considering the influence of disorder on V_1.

Although the complex behavior of the bandstructure in disordered MPCS is not yet completely understood, a simple picture allows to explain the reduced band separation for increasing uncorrelated disorder [17]. The band separation results from the strong coupling in this polaritonic system, which can be described by the coupling strength V_2 [13]. This V_2 can be interpreted as being caused by the spatial overlap of the electrical fields of particle plasmon and quasiguided mode [17]. In systems with positional disorder, this spatial overlap is reduced due to the shifted positions of the nanowires with respect to the electrical field of the quasiguided mode. As a result, the polariton coupling V_2 is reduced and hence the width of the stop-gaps in the bandstructure. This effect also appears for correlated disorder, however, it is not visible there due to the smearing of the bands.

The effect of band broadening in the case of correlated disorder can be explained when taking into account the results for disordered MPCS at normal light incidence (see Section 19.3). While uncorrelated disorder reduces the amplitude of the polariton branch that corresponds to the quasiguided mode, correlated disorder additionally excites multiple quasiguided modes at different energies. We can attribute these modes to be responsible for the smearing of the bands for correlated disorder. Each mode couples to the plasmon and hence forms a polariton with slightly different energy and slightly shifted dispersion.

The resulting dispersion of the sample leads to the complete vanishing of pronounced polariton branches. This is not the case for uncorrelated disorder with no additional quasiguided modes.

Similar results were obtained for the bandstructure in TE polarization [17]. Increasing uncorrelated disorder retains the bandstructure and reduces the width of the stop-gap continuously. Correlated disorder, however, shows a broadening of the bands which leads to the complete destruction of the bandstructure even at moderate disorder.

As pointed out above, the reduction of the bandsplitting in TE polarization due to uncorrelated disorder indicates a modified V_1 in Eq. (19.13). Again, this can be understood intuitively as in such MPCS symmetric and antisymmetric quasiguided modes have their nodes and antinodes under the nanowires, respectively [14]. Hence, both modes experience a different effective dielectric environment, which leads to different energies according to their dispersion [27]. Introducing positional disorder modifies the positions of the nanowires with respect to both modes. As a result, the difference in the effective dielectric environment of symmetric and antisymmetric mode decreases, which leads to a reduced energy difference and hence a reduced splitting in the bandstructure.

19.5 Conclusion

In conclusion, we discussed the optical properties of disordered metallic photonic crystal slabs. We fabricated samples with artificial disorder and considered disorder models with different next-neighbor correlations. The linear transmission is strongly modified, depending on the implemented disorder model. While uncorrelated disorder only reduces the amplitude of spectral peaks, correlated disorder additionally causes a strong broadening of the resonances. Both effects are explained by numerical simulations. The peak reduction originates from a reduced excitation efficiency of the corresponding resonance, the broadening is caused by the excitation of multiple resonances at slightly different energies. We also discussed the bandstructure of such disordered structures as determined from angle-resolved transmission measurements. It is found that uncorrelated disorder retains the bandstructure and reduces the separation of the bands. Correlated disorder, however, destroys the bandstructure already at small amounts of disorder. Again, this smearing of the bands is a consequence of the excitation of multiple resonances. Our results may have strong consequences for using metallic photonic crystal slabs as photonic devices (see e.g. [32]).

Acknowledgements

We thank Peter Thomas (Marburg) and Siegfried Dietrich (Stuttgart) for fruitful discussions and the teams of R. Langen and K. D. Krause for technical support. Financial support by DFG (SPP 1113) is greatfully acknowledged.

References

1 Yablonovitch, E. (1987) *Phys. Rev. Lett.*, **58**, 2059.
2 John, S. (1987) *Phys. Rev. Lett.*, **58**, 2486.
3 Krauss, T.F., De La Rue, R.M. and Brand, S. (1996) *Nature*, **383**, 699.
4 Busch K., Lölkes S., Wehrspohn R.B. and Föll H. (eds) (2004) *Photonic Crystals*, Wiley-VCH, Weinheim, Germany.
5 Joannopoulos, J.D., Villeneuve, P.R. and Fan, S. (1997) *Nature*, **386**, 143.
6 Vlasov, Y.A., Astratov, V.N., Baryshev, A.V., Kaplyanskii, A.A., Karimov, O.Z. and Limonov, M.F. (2000) *Phys. Rev., E*, **61**, 5784.
7 Koenderink, A.F., Lagendijk, A. and Vos, W.L. (2005) *Phys. Rev., B*, **72**, 153102.
8 Freilikher, V.D., Liansky, B.A., Yurkevich, I.V., Maradudin, A.A. and McGurn, A.R. (1995) *Phys. Rev., E*, **51**, 6301.
9 Bertolotti, J., Gottardo, S., Wiersma, D.S., Ghulinyan, M. and Pavesi, L. (2005) *Phys. Rev. Lett.*, **94**, 113903.
10 Sainidou, R., Stefanou, N. and Modinos, A. (2005) *Phys. Rev. Lett.*, **94**, 205503.
11 Brown, E.R. and McMahon, O.B. (1995) *Appl. Phys. Lett.*, **67**, 2138.
12 Linden, S., Kuhl, J. and Giessen, H. (2001) *Phys. Rev. Lett.*, **86**, 4688.

13 Christ, A., Tikhodeev, S.G., Gippius, N.A., Kuhl, J. and Giessen, H. (2003) *Phys. Rev. Lett.*, **91**, 183901.

14 Christ, A., Zentgraf, T., Kuhl, J., Tikhodeev, S.G., Gippius, N.A. and Giessen, H. (2004) *Phys. Rev., B*, **70**, 125113.

15 Nau, D., Christ, A., Linden, S., Kuhl, J. and Giessen, H. (2004) *CLEO/IQEC and PhAST Technical Digest on CDROM*, The Optical Society of America, Washington, DC, IThB6.

16 Nau, D., Schönhardt, A., Bauer, C., Christ, A., Zentgraf, T., Kuhl, J. and Giessen, H. (2006) *phys. stat. sol. (b)*, **243**, 2331.

17 Nau, D., Schönhardt, A., Chigrin, D.N., Kroha, H., Christ, A. and Giessen, H. (2007) *phys. stat. sol. (b)*, **244**, 1262.

18 Nau, D., Schönhardt, A., Bauer, C., Christ, A., Zentgraf, T., Kuhl, J., Klein, M.W. and Giessen, H. (2007) *Phys. Rev. Lett.*, **98**, 133902.

19 Maschke, K., Overhof, H. and Thomas, P. (1973) *phys. stat. sol. (b)*, **57**, 237.

20 Huang, K. (1964) *Statistische Mechanik III*, Bibliographisches Institut Mannheim.

21 Kittel, C. (1996) *Introduction to Solid State Physics*, John Wiley & Sons.

22 Zentgraf, T., Christ, A., Kuhl, J., Gippius, N.A., Tikhodeev, S.G., Nau, D. and Giessen, H. (2006) *Phys. Rev., B*, **73**, 115103.

23 Ziman, J.M. (1979) *Models of Disorder*, Cambridge University Press.

24 Zernike, F. and Prins, J.A. (1927) *Z. Phys.*, **41**, 184.

25 Laue, M.V. (1941) *Röntgenstrahlinterferenzen*, Akademische Verlagsgesellschaft, Leipzig.

26 Klein, M.W., Tritschler, T., Wegener, M. and Linden, S. (2005) *Phys. Rev., B*, **72**, 115113.

27 Barnoski, M.K. (1974) *Introduction to Integrated Optics*, Plenum Press, New York.

28 Yariv, A. and Nakamura, M. (1977) *IEEE J. Quantum Electron*, **13**, 233.

29 Fan, S. and Joannopoulos, J.D. (2002) *Phys. Rev., B*, **65**, 235112.

30 Marcuse, D. (1969) *Bell Syst. Tech. J.*, **48**, 3187.

31 Yablonskii, A.L., Muljarov, E.A., Gippius, N.A., Tikhodeev, S.G., Fujita, T. and Ishihara, T. (2001) *J. Phys. Soc. Jpn.*, **70**, 1137.

32 Nau, D., Bertram, R.P., Buse, K., Zentgraf, T., Kuhl, J., Tikhodeev, S.G., Gippius, N.A. and Giessen, H. (2006) *Appl. Phys., B*, **82**, 543.

20
Superfocusing of Optical Beams below the Diffraction Limit by Media with Negative Refraction

A. Husakou and J. Herrmann

20.1
Introduction

Focusing of light beams into an extremely small spot with a high energy density is an important issue in optics. High-numerical-aperture focusing is used for miniaturization in key technologies, such as lithography (see e.g. [1]), optical data storage (see e.g. [2]), laser material nanoprocessing, and nano-optics, as well as in confocal microscopy and numerous other fields. Pushing the spot size below the diffraction limit is of exceptional importance for many of these applications. The physical limit for the smallest spot size of focused light for conventional optical devices is determined by the fact that all evanescent waves of an electromagnetic wave are lost during propagation. The reason is that all the components $E(k_x, k_y, z)$ in the spatial Fourier presentation of the electric field $E(x, y, z, t) = \exp(-i\omega t) \int\int E(k_x, k_y, z) \exp(ik_x x + ik_y y)\, dk_x\, dk_y$ with transverse wavenumber $k_\perp = (k_x^2 + k_y^2)^{1/2}$ which satisfy the relation $k_\perp > 2\pi/\lambda$ (so-called evanescent waves) decay exponentially as $\exp(ik_z z)$ because for these components the longitudinal wavenumber $k_z = (k_0^2 - k_\perp^2)^{1/2}$ becomes imaginary. Here $\lambda = 2\pi c/\omega$ is the vacuum wavelength, $k_0 = 2\pi/\lambda = \omega/c$ and z the propagation length. Near-field optical methods can overcome this limitation due to the creation and interaction of evanescent components with the sample. One of the most common ways is the use of tapered or etched optical fibers [3] that confine a light spot to a nanometer area. Solid immersion lenses [4] can also produce subdiffraction spots by the application of high-refractive-index solids as lens material in the near-field range of the sample. However, the main problem for the application of these techniques is the required subwavelength proximity of the moving near-field element and the medium which is to be studied or modified. Although impressive results have been achieved in this field [5], the required spatial near-field control, low scanning speed and reliability, and small throughput remain formidable obstacles for applications.

Nanophotonic Materials: Photonic Crystals, Plasmonics, and Metamaterials. Edited by R.B. Wehrspohn, H.-S. Kitzerow, and K. Busch

ISBN: 978-3-527-40858-0

The present paper is devoted to the study of the possibility for superfocusing of scanning light beams below the diffraction limit, without requirement of spatial control within a subwavelength distance between a moving near-field element and the surface. Here we present an overview of our results obtained in this field, part of these results were already published before [6–8]. The studied method is based on recently discovered materials with a negative refractive index [9], like a metamaterial based on split-ring resonators and wires [10,11], strongly modulated photonic crystals [12,13], or parallel nanowires and nanoplates [14,15] and coupled nanocones [16]. The high interest in this topic is mainly connected with the observation [17] that evanescent components can be enhanced by a thin slab of a material with a negative refractive index, which thus forms a "superlens" with a resolution below the diffraction limit. The enhancement of evanescent components can also be achieved in photonic crystals with negative refraction in parameter regions without a definite refraction index (so-called all-angle negative refraction) due to coupling to bound photon modes [18]. Moreover a simple metal slab can also enhance evanescent components and act like a superlens [19,20]. Based on this work, up to now the superlensing effect was studied for imaging of subwavelength features [18,21–24]. In the case of imaging, near-field components of the object placed near the superlens are enhanced by the slab, which leads to a super-resolution of the image [see Figure 20.1 (a), (b)]. In contrast, in the case of focusing of a beam no evanescent components are present in the input beam which could be amplified by a superlens, as illustrated in Figure 20.1(d). Therefore an additional element placed directly before the superlens has to be introduced, which creates week seed evanescent components from the beam. This element can be a simple aperture as shown in Section 20.2 and illustrated in Figure 20.1(c), (d). However, for focusing of a scanning beam a fixed position of the aperture does not allow to create the spot in an arbitrary position. To avoid this problem we study in Sections 20.3 and 20.4 the application of

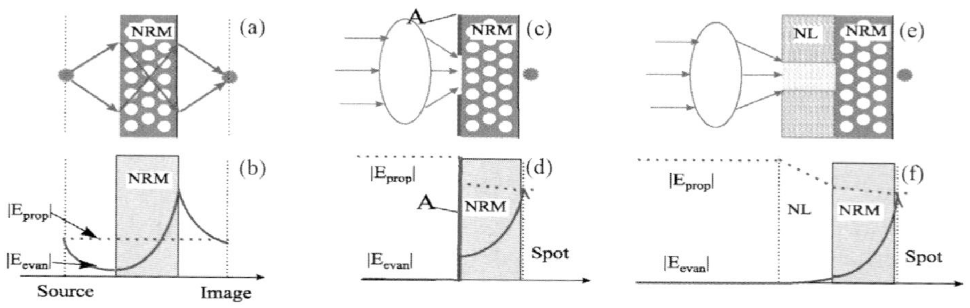

Figure 20.1 General scheme of superimaging by a negative-refraction media (NRM) (a)–(b), of superfocusing by a combination of an aperture and an NRM slab (c), (d), as well as of superfocusing of a scanning beam using a nonlinear layer and an NRM slab (e), (f). $|E_{prop}|$ and $|E_{evan}|$ denote the amplitudes of respectively propagating and evanescent components, A denotes an aperture, NL denotes a nonlinear layer which can be a saturable absorber or a Kerr-type nonlinear material, NRM is a slab of negative-refraction layer.

light-controlled nonlinear optical elements, as illustrated in Figure 20.1(e), (f), like a saturable absorber (SA) (Section 20.3) or an element with a nonlinear refraction index like a Kerr-type medium (Section 20.4) for the creation of week evanescent components from a beam.

20.2
Superfocusing of a Non-Moving Beam by the Combined Action of an Aperture and a Negative-Index Layer

20.2.1
Effective-Medium Approach

To get insight into the phenomenon of superfocusing we first describe the propagation of a monochromatic beam through a plane slab of a medium with a negative refraction index in the context of effective medium theory which models the artificial negative-refraction material (NRM) by a homogeneous material with $\mu < 0$ and $\varepsilon < 0$. The propagation of a beam through a homogeneous slab of a material with a negative refractive index can be described solving the Maxwell equation for the spatial Fourier components and using the boundary conditions on both sides of the slab. Then for arbitrary ε and μ the output field $\boldsymbol{E}_{\mathrm{s,p}}(k_x, k_y, L)$ is related with the input field $\boldsymbol{E}_{\mathrm{s,p}}(k_x, k_y, 0)$ by $\boldsymbol{E}_{\mathrm{s,p}}(k_x, k_y, L) = T_{\mathrm{s,p}}(k_x, k_y)\ \boldsymbol{E}_{\mathrm{s,p}}(k_x, k_y, 0)$, where $k_{x,y}$ are the transverse components of the wavevector and $T_{\mathrm{s,p}}(k_x, k_y)$ are the transfer functions [17]

$$T_{\mathrm{s,p}} = 4[(2+k_{\mathrm{s,p}})\mathrm{e}^{-iq_zL} + (2-k_{\mathrm{s,p}})\exp^{iq_zL}]^{-1}. \tag{20.1}$$

Here $\kappa_{\mathrm{s}} = \mu k_z/q_z + q_z/(\mu k_z)$ for S-polarized and $\kappa_p = \varepsilon k_z/q_z + q_z/(\varepsilon k_z)$ for P-polarized components with $q_z = \sqrt{\varepsilon\mu k_0^2 - k_\perp^2}$, $k_z = \sqrt{k_0^2 - k_\perp^2}$, $k_0 = 2\pi/\lambda$ and $k_\perp = \sqrt{k_x^2 + k_y^2}$. We neglect the back-reflection from the aperture. The spatial structure of the field at the output of the system is then found by the backward Fourier transformation. The input field $\boldsymbol{E}_{\mathrm{s,p}}(k_x, k_y, 0)$ here is given by the field after the circular aperture in a perfectly-conducting infinitely-thin film positioned immediately before the slab. In the case of a realistic (for example metallic) film of finite thickness, the diffraction will be more complicated due to various effects such as excitation of surface plasmons. However, for realistic film parameters and wavelengths in near IR or visible, coupling to the plasmons will be efficient only near the aperture. Thus, they will not significantly influence the field distribution after the film. The solution of the diffraction problem of a beam through a circular aperture with a diameter in the range of the wavelength [25], provides the diffracted field after the aperture $\boldsymbol{E}(k_x, k_y, 0)$. The crucial point here is that in difference to the incoming beam before the aperture, $\boldsymbol{E}(k_x, k_y, 0)$ contains propagating components as well as evanescent components, the latter of which can be amplified by the slab with the negative refraction index.

In Figure 20.2 (a) the surface I (red) represents the spatial transverse Fourier distribution after the aperture for a x-polarized input beam with parameters as given

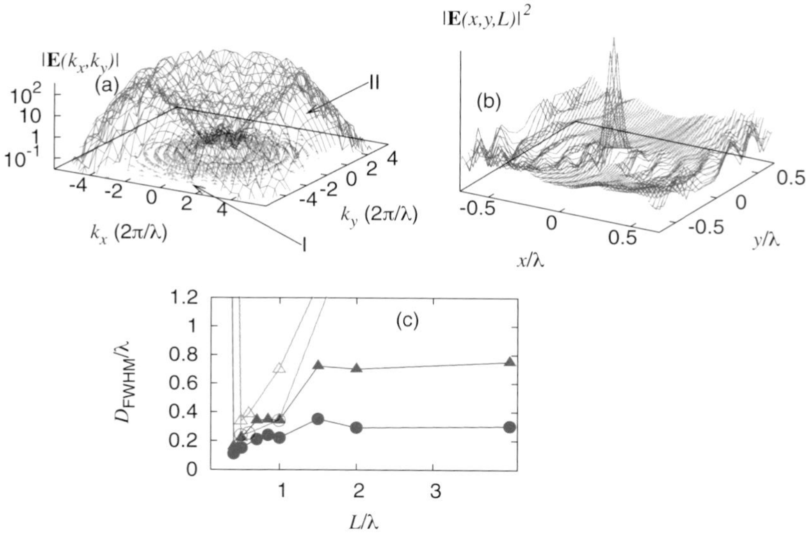

Figure 20.2 Superfocusing by an NRM and an aperture in the effective-medium theory. The spatial spectrum $|\boldsymbol{E}(k_x, k_y)|$ is shown in (a) after the aperture $|\boldsymbol{E}(k_x, k_y, 0)|$ (red, surface I) and after the slab $|\boldsymbol{E}(k_x, k_y, L)|$ (green, surface II). The spatial distribution of the energy density after the slab is shown in (b) and the dependence of the FWHM diameters along x (triangles) and y (circles) directions in (c). The parameters for (a), (b) are $L = 0.4\lambda$ and $\varepsilon = \mu = -1-0.0001(i+1)$. In (c), $\varepsilon = \mu = -1-0.0001(i+1)$ and $\varepsilon = \mu = -1-0.01(i+1)$ for solid (green) and open (magenta) points, respectively. Curves in (c) are guide for the eyes; the aperture diameter is 2λ.

in the caption, where weak evanescent components at the level 10^{-2} can be seen for $k_\perp \sim 2.0k_0$. During the transmission through the slab the evanescent components are strongly amplified. Though the thickness of the slab in this case is only 0.5λ, the amplification of the field for $k_\perp = 3k_0$ is around 10^4, which explains the appearance of the 'ring' of the amplified components around $k_\perp = 3.5k_0$. The corresponding transverse spatial distribution of the energy density including the longitudinal component with a strongly focused peak (superfocusing) is illustrated by Figure 20.2(b). The FWHM of this peak is $0.15\lambda/0.12\lambda$ in the x/y direction, respectively; the difference arises due to difference in diffraction and transmission for the different vector components of the field. The dependence of the output FWHM on the slab thickness is presented in Figure 20.2(c) by solid (green) and open (magenta) points for $\varepsilon = \mu = -1-0.0001(i+1)$ and $\varepsilon = \mu = 1-0.01(i+1)$, respectively. The deviations of ε and μ from the optimum value -1 limit the superfocusing (cf. Ref. [26]). The existence of an optimum slab thickness [minima of curves in Figure 20.2(c)] is caused by the fact that the increasing amplification of evanescent components with larger slab thickness is counteracted by the decrease of the transmission coefficients due to the deviations from the ideal value of $\varepsilon = \mu = -1$. For a higher level of deviations 10^{-2}, the optimum FWHM at the output are $0.23\lambda/0.33\lambda$, which is still below the diffraction limit.

20.2.2
Direct Numerical Solution of Maxwell Equations for Photonic Crystals

To explore the phenomenon of beam superfocusing in a real inhomogeneous NRM system, we now study this effect in one transverse direction for a two-dimensional (2D) photonic crystal. For certain design parameters and below the bandgap edge strongly modulated photonic crystals behave as a material having an effective negative refraction index [12]. Evanescent components are amplified in such photonic crystals analogous to the effective medium model. Besides it was shown that in the case of all-angle negative refraction [27] the amplification of evanescent waves in PCs can also be achieved due to resonant coupling with bound photon modes [18]. Here we will consider both superfocusing in a PC with an effective negative refractive index as well as in regions with all-angle negative refraction. We consider here the two-dimensional case, i.e. the input beam is focused only in one transverse direction, with larger FWHM in the other transverse direction (translation direction). The aperture and the structure of the photonic crystal are invariant in respect to the translation direction, as well as the electromagnetic field. Superfocusing in the translation direction is not considered in this scheme. Note that the same 2D geometry is also considered in Sections 20.3.2 and 20.3.4. The 1D aperture is considered infinitely thin and ideally-absorbing. The propagation of the TE-polarized beam (electric field perpendicular to the translation direction) through the 2D PC is described by the numerical solution of Maxwell equations, found by the decomposition of the field into a large ($\sim 10^7$) number of plane waves, including the reflected and transmitted ones outside the slab, as developed in Ref. [28]. The periodic structure determines the coupling between all of these waves, resulting in a large-order system of linear equations which is solved numerically using standard algorithms. First we consider a rods-in-air structure with a periodic arrangement of silicon rods (with $\varepsilon = 12.96 + 0.01i$) with diameter $0.7a$ where a is the lattice constant, with interfaces parallel to the $\Gamma - K$ direction. This PC possesses a negative refractive index for frequencies around $\omega = 0.56\ldots0.6 \times 2\pi c/a$ [12,29]; we have chosen $\omega = 0.56 \times 2\pi c/a$ with an effective refractive index close to -1. The transfer function into the forward direction (0th Bragg order) is presented in Figure 20.3 (a), and the amplitude of the Fourier components after the slab is presented by the solid (red) curve in Figure 20.3(b). As can be seen, evanescent components with $k_x > k_0$ are strongly amplified, and play a crucial role in the formation of the focused spot. The phase of the Fourier-transformed field [dashed (green) curve in Figure 20.3(b)] shows large changes in the whole presented k-range, however it only weakly varies inside the range with a large Fourier amplitude. Therefore the superposition of all Fourier components is mainly constructive. The spatial transverse field structure is presented in Figure 20.3(c) by the distribution of the $|H_z|^2$ component of the field (relative to the input field). The field at the output is weaker than at the input, because of the reflection at the slab interfaces caused by the refractive index contrast. The FWHM is 0.25λ in this case, however there exist a modulated broader background, and in Figure 20.3(d) it is shown that the range of the aperture diameters for superfocusing is very narrow.

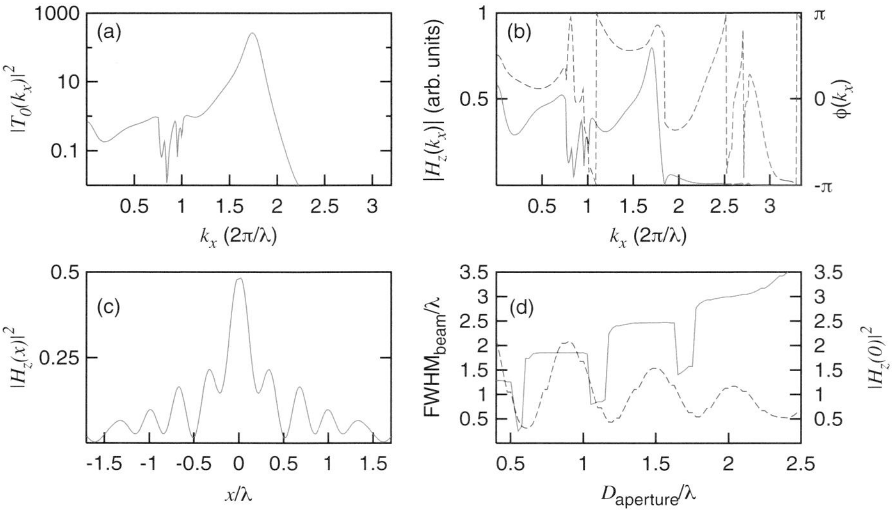

Figure 20.3 Superfocusing by a hexagonal rods-in-air lattice PC with an effective negative refraction index and an aperture. Transfer into the 0th Bragg order is shown in (a), the spatial spectrum (solid, red) and phase (dashed, green) of the field *Hz* are given in (b). The spatial distribution of the field is shown in (c), and the dependence of the FWHM of the beam at the output (solid, red) and the square of the maximum field $|H_z(0)|^2$ (dashed, green) on the aperture width in given in (d). The input beam has a Gaussian shape with FWHM of 3.2λ, the photonic crystal slab consists of 4 layers of circular rods with radius 0.35*a* with lattice constant *a*. The frequency of the field is $0.57 \times 2\pi c/a$, the dielectric permittivity of the rods $\varepsilon = 12.96 + 0.01i$. The aperture width in cases (b) and (c) is 0.55λ.

Now we study beam focusing by a 2D photonic crystal *without definite effective refractive index* but within the range of so-called all-angle negative refraction [18]. We consider a photonic crystal made of '+'-shaped air holes in a high-index material. Such a PC made from silicon ($\varepsilon = 12$) with geometric parameters as given in Figure 20.5 (b) exhibits all-angle negative refraction near the frequency $0.27 \times 2\pi c/a$ [18]. As can be seen in Figure 20.4(a), the evanescent components of the field are also amplified in this system. In difference to the case presented in Figure 20.3 now sharp peaks in the transfer function appear. These peaks influence the spatial Fourier distribution of the field at the output as shown in Figure 20.4(b). The phase presented by the dashed (green) line in Figure 20.4(b), shows a rather irregular behavior, however the phases are nearly matched in the dominating regions with large spectral amplitudes. Therefore the spatial distribution as shown in Figure 20.4(c) shows a sharply focused spot with FWHM of 0.25λ, and only weak background radiation. The peak of $|H_z(0)|^2$ is approximately 4 times higher than at the input. The dependence of beam FWHM at the output of the PC on the aperture width is illustrated by Figure 20.4(d) and shows a well-established minimum, a notable enhancement of the field, and superfocusing over a large parameter range. It was shown in the previous studies (see e.g. [12] and [26]), that

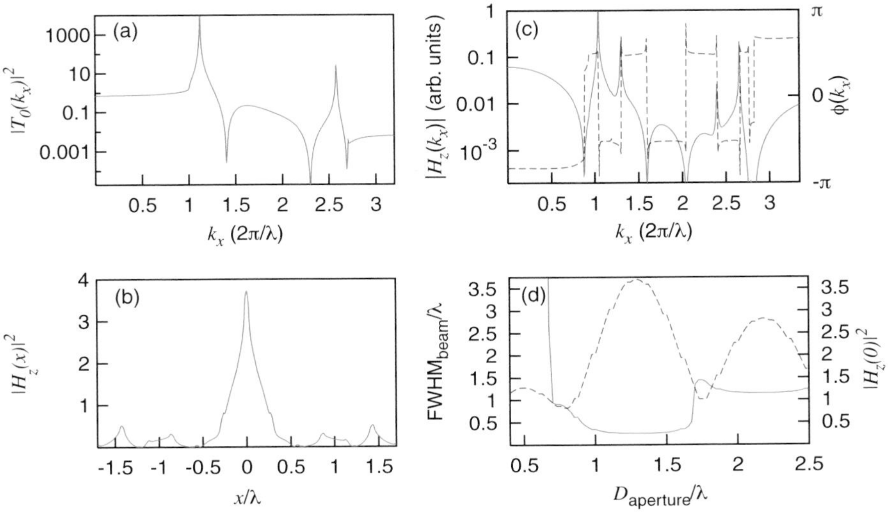

Figure 20.4 Superfocusing by a square air-holes lattice PC with an all-angle negative refraction and an aperture. Transfer into the 0th Bragg order is shown in (a), the spatial spectrum (solid, red) and phase (dashed, green) of the field H_z is given in (b). The spatial distribution of the field is shown in (c), and the dependence of the FWHM of the beam at the output (solid, red) and the square of the maximum field $|H_z(0)|^2$ (dashed, green) on the aperture width is given in (d). The input beam has a Gaussian shape with FWHM of 3.2λ, the photonic crystal slab consists of 4 layers of '+'-shaped air holes with parameters as given in Figure 20.5(b). The frequency of the field is $0.27 \times 2\pi c/a$, the dielectric permittivity of the bulk is $\varepsilon = 12 + 0.01i$. The aperture width in cases (b) and (c) is 1.15λ.

efficient amplification of the evanescent components is possible only in a narrow region of frequencies. We have observed a similar effect: even a small change of the optical frequency by ±0.5% results in approximately a twofold increase of the FWHM and the disappearance of superfocusing. Nevertheless, our calculations predict that optical beams with spectral width below 0.2% of the cetral frequency

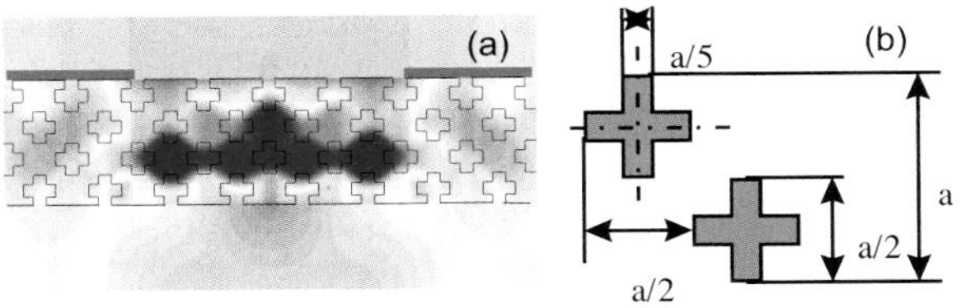

Figure 20.5 Distribution of the field of the beam (a) and the structure of the unit cell of the photonic crystal for the same system as in Figure 20.4. The value of $|\boldsymbol{H_z}|$ is illustrated by shadows of blue in (a), and the aperture is represented by green; the reflected beam is omitted for the sake of clarity.

can be focused below the diffraction limit using any of the setups considered in this paper. In the case of pulsed beams, this corresponds to a shortest pulse duration of about 2 ps.

To illustrate the process of superfocusing in more detail, we have calculated the spatial structure of the field inside and behind the PC as presented in Figure 20.5 (a). The input beam after passing the aperture (indicated by green) undergoes complex redistribution in the photonic crystal, and is strongly focused below the diffraction limit directly behind the PC. With further propagation, the FWHM of the beam quickly increases. We see that the considered PC with all-angle negative refraction is better suited for superfocusing than the PC with negative refractive index studied in Figure 20.3.

To summarize this section, we have shown that it is possible to achieve focusing of a light beam below the diffraction limit by using a combination of an aperture and a slab with a medium exhibiting negative refraction caused by the creation and the amplification of evanescent components of the field. Note that the fairly restricted region for the parameters μ and ε in the effective medium model is eliminated in real photonic crystals in parameter regions where the concept of effective negative-index is valid and in the region of all-angle negative refraction. Although the above analysis was done for a 2D photonic crystal, a 3D photonic crystals with all-angle negative refraction has been already designed and studied in Ref. [30] which allows superfocusing of light beams in both transverse directions similar to Figure 20.2.

20.3
Focusing of Scanning Light Beams Below the Diffraction Limit Using a Saturable Absorber and a Negative-Refraction Material

The aim of the present section is to show that a scanning light beam can be focused below the diffraction limit without the control of moving near-field elements using the combination of two main components: A light-controlled saturable absorber (SA) which creates seed evanescent components from the beam and a layer of negative-refraction material which amplifies the evanescent waves. Focusing to spots with a FWHM in the range of 0.2–0.3λ is predicted. The scheme for the studied method is presented in Figure 20.1(e), (f), in which a thin layer of a SA serves as the nonlinear layer NL and a negative-refraction material (NRM). The transmission of the SA for intensities in the range of the saturation intensity depends on the transverse coordinates, with larger transmission in the center and increasing loss in radial direction. If the input beam spot at the SA is already in the wavelength range, the action of the SA will lead to the creation of weak seed evanescent waves as shown in Figure 20.1(f), similar to a circular aperture since the intense beam creates the aperture-like transparency region by itself. However, now the aperture action is controlled by the input power of the scanning light beam and not by an aperture with fixed position [Figure 20.1(c)]. After the SA the evanescent components are significantly amplified by the layer of

NRM which results to a subdiffraction spot after the layer, indicated by the red point in Figure 20.1(e). Layers with semiconductors or semiconductor quantum dots are appropriate examples for a SA.

20.3.1
Effective-Medium Approach

First we describe beam focusing by a plane slab of an NRM in the context of effective medium theory. The propagation of a beam through a homogeneous slab of a material with a negative refractive index was already described in Section 20.1 and given by Eq. 20.1. The action of a saturable absorber is due to light absorption in bulk or confined semiconductors by inter-band transitions between the valence and the conduction bands. Under intense radiation the increasing population of the conduction band leads to the saturation of the absorption. In the case of quantum dots the discrete levels due to confinement allows the description of light-induced absorption in the frame of a two-level model. In bulk semiconductors interband levels and many-body aspects play a significant role, but the experimental results for a fixed frequency can also be described by the simple two-level model with the intensity-dependent absorption coefficient for the field $\alpha(I) = \alpha_0(1 + i\beta)/(1 + I/I_s)$, where I_s is the saturation intensity, α_0 is the low-intensity absorption coefficient, and β describes the modification of the phase due to detuning of the light frequency from the transition frequency. Neglecting diffraction and decay of the evanescent components, we derive the following implicit expression for the intensity $I(z)$ and the phase shift $\Delta\phi(z)$ after the slab with thickness z:

$$\frac{I(z) - I(0)}{I_s} + ln\frac{I(z)}{I(0)} = -2\alpha_0 z, \quad \Delta\phi(z) = \beta\left\{\alpha_0 z - \frac{[I(z) - I(0)]}{2I_s}\right\}. \tag{20.2}$$

Typical values near the bandgap for bulk semiconductors or dense quantum dot layers are in the range of $\alpha_0 \sim 10^5\,\text{cm}^{-1}$ and $I_s \sim 10^5\,\text{W/cm}^2$. With these parameters the required input intensities are in the range of $\sim 10^6\,\text{W/cm}^2$, which is satisfactory in practical applications. The amplitude of the evanescent components grows with increasing $\alpha_0 z$ and $I(0)/I_s$. To achieve sufficiently strong evanescent components, both of these parameters should be of the order of unity, as will be shown further. With the above parameters, $\alpha_0 z \sim 1$ is satisfied for $z < \lambda$, therefore neglecting the decay of evanescent waves in the SA is justified.

We consider the input beam as a modified Gaussian beam with a diameter around λ, with all evanescent components eliminated in the spatial spectrum before the SA. In Figure 20.6 (a) the surface I (red) represents the spatial transverse Fourier distribution after the SA for a x-polarized input beam with parameters as given in the caption, where weak evanescent components at the level 10^{-2} can be seen for $k_\perp \sim 2.0k_0$. During the transmission through the slab of the NRM the evanescent components are strongly amplified, by a factor of around 10^3 for $k_\perp = 2k_0$ which explains the appearance of the amplified components around $k_\perp = 2.5k_0$. The corresponding transverse spatial distribution of the energy density

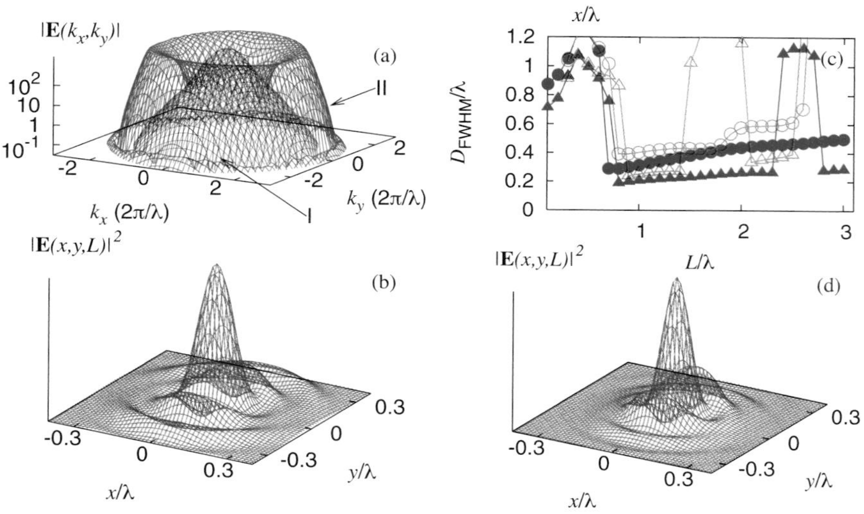

Figure 20.6 Superfocusing described by the effective-medium theory with an NRM and a saturable absorber (SA). The spatial spectrum $|\mathbf{E}(k_x, k_y)|$ is shown in (a) after the slab of the SA $|\mathbf{E}(k_x, k_y, 0)|$ (red, surface I) and after the slab of NRM $|\mathbf{E}(k_x, k_y, L)|$ (green, surface II). The spatial distribution of the energy density after the slab is shown in (b) and the dependence of the FWHM diameters along x (triangles) and y (circles) directions in (c). The SA parameters are $\alpha_0 z = 2$, $\beta = 0$ and $I(0)/I_s = 1$. The parameters for the NRM in (a), (b) are $L = 0.9\lambda$ and $\varepsilon = \mu = -1 - 0.0001(i+1)$. In (c), $\varepsilon = \mu = -1 - 0.0001(i+1)$ and $\varepsilon = \mu = -1 - 0.001(i+1)$ for solid (green) and open (magenta) points, correspondingly. In (d), all parameters are the same except $\beta = -1.5$ and $L = 0.7\lambda$. Curves in (c) are guides for the eyes; the input beam diameter is λ.

(including the longitudinal component of the field) with a strongly focused peak is illustrated by Figure 20.6(b). The FWHM of this peak is $0.3\lambda/0.2\lambda$ in the x/y direction, respectively; the difference is caused by different amplitudes of the S- and P-polarized components of the field. Note that the obtained spot width is much less than the width which is achieved due to the action of the SA only. The latter is $\sim 0.9\lambda$, thus the introduction of the NRM is crucial in achieving a subdiffraction spot. The dependence of the output FWHM on the slab thickness is presented in Figure 20.6(c) by solid (green) and open (magenta) points for $\varepsilon = \mu = -1 - 0.0001(i+1)$ and $\varepsilon = \mu = -1 - 0.001(i+1)$, respectively. The deviations of ε and μ from the optimum value -1 inhibit the amplification of components with high $k_\perp$ [26] and limit the superfocusing. For the higher level of deviations $\varepsilon = \mu = -1 - 0.001(i+1)$, an output FWHM of $0.39\lambda/0.23\lambda$ is achieved, which is still below the diffraction limit.

If the frequency of the beam is shifted from the resonance frequency, an intensity-dependent phase shift arises, which enhances the evanescent component after the SA. In Figure 20.6(d) this case is presented with $\beta = -1.5$. The achieved spot size is $0.23\lambda/0.16\lambda$, which is smaller than for the resonant case $\beta = 0$.

20.3.2
Direct Numerical Solution of Maxwell Equations for Photonic Crystals

Let us now consider a 2D photonic crystals as a real medium with negative refraction in an appropriately chosen parameter range, which enables superfocusing of a beam in one transverse direction. Since the effective medium theory provides only an approximate description for this system, we now solve the Maxwell equations for the 2D spatially inhomogenous periodic structure as described in Section 20.2.2. The field distribution after the layer of the saturable absorber obtained by the solution of the Eq. (20.2), is decomposed into spatial Fourier components, which are multiplied by the complex transmission coefficients of the photonic crystal slab. We consider the parameter region with all-angle negative refraction of a hexagonal lattice of circular holes in the material like GaAs with parameters as indicated in the caption of Figure 20.7 for a photonic crystal structure presented in Figure 20.1(e). In Figure 20.7 (a), the transmission into the 0th Bragg order is presented in dependence on the transverse wavenumber of the incoming wave, which shows amplification of

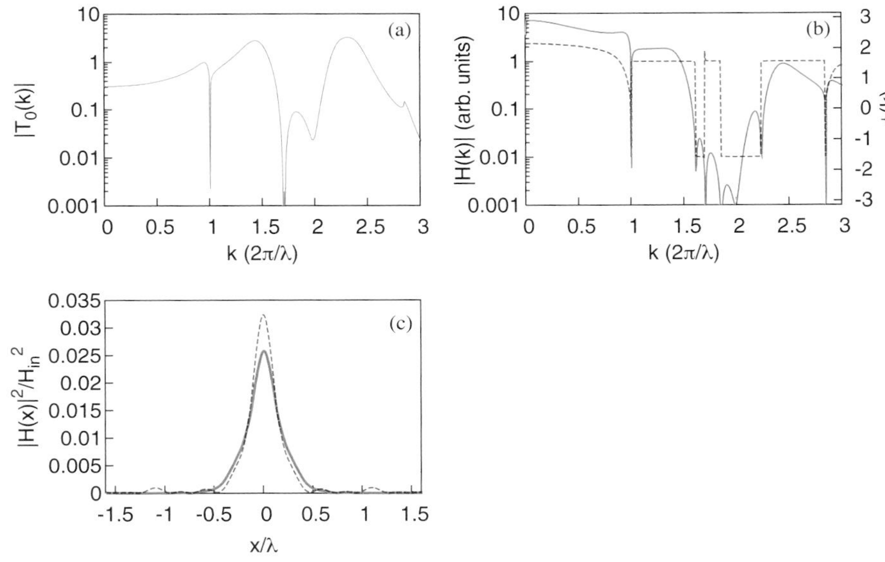

Figure 20.7 Superfocusing by a hexagonal air-hole lattice PC exhibiting all-angle negative refraction and a SA. Transfer into the 0th Bragg order is shown in (a), the spatial spectrum (solid, red) and phase (dashed, green) of the field H_z are given in (b). The spatial distribution of the field is shown in (c), H_{in} being the peak value of the input field. The input beam has a FWHM of 1.0λ, the photonic crystal slab consists of 3 layers of circular holes with radius 0.44*a* with lattice constant *a*, with additional layers of host material of thickness 0.42*a* above and below the layers. The frequency of the input TE-polarized light is $0.26\times 2\pi c/a$, the dielectric permittivity of the rods $\varepsilon = 12.5 + 0.01i$. The parameters of the SA in cases (b) and (c) are $\alpha_0 z = 3.0$, $I(0)/I_s = 4.5$ and $\beta = 0$. For $\beta = -1$ and $I(0)/I_s = 4.2$ and the same other parameters, the spatial distribution is shown by dashed green line in (c).

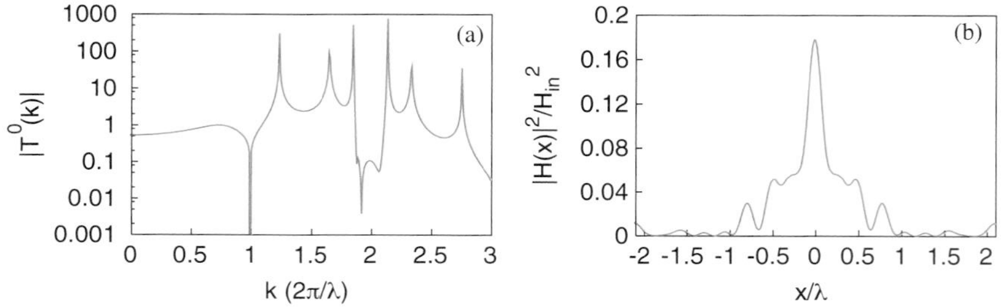

Figure 20.8 Superfocusing by a SA and a hexagonal air-hole lattice PC with the same parameters as in Figure 20.7, except for a different frequency $0.25 \times 2\pi c/a$ and $I(0)/I_s = 6.0$. Transfer into the 0th Bragg order is shown in (a), and the spatial distribution of the field is shown in (b).

evanescent waves for $k_\perp > 2\pi/\lambda$. The spectrum and phase at the output are shown in Figure 20.7(b), indicating a significant contribution of the evanescent components as well as an almost constant phase of the Fourier components with large amplitude. The broad spatial spectrum and the constructive interference result in a focused spot with a FWHM of 0.3λ remarkably below the diffraction limit, as illustrated by Figure 20.7(c). Similar as in the case of Figure 20.6, the SA alone does not lead to a subdiffraction spot. In order to achieve superfocusing, the small-signal loss $\alpha_0 z$ has to be at least in the order of unity, which results in a decrease of the intensity after the SA. As in the case of the homogeneous NRM presented in Figure 20.6, shifting the frequency below the resonance introduces an additional intensity-dependent phase shift and improves the results: as shown in Figure 20.7(c) by the dashed green curves, for $\beta = -1$ the strength of the evanescent components increases, and the output spot has a lower FWHM of 0.26λ. Due to the relatively strong absorption by the SA before the photonic crystal, the intensity at the output of the system is significantly lower than at the input. However, a higher output intensity can be achieved by the optimization of the parameters such as changing the wavelength of the input light from 3.84*a* to 4*a*. The occurrence of strong peaks in the transmission shown in Figure 20.8(a), compared to Figure 20.7(a) leads to a correspondingly increased maximum of the spatial intensity distribution, as can be seen in 8(b), with an output spot with FWHM of 0.23λ. However, the spatial intensity distribution shows side maxima in the wings and the spot quality is reduced. Although the considered 2D photonic crystal allows focusing in only one direction, 3D photonic crystals with negative refraction [30] should enable focusing below the diffraction limit in both transverse directions.

In conclusion of this section, we have shown that superfocusing of scanning light beams below the diffraction limit can be achieved without the necessity of subwavelength spatial control of a near-field element. We propose for this aim a combination of a saturable absorber and a layer of a negative-refraction medium, which creates and amplifies evanescent component of the beam. The smallest spot is in the range of λ/4.

20.4 Subdiffraction Focusing of Scanning Beams by a Negative-Refraction Layer Combined with a Nonlinear Kerr-Type Layer

As shown in Section 20.2 a beam with a fixed direction can be focused below the diffraction limit by using an aperture combined with an NRM. For focusing of scanning beams a fixed position of an aperture do not allow to focus the beam at an arbitrary point. Therefore in Section 20.3 a saturable absorber was studied as a light-controlled nonlinear element creating evanescent components from the beam. However, to achieve a sufficient magnitude of seed evanescent components, strong absorption was found to be necessary, leading to a relatively large loss with an intensity in the focus at the level of $10^{-2}-10^{-1}$ of the input intensity. In this section, we theoretically study subdiffraction focusing of scanning light beams by the combination of a Kerr-like nonlinear layer NL and a layer of NRM, as illustrated in Figure 20.1(e), (f). A Kerr-like nonlinear layer generates evanescent components due to a position-dependent nonlinear phase modulation which leads to transverse spectral broadening, analogous to the time-dependent self-phase modulation inducing the generation of new frequency components. In Figure 20.1(f), the creation of the evanescent components E_{evan} by a Kerr medium serving as nonlinear layer NL and its amplification by an NRM is schematically shown. The phases of both propagating and evanescent components are almost matched for optimized parameters, which allows constructive interference of all components and the formation of a subwavelength spot. Subdiffraction focusing is described in Section 20.4.1 in the context of the effective-medium description for quasi-homogeneous NRMs, such as metamaterials, and in 2.4.2 by direct numerical solution of Maxwell equations in a periodic medium for an NRM implemented by a photonic crystal.

20.4.1 Effective-Medium Approach

The modification of the refractive index by intense light is one of the fundamental effects in nonlinear optics. In many cases such modification is achieved due to the electronic Kerr effect, but there exist a number of other processes which can also lead to a high nonlinear refractive index. Although those processes with the highest nonlinear coefficients have typically a large response time, ultrafast operation is not a critical requirement for the purpose studied here. For the description of beam propagation through a nonlinear layer including the generation of evanescent components the paraxial approximation cannot be used. An exact treatment of this process is based on the general wave equation written for a monochromatic field in the form

$$\frac{\partial^2 E(k_x, k_y, z)}{\partial z^2} = k_z^2 E(k_x, k_y, z) + k_0^2 P_{\mathrm{NL}}(k_x, k_y, z), \tag{20.3}$$

where $k_z = \sqrt{n_0^2 k_0^2 - k_x^2 - k_y^2}$, $k_0 = \omega_0/c$, n_0 is the refractive index of the nonlinear layer, ω_0 is the frequency of the field, $P_{\mathrm{NL}}(k_x, k_y, z)$ denotes the Fourier transform of the nonlinear polarization $P_{\rm NL}(x,y,z) = \varepsilon_0\chi_3 E(x,y,z)^3$, $\chi_3 = (4/3)c\varepsilon_0 n_0^2 n_2$ is the nonlinear susceptibility. In the perturbation-theory approach the solution can be found in the form

$$E(k_x,k_y,z) = E(k_x,k_y,0)\exp(ik_z z) - \frac{k_0^2 P_{\rm NL}(k_x,k_y,0)}{2k_{z0}} \times \left[\frac{\exp(in_0k_0z)-\exp(ik_zz)}{n_0k_0-k_z} + \frac{\exp(in_0k_0z)-\exp(-ik_zz+ik_zL+in_0k_0L)}{n_0k_0+k_z}\right] \tag{20.4}$$

for $z < L$ where L is the thickness of the nonlinear layer. The first term in the square brackets corresponds to the forward-propagating waves, and the second to the backward-propagating waves which is zero for $z > L$. It can be shown that the forward-propagating part satisfies a simpler first-order equation:

$$\frac{\partial E(k_x,k_y,z)}{\partial z} = ik_z E(k_x,k_y,z) + i\frac{k_0^2}{2k_z} P_{\rm NL}(k_x,k_y,z). \tag{20.5}$$

Inside the nonlinear layer the nonlinear polarization is also influenced by the weak backward waves, but this effect is negligible for thin layers with a small relative modification of the refractive index. Eq. (20.6) is valid also without the perturbation theory, and can be derived using quite general assumptions as shown in Ref. [31]. This equation does not rely on the paraxial approximation and, as an additional feature, correctly describes the evolution of the components with $k_x^2 + k_y^2 > n_0^2 k_0^2$. Note that the nonlinear term has a divergence at $k_z = 0$ in the Fourier domain, but after Fourier transform to the space the field $E(x, y, z)$ remains finite. Besides, Fourier components with $k_z = 0$ are very weak and do not contribute to the formation of the spot in our calculation. Equation (20.5) was solved by the split-step Fourier method. Equation (20.5) is written in scalar form. In our case, this is justified, since the input beam is linearly polarized and relatively wide. The generated evanescent components are weak for the considered layer thicknesses, and their vectorial nature does not influence the nonlinear polarization.

For a homogeneous or quasi-homogeneous medium with negative refractive index the effective medium approach can be used to describe the propagation of the beam and the amplification of evanescent components in an NRM. Various negative-refraction metamaterials based on structures with a scale much smaller than the wavelength can be described by this method. In this approach the propagation of a beam through a homogeneous slab of an NRM with thickness L and effective parameters ε and μ is described by the transfer functions $T_{\rm s,p}(k_x, k_y)$ given in Eq. (20.1) in Section 20.2.1.

In Figure 20.9 superfocusing of a light beam by the combination of a Kerr-type nonlinear layer and an NRM is illustrated, with parameters given in the caption. We describe the profile of the field at the input surface of the nonlinear layer as modified Gaussian with evanescent components set to zero. In Figure 20.9(a) the spatial transverse Fourier distribution $E(k_x, k_y)$ after the nonlinear layer is shown by the red surface, and weak evanescent components at a level around 10^{-3} relative to the

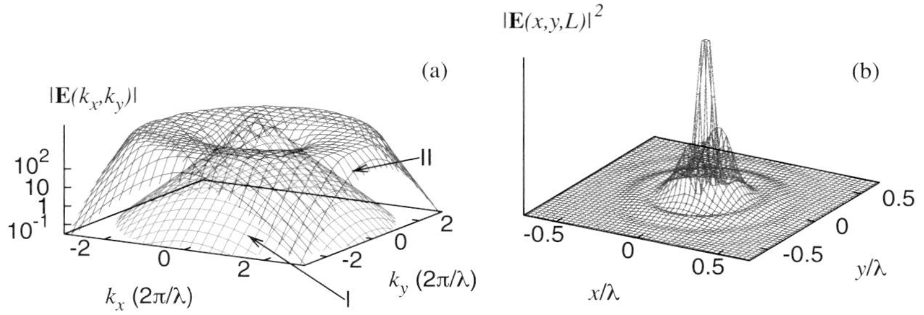

Figure 20.9 Superfocusing by an NRM and a Kerr-type medium in the effective-medium theory. The spatial spectrum $|\boldsymbol{E}(k_x, k_y)|$ is shown in (a) after the nonlinear layer (red, surface I) and after the NRM layer (green, surface II). The spatial distribution of the energy density after the slab is shown in (b). The maximum change of the refractive index is $\Delta n_{\text{Kerr}}^{\max}/n_0 = n_2 I_{\max}/n_0 = 0.033$, the linear refractive index of the Kerr medium is $n_0 = 3.0$, the thickness of the nonlinear layer is $L_{\text{kerr}} = 2\lambda$. The thickness of the NRM layer is $L_{\text{NRM}} = 0.6\lambda$, $\text{FWHM}_{\text{in}} = 1.5\lambda$ and $\varepsilon = \mu = -1 - 0.0001(i + 1)$.

maximum can be seen for $k_\perp > 2\pi/\lambda$. The transverse Fourier distribution after the slab of the NRM is shown in green, with clearly visible large amplification of evanescent components. The corresponding transverse spatial distribution is presented in Figure 20.9(b). The minimum beam diameter is $0.22\lambda/0.15\lambda$ in the x/y direction, respectively.

Note that for a nonlinear layer with a refractive index n_0 around unity, the evanescent components are lost over the distance $\lambda/2\pi$, and therefore the effective length over which they are generated is of the same order. That means that to achieve significant (of about $\pi/2$) phase modification necessary to generate seed evanescent components of sufficient intensity, over such a short distance, the maximum nonlinear contribution to the refractive index should be around unity. Since materials with such high nonlinear modification of the refractive index are not available, we cannot use a nonlinear film with $n_0 \sim 1$ and thickness around $\lambda/2\pi$ as the nonlinear layer. Instead, we considered in Figure 20.9 a nonlinear layer with relatively high linear refractive index $n_0 = 3.0$ and thickness of about λ, which allows to preserve and accumulate the evanescent components with $k_0 < \sqrt{k_x^2 + k_y^2} < n_0 k_0$. In this case the effective length of the nonlinear layer is limited only by the gradual increase of the spot size due to diffraction, which occurs after a propagation distance of several λ, and a relative nonlinear modification of the refractive index in the order of 0.03 is sufficient. There exist a large number of natural and artificial materials with strong optical nonlinearities. To give a few examples, the nonlinear refractive index of different semiconductors can have n_2 values in the order of 10^{-12} W/cm2. Due to near-resonant processes in the vicinity of the bandgap, values in the range of $10^{-8}\,\text{W/cm}^2$ can be achieved, for example in ZnSe in the wavelength range from 440 to 460 nm [32]. Metal-dielectric multilayer structures can yield a complex n_2 in the order of $|n_2| = 7 \times 10^{-8}\,\text{W/cm}^2$ [33]. Extremely high nonlinearities $n_2 \sim 10^0$–$10^3\,\text{W/cm}^2$ can be achieved in thin dye-doped liquid-crystal layers [34]. One has to take into account that for slow processes with a large response time

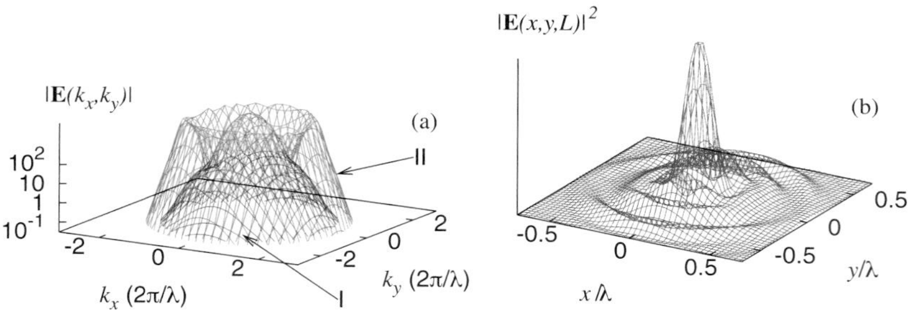

Figure 20.10 Superfocusing by an NRM and a Kerr-type medium in effective-medium theory, for a weaker nonlinearity. Denotations and parameters are the same as in Figure 20.9, except for $\Delta n^{\max}_{\text{Kerr}}/n_0 = n_2 I_{max}/n_0 = 0.003$, $L_{\text{NRM}} = 1.3\lambda$, and FWHMin $= 1.5\lambda$.

T_{res} the effective nonlinear refractive index is given by $n_2\tau/T_{\text{res}}$ where τ is the pulse duration. Depending on the chosen nonlinear material, the parameter $\Delta n^{max}_{\text{Kerr}}/n_0 = n_2 I/n_0 = 0.03$ which is assumed in Figure 20.9 can be achieved with intensities in the order of 10^7–10^{12} W/cm^2.

The nonlinear modification of the refractive index which was used in the above example corresponds to rather high value of $\Delta n^{max}_{\text{Kerr}}/n_0 = 3.3\%$. It is, fortunately, not an absolute prerequisite of superfocusing, and significantly lower values can be used, albeit with tradeoff of the spot size. In Figure 20.10 the superfocusing is illustrated for a one order of magnitude lower nonlinear modification of the refractive index $\Delta n^{max}_{\text{Kerr}}/n_0 = 3 \times 10^{-3}$, which can be achieved even by a fast Kerr nonlinearity.

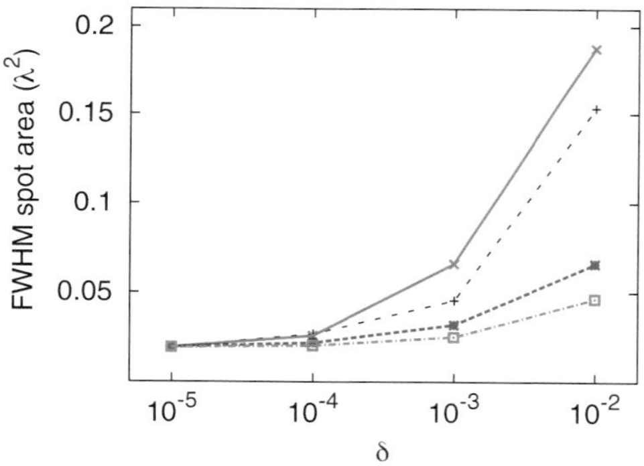

Figure 20.11 Dependence of the spot area on the deviations $\delta\varepsilon$ and $\delta\mu$ of the parameters ε and μ from the ideal case $\varepsilon = \mu = -1$. Cyan dash-dotted, green dashed, red solid, and black dotted curves correspond to deviations in Re (ε), Re (μ), Im (ε, μ), and all of the above parameters, correspondingly. The explicit parameters are for the cyan dash-dotted curve: $\delta\mu = -0.00001(i+1)$; $\delta\varepsilon - 0.00001i - \delta$; for the green dashed curve: $\delta\varepsilon = -0.00001(i+1)$; $\delta\mu = -0.00001i - \delta$; for the red solid curve: $\delta\varepsilon = \delta\mu = -0.00001 - \delta i$; and for the black dotted curve: $\delta\varepsilon = \delta\mu = -\delta(i+1)$. The NRM length corresponds to the optimum (smallest spot area) in each case. Other parameters are the same as in Figure 20.9.

The transverse spectrum is narrower than in Figure 20.9 due to a lower amplitude of the seed evanescent components, and the output spot is larger with a FWHM of $0.38\lambda/0.25\lambda$, but still exhibits focusing below the diffraction limit. The role of the deviations $\delta\varepsilon = \varepsilon + 1$ and $\delta\mu = \mu + 1$ from the ideal -1 values is described in Figure 20.11, where the FWHM spot area is plotted against the level of different deviations, as indicated in the caption.

It can be seen that the most sensitive parameter is the imaginary part of ε and μ (red solid curve) corresponding to loss, which thwarts superfocusing even faster than the simultaneous deviations of all parameters from the ideal case (black dotted curve). The reason is that although the spectrum is broader in the case when only loss is present, the phases of the additional spectral components can lead to destructive interference, yielding a larger spot. The dependence of the superfocusing on the Re $(\delta\varepsilon)$ is – somewhat surprisingly – the least sensitive. As can be seen, a very low level of deviations is required to achieve superfocusing, similar to subdiffraction imaging [26], however, the requirements are much less stringent in the case of photonic crystals considered in the next section.

20.4.2 Direct Numerical Solution of Maxwell Equations for Photonic Crystals

To quantitatively describe superfocusing by the combination of a nonlinear layer and a photonic crystals as an inhomogeneous negative-refraction material, we now study a 2D photonic crystal with geometric parameters appropriate for an experimental realization.

We consider the parameter region with all-angle negative refraction of a hexagonal lattice of circular holes in a material like Si with parameters and geometry as indicated in Figure 20.12 (a). In Figure 20.12(b), the transmission into the 0th Bragg order is presented in dependence on the transverse wavenumber of the incoming wave,

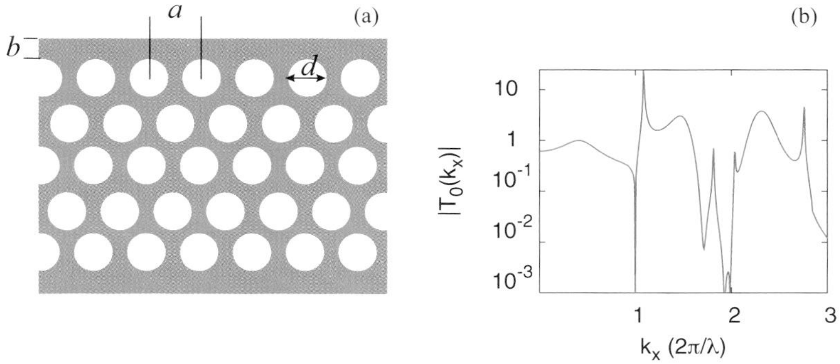

Figure 20.12 Photonic crystal structure (a) and transmission into 0th Bragg order as a function of the transverse wavevector (b). The considered design is a silicon substrate having 5 layers of holes as shown in (a) with interfaces parallel to $\Gamma - K$ direction. Parameters of the structure are $\varepsilon = 12.5 + 0.01i$ at $\lambda = 1000$ nm, $a = 260$ nm, $d = 229$ nm, $b = 110$ nm; a, b, d and λ can be scaled by a common factor.

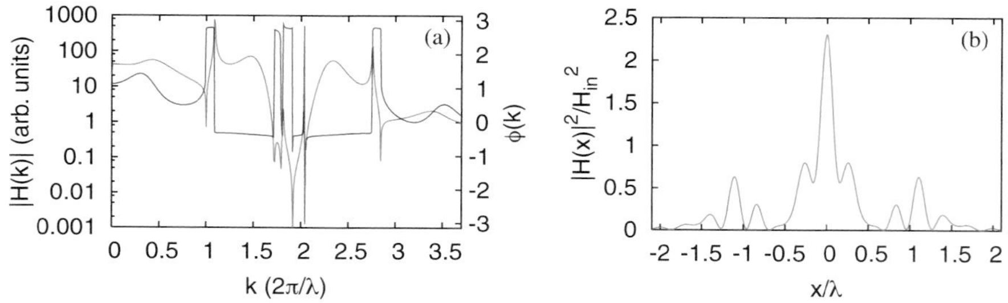

Figure 20.13 Superfocusing by a Kerr-type medium and a hexagonal air-holes lattice PC with all-angle negative refraction. The transverse spectrum (solid, red) and the phase (dashed, green) of the field H_z is given in (a), and the spatial distribution of the field is shown in (b). The input beam has a Gaussian shape with a FWHM of 1.0λ. The maximum nonlinear change of the refractive index is $\Delta n_{\mathrm{Kerr}}^{\max}/n_0 = 0.066$, the linear refractive index of the Kerr medium is $n_0 = 3.0$, the thickness of the nonlinear layer is $L_{\mathrm{kerr}} = 1.9\lambda$.

which shows several peaks for evanescent components with $k_{\perp} > 2\pi/\lambda$ with a transmission up to 10^2.

In Figure 20.13, superfocusing by the combination of a nonlinear layer and a photonic crystal slab is illustrated by the calculated transverse spectrum (a) and the spatial distribution (b), with parameters of the photonic crystal as given in Figure 20.12. The amplification of the evanescent components leads to the formation of a wide transverse spectrum with a phase which varies mostly in the range from 0 to $\pi/3$ over the whole range where the spectrum has significant magnitudes. This implies the formation of a localized spot in space, and indeed as presented in Figure 20.13(b) the FWHM of the spot is 0.18λ, significantly below the diffraction limit. The seed evanescent components are created mostly at the center of the input beam and are in phase with the input beam for a Kerr-like nonlinear layer. Therefore the phase-matching at the output depends mainly on the phase of the transmission coefficient, which can be made almost constant by optimizing the parameters of the system. The peak intensity is approximately two times higher that the input intensity. In comparison, superfocusing by the combination of a saturable absorber and a photonic crystal with negative refraction yields a relative intensity of the focused spot in the range of 10^{-2}–10^{-1}, and somewhat larger spot sizes (see Section 20.3.2). On the other hand, the required intensities for superfocusing are in that case lower, and the transmission in Section 20.3.2 is still significantly higher than that obtained using a tapered fiber as near-field element.

20.5
Conclusion

We have predicted that superfocusing of beams below the diffraction limit is possible using the combination of an optical element which generates seed evanescent components from the beam, and a layer with negative refraction which amplifies them. For focusing of a non-moving beam a wavelength-scale aperture can be used for the creation of seed evanescent components. For superfocusing of scanning beams a light-controlled saturable absorber or a nonlinear material with a Kerr-type nonlinearity enables focusing of beams to spots below the diffraction limit. This method do not require spatial near-field control of a moving near-field element and allows an arbitrary position of the focused spot of scanning beams due to the action of the light-controlled nonlinear layer. The creation and amplification of evanescent compnents by these two elements allow the formation of a 0.18λ-wide focused spot, with matched phases of the transverse spectral components implying constructive interference, and an intensity up to two times larger than the input intensity. Superfocusing is demonstrated both in the effective-medium theory for a metamaterial-based NMR and for photonic crystals with realistic parameters. In comparison with an arrangement using the combination of a saturable absorber for the creation of seed evanescent components and an NRM, the focused spot intensity for a Kerr-type nonlinear layer is more than one order of magnitude higher, but the required input intensity is also higher.

Although the results of our study using the effective medium theory, and the numerical calculation for an photonic crystal predict similar results concerning the spot size and the optimum geometrical parameters, there exist an important difference: the high sensitivity of superfocusing on the deviations of ε and μ from -1 in the effective medium theory does not arise for photonic crystals and superfocusing is even possible in the range of so-called all-angle negative refraction where an effective index cannot be defined. The physical mechanism here is that wide transmission peaks for evanescent components due to resonances with bound photon modes yield a sufficient amplification of the evanescent components which allow superfocusing. This means that not the negative index is the main physical requirement for the predicted effect, but the existence of wide transmission peaks much larger than unity for evanescent components.

Acknowledgements

We acknowledge financial support from the Deutsche Forschungsgemeinschaft in the framework of the SPP 1113 focus program.

References

1 Wong, A.K. (2001) *Resolution enhancement techniques in optical lithography*, SPIE Press, Bellingham, Washington.

2 McDaniel, T.W. (1997) *Handbook of magneto-optical data recording: materials, subsystems, techniques*, Noyes publishing, Westwood.

3 Betzig, E. *et al.* (1991) *Science*, **251**, 1468.

4 Mansfield, S.M. and Kino, G.S. (1990) *Appl. Phys. Lett.*, **57**, 2615.

5 Paesler, M.A. and Moyer, P.J. (1996) *Near-field optics: theory, instrumentation and applications*, John Wiley and Sons, New York.

6 Husakou, A. and Herrmann, J. (2004) *Opt. Express*, **12**, 6491.

7 Husakou, A. and Herrmann, J. (2006) *Phys. Rev. Lett.*, **96**, 013902.

8 Husakou, A. and Herrmann, J. (2006) *Opt. Express*, **14**, 11194.

9 Veselago, V.G. (1968) *Soviet Phys. Usp.*, **10**, 509–518. [Usp. Fiz. Nauk **92**, 517–526 (1967)].

10 Shelby, R.A., Smith, D.R. and Schultz, S. (2001) *Science*, **292**, 77.

11 Enkrich, C., Wegener, M., Linden, S., Burger, S., Zschiedrich, L., Schmidt, F., Zhou, J.F., Koschny, T. and Soukoulis, C.M. (2005) *Phys. Rev. Lett.*, **95**, 203901.

12 Notomi, M. (2000) *Phys. Rev. B*, **62**, 10696.

13 Cubukcu, E. *et al.* (2003) *Nature*, **423**, 604.

14 Dolling, G., Enkrich, C., Wegener, M., Zhou, J., Soukoulis, C.M. and Linden, S. (2005) *Opt. Lett.*, **30**, 3198–3200.

15 Shalaev, V.M., Cai, W., Chettiar, U., Yuan, H.K., Sarychev, A.K., Drachev, V.P. and Kildishev, A.V. (2005) *Opt. Lett.*, **30**, 3356–3358.

16 Grigorenko, N., Geim, A.K., Gleeson, H.F., Zhang, Y., Firsov, A.A., Khrushchev, I.Y. and Petrovic, J. (2005) *Nature*, **438**, 335–338.

17 Pendry, J.B. (2000) *Phys. Rev. Lett.*, **85**, 3966.

18 Luo, C. *et al.* (2003) *Phys. Rev. B*, **68**, 045115.

19 Fang, N. *et al.* (2005) *Science*, **308**, 534.

20 Melville, D. and Blaikie, R.J. (2005) *Opt. Express*, **13**, 2127.

21 Parimi, P.V. *et al.* (2003) *Nature*, **426**, 404.

22 Cubukcu, E. *et al.* (2003) *Phys. Rev. Lett.*, **91**, 207401.

23 Feise, M.W. and Kivshar, Yu.S. (2005) *Phys. Lett. A*, **334**, 326.

24 Lu, Z., Murakowski, J.A., Schuetz, C.A., Shi, S., Schneider, G.J. and Prather, D.W. (2005) *Phys. Rev. Lett.*, **95**, 153901.

25 Meixner, J. and Andrejewski, W. (1950) *Ann. Phys.*, **7**, 157–168.

26 Smith, D.R. *et al.* (2003) *Appl. Phys. Lett.*, **82**, 1506.

27 Luo, C., Johnson, S.G., Joannopoulos, J.D. and Pendry, J.B. (2002) *Phys. Rev. B*, **65**, 201104.

28 Sakoda, K. (2001) *Optical properties of photonic crystals*, Springer.

29 Foteinopoulou, S. and Soukoulis, C.M. (2003) *Phys. Rev. B*, **67**, 235107.

30 Luo, C., Johnson, S.G. and Joannopoulos, J.D. (2002) *Appl. Phys. Lett.*, **81**, 2352.

31 Kolesik, M. and Moloney, J.V. (2004) *Phys. Rev. E*, **70**, 036604.

32 Ding, Y.J., Guo, C.L., Swartzlander, G.A. Jr., Khurgin, J.B. and Kaplan, A.E. (1990) *Opt. Lett.*, **15**, 1431–1433.

33 Bennink, R.S., Yoon, Y.-K. and Boyd, R.W. (1999) *Opt. Lett.*, **24**, 1416–1418.

34 Lucchetti, L., Gentili, M. and Simoni, F. (2005) *Appl. Phys. Lett.*, **86**, 151117.

21 Negative Refraction in 2D Photonic Crystal Super-Lattice: Towards Devices in the IR and Visible Ranges

Y. Neve-Oz, M. Golosovsky, A. Frenkel, and D. Davidov

21.1 Introduction

Negative Refraction (NR) and left-handed metamaterials are of high interest today, particularly, for their ability to focus electromagnetic waves using flat lenses [1–8]. There is extensive research on negative refraction and focusing in the microwave and mm-wave ranges but only relatively few studies on metamaterial devices in the IR and in the visible ranges. To our best knowledge, no clear evidence for focusing using flat lenses in this higher frequency range is available today.

Left-handed metamaterials could be fabricated using two main routes: (i) periodic arrays of lumped elements or split-ring resonators; (ii) "conventional" photonic band-gap materials. In the former approach the real part of the permeability, μ', and the permittivity, ε', for each element in the array are negative leading to negative refractive index, n. In the latter approach the complicated dispersion relation results in the negative refractive index and the left-handed properties which become evident by plotting "equal frequency surfaces and contours" [9–18]. According to the present knowledge, the second approach is better suited for devices in the IR and the visible range although losses may be a major difficulty here. Insertion loss presents a difficulty even in the microwave range. For example, microwave transmission experiments on "conventional" metallic two-dimensional (2D) photonic crystals and for frequencies in the close vicinity of the band-gap edge show that negative refraction is associated with significant losses [9]. There are also very few reports on omni-directional negative refraction using photonic crystals; an essential requirement for far-field focusing.

Recently, our group in Jerusalem [13,14] have suggested a third approach to achieve negative refraction, namely, (iii) 2D-photonic crystals superlattice. The structural modulation in the superlattice results in a transmission subband at the middle of the stopband. This subband is characterized by negligible loss and negative refraction. Indeed, Saado et al. [14] have used 2D superlattice based on the array of dielectric rods with two different diameters but with very high dielectric constant ($\varepsilon \sim 90$) and in

Nanophotonic Materials: Photonic Crystals, Plasmonics, and Metamaterials. Edited by R. B. Wehrspohn, H.-S. Kitzerow, and K. Busch

ISBN: 978-3-527-40858-0

their microwave experiments demonstrated negative refraction and high transmittivity. However, materials with high dielectric constant are not feasible in the optical and the infrared ranges. Furthemore, the transmission band demonstrated by Saado *et al.* was rather narrow (~2% of the stopband width) and not omnidirectional. The purpose of the present paper is to extend the ideas presented in Ref. [13] and to demonstrate (mainly by simulations) photonic crystal superlattice made of the dielectric rods with low dielectric constant, such that the results will be directly scalable to the optical range. We demonstrate here that such PC superlattice made of low-ε dielectric rods may exhibit: (a) negative refraction for frequencies in the transmission subband (b) high transmitivity; and (c) large bandwidth (~15% of the stopband bandwidth).

21.2 Design

We consider two-dimensional arrays based on dielectric rods with two different diameters. For simplicity, we fix the ratio of diameters, $d_1/d_2 = 2$, and the dielectric constant, $\varepsilon = 4$, and consider periodic arrays with different unit cell that can be constructed from such rods. Figure 21.1 shows two such arrays where the large rods

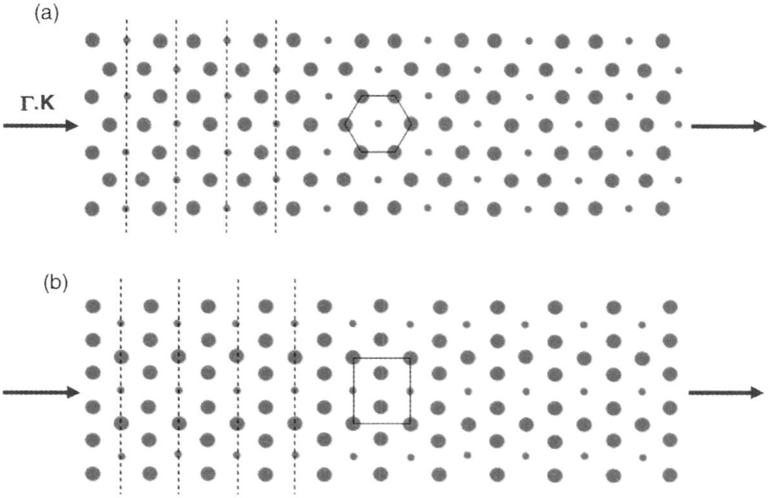

Figure 21.1 Design of the two-dimensional (2D) photonic crystal superlattice based on two types of dielectric rods with the dielectric constant $\varepsilon = 4$ and with the diameters: $d_1 = 2.5$ mm and $d_2 = 1.25$ mm. The large rods form a hexagonal lattice whereas the distance between their centers is $a = 5.4$ mm. (a) The small rods form a hexagonal sublattice and the ratio of the numbers of the large and small rods is 2:1. (b) The small rods form a tetragonal sublattice and the ratio of the numbers of the large and small rods is 3 : 1. Configuration (a) with the high-ε rods has been analyzed previously [13,14].

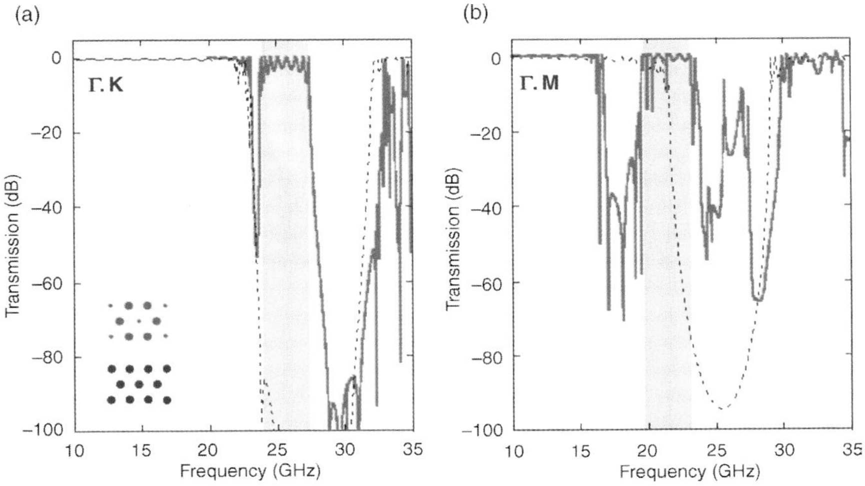

Figure 21.2 Computer simulation of the transmission through the superlattice shown in Figure 21.1a along the (a) Γ–K and the (b) Γ–M directions. The corresponding structures are shown in the inset of Figure 21.2a. The transmission subband inside the stopband is shown by blue hatched area. For the Γ–K direction the transmission band is between 24 GHz and 27.5 GHz. The dashed blue line shows transmission through a conventional 2D photonic crystal where all rods have the same dielectric constant and the same (large) diameter.

form a hexagonal lattice while the small rods form either (a) hexagonal lattice; or (b) tetragonal lattice.

In our previous studies we analyzed the configuration (b) whereas dielectric constants of the individual rods were either $\varepsilon \sim 90$ [14] or $\varepsilon = 12.9$ [13]. We found narrow transmission subband (bandwidth 1.5–2%) inside the stopband and negative refraction. Our search for wider bandwidth resulted in configuration (a) and other configurations, to be presented elsewhere.

Using ANSOFT-HFSS software we calculate the transmittivity of this superlattice with non-primitive unit cell (Figure 21.2a and b) for two directions Γ–K and Γ–M. We compare these results with the transmittivity of the conventional array with a primitive unit cell consisting only of the large rods with $\varepsilon = 4$.

Clearly, transmittivity of the conventional PC along the Γ–K direction demonstrates a wide stopband between 23.5 GHz and 32 GHz (Figure 21.2a; dashed blue line). The stopband edges of the superlattice PC are almost the same (red solid lines in Figure 21.2a), but there appears a new feature– a transmission subband between 24 GHz and 27.5 GHz (blue hatched area). A similar behavior is observed for the Γ–M direction, although due to structural modulation the low-frequency stopband edge is shifted down from 22 GHz to 17 GHz. The transmission subband (20–24 GHz) inside a photonic stopband is clearly visible here as well. The insertion loss is very low (few db). Note also that the width of the transmission band is of the order of ~15% of the total stopband, much wider than that for the 2D structures reported previously [13,14].

21.3
Simulations, Results and Discussion

21.3.1
Wave Transmission Through the Superlattice Slab: Evidence for Negative Phase Velocity

For a "meta-material" to be Left-Handed (LH), the group velocity, v_g, and the phase velocity, v_p, should be in opposite directions [8,9]. We demonstrate this by computer simulations of wave transmission through our PC superlattice slab. Figure 21.3 shows the magnitude and the phase of the electric field distribution before, inside and after the PC superlattice slab for the 25 GHz wave propagating along the Γ–K direction. Figure 21.3a shows that the electric field is concentrated mainly on the small rods (see the dashed vertical lines).

To understand phase propagation we consider a hypothetical uniform material with $n = -0.4$ (Figure 21.3c). The phase propagation in this material is marked in a sequence of colors. **Outside** this material the phase propagates from left to right and

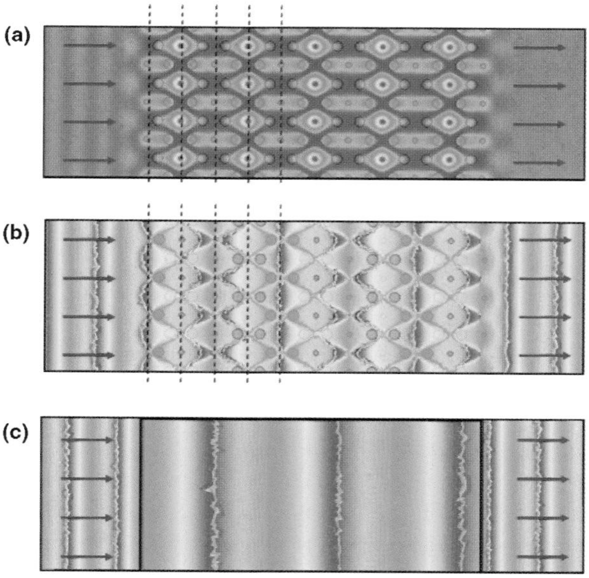

Figure 21.3 (a) Magnitude and (b) phase of the electric field distribution for the 25.2 GHz wave propagating along the Γ–K direction of the superlattice shown in Figure 21.1. (c) Electric field phase for the wave propagating inside the hypothetical uniform material with a negative refraction index $n = -0.4$. (See text for details).

the order of colors is red, yellow, green, blue. This corresponds to the positive phase velocity. Inside this material the order of colors is reversed, namely, blue, green, yellow and red, indicating that the phase propagates from right to left, as expected for the material with the negative refraction index. We observe a very similar reversed order of colors inside our superlattice (Figure 21.3b). Note that we consider the phase only in the regions where electric field is maximized, i.e. along the dashed lines connecting the small rods.

The intense electric field concentration on the small rods and the reverse direction of phase propagation can be regarded as a propagating coupled-cavity mode [19,20]. The phase reversal inside the superlattice PC slab suggests negative phase velocity. The phase velocity is defined as $v_p = \omega/\boldsymbol{k}$ and thus it has the same direction as the wave-vector, $\boldsymbol{k}$. Note however, that the radiation is propagating from left to right (see arrows), suggesting that the group velocity points in this direction and is "positive". Hence, the group velocity and the phase velocity are in opposite directions, that makes this superlattice a Left-Handed Material.

21.3.2 Refraction Through a Superlattice Prism

Negative refraction can be also demonstrated by considering wave propagation through a prism made of our superlattice. Figure 21.4a shows our simulations for the phase of the 25 GHz wave refracted by the prism. The incident plane wave enters

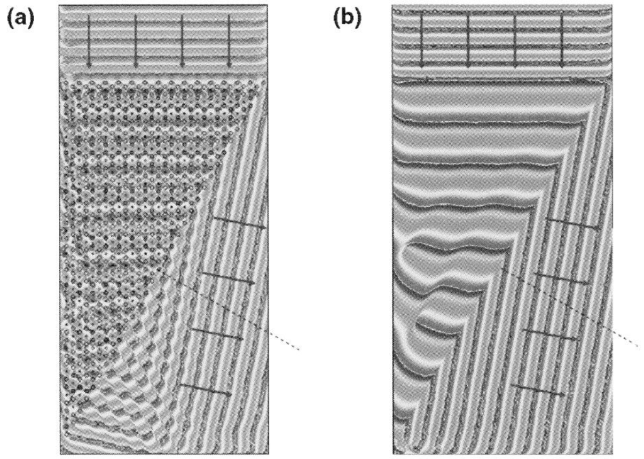

Figure 21.4 (a) Phase of the 25 GHz wave refraction through a superlattice prism. The phase is color coded. The direction of the beam propagation is orthogonal to the phase front. The incoming beam is shown by the arrows at the top of the prism. The refracted beam is shown by the arrows at the broad side of the prism. The dashed line shows the normal to the prism surface. Note the negatively refracted wave (outgoing arrows). (b) Negative refraction by a homogenous prism made of a uniform material with the negative refraction index, $n = -0.4$.

from the top of the prism, propagates there and exits through the broad face of the prism. The direction of the wave is orthogonal to the phase front. Figure 21.4a shows that the outgoing wave is negatively refracted (it makes a "negative angle" with a normal to the output face of the prism).

For comparison, Figure 21.4b shows refraction by a hypothetic uniform medium with the negative refraction index, $n = -0.4$. As clearly seen, the outgoing refracted wave propagates in the same direction as the outgoing wave refracted through the superlattice prism.

Figure 21.5 shows the far-field radiation pattern upon refraction through the prism shown in Figure 21.4. The calculations were performed at frequencies: 24–27 GHz all in the transmission subband. It is seen that lower frequencies are deflected stronger as compared to higher frequencies. The magnitude of the refracted wave (the length of the lobe) decreases towards the subband edges and it is the highest at the center of the transmission subband where it is comparable to the magnitude of the incident

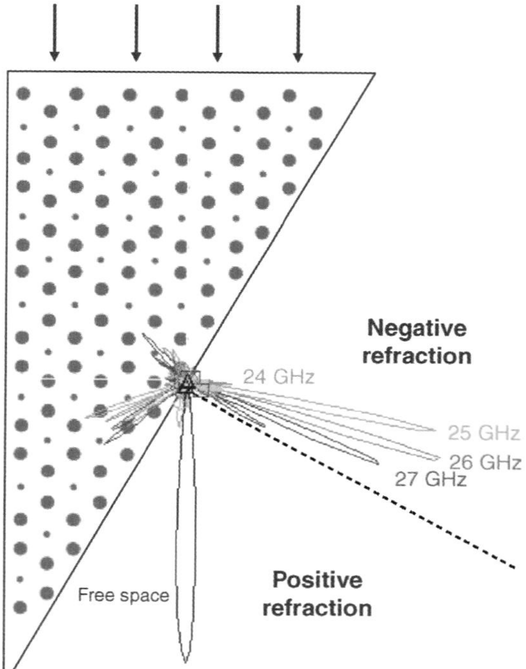

Figure 21.5 Far-field radiation pattern of a plane wave refracted through a superlattice prism for the frequencies corresponding to the transmission subband. The refracted beams form narrow lobes. The lobes with the maximum intensity correspond to the frequencies at the center of the transmission subband. The refracted beams direction and intensity strongly depend on the frequency. The intensity of the refracted beam is rather high (~80–90%) as compared to that of the incident wave. The dashed line shows the normal to the prism surface.

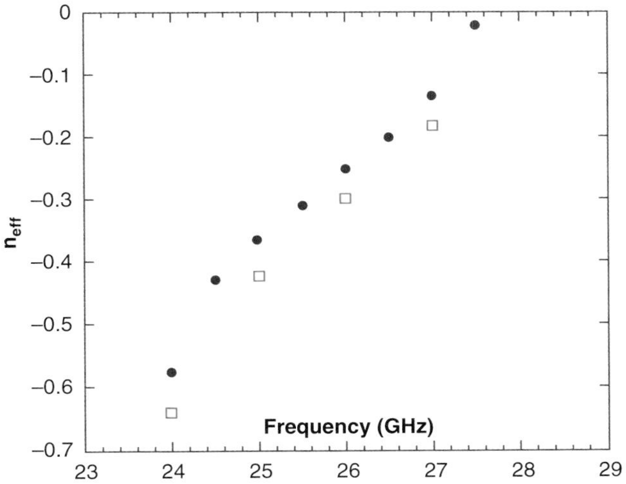

Figure 21.6 Effective refraction index, n_{eff}, as a function of frequency for the frequencies inside the transmission subband. Blue circles indicate n_{eff} derived from the far-field radiation pattern of the superlattice prism. Red squares indicate n_{eff} derived from the equal frequency contours of the infinite superlattice. At the lowest frequencies n_{eff} is -0.6 and at highest frequencies it approaches $n_{eff} = 0$.

wave (the lobe indicated as "free space"). This demonstrates that for frequencies in the transmission subband, the refracted wave exhibits minimal losses. We also simulated the refraction of a plane wave at a frequency outside the transmission band and we found positive refraction (see Ref. [14]).

We consider the angle of the refracted beam in Figure 21.5 and the Snell's law, and calculated the refraction index, n_{eff}, for several frequencies in the transmission band (Figure 21.6). It is negative and its magnitude varies from $n = -0.7$ at the lower subband edge to $n = 0$ at the upper subband edge.

21.3.3 Determination of the Refractive Indices Using the Equal Frequency Contours

Equal frequency contours provide a tool for graphical calculation of the direction of the refracted waves and thus allow independent estimate of the refraction index. Figure 21.7a shows the equal frequency contours ($\boldsymbol{k}_y$ versus $\boldsymbol{k}_x$ in $\boldsymbol{k}$-space) for free space (concentric circles–left panel) and for the superlattice (contours inside the hexagon–right panel). Note, that the gradient of the equal frequency surface points outwards in the former case and points inwards in the latter case. In Figure 21.8 we combine the equal frequency contours for the free space and for the superlattice, to show graphically the 25 GHz wave refraction through the superlattice prism.

The incident $\boldsymbol{k}$-vector (black arrow) and the incident group velocity (blue arrow) are pointing down (Figure 21.8). The $\boldsymbol{k}$-vector, say for the 25 GHz frequency, begins at

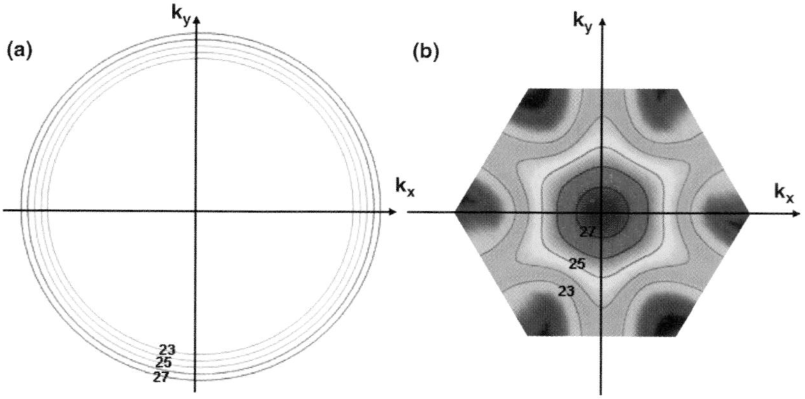

Figure 21.7 (a) Equal frequency contours in the free space (concentric circles in different colors) corresponding to different frequencies in the transmission band from 23 GHz to 27 GHz (marked on the contours). (b) Equal frequency contours in the superlattice prism (hexagonally shaped contours) for the same frequencies, shown in different colors.

the axis origin and ends at the equal frequency contour of 25 GHz in the free space. The group velocity vector, $v_g = \text{grad}_k\ (\omega)$ (blue arrow), begins where the ***k***-vector ends (because they refer to the same point in ***k***-space). The group velocity is orthogonal to the equal frequency contour. The free space group velocity, $\boldsymbol{v}_g$, points in

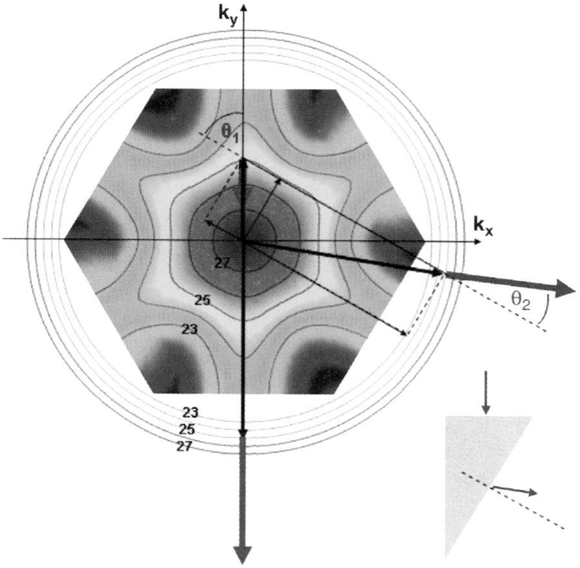

Figure 21.8 Refraction of a plane 25 GHz wave through a superlattice prism. The equal frequency contour presentation. Inset shows the refraction through the prism. See text for details.

the same direction as the $\boldsymbol{k}$-vector while the superlattice group velocity at 25 GHz points 'up-hill' (the gradient, $\text{grad}_k(\omega)$, points in the direction of the steepest slope). The $\boldsymbol{k}$-vector in the superlattice begins at the axis origin and ends where the group velocity vector begins. Note that the group velocity vector and the $\boldsymbol{k}$-vector are in the opposite directions, as expected for the left-handed material.

The dashed line is a normal to the output prism surface and it crosses the equal frequency contour of 25 GHz at the point where the $\boldsymbol{k}$-vector ends. The $\boldsymbol{k}$-vector component parallel to the prism surface (shown by a dashed line) is continuous everywhere. The outgoing $\boldsymbol{k}$-vector ends at the equal frequency contour of 25 GHz in free space. The outgoing group velocity in free space points in the same direction as the $\boldsymbol{k}$-vector and its magnitude is the same as for the incident wave. The refraction angle is the angle between the outgoing group velocity and the normal to the prism surface. This angle, together with the Snell's law, yields refraction index.

In Figure 21.6 we plotted the refractive index values for the different frequencies in the transmission subband as found from the analysis based on the equal frequency contours. The results agree quite well with the simulations above.

21.4 Conclusions and Future Directions

In this work we report a 2D superlattice made of dielectric rods with two different diameters but with relatively low dielectric constant $\varepsilon = 4$. The work is based on computer simulations. However, our recent experimental studies for the microwave range using the same concept but using rods with high dielectric constant [13] indicate that experiment and simulations lead practically to the same results. Here the appearance of transmission subband inside the stopband; for the Γ–K direction the stop-band is similar to that of a PC made of the large diameter rods. In the transmission-band frequency range we achieve Negative Refraction together with high to full transmission in the wide frequency range (15% of the total stopband width) and along different directions in the $\boldsymbol{k}$-space. The phase of the electric field reverses its direction inside the superlattice. These two phenomena indicate negative phase velocity and a demonstration of Left Handed Material. We show also refraction by a 60° prism made of the superlattice. This prism deflects the radiation in the direction that corresponds to negative refraction index between $n = -0.1$ to $n = -0.6$ for frequencies in the transmission band. These results are consistent with analysis using equal frequency contours. Observation of an efficient far-field focusing is the next step in our studies but this requires omni-directional refraction, which can be achieved by introducing some disorder (see Ref. [14]) and/or quasicrystalline structure.

We believe that the 2D superlattice presented here may have significant advantage with respect to other structures. Our work, we believe, is a step towards the observation of negative refraction and focusing at higher frequencies, particularly in the visible range. However, in the optical and IR ranges it is easier to fabricate arrays of holes in a dielectric substrate rather than arrays of rods. Simulations in this direction are in progress.

References

1 Luo, Chiyan, Johnson, S.G., Joannopoulos, J.D. and Pendry, J.B. (2003) *Phys. Rev. B*, **65**, 201104.

2 Luo, Chiyan, Johnson, S.G., Joannopoulos, J.D. and Pendry, J.B. (2003) *Opt. Express*, **746** (11), 7.

3 Pendry, J.B. (2000) *Phys. Rev. Lett.*, **85** (18), 3966.

4 Ziolkowski R.W. and Heyman, E. (2001) *Phys. Rev. E*, **64**, 056625.

5 Zengerle, R. (1987) *J. Mod. Opt.*, **34** (12), 1589.

6 Engheta N. and Ziolkowski, R.W. (2005) *IEEE Trans. Microw. Theory Tech.*, **53** (4), 1535.

7 Guven, K., Aydin, K., Alici, K.B., Soukoulis, C.M. and Ozbay, E. (2004) *Phys. Rev. B*, **70**, 205125.

8 Foteinopoulou S. and Soukoulis, C.M. (2003) *Phys. Rev. B*, **67**, 235107.

9 Parimi, P.V., Lu, W.T., Vodo, P., Sokoloff, J., Derov, J.S. and Sridhar, S. (2004) *Phys. Rev. Lett.*, **92**, 127401.

10 Notomi, M. (2000) *Phys. Rev. B*, **62**, 10696.

11 Berrier, A., Mulot, M., Swillo, M., Qiu, M., Thylén, L., Talneau, A. and Anand, S. (2004) *Phys. Rev. Lett.*, **93**, 073902.

12 Bulu, I., Caglayan, H. and Ozbay, E. (2006) *J. Phys.: Conf. Ser.*, **36**, 33.

13 Neve-Oz, Y., Golosovsky, M., Frenkel, A. and Davidov, D. (2006) *Proc. Eur. Microw. Assoc.*, **2**, 94.

14 Saado, Y., Neve-Oz, Y., Golosovsky, M., Davidov, D. and Frenkel, A. (2007) *phys. stat. sol. (b)*, **244** (4), 1237.

15 Richter, S., Hillebrand, R., Jamois, C., Zacharias, M., Gösele, U., Schweizer, S.L. and Wehrspohn, R.B. (2004) *Phys. Rev. B*, **70**, 193302.

16 von Rhein, A., Pergande, D., Greulich-Weber, S. and Wehrspohn, R.B. (2007) *J. Appl. Phys.*, **101**, 086103.

17 Padilla, W.J., Basov, D.N. and Smith, D.R. (2006) *Mater. Today*, **9**, 28.

18 Neff, C.W., Yamashita, T. and Summers, C.J. (2005) Lasers and Electro-Optics Society, 2005. LEOS 2005. The 18th Annual Meeting of the IEEE, p. 166.

19 Guven K. and Ozbay, E. (2005) *Phys. Rev. B*, **71**, 085108.

20 Ozbay, E., Bayindir, M., Bulu, I. and Cubukcu, E. (2002). *IEEE J. Quantum Electron.*, **38**, 7.

22
Negative Permeability around 630 nm in Nanofabricated Vertical Meander Metamaterials

Heinz Schweizer, Liwei Fu, Hedwig Gräbeldinger, Hongcang Guo, Na Liu, Stefan Kaiser, and Harald Giessen

22.1
Introduction

Dielectric metallic metamaterials [1–4] have been extensively studied to achieve material parameters of negative permittivity ε and permeability μ [5,6]. The advantages using materials with negative ε and μ were first pointed out in detail by Veselago [7]. He found that simultaneous negative ε and μ in such materials results in a negative refractive index and in backward wave propagation. This property justifies the name lefthanded (LH) materials or synonymously negative index materials (NIM). Some interesting optical imaging properties of NIM were already indicated by Veselago and later theoretically proven by Pendry [8]. The use of metamaterials for various technical applications is becoming increasingly interesting. For instance, a compact subwavelength thin cavity resonator can be realized by partial filling of the cavity with NIMs [9]. Implementation of NIMs in antenna gives the possibility of forward, broadside, or backward radiation designs [10]. Also there is a possibility for electronically scanned antenna beams [10]. Controlling the visibility of an object with respect to the electromagnetic transmission or reflection (cloaking) is one of the newest ideas [11]. With respect to fundamental physics, NIMs would be interesting for controlling the optical emission in atoms and molecules by electrical vacuum forces [12,13].

By using electrical engineering techniques for metamaterials, Smith [14] and Shelby [15] demonstrated first negative refraction in the GHz-range on the base of split-ring resonators (SRR) [5]. In their approach, the SRR structure produces the negative permeability, and a diluted wire grating produces the negative permittivity [5].

A major problem of resonant-structure based metamaterials is the principally restricted bandwidth of the negative refractive index effects. The reasons are outlined below. To overcome such a bandwidth restriction, several authors have proposed metamaterial structures on the basis of transmission line (TL) theory [10,16]. The

Nanophotonic Materials: Photonic Crystals, Plasmonics, and Metamaterials. Edited by R.B. Wehrspohn, H.-S. Kitzerow, and K. Busch

ISBN: 978-3-527-40858-0

idea of backward waves in TL theory dates back to the 1950's [1]. At that time, no ample application was found for backward waves due to their strong dispersion, which is normally not desired in a TL. The above mentioned groups [10,16] solved the problem of the structural transfer from a transmission line architecture with longitudinal capacitance and shunt inductance to metamaterials. They fabricated 1D [10] and 2D [10,16] structures to obtain a material response with simultaneous negative ε and μ.

In this contribution we combine the GHz TL approach for LH materials with the advantages of equivalent circuit analysis [17] of optical metamaterial structures [18] to obtain negative permittivity and negative permeability in the optical frequency range. From both the TL approach and Pendry's proposal [5], one finds that a negative ε can be achieved with ease. The plasmonic behavior of metals gives a dispersion relation of ε which has negative values below the plasma frequency. This frequency can be moved into the visible spectral range or even further into the GHz-range by small filling factors of metallic wires in a metamaterial (wire grating), which behave like a diluted metal [5] or an inductive impedance in the language of TL equivalent circuits [18].

The realization of a negative μ on the other hand is not easy. When trying to realize structures in the optical frequency range using the TL model for broadband LH materials, indications from Marcuvitz [17] equivalent circuit analysis of waveguide obstacles can be helpful. It is found that a narrowed conductive path along the propagation direction of an electromagnetic wave in a waveguide results in a longitudinal capacitance, while a lateral conductive path results in a shunt inductance. Based on this principle, we design a novel structure for negative μ using the TL analysis method [19].

We propose metallic meander strip and plate structures on pre-patterned dielectric substrates as 3D sublayers for realizing the longitudinal capacitance. In such structures, there are two degrees of freedom for the design. One is the conductive path along k, and the other is perpendicular to k. The path along k contributes to series impedance Z, and the other one to shunt admittance Y. The path along k, or the depth of the pattern can be varied up to 200 nm or even beyond, which is an appreciable parameter variation for a longitudinal capacitance.

In our theoretical analysis, we combine numerical simulations of a Maxwell equation solver (CST Microwave Studio) with the TL analysis. Numerical simulations yield scattering parameters S_{ij}, which are used for retrieval [20–22] of the effective material parameters ε and μ. However, these parameters do not give specific information about the relevant equivalent circuit. Therefore it is hard to get a direct idea for designing or optimizing a structure solely from the scattering parameters. On the other hand, TL analysis provides the possibility to map the Maxwell equations to circuit elements parallel and perpendicular to the light propagation vector $\boldsymbol{k}$. From this TL analysis, a direct and helpful relation (see Eq. (22.8) below) between μ and Y or between ε and Z can be obtained, and insight into the principal properties of the metamaterial structure can be acquired. An approach based on resonators (paired wire structures) is presented in [23,24] and appears also as a possible candidate to realize an optical magnetic response.

22.2 Theoretical Approach

22.2.1 Transmission Line Analysis

Mapping the longitudinal and transversal E- and H-fields onto their corresponding voltage and current modes, respectively, yields the transmission line equations of a metamaterial [10,16,17]. This enables a description of a metamaterial by an impedance matrix $\boldsymbol{A} = \boldsymbol{A}(Z, Y)$, where Z and Y are the series impedance and shunt admittance, respectively, when the unit cell of the material under investigation is much smaller than the operation wavelength. The matrix of the whole material is the matrix product of the single cells, constructed according to the standard rules of TL circuit analysis [1,25] (Eq. (22.1)).

$$\boldsymbol{A} = \begin{pmatrix} 1 + ZY & Z \\ Y & 1 \end{pmatrix} = \begin{pmatrix} A_{11} & A_{12} \\ A_{21} & A_{22} \end{pmatrix}. \tag{22.1}$$

The propagation in vacuum or medium can be represented by a transfer matrix which is not shown here [25]. From the impedance matrices $\boldsymbol{A}$ we can obtain scattering parameters S_{ij} (Eq. (22.2)), which can be compared with measured or simulated results. One obtains [10,25]:

$$S_{21} = \frac{2}{N} \tag{22.2}$$

and for homogeneous materials

$$S_{21} = |S_{21}|\exp(\gamma l_{\mathrm{p}}), \tag{22.3}$$

with γ is the propagation constant, l_{p} is physical length, and N is given by Eq. (22.4) from the elements of the $\boldsymbol{A}$ matrix [25]:

$$N = A_{11} + \frac{A_{12}}{Z_c} + A_{21}Z_c + A_{22}, \quad \text{and} \quad Z_c = \sqrt{\frac{Z}{Y}}. \tag{22.4}$$

The complex propagation constant γ obtained from the transmission line equation depends on the impedance and shunt admittance of the material (Eq. (22.5)):

$$\gamma = \frac{1}{l_{\mathrm{p}}}\sqrt{ZY} = a + jb. \tag{22.5}$$

The propagation constant β and the damping of the wave α can be obtained from the scattering coefficient of S_{21} (Eq. (22.2)), using its real and imaginary part (with $\phi_{S_{21}}$ = phase of S_{21}):

$$\beta = -\frac{\phi_{S_{21}}}{l_{\mathrm{p}}} \equiv \frac{1}{l_{\mathrm{p}}} \arctan\left(\frac{\mathrm{Im}(N)}{\mathrm{Re}(N)}\right) \tag{22.6}$$

and

$$\alpha = -\frac{1}{l_{\mathrm{p}}}\ln\left(\frac{2}{|N|}\right). \tag{22.7}$$

From known longitudinal impedances and transversal admittances, the material parameters μ and ε can be obtained in a homogeneous material (Eq. (22.8)):

$$\mu = -j\frac{1}{k_0 l_p}\frac{Z}{Z_0},$$

$$\varepsilon = -j\frac{1}{k_0 l_p}\frac{Y}{Y_0}, \qquad (22.8)$$

with $k_0 = \omega/c$, $Z_0 = \sqrt{\mu_0/\varepsilon_0}$, $Z_0 = 1/Y_0$, $c = 1/(\sqrt{\mu_0\varepsilon_0})$, and the refractive index $n = \sqrt{\varepsilon\mu} = (j/k_0 l_p)\sqrt{ZY} = \beta/k_0$.

As one can see from Eq. (22.8), for a negative μ a series capacitive impedance Z is required, and for a negative ε, a shunt inductive impedance Y is required. This is well known for microwave broad bandwidth metamaterials [10,16]. In the case of resonant structures, however, Z and Y always consist of a mixture of capacitive and inductive contributions. The application of the relations given in Eq. (22.8) to resonant Z and Y shows that positive μ and ε values would result, destroying the negative index effect.

22.2.1.1 Three Basic TL Circuits

Negative ε and μ result in a lefthanded vector triad (E, H, and k) [7], which gives rise to the definition of LH elements. For instance, a longitudinal (series) capacitance and a shunt inductance are lefthanded elements, while a shunt capacitance and a series inductance are righthanded elements. The relations between the effective material parameters and the impedances help us to construct metamaterials with desired dielectric properties. From Eq. (22.8), the required type of the series impedance Z and shunt impedance Y can be determined. An additional important connection to optical experiments is given by Eq. (22.2), which gives us the relation between the impedance matrix and the optical spectra. For general cases with multireflections, Eq. (22.3) is likely incomplete. A more detailed evaluation should consider the multiple-path of the transmission and reflection coefficients [20,21]. This case will be discussed in Section 2.2. To illustrate the principal relations between impedance matrix and optical spectrum we discuss here briefly the cases of purely right-handed, purely left-handed, and composite right-left-handed (CRLH) metamaterials. In Figure 22.1 we see a comparison of the transmission spectra $|S_{21}|^2$ in terms of frequencies.

As one can see, the RH material in Figure 22.1(a) starts with a high value in transmittance which is expected from an impedance network with a longitudinal RH inductance (L_{RH}) and a RH shunt capacitance (C_{RH}) (see inset in Figure 22.1(a)) and shows an optical low pass behavior. The characteristic material parameters are $\varepsilon \sim C_{RH}$, $\mu \sim L_{RH}$, and $n \sim \sqrt{L_{RH}C_{RH}}$. In the case of LH metamaterials, $|S_{21}|^2$ is small at low frequencies and is large at high frequencies. This is the typical behavior expected from a circuit with a longitudinal (series) capacitance and a shunt inductance (see the inset in Figure 22.1(b)). The material parameters are $\varepsilon \sim -1/L_{LH}\omega^2$, $\mu \sim -1/C_{LH}\omega^2$, and $n \sim -1/\sqrt{L_{LH}C_{LH}}\,\omega^2$. In real structures, however, parasitic LH-impedances in purely RH-materials and vice versa RH-impedances in purely

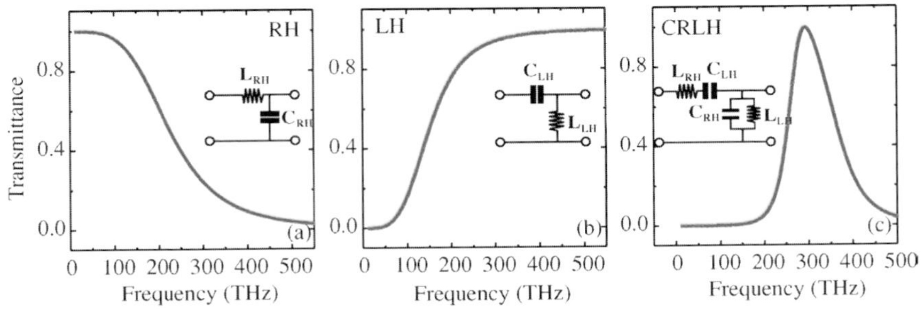

Figure 22.1 Comparison of transmission spectra ($|S_{21}|^2$) of (a) purely right-handed material, (b) purely left-handed material, and (c) composite right- and left-handed material. The parameter values are: $L_{RH} = \text{pH}$, $C_{RH} = 10^{-19} F$, $C_{LH} = 10^{-19} F$, $L_{LH} = \text{pH}$.

LH-materials are unavoidable. Therefore a CRLH-structure in a realistic metamaterial structure is expected. This is demonstrated in Figure 22.1(c). Here we see that a CRLH-metamaterial shows an optical bandpass behavior with characteristic frequencies of ω_{se} and ω_{sh} given by $\omega_{se}^2 = 1/(L_{RH}C_{LH})$ and $\omega_{sh}^2 = 1/(L_{LH}C_{RH})$, where $\varepsilon(\varepsilon \sim C_{RH} - 1/L_{LH}\omega_{sh}^2)$, and $\mu(\mu \sim L_{RH} - 1/C_{LH}\omega_{se}^2)$ change their sign from negative to positive values. A further important characteristic parameter is the phase-advance dispersion curve $[\phi = \phi(\beta)]$ as function of frequency or simply the dispersion relation of $\omega = \omega(\beta)$. From these curves we also obtain the phase velocity ($v_{ph} = \omega/\beta$) and the group velocity ($v_{gr} = \partial\omega/\partial\beta$), which gives us a possibility to define LH- or RH-metamaterials. In particular, for purely RH metamaterials, we have

$$\mathbf{v}_{ph} \uparrow\uparrow \mathbf{v}_{gr} \tag{22.9}$$

and for purely LH metamaterials we have

$$\mathbf{v}_{ph} \uparrow\downarrow \mathbf{v}_{gr}. \tag{22.10}$$

22.2.1.2 Role of the Series Capacitance

Previously, we have shown that a pure series capacitance results in a parabolic negative $\mu(\omega)$ spectrum of a metamaterial. From the analysis of the magnetic resonances, it turns out that the series capacitance has always a parallel parasitic inductance as shown in the inset of Figure 22.2 as in SRR-structures [16,26]. A metamaterial with a unit cell much smaller than the operating wavelength $l_{el} < \lambda/4$ can be described by a TL equivalent circuit model. The circuit model shown in the inset of Figure 22.2(b) is calculated using the TL theory described in the former section. The series L and C which form Z are chosen to keep the resonance frequency at 216 THz. The vacuum shunt C_{vac} and series L_{vac} are not shown in the circuit. From the scattering parameters S_{ij} of the circuit, effective material parameters can be retrieved [14] when an electrical length is given.

In Figure 22.2, the retrieved real parts of the permeability and permittivity for different values of C in the circuit model shown in the inset of Figure 22.2(b) are

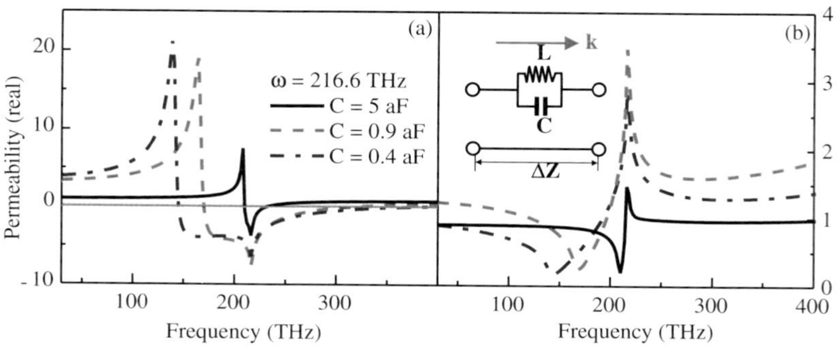

Figure 22.2 (a) Influence of the ratio C/L on the bandwidth and amplitude of the negative permeability while keeping the resonance frequency constant. The inset to (b) shows the circuit model for the magnetic resonance as in SRRs.

compared. In the calculation, an electrical length of 260 nm was used (the length which influences only the absolute amplitude of the retrieved parameters). It is found that with smaller C, both the amplitude and the bandwidth of the negative μ is larger. From the simple relation of $Z' = j\omega\mu$ for a homogeneous medium this effect can also be understood. At low and high frequencies, taking into account the additional condition of fixed resonance frequency $\omega_0 = 1/\sqrt{LC}$, μ approaches the asymptotic values $\mu \sim L \sim 1/C$ for $\omega \rightarrow 0$ and $\mu \sim -1/\omega^2 C$ for $\omega > \omega_0$. This shows that structures with smaller capacitance values are favorable.

22.2.2 Numerical Simulations and Syntheses with TL Analysis

22.2.2.1 Metamaterials with Different Unit Cells

In this section, we analyze metallic structures by the combination of TL analysis and numerical simulations. The geometry of the metallic structure in a metamaterial unit cell has important consequences on the material parameters of ε and μ [26] and on the bandwidth of the negative refractive index. Up until now, different geometrical structures have been reported for obtaining negative permeability: Split-ring resonators (SRRs) [5,6,14,15,27,28], double-bars [29–31], and cross-bars [32–34], for instance. According to the TL analysis for metamaterials discussed in the previous section, we study the evolution of the serial impedance in metamaterials, especially concentrating on the series capacitance.

Numerical Results for Several Structures with Different Series Capacitances The numerical results are obtained by commercial software (CST Microwave Studio) for solving Maxwell's equations using the finite integration technique (FIT). The following structures shown in Figure 22.3(A–D) were numerically simulated to study the variation of the series C in different structures. The structure parameters

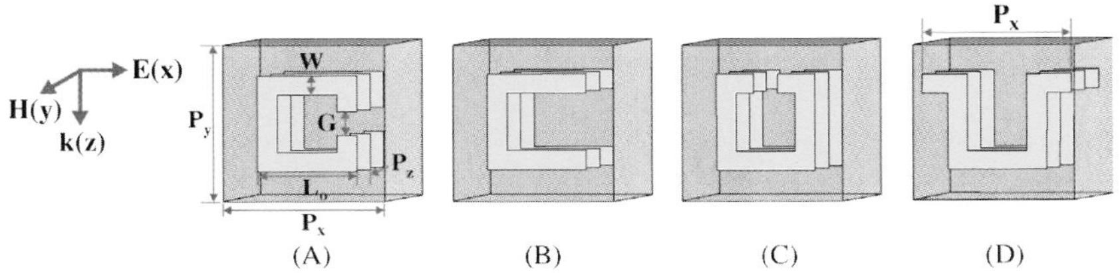

Figure 22.3 Schematics of SRRs and SRR-like metallic structures. In (A–B), the E-field is along the SRR-arms, and the H-field is coupled into the ring. In (C–D), the E-field is along the gap-bearing arm, and the H-field is coupled into the ring. (A, C) are conventional SRRs for achieving negative effective permeability. (B) shows U-structures with a fully opened gap in the ring. (D) shows the SRR-structures with outwards opened arms. The arms are extended so that they are connected with the neighboring rings. The structure is continuous like a meander and periodic along the E-field.

of the SRR and the SRR-like structures are given in Figure 22.3. We assume three unit cells along the $H(y)$-direction as shown in Figure 22.3(A–D). Along the k-direction there is only one unit cell. In the simulation, a Drude model was used for the material dispersion of Au with plasma frequency $\omega_p = 1.37 \times 10^{16}$ Hz and scattering frequency $\omega_c = 1.2 \times 10^{14}$ Hz [35]. The effective material parameters are retrieved within the unit cell along k-direction. The corresponding numerical results are shown in Figure 22.4. Structure A is the most common light incidence configuration for obtaining negative permeability. From curve A, in Figure 22.4, we see that the magnetic resonance is around 90 THz with a small amplitude of negative μ. The U-structure in Figure 22.3(B) with a completely opened gap

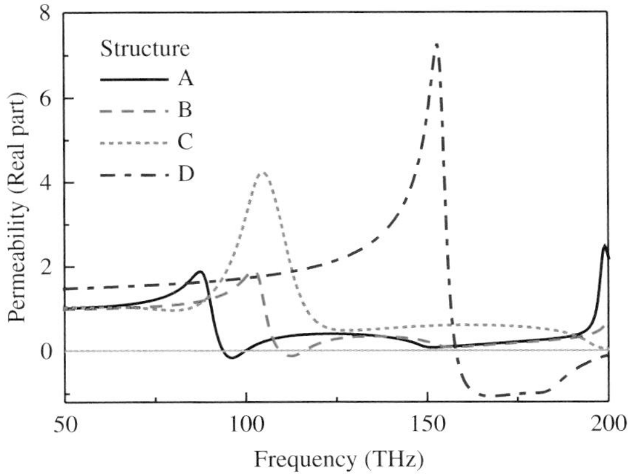

Figure 22.4 Real part of the retrieved effective permeability spectra with the corresponding structures and light incidence configurations as shown in Figure 22.3. The structures have an outer length L_0 of 370 nm × 370 nm in a unit cell of $P_x \times P_y \times P_z = 600 \times 600 \times 100\ \text{nm}^3$. The line width W of the ring is 70 nm. The thickness along the H-field is 20 nm.

implies a decrease in C. From curve B in Figure 22.4 we see that this is not effective for increasing the negative μ, although the resonance frequency is blue-shifted due to the reduced LC product. In Figure 22.3(C) the SRRs are rotated by 90° so that the E-field is along the gap-bearing arm. Curve C in Figure 22.4 shows that the negative μ disappears in the investigated spectral range, although the H-field is fully coupled into the rings. This is due to the strong electrical coupling into the ring [36]. However, when the direction of the open arms are changed towards the outside of the rings, the metallic structure along the E-field becomes continuous, as is shown in Figure 22.3(D). We call it "meander structure" as it has a meander shape in the transverse x-direction. This structure yields a very large bandwidth of negative μ (400 nm), in conjunction with a rather large amplitude as we can see in curve D in Figure 22.4, although other parameters are kept the same as in the SRRs shown in (C).

TL-Models to Interpret the Results The excellent result using structure (D) in Figure 22.3 can be explained by a reduction of the series C in the magnetic resonance. In Figure 22.3, although not shown in the schematics, all the structures are in fact periodic along both the H-field and the E-field. This implies additional coupling with the neighboring SRRs along the E-field for structure (A–C). This coupling is represented by a series capacitance C_s [37], parallel to capacitance C_1 of the ring (see also the inset in Figure 22.5). This is also the case when the ring is arranged as in (C). On the other hand, when the arms of the ring are connected together and there is no vacuum gap between the rings along the E-direction, C_s vanishes or is reduced. The longitudinal C_1 is then decreased dramatically. This leads to an increase in the

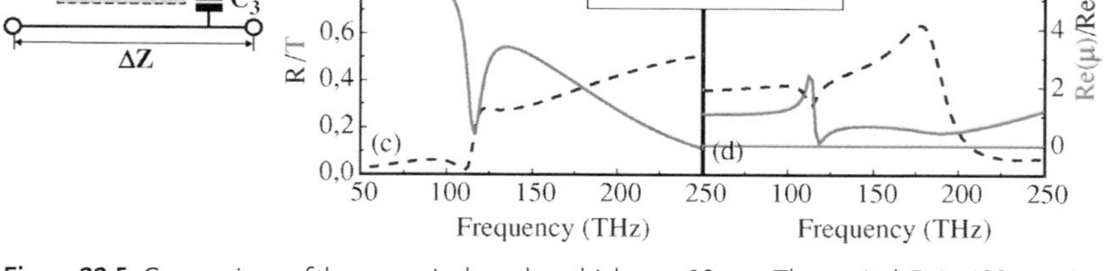

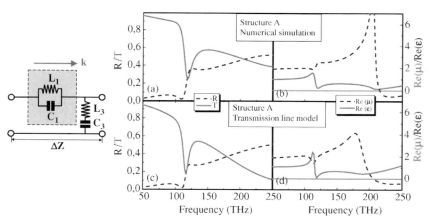

Figure 22.5 Comparison of the numerical results with the calculated results using the TL model, shown here for structure A. The used parameters of the lumped elements in the TL calculation are shown in Table 22.1. The structure parameters are: outer length 230 nm, line width 30 nm, thickness 20 nm. The period P_x is 400 nm, P_z is 100 nm, and the gap in the SRRs is 80 nm. The dashed lines represent reflectance and Re (ε), and the solid lines represent transmission and Re (μ), respectively.

resonance frequency $\omega_1 = 1/\sqrt{L_1 C_1}$. Furthermore, the induced AC current in the upward arm and the downward arm in one unit cell produces magnetic fields with the same direction which further enhance the magnetic resonances.

TL analysis [31] can justify this explanation further. TL models are constructed for the SRRs and the meander shown in Figure 22.3(A) and (D). Their scattering parameters and retrieved material parameters are shown in Figures 22.5 and 22.6, respectively.

The series impedance of L_1 and C_1 in Figures 22.5 and 22.6 represents the magnetic resonance (SRR resonance or meander loop resonance). The shunt impedance of L_3 and C_3 represents side impedance contributions along the E-field direction in a plane perpendicular to the $\boldsymbol{k}$-vector. The shunt impedance L_2 (see circuit depicted in the inset of Figure 22.6) accounts for the continuous wires in the meander matrix along the E-field direction. Only the elements in the shaded boxes contribute mainly to the retrieved permeability material parameters because the resonance frequencies of the shunted elements shown in Figures 22.5 and 22.6 are far away from the magnetic resonance frequencies. The shunt elements L_3 and C_3 are necessary for obtaining a satisfactory description on the high frequency side of the R/T-spectra. Here we concentrate only on the magnetic resonances.

Through fitting the numerical reflectance and transmittance spectra with the equivalent circuit models, the values of L and C in the equivalent circuit are determined. The values of L and C are listed in Table 22.1. In both figures, Figures 22.5 and 22.6, parts (a) and (b) show the calculated numerical results and parts (c) and (d) show the calculated TL results, respectively.

Through comparison of the reflectance/transmittance-spectra calculated by both methods, we see that the two models describe the two structures very well.

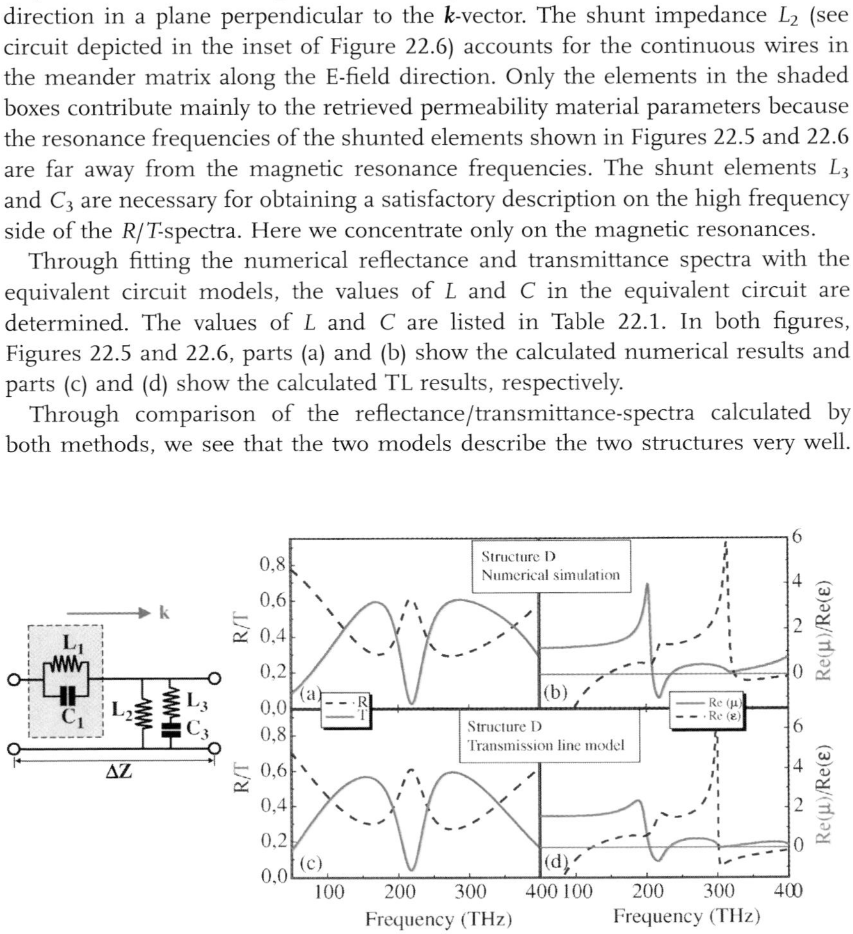

Figure 22.6 Comparison of the numerical results with the calculated results using the TL model shown here for structure D. The used parameters of the lumped elements in the TL calculation are shown in Table 22.1. The structure parameters are: outer length 230 nm, line width 30 nm, and thickness 20 nm. The period P_x is 400 nm, P_z is 100 nm. The dashed lines represent reflectance and Re (ε), and the solid lines represent transmission and Re (μ), respectively.

Table 22.1 Values of lumped C and L used to fit the numerical curves shown in Figures 22.5 and 22.6. Ohmic resistances (not shown in Figures 22.5 and 22.6) are in series with the corresponding L and account for loss. The unit for C is $10^{-19}F$, for L is $10^{-13}H$, and for R is Ω.

Lumped element	C_1	L_1	R_1	L_2	R_2	C_3	L_3	R_3
TLM for SRR	500	0.38	1.2			30	1.0	50
TLM for meander	53	1.0	6.9	2.7	10	13	0.86	30

Furthermore, the retrieved material parameters agree also very well with each other. This confirms that the TL models are reasonably constructed.

From Table 22.1, it is obvious that C_1 in the circuit model for meanders is one magnitude smaller than the one in the circuit model for SRRs. This explains the broad bandwidth and the large amplitude with negative μ in the meander structure.

22.2.2.2 Numerical Simulation of Meander Structures

As we already discussed in Section 22.2.2.1, the meander structure can induce a very strong magnetic resonance due to its reduced longitudinal capacitance [19]. Furthermore, this structure is more suited for fabrication than the SRRs for the negative μ configuration. Figure 22.7 shows a realistic meander structure with newly defined structure parameters and a suitable orientation of the electromagnetic field. It can be fabricated on ridge patterned substrates [19,38]. Through planarization of the ridge surface and choosing a proper ridge width W_r, the structure can be realized symmetrically along the k-direction, neglecting the presence of the substrate. Every parameter shown in the design can be changed freely within some range using E-beam lithography. However, in this section we show only two variations numerically to demonstrate the advantages with this novel structure for negative μ. The structure is buried in SiO_2 (or a similar medium) with a refractive index near 1.5.

In retrieving the effective material parameters, $T+d$ was used as the physical length. A Drude model was used for the material dispersion of Ag with plasma frequency $\omega_p = 1.37 \times 10^{16}$ Hz and scattering frequency $\omega_c = 0.85 \times 10^{14}$ Hz.

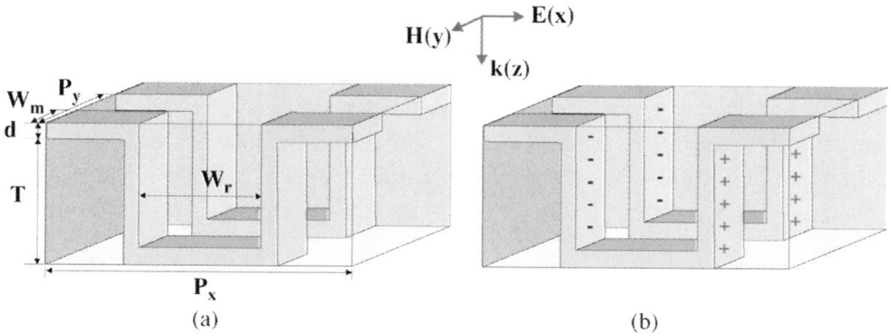

Figure 22.7 Schematic of a meander strip structure buried in SiO_2 with a refractive index of 1.5 [38].

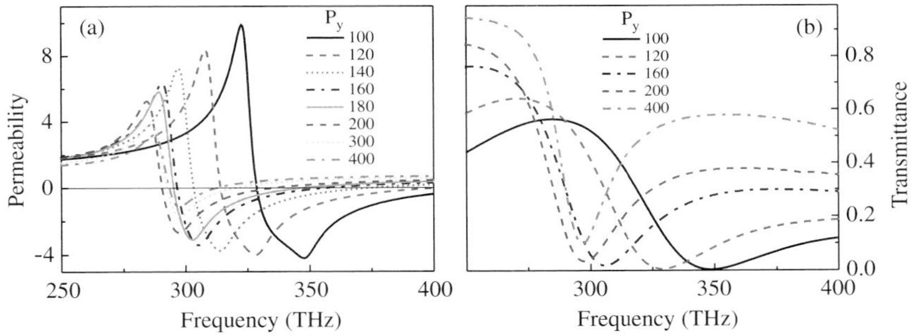

Figure 22.8 Numerical results of Ag meander structures buried in SiO_2 without any substrate. P_y varies from 100 nm to 400 nm. The other parameters are kept constant. $P_x = 200$, $T = 100$ nm, $W_m = 100$ nm, $W_r = 80$ nm, and $d = 20$ nm. (a) Retrieved effective permeability spectra with different P_y; (b) Selected transmittance spectra corresponding to the curves in (a) with different P_y.

Figure 22.8 shows the retrieved permeability from a structure with varying P_y and with constant $P_x = 200$ nm, $T = 100$ nm, $W_m = 100$ nm, $W_r = 80$ nm, $d = 20$ nm. We find that with larger P_y above 200 nm, in particular with larger distance along the H-field direction, the interaction between the meander strips is small and the resonance frequencies are almost independent of P_y. Only the amplitude of the negative μ is changed due to the changed filling factor of the meander. However, when P_y is decreased below 200 nm, the interaction becomes stronger and shifts the resonance to higher frequencies. Both the bandwidth and the amplitude of the negative μ are increased. This can be explained mainly by the reduced longitudinal C_1. When the distance between the meanders along y-direction is large, each meander ring has its independent resonance behavior. Along the E-field direction the charge distribution is also the same, as schematically shown in Figure 22.7(b). When P_y is decreased, the interaction of the charges with the same sign between neighboring meanders will weaken the E-field strength in every meander and decrease the longitudinal C_1. This shifts the resonance directly to higher frequencies. At high frequencies above the resonance, μ approaches the asymptotic value $\mu \sim -(1/l_p)1/(C_1\omega^2)$, while at low frequencies below the resonance we have $\mu \sim L_1/l_p$. For $P_y = 100$ nm, the meander strips become continuous along the H-field direction. This means the whole surface of the structure is covered with continuous metal (meander plate), which results in the broadest bandwidth and the largest amplitude of the negative μ. The results shown in Figure 22.8(a) indicate that a continuous metal on the ridge patterned surface leads to a broad and strong magnetic resonance. The structure can be regarded as a corrugated surface covered with a metal film. It is interesting to know how far the magnetic resonance frequency of the meander plate can be shifted towards higher frequencies. In further numerical simulations the structure is kept the same as that for Figure 22.8, but with $P_y = 100$ nm and a variation of T. We have also analyzed the retrieved real part of the permeability with respect to frequency. It is amazing to see that the magnetic resonance with 20 nm ridge height can reach a frequency of 700 THz, which is around 430 nm with a still rather large magnitude of negative μ (μ = −3 at 700 THz). This is below the absorption edge of Al or Ag and far beyond the saturated resonances due to kinetic inductance of the SRRs [37]. From the calculated resonance frequency versus $1/T$ we find that with the downscaling of the structure the magnetic resonance saturates only very slowly. For large periods of $P_x = 300$ nm, a frequency as high as 600 THz can also be achieved. Similar results of the resonance frequency with downscaling of the unit cell can be found in Refs. [26,37,39], which demonstrate that the kinetic energy of the electrons contributes to some degree to the inductance of a metamaterial.

22.3 Experimental Approaches

22.3.1 Fabrication Technologies

For the realization of metamaterial structures, different fabrication processes have been investigated. For plane metallic matrices, lift-off processes were developed. For the novel meander strip structures, a grating structure etched on a ridge patterned dielectric substrate was developed.

22.3.1.1 Plane Metallic Matrices

Pure metallic matrix structures on a dielectric substrate (mostly glass) have been realized using a high-resolution E-beam lithography process. The fabrication steps are as follows: After cleaning the SiO_2-substrate, a metallic adhesion layer (Cr) and then a metallic layer (Au or Ag) for the matrix elements was evaporated onto the substrate. Then a positive tone resist (960 K PMMA, 80–100 nm) was spin coated onto the wafer and the desired metallic matrix structure was written directly by E-beam lithography. After E-beam writing, an ultrasonic-wave supported lift-off process in MIBK solution was applied to achieve the matrix of the metallic elements. The single element as described in Section 22.2.2 was varied from SRRs to meander geometries. The metallic elements in the matrix were arranged periodically in periods of P_x from 200 nm to 400 nm and P_y from 200 nm to 400 nm. The smallest structure size (wire or gap) obtained was 30 nm for thicknesses of the metallic layer in the range between 20 nm and 30 nm. An example of a 2D array of double split rings is shown in Figure 22.9. The smallest SRR in Figure 22.9 has a diameter of 100 nm. Writing times of a few hours result in matrices in the range of $2 \times 2\,mm^2$ [27,28,40].

An alternative route to realize plane metallic matrix structures was based on the negative tone resist AR–N7500.18. In this route a negative tone resist is spin coated on the metallic layer (Cr, Au or Ag). After E-beam writing, developing, and fixing with AR-N-developer, the negative tone resist structure serves directly as etching mask in an ECR-RIE or RIBE Ar-plasma dry etching step.

In the case of meander structures we used the photo polymer (PC403). The lithography steps in this variant were nearly identical to that previously discussed,

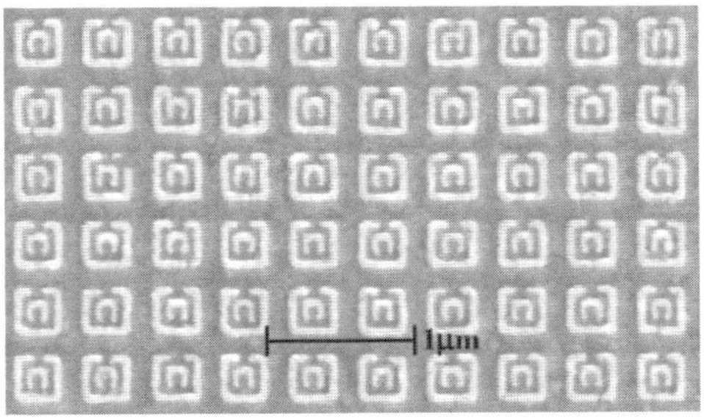

Figure 22.9 Section of a fabricated double SRR matrix structure on planar SiO_2 substrates. The whole array spans 2×2 mm^2. The smallest SRR sizes (inner SRR) are 100 nm [40].

and pattern transfer was carried out with a reactive (RIBE or ECR-RIE) dry etching step in an Ar/O_2 plasma. 2D chess board mesas were also fabricated.

22.3.1.2 Novel Meander Structure

As discussed in Section 22.2.1 a metamaterial should have reduced RH-circuit element contributions to improve LH-properties and to reduce damping. For that purpose, a new geometrical arrangement is necessary to overcome the restrictions of plane matrices. In plane matrices of metallic elements, especially in the case of resonant elements, in addition to the desired LH-circuit elements, always RH-circuit elements appear, reducing the bandwidth and increasing the damping of a metamaterial. The new approach here is the realization of a vertical metallic meander matrix [41,42] on a pre-patterned dielectric substrate which is patterned in ridges or chessboard-like mesas. The pattern depth (z-direction in this paper) serves as an additional degree of freedom in addition to the two other design degrees in x/y-direction. This extra degree of freedom gives the possibility for realization of one type of impedance – in our case the desired longitudinal capacitance – as LH-circuit element with suppressed contributions of parasitic RH-circuit elements (series inductance). The configuration is shown in Figure 22.7 and shows perfect coupling of the H-field into the meander loop. Furthermore, for tilted directions of the propagation vector $\boldsymbol{k}$ with respect to the average meander matrix plane, the $\boldsymbol{H}$-vector remains always fully coupled. In comparison to GHz-meander approaches [43] this approach allows for an additional degree of freedom: The height of the meander line (y-direction in Figure 22.7) can be freely chosen without a limitation by the metal layer thickness. Moreover, the depth and the width of the ridge or mesa in conjunction with the height (y-direction) of the meander gives a certain degree of independent parameter variation to design Z- and Y-contributions of a metamaterial. The stacking of each 3D-patterned sublayer completes the process for the metamaterial structure. The fabrication steps are as follows: We start with a profound cleaning of the SiO_2-substrate and deposit a layer sequence of 30 nm MgF_2 and up to 40 nm of Au or Ag. After a negative tone resist E-beam step, the negative resist is used as etching mask in an IBE (Ar plasma) etching step to pattern this first metallic layer. We continue with a MgF_2-spacer layer (50 nm to 150 nm) and a PC403 dielectric substrate layer (different thicknesses 80 nm to 150 nm). The PC403 is patterned with a RIBE: Ar/O_2 dry etching step using a negative tone resist etching mask. After pattern transfer, a metallic Au or Ag layer is deposited by E-beam evaporation on the ridge-/mesa-patterned substrate and is subsequently patterned in a linear or a 2D grating, forming the 1D or 2D meander line matrix. The AFM picture shown in Figure 22.10 depicts a meander array of this process step. The picture shows Au meander lines on PC403 ridges. The horizontal direction in Figure 22.10 corresponds to the x-direction in Figure 22.7. After planarization with PC403, a top wire grating is formed identical to the metallic grating on the bottom. Finally one obtains a sequence of planar grating, meander line grating, and planar grating which serve as building blocks for the metamaterial structure. The stacking of metamaterial structures using planarization techniques will be forthcoming [44].

22.3.2
Characterization of Fabricated Structures

In this section we discuss a selection of experimental realizations of sublayer structures which can serve as unit cells of a metamaterial. We primarily analyze two types of meander structures: grating meander structures as shown in Figure 22.7, and a meander structure with closed metallic surface (closed gaps in y-direction of Figure 22.7). The optical measurements of the reflection and transmission spectra were carried out using a Fourier spectrometer (Bruker IFS 66v/S) equipped with an IR microscope. The transmitted and reflected light was detected by a N_2-cooled Mercury Cadmium Telluride detector. The system (detector, spectrometer, and polarizer) has a detection range between 30 THz and 500 THz, corresponding to a wavelength of 10 μm and 600 nm, respectively. A silver/gold mirror was used as reference for reflectance. The bare glass substrate served as reference for the transmission.

Figure 22.10 Section of a fabricated meander matrix structure on patterned dielectric substrate with a period of 200 nm, a ridge width of 100 nm, and a ridge depth of 80 nm [19]

22.3.2.1 Experimental Results of Meander Strips

The meander matrix structure simulated in chap. 22.2.2.2 (Figure 22.6) was realized on a dielectric photopolymer (PC403) on glass substrate in different variations of the period P_x and P_y to vary the serial capacitance C_1 (inset in Figure 22.6). Experimental reflection spectra of meander gratings on PC403 are shown in Figure 22.11(a) for meander periods $P_x = 200$ nm and 350 nm. The reflection spectra show an overall low-pass characteristics (high-pass in transmittance) with magnetic resonances at 710 nm and at 860 nm for $P_x = 200$ nm and 350 nm, respectively. As expected from the simulation (Figure 22.11(b)), the magnetic resonance shifts to shorter wavelengths if the period P_x is decreased. The deviations of the resonance amplitude between experiment and simulation are explained by some fabrication imperfections when forming the geometrical meander loop shape. An additional fitting of the permeability as carried out in [23,24] was not performed because percolation threshold of the metal layers on polymer mesa ridges starts on larger metal thicknesses than on flat surfaces. Therefore the calculated spectra appear sharper despite the use of the experimentally determined scattering rate [35] for evaporated gold metal layers. Due to E-beam writing the structural periodicity is kept precisely as calculated, and the calculated frequencies agree without fitting parameters satisfactorily [compare Figure 22.11(a) and (b)] with the measured ones. After application of a retrieval process [20–22] we obtain the optical constants shown in Figure 22.11(c). The retrieved μ shows a negative value around 700 nm for the meander wires. The asymptotic parabolic behavior for smaller wavelengths (higher frequencies) can also be recognized, however, an additional approximately constant contribution of a RH-element appears. The asymptotic value of μ for large wavelength values is mostly given by the inductance $L_1 (\mu \sim L_1/l_p)$, bridging the series capacitor C_1 of the material. C_1 is depicted in the circuit shown in the inset of Figure 22.6. The crossover from negative to positive values of ε is related to the shunt inductance Ymainly depending on L_3 and C_3, where the inductance L_3, arising from interactions between elements perpendicular to the light propagation constant, determines together with L_2 the amplitude of ε, $[\varepsilon \sim (1/l_p)\ (\text{const.} - (1/L_2 + 1/L_3)\ \omega^{-2})]$. From these results we infer that the meander grating structure has promising capacitive properties but needs

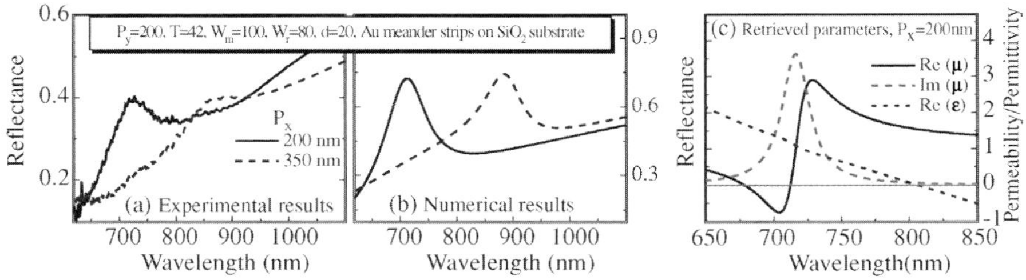

Figure 22.11 Reflection spectra of fabricated gold meander wire structures with a linewidth of 30 nm and a period of 200 nm on a SiO_2 substrate. In part (a) and (b) solid and dashed lines represent periods of 200 nm and 350 nm, respectively.

additional inductive degrees of freedom to exhibit negative n. This can be realized by an additional plane grating layer which adds an additional shunt inductance [17–19] to the meander material TL-circuit.

22.3.2.2 Experimental Results of Meander Plates

A closed metallic layer on a ridge patterned substrate forming a metallic meander surface would reduce strongly the capacitance in the metamaterial when the contribution of a serial inductance is kept low. According to the discussion in Section 22.2.2.2 (Figure 22.8), the highest operation frequencies and largest μ amplitudes can be expected. For the smallest periods (200 nm) of the meander plate structure a minimum of the transmission (around 800 nm) was found which marks the beginning of a high pass region which is also the region of negative μ. With the variation of period P_x from 350 nm to 200 nm, the beginning of the negative μ region shifts from 790 nm (380 THz) to 600 nm (500 THz). The retrieval process shows for the smallest period P_x (200 nm), a resonance at 675 nm for the permeability where μ changes sign from positive to negative values. The negative μ ranges from 550 nm to 665 nm with the largest amplitude of $\mu = -4.5$ at 650 nm. The μ value of -1 is present within a bandwidth of 50 nm centered at 630 nm. For shorter wavelengths (higher frequencies), μ remains negative and shows the parabolic behavior $\mu \sim -(1/l_p C_1)\lambda^2$ as expected from the equivalent circuit. In the limit of long wavelengths, μ becomes positive due to the parasitic inductance L_1 ($\mu \sim L_1/l_{el}$). These results show that metallic meander surfaces are very promising candidates to obtain negative permeability at very broad bandwidths in the visible wavelength range. We also calculated the ratio of the real and imaginary part of the permeability to determine a figure of merit for the permeability ($FOM_\mu := -\Re\{\mu\}/\Im\{\mu\}$) (Figure 22.12). One obtains a maximum value of 2.2 for meander structures, whereas for SRR-structures a value of 0.22 is obtained. These results also show that a resonant characteristic due to L_1 limits the wavelength range of negative μ, as an oscillator always consists indispensably of a left-handed and a right-handed element leading to composite right-left-handed

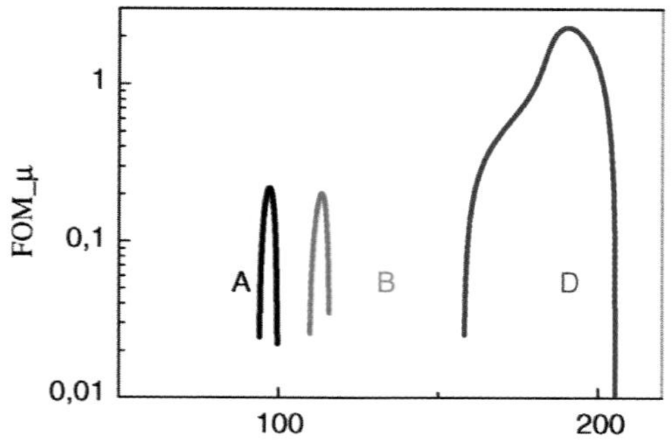

Figure 22.12 Comparison of the calculated figure of merit for the permeability (FOMμ) for different unit cell elements A, B, D. The element structures A, B, D are shown above in Figure 22.3.

materials. A non-resonant material, however, can consist of purely left-handed circuit elements, leading in principal to a broadband negative index material.

22.4 Conclusion

We have shown that the combination of the transmission line analysis together with the numerical solution of the Maxwell equations yields an efficient way to determine the electrical nature of a metamaterial structure. A key element in realizing the negative μ is the longitudinal capacitance in a metamaterial. This requirement lead us to a new structural design of vertical meander metamaterial structures to control the longitudinal capacitance in a large parameter range. Different types of meander metamaterial structures were fabricated. Broadband negative permeability at near infrared and visible frequencies were obtained in meander plate structures. This makes the metallic meander structures very promising candidates for broadband optical NIMs.

Acknowledgements

The authors acknowledge support by the Deutsche Forschungsgemeinschaft DFG (SPP 1113, contract Schw 470/19-1), Landesstiftung Baden-Württemberg, and Bundesministerium für Bildung und Forschung BMBF (13N9155). We would also like to thank for kind support from J. Pflaum and M. Dressel. For their support in lithography and etching we thank E. Frank and M. Ubl.

References

1 Collin, R.E. (1991) *Field Theory of Guided Waves*, IEEE Press, Oxford University Press, Oxford. Chap. 8 and 12.
2 Kock, W. (1946) *Proc. IRE*, **34**, 828–836.
3 Kock, W. (1948) *Bell Syst. Tech. J.*, **27**, 58.
4 Shivola, A. (2007) *Metamaterials*. **1**, 21.
5 Pendry, J.B., Holden, A.J., Stewart, W.J. and Youngs, I. (1996) *Phys. Rev. Lett.*, **76**, 4773.
6 Linden, S., Enkirch, C., Wegener, M., Zhou, J., Koschny, T. and Soukoulis, C. (2004) *Science*. **306**, 1351.
7 Veselago, V.G. (1968) *Sov. Phys. -Usp.*, **10**, 509.
8 Pendry, J.B. (2000) *Phys. Rev. Lett.*, **85**, 3966.
9 Engheta, N. (2002) *IEEE Antennas Wirel. Propag. Lett.*, **1**, 10.
10 Caloz, C. and Itoh, T. (2006) *Electromagnetic Metamaterials. Transmission Line Theory and Microwave Applications*, Wiley-Interscience, John Wiley & Sons, Inc., Hoboken, New Jersey.
11 Schurig, D., Mock, J.J., Justice, B.J., Cummer, S.A., Pendry, J.B., Starr, A.F. and Smith, D.R. (2006) *Science Express*, **19**, 1133628.
12 Casimir, H. and Polder, D. (1948) *Phys. Rev.*, **73**, 360.
13 Leonhardt, U. and Philbin, T. (2007) *New J. Phys.*, **9**, 254.

14 Smith, D., Padilla, W., Vier, D., Nemat-Nasser, S. and Schultz, S. (2000) *Phys. Rev. Lett.*, **84**, 4184.

15 Shelby, R.A., Smith, D.R. and Schultz, S. (2001) *Science*, **292**, 79.

16 Eleftheriades, G.V. and Balmain, K.G. (2005) *Negative-Refraction Metamaterials: Fundamental Principles and Applications*, IEEE Press Wiley-Interscience, Hoboken, NJ.

17 Marcuvitz, N. (1951) *Waveguide Handbook, M.I.T. Rad. Lab. Ser.*, McGraw-Hill, Inc., Exeter.

18 Ulrich, R. (1967) *Infrared Phys.* **7**, 37.

19 Schweizer, H., Fu, L., Gräbeldinger, H., Guo, H., Liu, N., Kaiser, S. and Giessen, H. (2007) *phys. stat. sol. (b)*. **244**, 1243.

20 Boughriet, A.H., Legrand, C. and Chapoton, A. (1997) *IEEE Microw. Theory Techn.*, **45**, 52.

21 Nicolson, A.M. and Ross, G.G. (1970) *IEEE Trans. Instrum. Meas.*, **IM-19**, 377.

22 Weir, W.B. (1974) *IEEE Proc.* **62**, 33.

23 Yuan, H.K., Chettiar, U., Cai, W., Kildishev, A., Boltasseva, A., Drachev, V. and Shalaev, V. (2007) *Opt. Express*, **15**, 1076.

24 Cai, W., Chettiar, U., Yuan, H.K., de Silva, V., Kildishev, A., Drachev, V. and Shalaev, V. (2007) *Opt. Express*, **15**, 3333.

25 Pozar, D. (2005) *Microwave Engineering*, John Wiley & Sons, Inc., Hoboken, NJ.

26 Tretyakov, S. (2007) *Metamaterials*. **1**, 40.

27 Guo, H., Liu, N., Fu, L., Schweizer, H., Kaiser, S. and Giessen, H. (2007) *phys. stat. sol. (b)*. **244**, 1256.

28 Liu, N., Guo, H., Fu, L., Schweizer, H., Kaiser, S. and Giessen, H. (2007) *phys. stat. sol. (b)*. **244**, 1251.

29 Shalaev, V., Cai, W., Chettiar, U., Yuan, H.K., Sarychev, A., Drachev, V. and Kildishev, A. (2005) *Opt. Lett.*, **30**, 3356.

30 Zhou, J., Zhang, L., Tuttle, G., Koschny, T. and Soukoulis, C. (2006) *Phys. Rev.*, **B73**, 041101.

31 Fu, L., Schweizer, H., Guo, H., Liu, N. and Giessen, H. (2007) *Appl. Phys. B*, **86**, 425.

32 Dolling, G., Wegener, M., Soukoulis, C. and Linden, S. (2007) *Opt. Lett.*, **32**, 53.

33 Zhang, S., Fan, W., Mallory, K. and Brueck, S. (2005) *Opt. Express*, **13**, 4922.

34 Soukoulis, C., Linden, S. and Wegener, M. (2007) *Science*, **315**, 47.

35 Dolling, G., Enkrich, C., Wegener, M., Soukoulis, C. and Linden, S. (2006) *Science*, **312**, 892.

36 Katsarakis, N. (2004) *Appl. Phys. Lett.*, **84**, 2493.

37 Zhou, Z., Koschny, T., Kafesaki, M., Economou, E.N., Pendry, J.B. and Soukoulis, C.M. (2005) *Phys. Rev. Lett.*, **95**, 223902.

38 Schweizer, H., Gräbeldinger, H., Giessen, H., Fu, L., Berroth, M., Rumberg, A. Patent application DE 10 2005 052 208. 4-34. Breitband NIM.

39 Klein, M., Enkrich, C., Wegener, M., Soukoulis, C. and Linden, S. (2006) *Opt. Lett.*, **31**, 1259.

40 Schweizer, H., Gräbeldinger, H., Zentgraf, T., Kuhl, J., Loa, I., Syassen, K. and Giessen, H. (2005) in: Proceedings of EPFL Latsis Symposium 2005, Negative refraction: revisting electromagnetics from microwaves to optics p. 45.

41 The fish net structure [42] appears very similar to the meander structure on pre-patterned substrates, however, in [42] the sructure is planar whereas an important difference of the pre-patterned meander structure is the 90 degree tilt of the loops with respect to the surface normal vector for full coupling of the magnetic field at arbitrary angles of incidence.

42 Fedotov, V., Mladyonov, P., Prosvirnin, S. and Zheludev, N. (2005) *Phys. Rev.*, **E72**, 056613.

43 Chen, H., Ran, L., Huangfu, J., Zhang, X., Chen, K., Grzegorczyk, T. and Kong, J. (2004) *J. Appl. Phys.*, **96**, 5338.

44 Liu, N. (2007) *et al.* to be publ.

Index

Nanophotonic Materials: Photonic Crystals, Plasmonics, and Metamaterials. Edited by R.B. Wehrspohn, H.-S. Kitzerow, and K. Busch

ISBN: 978-3-527-40858-0